HOW THINGS WORK

FLIGHT

TIME LIFE BOOKS ®

Other publications:

COVER

A Lockheed TriStar jet touches down at Palmdale, California at dusk. The time exposure, taken by a camera mounted on the airplane's tail fin and triggered by remote control, captures ribbons of runway lights streaming from the jet's nose.

HOW THINGS WORK

FLIGHT

TIME-LIFE BOOKS

ALEXANDRIA, VIRGINIA

Library of Congress Cataloging-in-Publication Data

Flight
      p. cm. – (How things work)
  Includes index.
  ISBN 0-8094-7850-1 (trade)
  ISBN (invalid) 0-8094-7851-1 (lib)
  1. Flight—Popular works.
  I. Time-Life Books. II. Series.
  TL546.7.F53 1990
  629.132—dc20   90-37365
                                CIP

How Things Work was produced by
ST. REMY PRESS

PRESIDENT	Pierre Léveillé
PUBLISHER	Kenneth Winchester

Staff for FLIGHT

Editor	Elizabeth Cameron
Senior Art Director	Diane Denoncourt
Art Director	Francine Lemieux
Contributing Editors	George Daniels, Peter Pocock, Bryce S. Walker
Assistant Editor	Mitchell Glance
Research Editor	Fiona Gilsenan
Researcher	Hayes Jackson
Picture Editor	Chris Jackson
Designer	Chantal Bilodeau
Illustrators	Maryse Doray, Nicolas Moumouris, Robert Paquet, Maryo Proulx
Index	Christine M. Jacobs
Administrator	Denise Rainville
Accounting Manager	Natalie Watanabe
Production Manager	Michelle Turbide
Systems Coordinator	Jean-Luc Roy

Time-Life Books Inc. is a wholly owned subsidiary of
THE TIME INC. BOOK COMPANY

President and Chief Executive Officer	Kelso F. Sutton
President, Time Inc. Books Direct	Christopher T. Linen

TIME-LIFE BOOKS INC.

EDITOR	George Constable
Director of Design	Louis Klein
Director of Editorial Resources	Phyllis K. Wise
Director of Photography and Research	John Conrad Weiser
PRESIDENT	John M. Fahey Jr.
Senior Vice Presidents	Robert M. DeSena, Paul R. Stewart, Curtis G. Viebranz, Joseph J. Ward
Vice Presidents	Stephen L. Bair, Bonita L. Boezeman, Mary P. Donohoe, Stephen L. Goldstein, Juanita T. James, Andrew P. Kaplan, Trevor Lunn, Susan J. Maruyama, Robert H. Smith
New Product Development	Trevor Lunn, Donia Ann Steele
Supervisor of Quality Control	James King
PUBLISHER	Joseph J. Ward

Editorial Operations

Production	Celia Beattie
Library	Louise D. Forstall
Correspondents	Elisabeth Kraemer-Singh (Bonn); Christina Lieberman (New York); Maria Vincenza Aloisi (Paris); Ann Natanson (Rome).

THE WRITERS

Walter Boyne enlisted in the U.S. Air Force in 1951 and retired as a colonel in 1974 with more than 5,000 flying hours in a score of aircraft. He is the prize-winning author of fifteen books on aviation and automotive subjects, including three novels. Among his non-fiction works are *Boeing B-52—A Documentary History*, *The Smithsonian Book of Flight* and *The Leading Edge*. His novels include *The Wild Blue* (with Steven L. Thompson), *Trophy for Eagles* and the second book of a trilogy, *Eagles at War*.

Terry Gwynn-Jones served for 32 years as a fighter pilot and flight instructor with the British, Canadian and Australian air forces. In 1975, he teamed with pilot Denys Dalton to set an around-the-world speed record—122 hours and 17 minutes—for piston-engined aircraft. His books include *True Australian Air Stories*, *On a Wing and a Prayer* and *Farther and Faster*.

Valerie Moolman, a former editor for Time-Life Books, is the author of 45 books, documentary film scripts and dramatic shows on subjects ranging from aviation to social history. She is the author of two books in Time-Life's *Epic of Flight* series and was a contributor to *Understanding Computers* and *Mysteries of the Unknown*.

THE CONSULTANTS

Dr. Tom D. Crouch is Chairman of the Department of Aeronautics at the National Air and Space Museum in Washington, D.C. He holds a Ph.D from Ohio State University and has written several books and articles on the early history of aviation. He is also an avid balloonist.

Dr. Howard S. Wolko is Special Advisor for Technology to the Department of Aeronautics of the National Air and Space Museum. He holds a D.Sc. in Theoretical and Applied Mechanics from George Washington University and has extensive experience in industry, the academics and the Federal Government. He also participated in the development of the X-series of experimental research aircraft.

For information about any Time-Life book,
please write:
Reader Information
Time-Life Customer Service
P.O. Box C-32068
Richmond, Virginia
23261-2068

CONTENTS

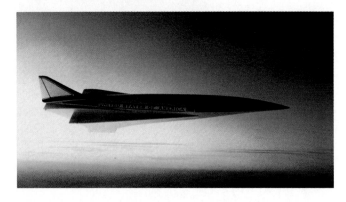

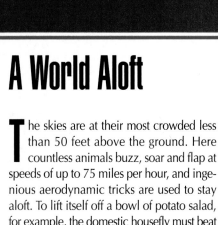

A World Aloft

The skies are at their most crowded less than 50 feet above the ground. Here countless animals buzz, soar and flap at speeds of up to 75 miles per hour, and ingenious aerodynamic tricks are used to stay aloft. To lift itself off a bowl of potato salad, for example, the domestic housefly must beat its wings furiously—up to 200 times each second. Likewise the American woodcock, a forest-dweller that would rather hide than fly, bounds into the air like a Harrier jump jet. To avoid predators, the "flying" fish manages to launch itself into the air and sail for short distances on its long, wing-like pectoral fins. Humans entered the crowded skies in a modest way on December 17, 1903, when the first successful flying machine, the Wrights' *Flyer*, covered 852 feet in 59 seconds.

Dandelion seedhead
Borne on the wind
The parachute-like seedhead of the dandelion weed is actually lighter than the air itself. When the relative humidity of the air rises above 70 percent, the seedhead gets heavier and floats to the ground.

Woodcock
5 miles per hour
When threatened, the woodcock may reluctantly take to the air, but is the world's slowest flying bird.

Housefly
4 miles per hour
A deft and acrobatic flier, the domestic housefly can even land upside down on the ceiling.

Blimp
35-40 miles per hour
Although impractical for commercial air travel, modern helium-filled dirigibles have found a role as flying billboards.

Wrights' *Flyer*
9.8 miles per hour
As his confidence grew with successive flights, Orville Wright pushed the world's first powered airplane to 30 miles per hour.

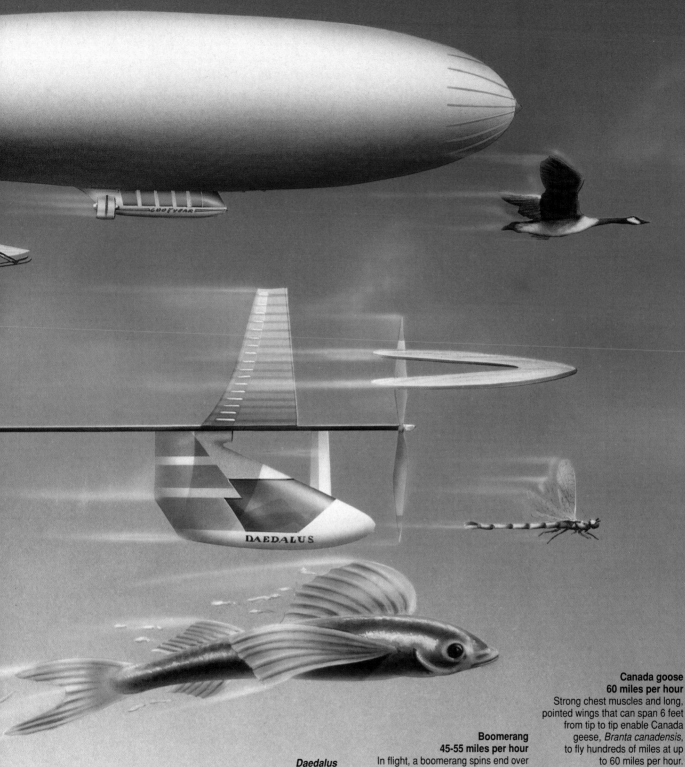

**Canada goose
60 miles per hour**
Strong chest muscles and long, pointed wings that can span 6 feet from tip to tip enable Canada geese, *Branta canadensis*, to fly hundreds of miles at up to 60 miles per hour.

**Boomerang
45-55 miles per hour**
In flight, a boomerang spins end over end 10 times per second. Its elliptical flight seldom exceeds 12 seconds.

**Dragonfly
60 miles per hour**
Engineers studying the dragonfly's use of airflows have clocked the insect at 60 miles per hour. This remarkable flier has inhabited the planet for 250 million years and, for its size, generates three times the lift of the most efficient aircraft.

Daedalus
18 miles per hour
This 70-pound aircraft, piloted by an Olympic cyclist, holds the distance record for human-powered flight: 72 miles.

**Flying fish
35 miles per hour**
Flying fish skim along the surface of the water to gain momentum, then lift off a wave and glide up to 150 feet on wing-like pectoral fins. They have been clocked aloft for 42 seconds.

Beyond 90 miles per hour, only the peregrine falcon remains from the animal kingdom. At these speeds, for both birds and flying machines, lift is produced by thin, tapered wings called airfoils. Air must travel farther and faster over the curved upper surface of the airfoil, creating two different pressure areas that lift the wing. A helicopter looks different from an airplane, but its whirling rotor blades are in fact spinning airfoils.

Flying disk
74 miles per hour
The familiar spinning plastic disk is a sophisticated aerodynamic toy.

Hockey puck
85-90 miles per hour
When a professional hockey player winds up and takes a slapshot, the vulcanized rubber disk streaks toward the goal at up to 90 miles per hour.

Helicopter
170 miles per hour (cruising)
Helicopters usually sacrifice speed for maneuverability and are more difficult to fly than a plane.

Cessna Twin
163 miles per hour (cruising)
This popular business plane has a top speed of 245 miles per hour and a range of more than 1,200 nautical miles.

Voyager
122 miles per hour (cruising)
Voyager entered aviation history in 1986 by flying nonstop around the world on a single tank of fuel.

Boeing 747
580 miles per hour (cruising)
Four turbofan engines and a wing surface
area greater than a basketball court are
necessary to lift this 390-ton jumbo jet.

Peregrine falcon
217 miles per hour (diving)
Fastest animal in the world, the peregrine falcon folds
its wings to dive literally at breakneck speed—the
force of impact breaks the neck of its prey.

Executive jet
509 miles per hour (cruising)
The Gulfstream III carries eight passengers and a crew of three at speeds close
to those of commercial jetliners.

F-86E Sabre
690 miles per hour
In the hands of a hot pilot, this Korean War-
era fighter could approach the speed of sound in
steep dives. A classic fighting jet, the Sabre
was powered by a turbojet engine that
produced 5,900 pounds of thrust.

Golf ball
170 miles per hour
Hit off the tee with a No. 1 wood, a 1.6-
ounce golf ball can reach a speed of
170 miles per hour. Its dimpled cover
increases the ball's distance and accuracy. Professional golfers can drive the
ball more than 300 yards.

Concorde
Mach 2
The world's first commercial supersonic flier, the Concorde can fly 3,050 miles without refueling. Cruising at a maximum altitude of 60,000 feet, it makes the trip between New York and London in three hours and forty-five minutes.

Bell X-1
Mach 1.07
Legendary pilot Chuck Yeager broke the sound barrier aboard "Glamorous Glennis" on October 14, 1947. His top speed was Mach 1.45, or 967 miles per hour.

McDonnell Douglas F-15 Eagle
Mach 2.5 (maximum)
With an airframe able to withstand 9 Gs, the 81,000-pound Eagle is powered by two engines with a combined thrust of 48,000 pounds.

.22 bullet
Mach 2.4
Muzzle velocities vary with ammunition load and muzzle length; the 22-250 Remington rifle bullet reaches speeds of Mach 3.5.

Fighter pilots have known for decades that shock waves begin to form around any object traveling near the speed of sound *(page 122)*. But that speed is not a constant—at sea level it is about 760 miles per hour, while at 40,000 feet it drops to about 660 miles per hour.

To avoid ambiguity, scientists adopted the Mach number, named for the 19th-Century Austrian physicist who first measured the speed of sound. An object's Mach number is equivalent to its speed divided by the speed of sound at the object's altitude.

At this threshold, high temperatures and pressures make a conventional aircraft difficult to handle and put dangerously high stresses on its structural materials. Beyond Mach 1.05, the shock waves fold back over the aircraft and the "ride" smooths out as the plane reaches supersonic speed.

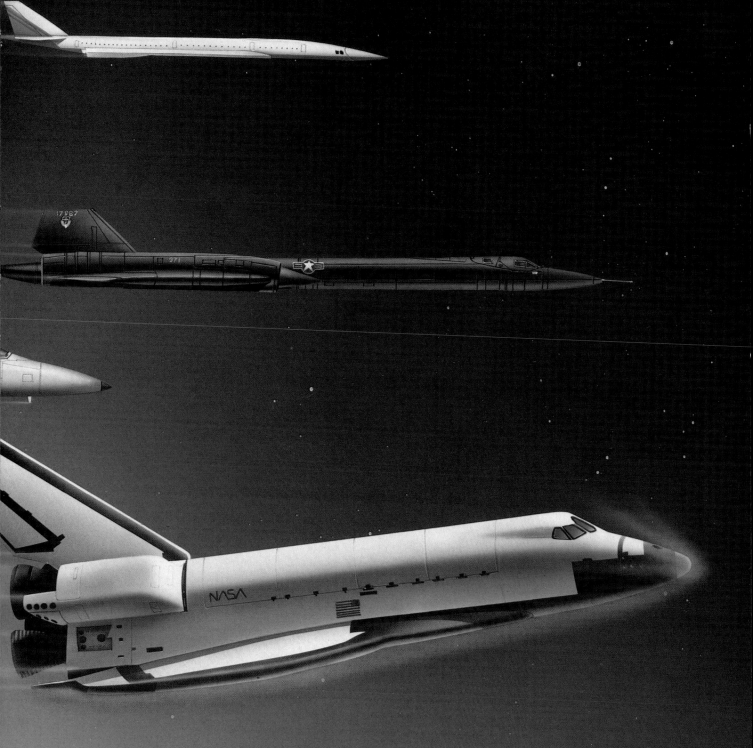

High-speed civil transport plane
Mach 3+ (cruising)
Scheduled for service by the year 2000, the so-called
Orient Express will fly from Los Angeles to Sydney,
Australia, in 2½ hours and have a nonstop fuel
range of 7,000 miles—twice that of the Concorde.

Lockheed SR-71
Mach 3.2 (cruising)
Featuring an airframe of titanium and
heat-resistant plastic, this reconnaissance plane is the world's fastest jet.

Space shuttle
Mach 21 (reentry)
The shuttle reenters the Earth's atmosphere at more than
15,000 miles per hour. Protective tiles on the belly of the
craft absorb the intense heat created by this friction.

THE WONDER OF FLIGHT

The horizon tilts and the green Earth far below swirls in response to the aircraft's sprightly movements. Rushing air snatches at the pilot's clothes and streams over wings of colorful sailcloth. At a touch of the throttle, the flying machine leaps ahead under the impetus of its spinning propeller. This ultralight aircraft, an ungainly apparatus made from 200 pounds of fabric and aluminum tubing, is one modern expression of the age-old human dream of flying. High above—barely visible as a silver speck creeping across the sky—is another: a huge passenger jet, seven miles up and bound for another continent at 600 miles per hour.

And there are many more. Humans have only recently puzzled out the principles of aviation and created devices that climb into the skies, but those devices have evolved rapidly and taken innumerable forms. Some, such as balloons, are lighter than air, which buoys them up as water supports a floating cork. Others, such as gliders and airplanes, are heavier than air, and stay aloft because of a force, called lift, generated by their movement through the air.

A means of staying in the air is only one of the essential elements of human flight; the development of flying machines has been shaped by other needs as well. The demand for ever-greater speeds has led to powerful, lightweight engines. Maneuvering these faster aircraft, in turn, calls for increasingly complex control systems: Where early pilots shifted their weight to change course, modern fliers activate guidance computers with the slightest of hand movements. Navigation, once a matter of following railroad tracks from town to town, is now possible without a glance at the ground; pilots instead rely on sophisticated electronic equipment and signals from satellites.

Nor is navigation simply wayfinding: The vast expanses where aviators once found only birds and clouds are now so crowded with traffic that their use must be rigorously regulated to prevent collisions.

Cruising about 1,000 feet above the Swiss countryside, the pilot of an ultralight aircraft surveys the drifting passage of a hot-air balloon. A 30-horsepower engine, with a cruising speed of about 40 miles per hour, powers this "lawn chair with wings." Strong, lightweight materials contribute to its safety and economy.

AIR AND ITS MOVEMENT

All this activity takes place in a realm that is itself highly active. The regions where aircraft operate are filled with winds and weather systems, ranging from gentle zephyrs to vast weather fronts, roaring arctic expresses, and formidable jet streams. All result from the turbulent motions of the molecules that make up the Earth's atmosphere. Countless air molecules—about 77 percent nitrogen and 21 percent oxygen with traces of other substances, including water vapor—move about freely, continuously jostling each other and occupying any space open to them. Everything that flies, from bird to balloon to jet plane, relies on interaction with this sea of air to stay aloft.

An understanding of the true nature of the atmosphere, and some of the forces involved in flight, began to emerge in the 17th Century with the work of Evangelista Torricelli, an Italian scientist who studied movement and pressure in fluids—a category that includes gases as well as liquids. In 1643, Torricelli invented the barometer, a device for measuring air pressure. Using simple glass tubes and bowls of mercury, he was able to determine the weight of a narrow column of air reaching from the Earth's surface to the outer edge of the atmosphere—about 14.7 pounds for a column one inch square. Compacted by the weight of overlying air, the atmosphere is most dense at the bottom. The pressure diminishes rapidly with altitude, falling to half its sea-level value about three-and-a-half miles above the surface of the Earth; 90 percent of the atmosphere is below the height of ten miles.

This relatively air-rich layer of the atmosphere, called the troposphere, is the region where clouds form and weather conditions develop. It is also the arena for natural flight, ranging from insects that flutter no more than a few feet from their place of birth, to arctic terns that make an annual round-trip migration of 22,000 miles, and bar-headed geese that soar as high as 30,000 feet as they pass over the Himalayas. Human flight, too, is generally limited to the troposphere, although long-distance jets sometimes venture into the next layer, known as the stratosphere. Manned balloons have explored the upper reaches of the stratosphere, which extends to a height of 30 miles; rocket-powered aircraft have gone beyond it. For the most part, though, flight is confined to altitudes below 80,000 feet, where the atmosphere is thick enough to feed engines and provide lift to the wings.

BORNE ON THE WIND

The lift that keeps airplanes aloft is only one of the aerodynamic forces that act on a body moving through the air. Simple air resistance, for example, slows any moving object by imposing a force in the direction opposite to the motion. In the case of a falling body, lift partly counteracts the force of gravity, which causes the object to accelerate—gaining speed at a steady rate. Air resistance

A SLICE OF THE SKY

As altitude increases, air becomes more rarefied and temperature drops, affecting both natural and mechanical flight. Insects and birds generally fly below 1,000 feet where oxygen is abundant and ambient temperature tolerable. Humans can fly up to 10,000 feet without specialized equipment. Above that height, planes must be equipped with pressurized, temperature-controlled cabins and supercharged engines that can suck in enough of the thin air to power the plane.

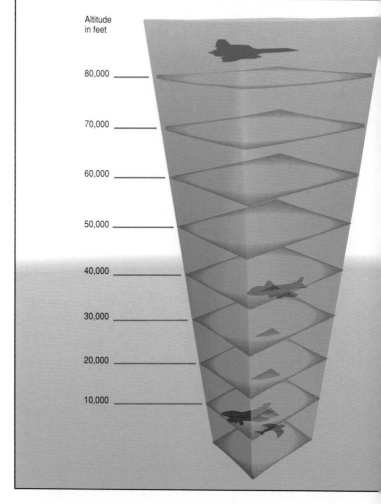

Altitude
in feet

80,000
70,000
60,000
50,000
40,000
30,000
20,000
10,000

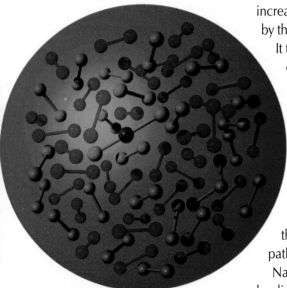

A GASEOUS MIX

Gas molecules dance about in this rendering of an iota of air, magnified more than a millionfold. Nitrogen (purple) and oxygen (green) predominate, with water vapor, carbon dioxide and other gases making up the rest. Objects moving through this dense soup of molecules create currents and eddies that produce a variety of aerodynamic forces.

increases with the object's speed, so that eventually the upward force generated by the air matches the downward pull of gravity, and the object stops accelerating. It then falls at a constant speed, called terminal velocity. The terminal velocity of a stone is relatively high, since it begets relatively little air resistance compared to its weight; a feather, on the other hand, falls more slowly because its broad surface creates a far greater air resistance in proportion to its negligible weight.

Lift is a more complex phenomenon. It is produced by the motion of a specially shaped surface through the air, and acts in a direction perpendicular to the motion. Thus if a lift-generating body moves horizontally, its lift works to counteract gravity. Without any motive power, a body cannot produce enough lift to overcome gravity and stay aloft indefinitely, but the force does act to slow the descent and allow the body to follow a sloping path to the ground—the graceful action known as gliding.

Nature exploits these forces in myriad ways. Many plants extend their range by distributing seeds that have evolved into aerodynamic shapes. The tiny seeds of the dandelion, suspended from downlike clusters, fall to the ground so slowly that the least breeze can sweep them up and carry them for miles. Other seeds have true wing shapes that generate lift, allowing them to glide long distances. The seed of the tropical palm *Zanonia macrocarpa*, for instance, is so well formed that it inspired the wing shape of an early biplane; it bears a remarkable similarity to the design of an advanced military aircraft, the Northrop B-2 Stealth bomber. The familiar maple seedpod spins as it descends like a tiny helicopter rotor, its wings generating lift that slows its fall and gives the wind time to carry the seed away from its parent tree.

Animals, too, use aerodynamic forces in their quest for survival: flying squirrels, lemurs, snakes, lizards, and—perhaps best-known of all—flying fish. All of these creatures have one thing in common. Though frequently airborne, they do not really fly—at least, not as birds, bats and insects fly, staying aloft indefinitely under their own power. Instead, they use specialized surfaces of their body to glide. The North American flying squirrel, for example, spreads a sail-like membrane between its front and back legs, enabling it to soar from lofty perches for distances of up to a hundred feet at a time. To prepare for landing, the squirrel arches its body, letting the membrane balloon outward to catch air and slow its fall.

CANVAS BRAKES TO BREAK A FALL

The nylon canopy that blossoms over the head of a parachutist performs the same function as the arched body of the flying squirrel, creating additional air resistance to cut the speed of descent. Expanded by the rushing air as it opens, the parachute initially develops a braking force that exceeds the pull of gravity, thus causing an actual deceleration. As the velocity dwindles, so does the strength of the air resistance, until it is finally in precise balance with the force of gravity. At this point, with no overall force acting on the parachute and its passenger, the descent continues at about 15 miles per hour without speeding up or slowing down.

The first parachutes, in the late 18th Century, were made like umbrellas. Thin wooden ribs supported vast expanses of canvas, some measuring as much as 195 feet across. The first person to entrust himself to one of these monumental con-

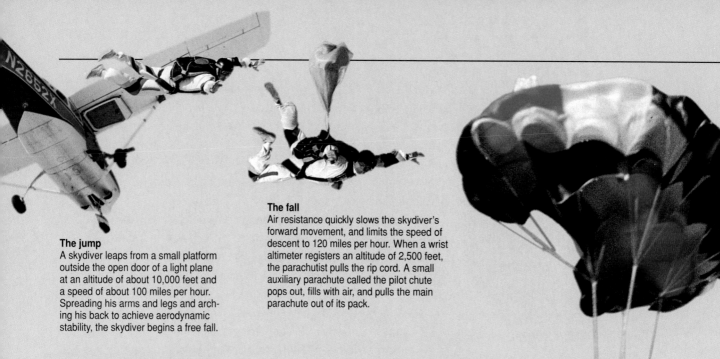

The jump
A skydiver leaps from a small platform outside the open door of a light plane at an altitude of about 10,000 feet and a speed of about 100 miles per hour. Spreading his arms and legs and arching his back to achieve aerodynamic stability, the skydiver begins a free fall.

The fall
Air resistance quickly slows the skydiver's forward movement, and limits the speed of descent to 120 miles per hour. When a wrist altimeter registers an altitude of 2,500 feet, the parachutist pulls the rip cord. A small auxiliary parachute called the pilot chute pops out, fills with air, and pulls the main parachute out of its pack.

The shock
The parachutist experiences a tooth-rattling jolt called opening shock as the main canopy fills with air, abruptly cutting the descent velocity from 120 to about 15 miles per hour. If the main parachute malfunctions, the skydiver quickly deploys a reserve chute; it can be effective if it opens as little as 200 feet above the ground.

traptions was Louis-Sebastien Lenormand, a French physicist, who drifted safely to the ground from the tower of the Montpellier Observatory, France, in 1783. Two years later another Frenchman, Jean-Pierre Blanchard, used a hot-air balloon as the jumping-off point for the first test of a parachute of his own design. The prudent Blanchard, however, delegated the actual descent to a dog. The wisdom of this decision was borne out in 1793, when Blanchard finally undertook his first parachute jump and broke his leg.

These early, unwieldy parachutes were little more than curiosities, especially in the absence of a way to get into the sky in the first place. As more flying machines like Blanchard's balloon began to appear, their very existence stole some of the mystique of the parachute. Lighter-than-air flight, after all, was more sustainable than a parachute drop, and held the potential for travel as well. The advent of powered flight, however, did create a practical application for the parachute, as a potential lifesaver in the event of midair mechanical failures. The continued development of powered aircraft eventually led to other parachute applications: The chute became a useful instrument for dropping supplies and people—particularly troops—into otherwise inaccessible areas. By the late 20th Century it also evolved into a piece of sporting equipment (*above*). In contrast to its gargantuan forebearers, a modern sport parachute stows neatly into a small pack; even when open, its nylon canopy is only about 25 feet across.

FROM FALLING TO SOARING

The flying machines that would eventually overshadow such novelties as the parachute passed through a long and arduous evolution. Early experimenters had to solve the fundamental problem of obtaining lift, which is only available when air is moving over a lifting surface, such as a wing. One way to achieve this is to equip the craft with a power source that will move it through the air. But as long as engines remained too heavy—as they did until the Wright Brothers' historic flight at the beginning of the 20th Century—powered flight remained an attribute unique to the animal kingdom. Two other lifting techniques were at hand, though. Would-be fliers could tap the power of gravity by climbing to a high place, like a mountain or hillside, then jumping off and using the speed of the fall to get the

This sequence of photographs displays a typical jump of a sport parachutist. Novices do not open the parachute themselves; a static line attached to the aircraft deploys it automatically.

The ride
An apex vent in the top of the parachute allows compressed air to escape and permits a steady descent. To achieve the control over direction and speed of descent needed for an accurate landing, the skydiver uses steering lines that spill air out of slots or gores in the chute.

The touchdown
Watching the approaching ground intently, the parachutist bends his knees just before landing. Even after a 10,000-foot descent, the impact is about the same as jumping off a wall eight feet high. At touchdown, the skydiver rolls with the momentum, absorbing most of the shock with his leg muscles, and immediately starts to gather in his chute.

air moving over the wing. Alternatively, they could seek out updrafts—masses of upward-moving air—and other wind patterns that would provide lift to keep the craft in the air.

Birds use both of these techniques to stay aloft with a minimum expenditure of energy. The Andean condor, for example, follows a carefully selected flight path that takes it into winds rushing up steep mountain slopes. After rising effortlessly with the moving air, the condor can then move away from the mountain, gliding at a shallow downward angle and trading altitude for lift-generating speed. Eventually the condor returns to the region of the updraft to take another ride to the top. Other birds, from albatrosses to vultures, use similar tactics in their continuous aerial hunt for food *(right)*. Migratory fowl, too, take advantage of weather patterns; often their flight paths allow them to soar on winds deflected upward by ridges. Some scientists suggest that the characteristic "V" formation of geese is also an energy-saving stratagem that allows every bird but the leader to rest its wing tip on the rising vortex of air displaced by the wing of the bird in front. The hard-working goose at the apex of the "V" periodically falls back into the formation, to be replaced by another, so that the free rides are shared equitably.

Lessons learned from observations of soaring birds found their way long ago into the construction of kites, the first man-made objects that exploited the forces in moving air. Appearing in China more than 2,000 years ago, the tethered fliers were originally used by adults in religious ceremonies and in war, as well as for entertainment. Chinese generals hoisted soldiers aloft on giant kites to observe enemy troops, and the first concrete evidence of kites in Europe is a book written in the 14th Century depicting three soldiers using a kite to drop a bomb on a castle. Only later did kites assume their modern identity as children's toys, although they were still employed at tasks as diverse as fishing in hard-to-reach waters and experiments with electricity.

As varied in use and size as they have been since antiquity, all kites share a reliance upon the simplest aerodynamic principles. The force that holds a kite aloft comes from the resistance of its surface area to the wind. Held at an angle by its string, the kite deflects the moving air downward; in a breeze of constant strength, the forces applied by the wind and the string are in perfect balance, and the kite hangs motionless in the air.

By the late 19th Century these forces were the subject of rigorous experimentation in Europe, where several inventors hoped to bring new generations of kites to war. One leading proponent of man-carrying kites was B. F. S. Baden-Powell, a British Army officer and brother of the founder of the Boy Scouts. His intention was to provide the army with a means of aerial observation. Baden-Powell's first success was in 1894, with a giant kite made of cotton cambric stretched over a bamboo frame. Thirty-six feet high, the kite had enough power to lift a man, although safety considerations limited flights to no more than ten feet above the ground (Baden-Powell usually took along a parachute on his own test flights). The next year Baden-Powell patented a new design, which used several kites about ten feet square, linked one above the other by ropes. The apparatus, which Baden-Powell called a Levitor, used four to seven small kites at a time, depending on the wind conditions. In moderate to strong winds, the Levitor proved capable of lifting the inventor as high as 100 feet.

SOARING AT SEA
The swooping flight of long-winged seabirds takes advantage of slower wind speeds near the surface of the open sea. With the wind at its back, an albatross descends from a height of about 50 feet, gaining speed and distance. Mere inches from the wave tops, it wheels into the weaker surface breeze, then uses its momentum to climb back through stronger winds before turning for another downwind dive. Using this technique, the albatross can patrol the ocean for hours without flapping its wings.

COASTING ON CURRENTS

The broad wings of eagles and condors allow the heavy birds to ride aloft on currents of air sweeping upward over cliffs and hillsides. The gliding birds tack across the updraft, steadily gaining altitude as they go.

RIDING THERMALS

Land birds with broad, rectangular wings frequently use thermals—columns of hot, rising air—to gain altitude. Entering the thermal near the ground, a vulture circles within the column, gaining several thousand feet with very little effort.

Big kites soon found other functions. In 1901, Italian inventor Guglielmo Marconi used a Levitor to raise the receiving antenna for his first transatlantic wireless test at St. John's, Newfoundland. A strong wind tore apart one kite, but a second fared better, carrying the antenna to a height of 400 feet, and Marconi received the message he was listening for. For a brief period in the 1900s, the British Army adopted a system of observation kites (using a design that was safer and more stable than Baden-Powell's); during World War I, the German navy sent similar man-lifting devices to sea with their submarines, vastly increasing the U-boats' chances of discovering distant prey. Like parachutes, however, kites were suited only for a few special jobs. And while kite-builders struggled to perfect their tethered creations, other inventors were pouring their energy into machines that would not merely resist the wind, but ride upon it.

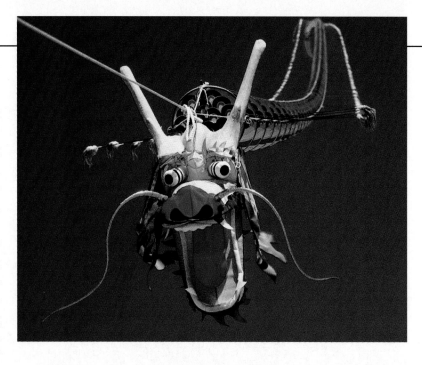

The fearsome face of a dragon supports a long, undulating tail on a classic Chinese kite. Made of hand-painted paper covering a split-bamboo frame, it is little different from kites that flew 2,000 years ago.

BREAKING THE TETHER

Otto Lilienthal was one of this new breed of experimenters. A distinguished German engineer, Lilienthal was captivated by the sweeping flight of seagulls he encountered while installing a foghorn of his invention in German lighthouses. He studied birds in meticulous detail, eventually publishing an authoritative book drawing connections between natural and artificial flight. His aviation research was equally painstaking, beginning with a series of kite experiments in the 1870s and progressing to free-flying machines with wings modeled after those of soaring birds. In the course of five years in the 1890s, Lilienthal flew 18 different kinds of gliders, taking careful notes on their aerodynamic qualities.

He launched his flights from the top of a 50-foot hill, built specially for the purpose in an open area near Berlin. The hill was conical, which allowed Lilienthal to fly directly into the wind no matter which direction it blew, thus increasing the lift-generating movement of air over the curved, fabric-covered wings of his fixed-wing gliders. Half-sitting on a trapeze so that his head and shoulders were above the wings, Lilienthal would step off the hill to begin a long, gentle glide to the ground. Maneuvering the craft by shifting his weight, the inventor regularly flew for distances up to 750 feet, becoming the first man to achieve sustained, controlled flight in a heavier-than-air machine.

Lilienthal paid dearly for his achievement. He was piloting one of his most reliable gliders on a summer's day in 1896 when a sudden gust of wind brought him to a standstill in midair. One wing lost its lifting power, dropped sharply, and the glider sideslipped to the ground. Lilienthal, his spine broken in the crash, died the next day. His work outlived him, however. The body of aerodynamic data generated by his experiments was an important step in the development of powered flight, which became a reality less than a decade after his death. And even after aircraft with engines began winging their way through the skies, the memory of

Eighty feet from tip to tip, the slender wings of a sailplane carry it on air currents at play above the Rocky Mountains.

Lilienthal's graceful gliders lived on in a quieter, more contemplative style of flying, embodied in the modern sailplane.

This sporting craft, with its long, tapered wings and streamlined, glassy-smooth fuselage, is a paragon of specialized efficiency in design, structure and materials. Built to produce maximum lift, it can soar for hours when conditions are right, covering hundreds of miles at speeds up to 150 miles per hour in a dive. Every flight, however, is a battle of human wits against nature; to stay aloft, a pilot must find an updraft and use it as an elevator to gain altitude. The pilot looks for visual cues—birds spiraling, leaves blowing upward, rising clouds—and then steers for them. But updrafts are not always available, and they sometimes disappear; the prudent pilot watches the rate-of-climb indicator to monitor lost altitude, while keeping an eye on the home field or an alternate landing spot within gliding range.

LIGHTER THAN AIR

The rising masses of warm air that prove useful to soaring birds and glider pilots are instrumental to another, completely different kind of flight. Captured in a leakproof bag, warm air retains its tendency to rise, and can be harnessed to lift weight as long as it is warmer, and thus lighter, than the air around it.

This simple principle occurred to French paper-maker Joseph Montgolfier one evening in 1783 as he watched the fire in his hearth. Struck by the potential lifting power of heated air, Montgolfier made a bag of fine silk and lit a fire under it. The bag filled up with hot air and rose to the ceiling. In June of that year, aided by his brother, Montgolfier demonstrated the effect in public by raising a huge cloth bag filled with hot, smoky air rising from burning wood and straw. When this primitive balloon actually proved workable, the brothers attached a basket and selected the world's first balloonists—a duck, a sheep and a cockerel. The animals landed unharmed after an eight-minute voyage.

The following month—November 1783—two human volunteers ascended in a Montgolfier balloon equipped with a brazier. The fledgling fliers took along a bucket of water and sponges in case the fire got out of control, but the first manned balloon ride was a complete triumph: The balloon rose 500 feet and floated above the rooftops of Paris for 25 minutes.

Even as the Montgolfier brothers were developing their hot-air balloon, another experimenter was working on a completely different approach to lighter-than-air flight. Professor Jacques A. C. Charles filled a small bag of rubberized silk with hydrogen, a gas only seven percent as heavy as air. Launched in Paris, it flew for about 45 minutes before coming down in the village

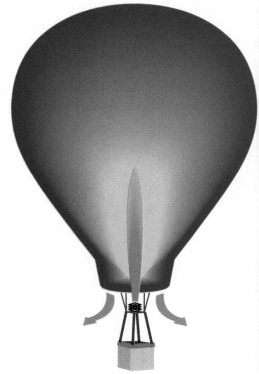

ARCHIMEDES' PRINCIPLE

The ascent of a hot-air balloon illustrates Archimedes' Principle, which states that any object suspended in a fluid experiences a buoyant force equal to the weight of the fluid it displaces. A typical balloon is subjected to an upward force of about 8,000 pounds—the weight of the air it displaces. When the air inside the envelope is heated, it expands and about one-quarter escapes through the bottom opening, reducing the overall weight of the balloon. Now the buoyant force exceeds the weight, and the balloon rises.

of Gonesse, about 15 miles away. The balloon met an unfortunate end: It was torn to shreds by pitchfork-wielding villagers, intent on destroying what they believed was a device of the devil. Undeterred, Charles built a larger, man-carrying balloon, which he flew just a month after Montgolfier's flight. The enthusiasm for ballooning had grown so great that Charles's takeoff from the Tuileries Gardens was witnessed by an estimated 400,000 people. The 28-foot-wide balloon, remarkably similar to modern gas-filled designs in terms of construction, valves and ballast, was a complete success, carrying Charles and a friend 27 miles in a two-hour flight.

STEERING IN THE SKIES

While balloons quickly proved their ability to carry people into the sky and over distances, they suffered the drawback of almost complete unpredictability. The only possible destination for any given balloon trip was somewhere directly downwind from the launch site. Hence inventors immediately turned to the problem of creating a balloon that could be steered under its own power: a dirigible. They were thwarted, however, by the materials and equipment available. The first airships were essentially elongated balloons, filled with hydrogen, trussed into shape like corseted whales, and equipped with primitive engines and rudimentary steering devices. These early airships were particularly prone to losing their aerodynamic shape, sagging limply with any decrease in the pressure of the gas that kept them inflated.

An inkling of a solution to this problem appeared in 1900, with the flight of the first rigid airship. Designed by Count Ferdinand von Zeppelin, a retired German cavalry officer, the dirigible boasted a rigid aluminum frame and a streamlined fabric envelope surrounding multiple gas-filled lifting compartments inflated with hydrogen. At 420 feet long and almost 40 feet in diameter, it was far larger than previous airships, and held enough hydrogen to lift 27,000 pounds. Structure, engines and ballast weighed so much, however, that the payload for the giant's first flight was just 660 pounds.

Count von Zeppelin spent the next decade developing his idea, overcoming business setbacks and a series of disastrous wrecks and fires. Eventually his ships proved powerful and reliable enough for a passenger airline serving a number of German cities to be inaugurated. During World War I, Zeppelins in the service of the German military even carried out bombing raids over England. But dirigible technology reached its peak of development after the war (and after the death of von Zeppelin in 1917), when a new generation of the giant dirigibles established high standards for comfort and elegance in travel. As fast as contemporary airplanes, these rigid airships had far greater range and carrying capacity than heavier-than-air machines. In the late 1920s, the *Graf Zeppelin*, most successful of the German fleet, routinely carried up to 20 passengers across the Atlantic in about 66 hours, pampering the travelers en route with superb cuisine and unmatched vistas of ocean and sky.

The glory days of the giant airships were numbered, however. By the early 1930s many of the dirigibles' performance advantages were surpassed by aircraft. Furthermore, structural deficiencies still lurked in the design of the enormous craft. Their aluminum frames were so large (the *Hindenburg*, star of the fleet in the 1930s, was 804 feet long and 134 feet in diameter) and so lightly built that the ships were extremely vulnerable to bad weather; turbulence could easily stress and distort the fragile structures.

An even greater hazard, however, was the flammable hydrogen used as a lifting gas in most dirigibles. In practical terms, the age of the rigid airship ended on May 6, 1937, at Lakehurst, New Jersey, when, for reasons unknown, the hydrogen aboard the *Hindenburg* suddenly burst into flame. The popular ship was just coming in for a landing, and the disaster, which claimed 35 lives, was reported to a stunned world in a live radio broadcast. Newsreel film showed the pride of the skies reduced to a bent, blackened skeleton.

The grim demise of the *Hindenburg* closed a chapter in the history of air travel, although airships continued to hold their specialized place in the lineup of flying machines. Sophisticated gas-control systems have made nonrigid dirigibles more practical; nonflammable helium has made them far safer. Today, airships are routinely employed to patrol vast expanses of ocean, monitor changing weather conditions, haul cargo in and out of otherwise inaccessible areas and, in their most public role, to carry advertising messages as flying billboards.

But airships, like balloons, kites and gliders, have proved to be peripheral to the main business of human flight. Each in its way helped give people a taste of flight, but none can deliver the combination of speed, power and mass access offered by aircraft with engines. With these, humans would conquer the skies.

Preparing a hot-air balloon for flight, a helper holds the envelope open for the flame that heats the air trapped inside. The expanding envelope is already beginning to rise from the ground.

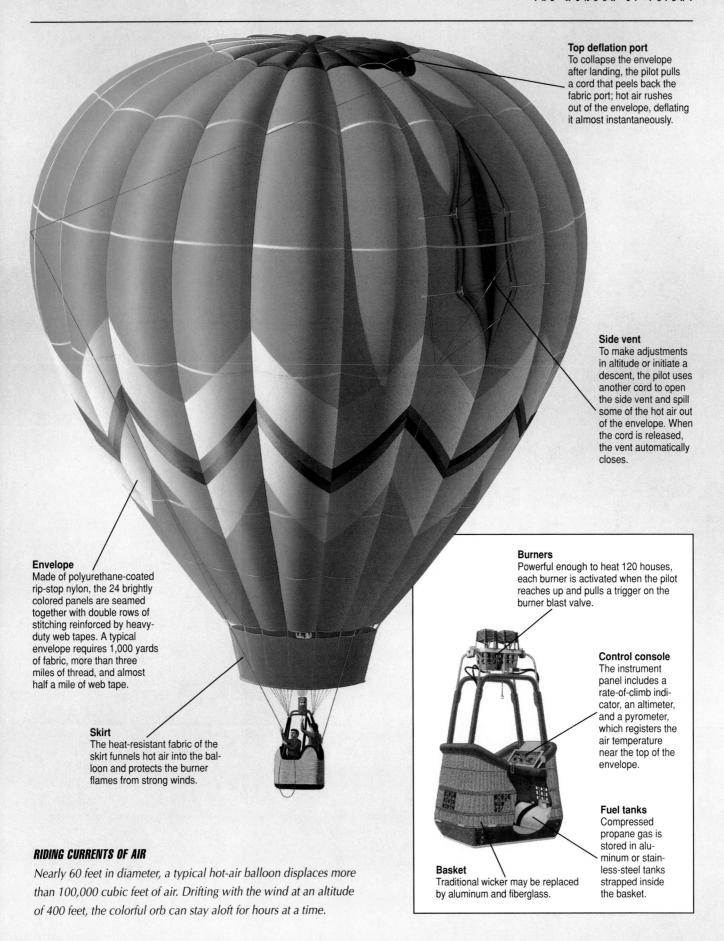

Top deflation port
To collapse the envelope after landing, the pilot pulls a cord that peels back the fabric port; hot air rushes out of the envelope, deflating it almost instantaneously.

Side vent
To make adjustments in altitude or initiate a descent, the pilot uses another cord to open the side vent and spill some of the hot air out of the envelope. When the cord is released, the vent automatically closes.

Envelope
Made of polyurethane-coated rip-stop nylon, the 24 brightly colored panels are seamed together with double rows of stitching reinforced by heavy-duty web tapes. A typical envelope requires 1,000 yards of fabric, more than three miles of thread, and almost half a mile of web tape.

Skirt
The heat-resistant fabric of the skirt funnels hot air into the balloon and protects the burner flames from strong winds.

Burners
Powerful enough to heat 120 houses, each burner is activated when the pilot reaches up and pulls a trigger on the burner blast valve.

Control console
The instrument panel includes a rate-of-climb indicator, an altimeter, and a pyrometer, which registers the air temperature near the top of the envelope.

Fuel tanks
Compressed propane gas is stored in aluminum or stainless-steel tanks strapped inside the basket.

Basket
Traditional wicker may be replaced by aluminum and fiberglass.

RIDING CURRENTS OF AIR

Nearly 60 feet in diameter, a typical hot-air balloon displaces more than 100,000 cubic feet of air. Drifting with the wind at an altitude of 400 feet, the colorful orb can stay aloft for hours at a time.

Vessels of the Air

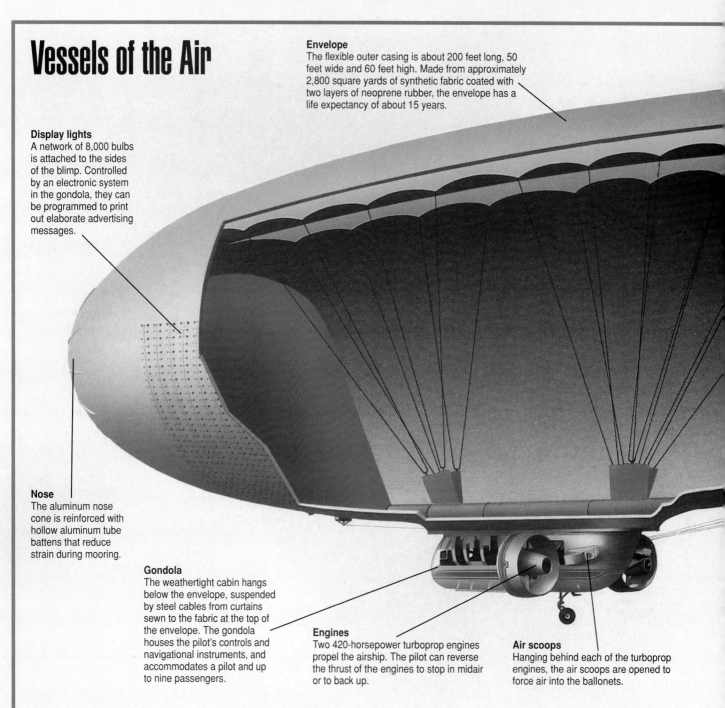

Envelope
The flexible outer casing is about 200 feet long, 50 feet wide and 60 feet high. Made from approximately 2,800 square yards of synthetic fabric coated with two layers of neoprene rubber, the envelope has a life expectancy of about 15 years.

Display lights
A network of 8,000 bulbs is attached to the sides of the blimp. Controlled by an electronic system in the gondola, they can be programmed to print out elaborate advertising messages.

Nose
The aluminum nose cone is reinforced with hollow aluminum tube battens that reduce strain during mooring.

Gondola
The weathertight cabin hangs below the envelope, suspended by steel cables from curtains sewn to the fabric at the top of the envelope. The gondola houses the pilot's controls and navigational instruments, and accommodates a pilot and up to nine passengers.

Engines
Two 420-horsepower turboprop engines propel the airship. The pilot can reverse the thrust of the engines to stop in midair or to back up.

Air scoops
Hanging behind each of the turboprop engines, the air scoops are opened to force air into the ballonets.

A blimp is a steerable, powered balloon that uses internal gas pressure to maintain its shape; the buoyancy of the gas keeps it aloft. A blimp like the one shown here (modeled after the Goodyear blimps) weighs about 12,000 pounds, not including pilot, passengers and cargo, and holds about 250,000 cubic feet of helium—a nonflammable gas seven times lighter than air.

At the takeoff command of "Up ship," the two engines roar to full power. The airship noses up at a startling angle, beginning a laborious ascent to its cruising altitude, about 1,000 feet. Even after the airship levels off, the wind causes it to pitch gently up and down, like a ship riding ocean waves. Throttled back, the engines are very efficient: An airship can fly eight hours a day for almost a week on the amount of fuel it takes to taxi a jumbo jet from ramp to departure gate.

To land, the pilot points the nose down and brings the engines back to full power, driving the airship toward the ground at 30 miles per hour. The course is squarely into the wind, to prevent the nose from being blown sideways as the blimp loses speed. The ship slows almost to a stop, ground crew quickly load on ballast, then a spindle on the airship's nose engages with latches on a tall mooring mast.

Anchored only at this point, the airship is free to rotate 360 degrees around the mast as the wind changes direction.

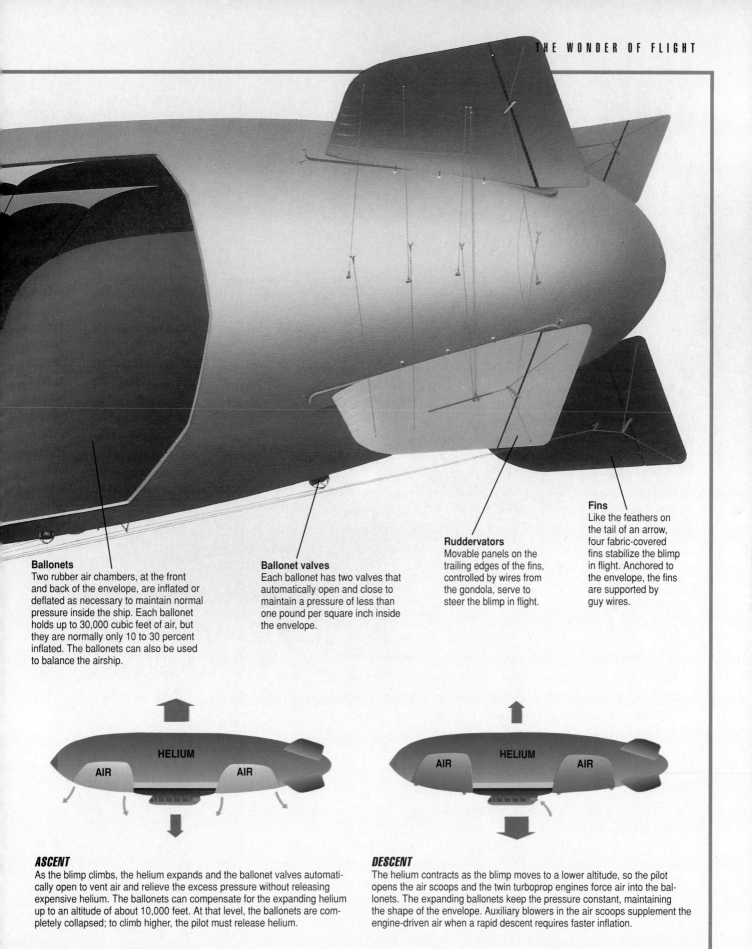

Ballonets
Two rubber air chambers, at the front and back of the envelope, are inflated or deflated as necessary to maintain normal pressure inside the ship. Each ballonet holds up to 30,000 cubic feet of air, but they are normally only 10 to 30 percent inflated. The ballonets can also be used to balance the airship.

Ballonet valves
Each ballonet has two valves that automatically open and close to maintain a pressure of less than one pound per square inch inside the envelope.

Ruddervators
Movable panels on the trailing edges of the fins, controlled by wires from the gondola, serve to steer the blimp in flight.

Fins
Like the feathers on the tail of an arrow, four fabric-covered fins stabilize the blimp in flight. Anchored to the envelope, the fins are supported by guy wires.

ASCENT
As the blimp climbs, the helium expands and the ballonet valves automatically open to vent air and relieve the excess pressure without releasing expensive helium. The ballonets can compensate for the expanding helium up to an altitude of about 10,000 feet. At that level, the ballonets are completely collapsed; to climb higher, the pilot must release helium.

DESCENT
The helium contracts as the blimp moves to a lower altitude, so the pilot opens the air scoops and the twin turboprop engines force air into the ballonets. The expanding ballonets keep the pressure constant, maintaining the shape of the envelope. Auxiliary blowers in the air scoops supplement the engine-driven air when a rapid descent requires faster inflation.

SHAPE AND POWER

I t is a moment every U. S. Navy carrier pilot must learn to face—the blurred, gut-wrenching instant of takeoff, when his F-14 Tomcat leaves the deck as a 30-ton unguided ballistic missile, hurled skyward by the world's most powerful slingshot.

The flight deck of an aircraft carrier is a maelstrom of organized confusion and violent sound. Planes land and lift off at split-second intervals, and the blast of their jets is so shrill that deck crews must wear ear protectors. Two steam-powered launch catapults shudder and hiss. The grinding noise of arresting-gear engines that power the landing-gear mechanism reverberates through the steel deck. An officer orchestrates traffic toward the catapults; he must launch a squadron of 20 fighter planes, at the rate of one every 45 seconds. From a control tower 140 feet above the rolling sea, an air boss monitors the entire operation, his finger poised to hit the abort button at the first hint of trouble.

The first F-14 is hitched to the forward catapult. A catapult officer checks that its wing flaps are set and signals the pilot to apply full power. The pilot revs up, makes a final check of his engine gauges, braces against his seat and throws a smart salute to the catapult officer.

The catapult is preset to fling the F-14 from standstill to 200 miles per hour, virtually instantaneously. So sudden is the acceleration that it confuses the senses. Pilots swear that their aircraft slow down upon leaving the deck, a misconception that arises because the brain takes several seconds to recover from the "catstroke." Even flight instruments lag behind. Battling gravity, the F-14 dips momentarily toward the waves. Then its 40,000 pounds of jet thrust bite hard, and the critical transition occurs: from uncontrolled missile to flyable aircraft. The pilot, now in full command, points the fighter's needle nose upward. The Tomcat rockets into the sky at a 60-degree angle.

With both throttles jammed full forward, the F-14's afterburners kick in, punching it through the sound barrier. Climbing almost vertically, faster than a .45 caliber bullet, the Tomcat reaches 50,000 feet in two minutes. Leveling off, it passes

A carrier-based F-14 Tomcat revs its engines to full screaming power just seconds before takeoff.

Mach 2—twice the speed of sound—accelerating toward its maximum velocity of 1,650 miles per hour.

What enables this amazing fighter to fly? Few people can gaze at the stream-lined beauty of a modern jet as it roars overhead without asking the same question. The answer—or at least part of the answer—can be traced back to the Wright Brothers who, nearly a century ago, took a close, discerning look at the world of nature, and so discovered the secret to powered flight.

THE FLIGHT OF BIRDS

Ever since humans first looked skyward, they have envied the flight of birds. Even in the era of transatlantic flights and space shuttles, aviation scientists still study them. Birds remain the lords of the air—the world's purest and most aerodynam-ically efficient fliers.

These remarkable creatures are capable of reaching speeds in excess of 100 miles per hour, of making migratory flights halfway around the globe, of soaring effortlessly on the wind for hours, and of performing aerobatics and dogfighting maneuvers that would cause a fighter pilot's head to spin. They serve as living blueprints of the theory of flight. Even the leading-edge flaps on contemporary jets find their counterpart in birds' wings. Indeed, without their example, humankind might still be searching for ways to conquer the air.

Like airplanes, birds were not always efficient fliers. Nature has taken millions of years to adapt them into the flying machines they are today. The discovery of a 150-million-year-old fossil in a Bavarian limestone quarry in 1861 gave science its first known bird. Christened *Archaeopteryx*—old wing—it was as much a feath-ered reptile as a bona fide fowl, and it provided convincing evidence for Charles Darwin's theory of evolution.

Within the next 150 million years, impelled by the needs of feeding, breeding and survival, birds developed into their modern form. Scientists believe that their

A multiflash photograph of an owl in flight reveals the powerful downstroke of the wings: The tips of the primary feathers are bent upward, flaring out almost at right angles to the wing. As the wing descends, these feathers bite into the air and pull the bird forward. At the end of each down-stroke, the wings are quickly brought back and tucked slightly in an oar-like move-ment that provides additional thrust. On the upstroke, the primary feathers separate to reduce air resistance. Then the whole cycle begins again.

light, hollow wing feathers evolved from thick reptilian scales and their stiff tail feathers, which act as in-flight stabilizers from the bony whiplash tail of their reptilian ancestors. To reduce weight further, sturdy bones suitable for land dwelling evolved into hollow, buoyant structures with paper-thin walls, some braced with cross-struts like the wings of airplanes.

The earliest experimenters with flight tried to mimic the flight of birds by strapping on homemade wings, but a human's bulky, ponderous physique is simply not designed for the task of self-propelled flight. Nor could the human heart cope with the aerobic demand of wing flapping—which, in the humble sparrow, is an astonishing 800 heartbeats per minute.

The pioneer birdmen were victims of a common misconception. For centuries, it was believed that birds "swam" across the sky, propelled by a backward and downward wing stroke. Modern high-speed photography shows quite a different process. A bird's forward thrust comes from the primary feathers at the wing tips, which act almost like propellers. As each downstroke begins, the tips of these primaries twist up at an angle to the rest of the wing. The descending tips thus bite into the air in much the same way as the angled blades of a propeller, pulling the bird forward. At the same time, the rest of the wing provides the lifting surface that keeps the bird airborne.

THE WRIGHTS AND THE TURKEY VULTURE

It was only when early aviators began to focus on gliding rather than flapping that they learned how the curved airfoil shape of a bird's wing generates aerodynamic lift. The first person to light upon this phenomenon was the British visionary Sir George Cayley, often called "The Father of the Airplane." In 1810 Cayley issued a pioneering report, *On Aerial Navigation*, which revealed his studies of crows and other soaring birds. In effect, it marked the birth of the science of aerodynamics. Cayley soon had more to say—on the shapes of airfoils, on movable control

Alula
A small tuft of feathers, originating from the thumb, the alula sweeps forward and separates when a bird lands.

Hand section
Similar to a human hand but with longer and narrower bones in relation to the rest of the arm, the bird's "hand" has just two fused fingers and a thumb that can open and close.

Bones
Filled with air, the honeycombed long bones of flying birds are reinforced by a crisscross of internal struts, much like the frame of an airplane or the stem of many plants.

Wings
Composed of an upper arm, a forearm and a hand section, the wing is attached to breast muscles through a triple-jointed support that lends flexibility and allows the shoulder to be twisted into different positions during flight.

Primary flight feathers
Stemming from the finger bones of the hand, they are largely responsible for the power for flight.

Secondary flight feathers
These flight feathers stem from the outside bone of the forearm and supplement the primary flight feathers.

Tail feathers
Help the wing balance, steer and brake the bird's body in flight.

Breastbone
The breastbone has a deep keel jutting out of its center, anchoring the bird's huge flight muscles.

Feathers
More than a million very fine projections, branching off a central shaft, are meshed together by miniscule hooklets, too small to be seen by the naked eye. If the hooklets disengage, the bird simply draws the feather through its beak several times and the projections automatically rehook, just like a zipper.

Built for Flight

Behold the bird—nature's perfect flying machine and the envy of all aircraft designers. Their wings are driven by muscles that are amongst the most powerful in the animal world.

Muscle-powered flight requires a high metabolic rate and a very efficient respiratory system. A bird's unusually large heart pounds at a fantastic rate—up to 1,000 beats per minute in a hummingbird. Its lungs are supplemented by a series of air sacs, spread throughout the body, that supply a continuous flow of oxygen to the powerful flight muscles, and help keep the bird's body temperature at a reasonable level.

To keep weight to a minimum, evolution has rid the bird of excess baggage. For instance, lightweight beaks have replaced heavy jaws laded with teeth. The bones are literally filled with air, crisscrossed by truss-like supports that provide both strength and resilience. The skeleton of a three-pound frigate bird, despite its seven-foot wingspan, weighs just four ounces. This efficiency of form is also apparent in the bird's most important structure. Feathers are one of the lightest yet strongest materials formed by any creature. They serve a dual purpose: providing the surfaces on the wings and tails for lift and propulsion, and insulating the bird against loss of heat and protecting it from the weather.

Although man has perfected designs and developed powerful engines that enable him to travel far faster than any bird, in aircraft that incorporate the latest in digital technology, he has still to attain the refined simplicity and efficiency of two flapping wings.

ANATOMY OF A POWERFUL FLIER

The ubiquitous Columba livia, *commonly known as the rock dove or pigeon, is about 13 inches long and weighs between 10 1/2 to 15 ounces. A strong flier, it is a very familiar sight to city dwellers worldwide.*

Air sacs
Connected to the lungs and interconnected by tubes, the air sacs deliver oxygen to almost every part of the body—even into the hollow cavities of the long bones. The air sacs also contribute to the buoyancy of the bird.

Breast muscles
Anchored between the breastbone, the *pectoralis major* is the largest flight muscle and drives the powerful downstroke of the wing. Beneath is the smaller *pectoralis minor* which pulls the wing up. In a strong flier like the pigeon, these breast muscles may account for one-half of the bird's total weight.

Wing muscles
Bundles of wing muscles pull the wing forward and back and rotate the upper arm bone. Others extend or fold the wing in much the same way as human triceps and biceps.

Legs
Probably the most effective shock-absorbing mechanism found in nature, each leg consists of three rigid bones with joints that work in opposite directions to cushion the landing.

surfaces such as the elevator and rudder, and on proposed configurations for airplane wings. Then in 1853, after a lifetime of cautious experimentation, Cayley built the world's first successful glider. Crewed by his reluctant coachman, the cloth-winged apparatus was reportedly seen wafting across a valley on Cayley's Yorkshire estate.

Four years later, French sea captain Jean-Marie Le Bris, who studied albatrosses while sailing around Cape Horn, constructed a 46-foot glider closely resembling the majestic seabird. He equipped it with an adjustable, pedal-operated tail and, as an artistic flourish, a long artificial beak. Racing downhill atop a horse-drawn cart, Le Bris apparently gained enough windspeed to ride his *Albatross* to a height of 300 feet. Unfortunately, he crashed on a second flight—shattering both his whimsical glider and his leg.

Such mishaps were all too frequent in flying's early days. Even the greatest of the glider pioneers, the German engineer Otto Lilienthal, who in six years of intensive experimentation made nearly 2,000 flights, took his share of tumbles, the last one fatal. Lilienthal's legacy was vital, however, for it inspired Orville and Wilbur Wright, joint owners of a bicycle repair shop in Dayton, Ohio, to turn their mechanical skills to aviation. The Wrights' first goal was to develop an efficient control system that would help prevent the kind of crashes that had killed Lilienthal. The brothers became dedicated bird-watchers, and it was Wilbur's observation of turkey vultures that provided the necessary clue. He noted that, when tipped sideways by a gust of wind, the bird righted itself by a slight twisting of its wing tips; the leading edge of one wing tip would turn down, while the other turned up. Wilbur reasoned that a system enabling pilots to twist, or warp, a glider's wings in this manner would provide adequate control. After successfully testing the theory on a kite, the brothers incorporated wing warping into their first glider, built in 1900.

By 1902 they had made nearly 1,000 flights over the sand dunes at Kitty Hawk, North Carolina, a site they had chosen for its rolling terrain and steady wind. They were now experimenting with their third glider, and had perfected a flight-control system that included an elevator to regulate the glider's nose position, and a rudder to help control the tail. To measure the performance of various wing designs, they had even constructed a primitive wind tunnel—an open-ended wooden box with a glass window and a steel fan driven by a one-horsepower gasoline engine—that forced air past model wings. Soon they had a stable, maneuverable aircraft. To achieve powered flight, all they needed was a lightweight gasoline engine. Finding none that suited them, the remarkable brothers designed their own, then carved its twin wooden propellers.

On December 17, 1903, the Wrights' *Flyer* made its wobbly maiden voyage—120 feet across the sand dunes of Kitty Hawk at shoulder height, with Orville piloting. Twelve glorious seconds of powered flight christened a new era in aviation. The brothers made three more flights that day, achieving in one of them a distance of 852 feet, and staying aloft for 59 seconds. Paying tribute to their feathered teachers, Orville wrote: "Learning the secret of flight from a bird was a good deal like learning the secret of magic from a magician. After you once know the trick, you see things that you did not notice when you did not know exactly what to look for."

As late as 1912, pioneer airmen were still trying to imitate the flapping flight of birds. Here a rare photograph captures an Austrian tailor, Franz Reichelt, as he prepares to leap off the Eiffel Tower. His billowing, parachute-like suit barely slowed his fall, and he plunged 190 feet to his death.

GETTING AN ANGLE ON LIFT

One secret the Wright Brothers explored was the phenomenon of lift—a basic aerodynamic effect that allows an aircraft to overcome the pull of gravity and so stay aloft.

Despite appearances, airplanes do not ride on a cushion of air; the process is more one of suction. The principle was first explained by the Swiss scientist Daniel Bernoulli, who in 1783 discovered that increasing the velocity of a fluid, as it moves through a pipe or other conduit, will diminish the pressure it exerts against the conduit's sides. For example, when a wide river is forced to negotiate a narrow canyon, the water pressure on the riverbed is measurably lower than in the gentler stretches upstream and downstream. A pair of pressure gauges, one placed in the canyon's swirling rapids, the other lodged in a quiet pool well below them, gives dramatic proof of this.

In the world of aeronautics, air is as much a fluid as water, and it behaves in just the same way. The shape of an aircraft's wing, called an airfoil, is specially designed to create lift, the force that makes heavier-than-air flight possible. A wing's upper surface is curved, while its bottom is straight. As the oncoming airstream divides to flow around the wing, it must travel faster over the curved upper surface (which acts much like the canyon) than across the straight bottom surface—since the distance is greater. According to Bernoulli's Principle, the pressure of the upper

A replica of the Wright Brothers' 1902 glider soars over the dunes of Kitty Hawk, North Carolina. By changing the angle of the wing tips—a technique known as wing warping—the Wrights could control their craft in an unsteady wind and bank it from side to side. With the addition of a front elevator to vary the glider's pitch, and a rear rudder for steering, the brothers had invented a trustworthy, controllable aircraft. All that it lacked was a power plant.

surface will be less than the pressure on the bottom surface. As a result of this difference, the wing lifts.

Reinforcing this basic flow pattern is a secondary sequence, which further augments the difference in pressure. As the airstream flows over the wing, it comes off the wing's trailing edge with a spin, known as the starting vortex. A law of aerodynamics states that a vortex always produces a counter vortex of equal strength that rotates in the opposite direction. Underneath the wing, air circulating in the counter vortex clashes with the main under-the-wing airstream. Velocity diminishes, and air pressure below the wing consequently increases. Above the wing, both airflows move in the same direction, and their combined velocity boosts the above-wing airflow. The result is an increase in lift.

Other factors affect the amount of lift a wing generates, from the manipulations of the pilot to the wing's basic design. Simply pushing an airplane's throttle increases lift; as airspeed rises, so does the pressure differential. Another aspect controlled by the pilot is the angle at which the wings meet the oncoming airstream. By inclining the wings at a slight angle—called the angle of attack—the air meets the undersurface of the wing at a steeper angle, and creates an even greater pressure difference above and below the wing. Generally the most efficient angle of attack for airplanes is about four degrees to the oncoming airstream.

The main design factors affecting lift are the wing's surface area and its camber—the degree of curvature of its upper surface. More camber, and more surface area, each produces more lift. As a rule of thumb, aircraft designed to fly at low speeds, like gliders, need extra lift and so carry large, highly cambered wings. Since velocity also substantially increases lift, high-speed jets can get by with smaller, flatter wings.

Aeronautical engineers express lift in pounds per square foot, and the heavier an aircraft the more lift it obviously requires from its wings. A lightweight glider can soar with ease on the miserly two pounds of lift supplied by each square foot of its long, fully cambered wings. A heavier Piper Tomahawk trainer moves at a

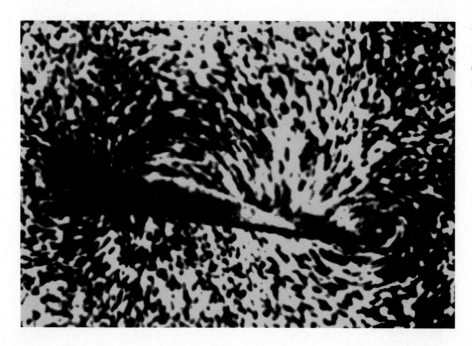

A pattern of oil droplets, sprayed into a windstream moving at about 100 miles per hour, reveals the circulation of a fluid around a stationary airfoil being tested inside a wind tunnel.

faster pace, and so its shorter, less-cambered wings provide all the 13.5 pounds of lift it needs to sustain level flight. The fire-breathing, 50-ton F-14 Tomcat relies on its sensational speed to eke 82 pounds of lift from each square foot of its razor-thin wings.

THE WEIGHT PROBLEM

"Why does fuel have to be so heavy? If gasoline weighed only a pound per gallon instead of six, there'd be no limit to the places one could fly." So wrote Charles Lindbergh about his most troublesome problem as he prepared, in the spring of 1927, to make the world's first nonstop transatlantic flight.

Lindbergh's complaint was understandable. His 2,150-pound monoplane, the *Spirit of St. Louis*, would need to burn an estimated 425 gallons of fuel as it flew the 3,600 miles between New York and Paris. But to sustain this weighty load, given the technology of the day, the plane would need the extra lift of an impossibly large wingspan. So designer Donald Hall took a calculated risk. He gutted the *Spirit* of every nonessential, until it was little more than a flying gasoline tank. This allowed him to cut the wingspan to a manageable 46 feet—and Lindbergh to lift off to aviation glory.

Today's designers face similar dilemmas of weight, lift and structural integrity. In engineering terms, weight is the downside force opposing lift; but unlike lift, it cannot be aerodynamically manipulated. The only solution is to reduce it as much as possible. To do so, designers have turned to a variety of composite materials, light in weight but of great strength, for use in airframe construction.

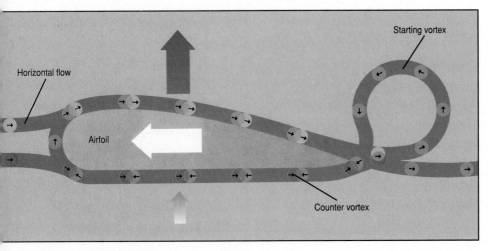

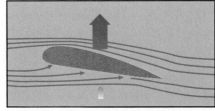

THE SECRET OF LIFT

This cross section of an airplane's wing moving through the air reveals its airfoil shape, and the activity of the air as it sweeps past. As the horizontal airstreams above the wing (yellow) and below it (red) join at the trailing edge, they form a circular eddy, called the starting vortex, and produce a counter vortex that rotates counterclockwise around the airfoil (orange). Underneath the wing, the counter vortex meets the horizontal flow and checks its speed. Above the wing, both airflows move in the same direction, and their combined velocity increases. The result is a substantial difference in pressure above and below the wing that sucks it upward.

Raising the angle of attack—the angle at which the wings meet the oncoming airstream—increases lift. For example, steepening the angle of attack from 4° to 8° (top) doubles the total lift force. When the angle of attack reaches about 14°, air flowing over the wing becomes turbulent; the low-pressure area above the wing starts to dissipate and destroys lift (bottom).

The first airframe material was wood—the Wrights' *Flyer*, for example, used it—and wood itself is a marvelous organic composite. With its long cellulose fibers locked in a rigid matrix of lignum, it is both flexible and sturdy. Wood has its limits, however, and designers soon began using metals such as steel, iron and aluminum. Now, with the physical stresses of today's high-speed aviation, even better materials are needed. So scientists are designing them, molecule by molecule, often mimicking the structure of natural objects. For example, a carbon composite, considerably lighter and six times stiffer than aluminum, draws most of its strength from micro-thin carbon fibers set in an epoxy matrix. And the strength is extraordinary. A filament of carbon or graphite can be spun out to miles in length, made finer than a human hair, and it will not break until the stresses upon it reach up to 20 times the breaking point for metals.

Structure is key in the new materials, and nature provides clues here as well. The interlocking cells of a honeycomb makes it extremely rigid in one direction and resilient in the other. This property is useful in airframes, so sheets of honeycomb made of aluminum, graphite epoxy, and even paper impregnated with plastic, are now being used in airliners and military jets.

Besides strength and lightness, another sought-after quality is heat resistance. Often, it is provided by metals—superalloys—blended of exotics such as tungsten, columbium, lithium and titanium. Exceptionally strong at high temperatures, superalloys are now being used in the compressors of jet engines, where temperatures can reach up to 2,500° F. They also find application in supersonic aircraft, where aerodynamic heating caused by high speeds can expose an aircraft's surface to temperatures as high as 1,000° F. Ceramics, capable of withstanding even greater heat than metals, may soon find an application in aero engine parts. And lightweight ceramic-reinforced polymers are already being tested as a material for the flight surfaces of a plane.

In 1986, composite materials were used to build an airplane that set the ultimate distance record. Two American pilots, Dick Rutan and Jeana Yeager, accomplished a nonstop voyage around the world in a specially designed craft called the *Voyager*. As with Lindbergh's *Spirit of St. Louis*, the design yardstick for *Voyager* was its fuel load. Designer Burt Rutan, Dick's brother, calculated that an ordinary aluminum-frame plane, to contain the fuel it would need for a round-the-world flight, would have to be as big as an aircraft carrier. His solution was to build the plane almost entirely of composites. He glued together layers of carbon-fiber cloth and honeycombed paper, sandwiched between a tough skin of graphite fibers, creating a material seven times stronger than aluminum. The result was a 2,860-pound aircraft—just 100 pounds heavier than Lindbergh's—capable of carrying

The lightweight, spider-slim Voyager *ghosts across the Mojave desert en route to its record-breaking, 25,000-mile nonstop circumnavigation of the globe. Pilots Dick Rutan and Jeana Yeager—Jeana flying and Dick napping—huddle in* Voyager's *tiny unpressurized cabin. Measuring just 3 1/2 by 7 1/2 feet, it was little larger than a phone booth; shifting places without knocking the controls was a minor feat of acrobatics.*

two passengers 25,000 miles nonstop, without in-air refueling. A consumate fuel miser, *Voyager* averaged 36 miles to the gallon.

Weight still remains an aircraft designer's toughest adversary. Jumbo jet or Mach 2 fighter, the airframe must be light and sturdy. As an industry saying puts it, an aeronautical engineer is a man "who must build for one pound of weight what any fool could do for two."

THE STRUGGLE FOR STABILITY

An experienced jet test pilot who recently flew a replica of Louis Blériot's famed 1909 monoplane, the first heavier-than-air machine to cross the English Channel, was amazed that the French inventor managed to fly at all. He found that the mental concentration, and continuous control adjustments required just to stay on an even keel, were unspeakably difficult and tiring. What most airplanes of the Blériot era lacked was stability.

Stability is the ability of an airplane to remain in the flight attitude the pilot selects, and to self-correct when it is upset by air turbulence. Slight air disturbances, such as updrafts or gusts of wind, tend to send it out of kilter in any of three dimensions, as shown in the diagrams at right. So aviation designers soon developed control surfaces to help preserve equilibrium: tail fins to keep the aircraft from weaving back and forth, horizontal stabilizers to discourage pitching, and wing alignments that prevent it from rolling from side to side.

DESIGNS FOR STABILIZING FLIGHT

An object traveling through the air is vulnerable to movement along its horizontal, vertical and longitudinal axes. The airplane pictured here displays the built-in design characteristics that automatically correct for movement along these axes.

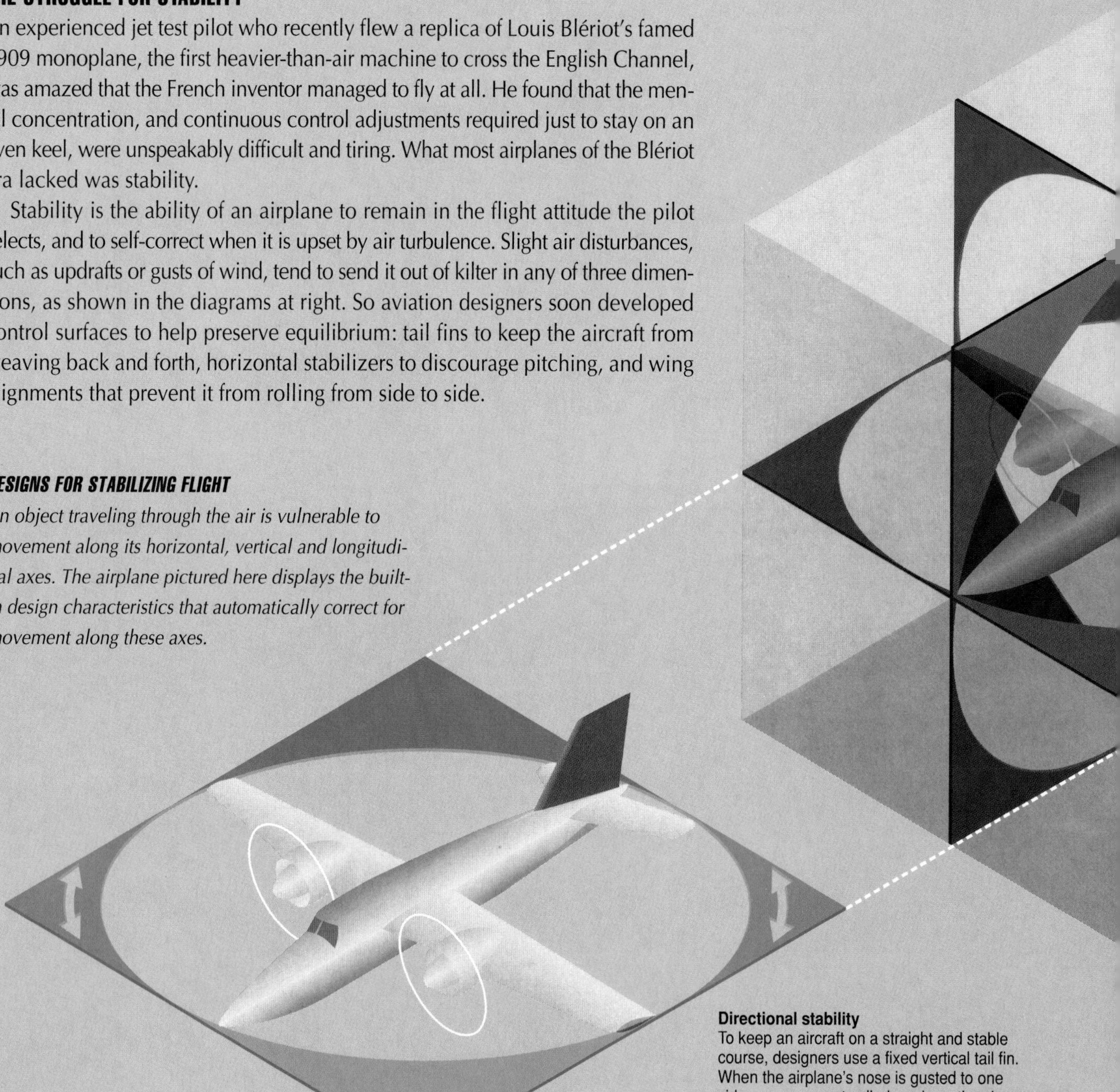

Directional stability
To keep an aircraft on a straight and stable course, designers use a fixed vertical tail fin. When the airplane's nose is gusted to one side—a movement called yawing—the change in airflow around the fin sets up an opposite, correcting force. Like the tail of a weathercock, it pulls the nose back into line with the airflow.

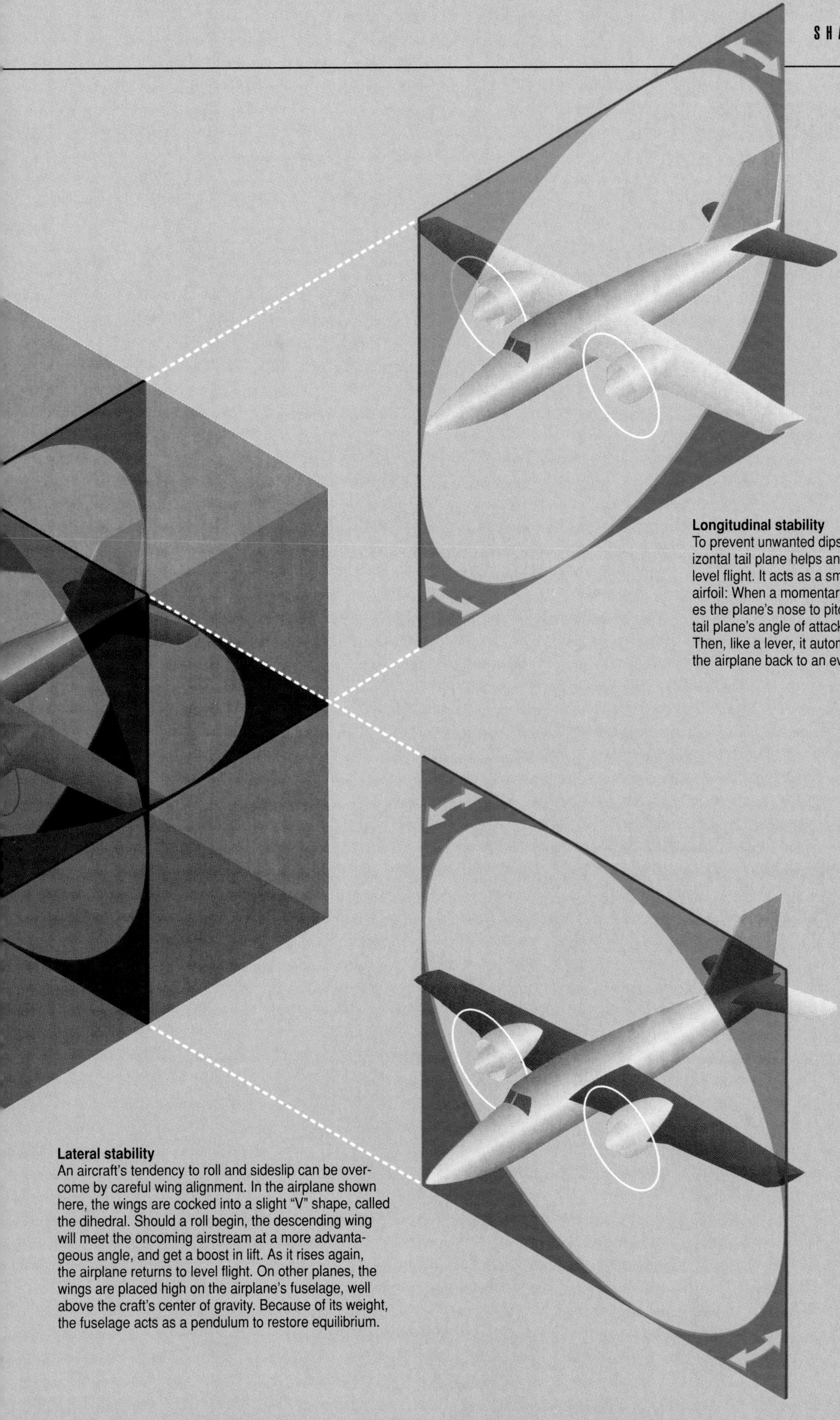

Longitudinal stability
To prevent unwanted dips or nose-ups, a horizontal tail plane helps an airplane maintain level flight. It acts as a small supplementary airfoil: When a momentary disturbance causes the plane's nose to pitch up or down, the tail plane's angle of attack also changes. Then, like a lever, it automatically hoists the airplane back to an even keel.

Lateral stability
An aircraft's tendency to roll and sideslip can be overcome by careful wing alignment. In the airplane shown here, the wings are cocked into a slight "V" shape, called the dihedral. Should a roll begin, the descending wing will meet the oncoming airstream at a more advantageous angle, and get a boost in lift. As it rises again, the airplane returns to level flight. On other planes, the wings are placed high on the airplane's fuselage, well above the craft's center of gravity. Because of its weight, the fuselage acts as a pendulum to restore equilibrium.

Testing Designs for Flight

"Test first, fly later": This has always been the order of business in the history of aviation research. The wind tunnel is the primary testing tool of aeronautical engineers. Its invention dates to 1871 when Francis Wenham, a marine engineer fascinated by the idea of powered flight, designed the first crude test chamber. He used a steam-powered fan to force air past the model of a wing design.

A modern wind tunnel employs the same principles as its 19th-Century predecessor but it yields far more accurate results. Inside the tubelike structure, a carefully controlled airsteam, usually produced by a huge fan, flows over scale models of airplanes or their components in the tunnel's test section. Computers, connected to the suspended model, measure and record airflow and aerodynamic forces.

Tests are conducted with models by relying on a formula, developed in 1883 by British physicist Osborne Reynolds. This formula takes into account the size and shape of the proposed design and the characteristics of the air flowing over it (its speed, density and viscosity). The equation results in a reynolds number. Next a scale model is tested in a wind tunnel, under conditions that yield the same reynolds number and the model's performance dictates the success or failure of the design.

A small, slow-speed plane such as a Cessna Skymaster has a relatively low reynolds number that is easy to simulate in a wind tunnel. But testing models of large, high-speed planes poses problems. There are two solutions—larger tunnels and larger models or manipulation of the properties of the air flowing over the model. Air in the tunnel can be pressurized, substituted with a denser gas such as nitrogen, or cooled to subzero temperatures to increase viscosity.

An electronic version of a wind tunnel is emerging as a useful tool of aerodynamicists. Called Computational Fluid Dynamics (CFD), it uses high-speed computers to generate mathematically the flow of fluid over a computer-designed model. The computer can analyze the aerodynamic forces on the aircraft's surfaces, quickly sort through a large number of possible design modifications and present the best solution.

Energy conservation is one of the main goals of wind-tunnel testing and CFD research. Commercial airliners in the United States consume more than ten billion gallons of fuel every year; even a one-percent improvement in fuel efficiency would save millions of gallons.

Computational fluid dynamics provides aerodynamicists with a time- and money-saving testing tool. Shown below, the computer-generated wind flow onscreen appears as multicolored trace lines over an F-16's wings and fuselage. The colors represent air pressure; red depicts the high-pressure areas, and blue the low.

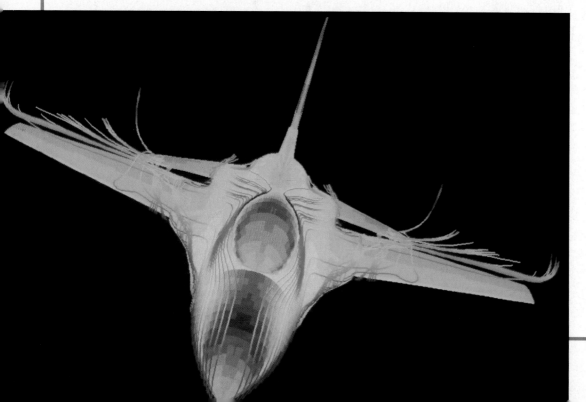

The National Transonic Facility (right) is a new kind of wind tunnel that uses cryogenic or extremely cold nitrogen gas at high pressure to test small-scale models of advanced aircraft or spacecraft as they fly through the sound barrier. Cryogenic temperatures as low as minus 300° F. reduce the viscosity and increase the density of the fluid flowing through the tunnel.

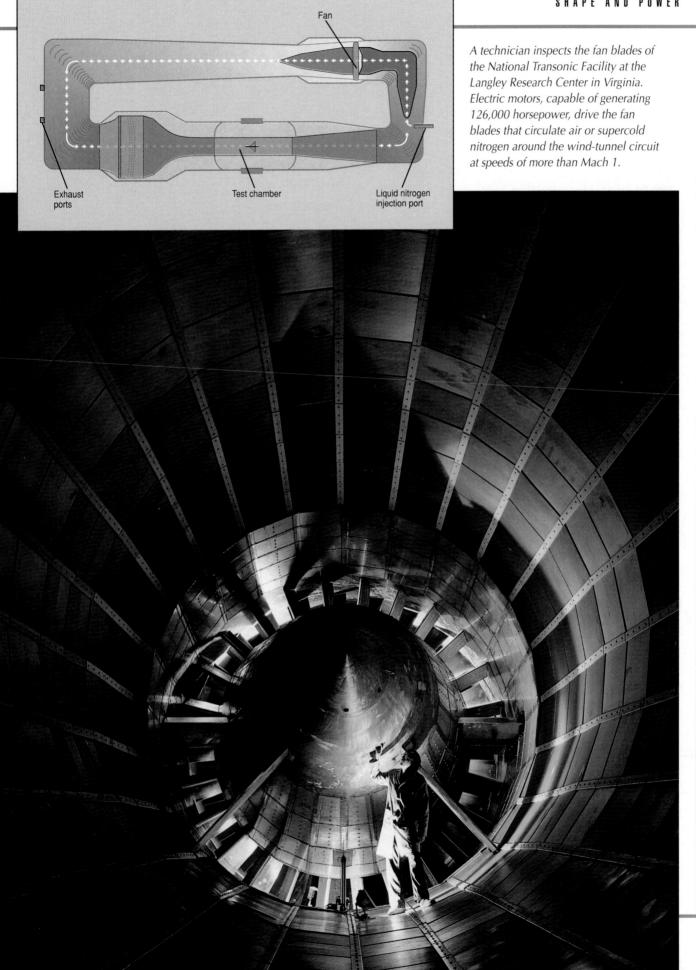

Fan

Exhaust
ports

Test chamber

Liquid nitrogen
injection port

A technician inspects the fan blades of the National Transonic Facility at the Langley Research Center in Virginia. Electric motors, capable of generating 126,000 horsepower, drive the fan blades that circulate air or supercold nitrogen around the wind-tunnel circuit at speeds of more than Mach 1.

Along with stability comes balance, for unless an airplane's weight is properly distributed it will soon tumble out of the sky. Aircraft fly best when their center of gravity is located just ahead of the point where the three dimensions of stability intersect. This means that whenever an airplane is made ready for takeoff, care must be taken that fuel and payload are correctly placed—like trying to position children of different weights on a seesaw, but considerably more complicated. Computers are now replacing the graphs and calculators commonly used by pilots to perform their vital preflight loading calculations. Some modern airliners, such as the supersonic Concorde, have onboard computers that feed information automatically to a center-of-gravity indicator in the cockpit.

THE POWER PLANT

Even while the early glider pioneers were searching for a flyable airframe, a number of farseeing airmen were attempting to solve the next problem: power. All kinds of bizarre power plants have been tested over the years, driven by everything from carbonic acid to gunpowder to elastic bands. One early favorite was steam, and in 1894 Britain's Sir Hiram Maxim, inventor of the machine gun, used a pair of 180-horsepower compound steam engines to lift his galumphing four-ton *Leviathan* momentarily from its launching rails. What was needed for true powered flight—beyond a suitable airfoil—was an engine that could deliver generous amounts of horsepower for relatively little weight. Unless an engine can move its airplane forward at sufficient speed, it will not create the lift required to raise the airplane, the payload, and the engine itself against the pull of gravity.

The earliest airborne power plants met none of these requirements. Heavy steam-driven devices of ponderous inefficiency, they were carried aloft on balloons and dirigibles—which, being lighter than air, had no need of aerodynamic lift. The first such machine to make the ascent was a 100-pound steam-puffing engine that French engineer Henri Giffard installed on an 88,000-cubic-foot airship in 1852. Weighing 117 pounds for every unit of horsepower it developed, Giffard's engine could barely push itself against a gentle breeze.

A leaner and more muscular type of power was clearly needed. The inspiration came from the automobile, where gasoline-burning, piston-driven, internal combustion engines were proving increasingly effective.

The Wright Brothers used what was basically a modified automobile engine to propel their *Flyer* on its historic leap at Kitty Hawk. Their mechanic, Charlie Taylor, built a four-cylinder gasoline engine that weighed a mere 180 pounds, including its flywheel, and generated about 13 horsepower. It gave the 605-pound *Flyer* just enough lift to become airborne.

Besides building an engine, the Wrights also designed and constructed their own propellers—necessary to convert the power of the engine into propulsive thrust. Propellers of a sort were already being used on the early airships; but these were huge paddles, often more than 20 feet in diameter, that turned lazily to push the balloon-like craft along at a walking pace. The Wrights needed a propeller that would withstand the stresses of spinning at 350 revolutions per minute—a rate reflecting their vision that a propeller should not be a paddle, but an airfoil—literally, a wing spinning through the air to generate lift in a forward direction, just as the airfoil shape of a fixed wing creates upward lift. As a propeller rotates, air

A piston engine
Power from six firing cylinders rotates the crankshaft at 5,300 rpms, which in turn drives the two-bladed propeller at 2,340 rpms. It has an enviable power-to-weight ratio of 240 horsepower to 400 pounds of engine weight. With electronic ignition, automatic fuel injection, and a specially designed muffler for dampening noise, the engine burns a frugal 12 to 14 gallons of fuel per hour at cruising speed.

An airfoil that spins
The twin-bladed propeller is made of molded fiberglass reinforced with carbon fibers. To increase the prop's effectiveness at varying speeds, the pitch of the blades can be adjusted by the pilot. Thus, in cruising flight, the blades are set at a fairly sharp angle—in effect, shifting the propeller into high gear while keeping engine speed and fuel consumption low. But when extra thrust is needed for takeoff and landing, the pitch is reduced so that the curved upper surface of the blades faces the airflow, thus gearing down for maximum lift at low airspeeds.

flows around the blade, moving faster over the curves of the blade's leading edge. The motion lowers the air pressure in front of the blade and pulls the aircraft forward. Figuring these forces in painstaking tests and calculations, the Wrights designed and handcarved a pair of elegant propellers from blocks of laminated spruce. They linked them to their engine with a system of chains and sprockets much like those on a bicycle.

During the next five years, the Wrights boosted their engines to 30 horsepower. But by then Europe had taken the lead in engine technology. The great leap forward came in 1905, when France's Léon Levasseur, reworking a speedboat engine, devised a V-8 power plant that delivered 50 horsepower and weighed only 110 pounds. Its remarkable power-to-weight ratio—one horsepower for each 2.2 pounds of engine weight—would not be exceeded for 20 years. But Levasseur's engine had problems with its fuel injection and cooling systems, and frequently broke down. It never achieved the wide commercial success of another innovative French design—the legendary Gnome rotary.

The Gnome was one of the first production engines manufactured exclusively for airplanes. The first model appeared in 1907, and its design was truly revolutionary. Five cylinders, rated at 10 horsepower each, were placed like wheel spokes around a fixed crankshaft. When the cylinders fired, the entire engine would turn, imparting its spin to a propeller bolted directly to it. Because of this rotary action, the engine was self-cooling; thin metal fins on each cylinder allowed the heat of combustion to dissipate directly into the passing air. It also acted as its own flywheel, with a marked savings in weight. The Gnome was not perfect: It was a glutton for fuel, and the mass of whirling cylinders tended to make the aircraft attached to it difficult to control. It also sprayed lubricating fluid—castor oil—back over the pilot. But the Gnome was light, peppy, smooth-running and reliable, and airplane designers loved them. By 1916, almost 80 percent of all aircraft ran either on Gnome rotaries or on engines derived from them.

This was the decade of World War I, when aircraft were first enlisted as instruments of battle, and designers turned every effort to improve both power and performance. Car manufacturers—Rolls-Royce, Mercedes and Hispano-Suiza—produced aerial power plants of ever-increasing horsepower. The rotary design gave way to other configurations, where the cylinders were fixed and the crankshaft turned. The result was a welcome cut in fuel and oil consumption, and greater efficiency on long-range flights. By 1918, water-cooled engines with banks of 12 cylinders were developing 300 to 400 horsepower. Nor did the rate of innovation decrease with the Armistice. The 1920s saw such refinements as variable-pitch propellers that effectively gave airplanes gearshifts; improved carburetors that could be adjusted for various altitudes and power settings; and starter motors that removed the dangerous task of hand-swinging the prop before each flight.

The United States became world leader in engine design following Charles Lindbergh's epoch-making Atlantic flight. The 500-pound, 220-horsepower air-cooled Wright Whirlwind that powered the *Spirit of St. Louis* set new standards of power for weight, economy and reliability. The Whirlwind was an air-cooled radial, with stationary cylinders placed in a collar around a spinning crankshaft. Its main drawback was wind resistance from the protruding cylinders, but this was solved by covering the engine with a streamlined cowling. Within ten years, Wright and the

THE FORCE WITHIN A TOY BALLOON
Inflated and tied shut, a toy balloon is a miniature engine waiting to start up. The compressed air within it pushes equally on all sides; but until the air finds an outlet, the balloon remains stationary.

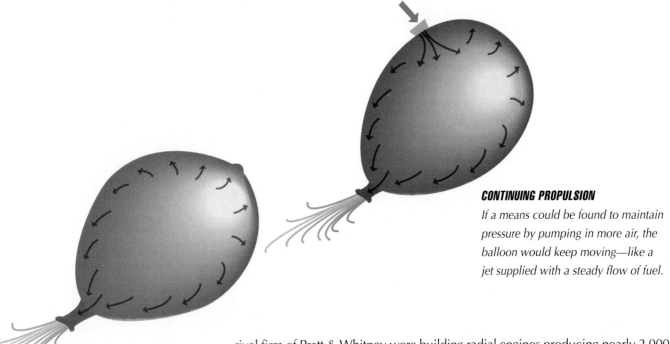

CONTINUING PROPULSION

If a means could be found to maintain pressure by pumping in more air, the balloon would keep moving—like a jet supplied with a steady flow of fuel.

THE FORCE IN ACTION

When the balloon's neck is opened, the trapped air rushes out. As the air escapes, it upsets the balance of forces inside the balloon, causing an equal and opposite reaction that propels the balloon forward like a squid in water—or a jet through the sky.

rival firm of Pratt & Whitney were building radial engines producing nearly 2,000 horsepower. Some of them are still in use today.

By the end of World War II, monstrous piston engines sprouting 28 cylinders were capable of developing 3,500 horsepower. Even larger engines were on the drawing boards, as airline companies prepared for a postwar boom in commercial aviation. Yet the days of roaring pistons and flashing propellers were numbered. In the last months of the war, a new phenomenon streaked across the sky: the jet.

Both sides in the conflict had been working to perfect this device. For well over a decade, a few visionary engineers had been toying with the idea that a stream of hot exhaust from a gas turbine might be used to drive an airplane. The first practical test of such a turbine, in a machine shop in 1937, almost ended in disaster. Recalled its British designer, Frank Whittle: "With a rising scream the engine began to accelerate out of control. Everyone took to their heels. I was paralyzed to the spot." Luckily for Whittle, the engine did not blow up. But problems continued to plague designers. Then, in the spring of 1944, the first fully operational jet aircraft, Germany's Messerschmitt 262 thundered into combat at 540 miles per hour—70 miles per hour faster than its swiftest propeller-driven opponent. Though its appearance came too late to alter the course of the war, the Me-262 swept jet technology into the modern age.

The action of a jet engine could not be more simple or direct. Air swallowed in through the front is compressed by whirling fans, mixed with kerosene-like fuel, and ignited. The gases of combustion expand with tremendous force and rush out the jet's tailpipe at speeds approaching 1,700 feet per second. The force of reaction thrusts the engine, and the airplane, forward.

The principle of reaction is one of the oldest in mechanics. It was demonstrated as long ago as the first century A.D., when Hero of Alexandria used it to operate the earliest-known jet engine—a bronze globe spun by jets of steam. Isaac Newton set it down as a basic law of physics in the 17th Century: For every action there is an equal and opposite reaction. One example is an inflated toy balloon racing across a room, a jet of air rushing out its open neck and creating a reaction force that propels the balloon forward.

In the natural world, the squid uses the same principle for quick movement in water. Water is drawn into the body, or mantle, where it is stored until needed. When the squid wants to move, it squirts the water backward through a funnel-like opening in the mantle's rear end. Octopi use the same technique, and some species of octopus expel the water with such force that they shoot above the surface and glide a short distance in the air.

Jet engines brought a marked improvement in power and thrust. They replaced the pounding pistons and intermittent explosions of conventional power plants with smooth-spinning turbines and steady combustion. The result was far more horsepower for each pound of engine weight. Whereas piston engines typically produced one-half pound of thrust for a pound of engine weight, today's jumbo jet engine produces four pounds of thrust for each pound of engine weight. Even so, the early jet engines were spendthrifts of fuel, especially at low speeds, and so were better suited to the performance demands of military flying.

Cost-conscious commercial airlines compromised by using turbo-props, in which the superior power of a jet turbine is used to drive a conventional propeller. But while this system saved fuel and gave pilots greater control during takeoff, landing, and other slow-speed maneuvers, it had dire limitations. The main problem was the propeller itself. At higher throttle settings, the propellers would start spinning so fast that their tips would approach the velocity of sound. The result was a dangerous increase in air turbulence. In fact, at about 450 miles per hour, the efficiency of the propellers dropped drastically and severe vibrations set in due to the enormous increase in drag.

Designers soon took another tack, focusing on jet exhaust. Various tests and calculations showed that the high-speed exhaust from the early turbojets was inherently inefficient; much more thrust can be obtained from a larger stream of slower-moving air. So in the early 1960s engineers began searching for ways to simultaneously slow down the jetstream and expand its volume. Large fans were installed in front of the jets' compressors, supplying extra air that bypassed the combustion chamber. This supplementary airstream, when mixed with the hot exhaust gases, slows them down and provides the sought-after extra boost of thrust.

ANATOMY OF A ROLLS-ROYCE

This Rolls-Royce RB211 turbofan displays its jet-powered efficiency. A pair of large whirling fans sucks air into the engine. Some of the air is directed to the compressors, which force it under tremendous pressure into the combustion chamber; mixed with fuel, ignited and expelled, it drives the turbines that turn the fans and compressors. A larger stream of air, pushed back by the fan blades, moves through bypass ducts to join with the exhaust gases as they spurt through the tail cone—cooling them, quieting them, and delivering most of the thrust.

Fan blades

Engine cowling

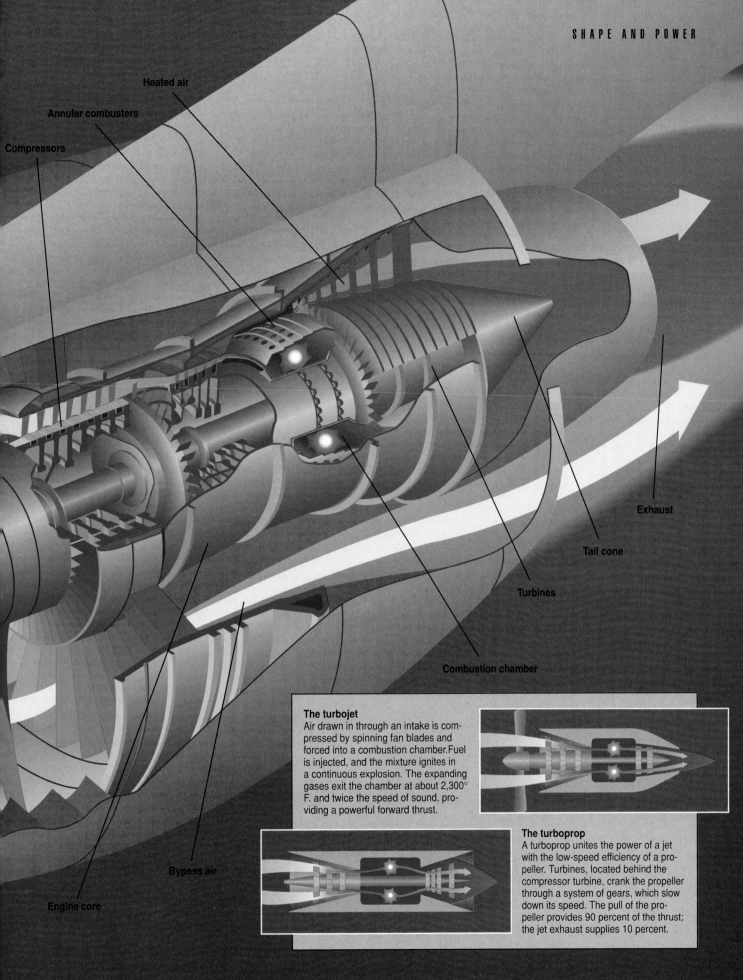

Heated air

Annular combusters

Compressors

Exhaust

Tail cone

Turbines

Combustion chamber

Bypass air

Engine core

The turbojet
Air drawn in through an intake is compressed by spinning fan blades and forced into a combustion chamber. Fuel is injected, and the mixture ignites in a continuous explosion. The expanding gases exit the chamber at about 2,300° F. and twice the speed of sound, providing a powerful forward thrust.

The turboprop
A turboprop unites the power of a jet with the low-speed efficiency of a propeller. Turbines, located behind the compressor turbine, crank the propeller through a system of gears, which slow down its speed. The pull of the propeller provides 90 percent of the thrust; the jet exhaust supplies 10 percent.

51

Pedal Power

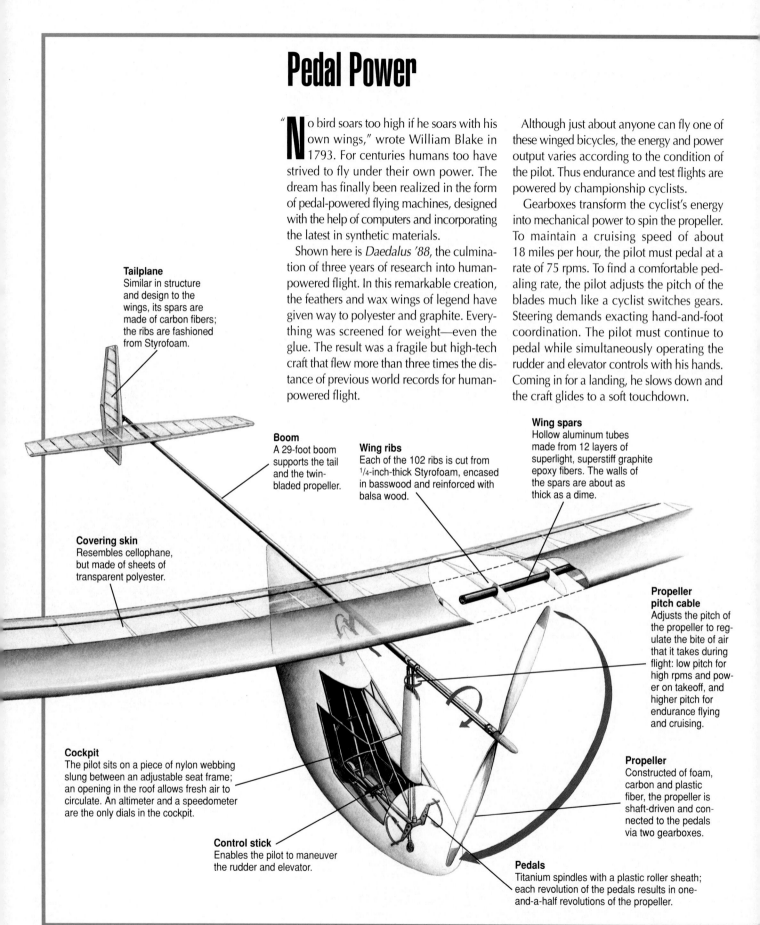

"No bird soars too high if he soars with his own wings," wrote William Blake in 1793. For centuries humans too have strived to fly under their own power. The dream has finally been realized in the form of pedal-powered flying machines, designed with the help of computers and incorporating the latest in synthetic materials.

Shown here is *Daedalus '88*, the culmination of three years of research into human-powered flight. In this remarkable creation, the feathers and wax wings of legend have given way to polyester and graphite. Everything was screened for weight—even the glue. The result was a fragile but high-tech craft that flew more than three times the distance of previous world records for human-powered flight.

Although just about anyone can fly one of these winged bicycles, the energy and power output varies according to the condition of the pilot. Thus endurance and test flights are powered by championship cyclists.

Gearboxes transform the cyclist's energy into mechanical power to spin the propeller. To maintain a cruising speed of about 18 miles per hour, the pilot must pedal at a rate of 75 rpms. To find a comfortable pedaling rate, the pilot adjusts the pitch of the blades much like a cyclist switches gears. Steering demands exacting hand-and-foot coordination. The pilot must continue to pedal while simultaneously operating the rudder and elevator controls with his hands. Coming in for a landing, he slows down and the craft glides to a soft touchdown.

Tailplane
Similar in structure and design to the wings, its spars are made of carbon fibers; the ribs are fashioned from Styrofoam.

Boom
A 29-foot boom supports the tail and the twin-bladed propeller.

Wing ribs
Each of the 102 ribs is cut from 1/4-inch-thick Styrofoam, encased in basswood and reinforced with balsa wood.

Wing spars
Hollow aluminum tubes made from 12 layers of superlight, superstiff graphite epoxy fibers. The walls of the spars are about as thick as a dime.

Covering skin
Resembles cellophane, but made of sheets of transparent polyester.

Propeller pitch cable
Adjusts the pitch of the propeller to regulate the bite of air that it takes during flight: low pitch for high rpms and power on takeoff, and higher pitch for endurance flying and cruising.

Cockpit
The pilot sits on a piece of nylon webbing slung between an adjustable seat frame; an opening in the roof allows fresh air to circulate. An altimeter and a speedometer are the only dials in the cockpit.

Propeller
Constructed of foam, carbon and plastic fiber, the propeller is shaft-driven and connected to the pedals via two gearboxes.

Control stick
Enables the pilot to maneuver the rudder and elevator.

Pedals
Titanium spindles with a plastic roller sheath; each revolution of the pedals results in one-and-a-half revolutions of the propeller.

The sun rises over Daedalus '88 as it takes wing across the Aegean, following the route of its namesake. Greek cycling champion, Kanellos Kanellopoulos, piloted the aircraft more than 70 miles from Crete to the Isle of Santorini—three times the distance of the previous world record for human-powered flight. As he prepared to touch down, a sudden gust of wind shattered the tail boom, and the craft crash-landed just ten yards short of its destination.

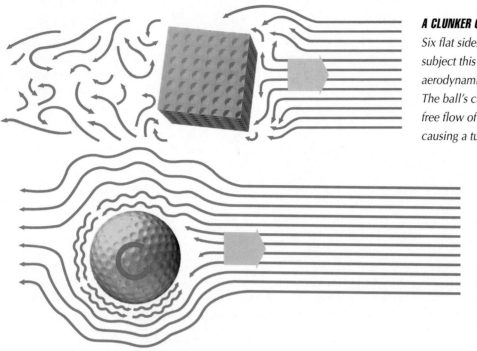

Six flat sides and squared-off corners subject this hypothetical golf ball to aerodynamic drag at its most extreme. The ball's cube shape obstructs the free flow of the passing airstream, causing a turbulent chaos downwind.

A SMOOTH-SPINNING SPHERE
Rounding out the ball's corners allows the airstream to flow past with only minor disruption. In addition, as the ball spins, it generates a degree of lift that increases its trajectory. The effect is further enhanced by the dimples on the ball's surface, which trap a layer of air and set it moving in the direction of spin. The trapped air on top moves with the airstream, while the air below collides with it—much like the counter vortex that helps create lift in an airplane's wing.

So was born the turbofan, the workhorse of modern aviation. Twenty-five percent more fuel efficient than the original turbojet power plants, the turbofan offers better range and more efficient propulsion. It is also quieter and smoother-running, and is capable of delivering effective thrust at both high and low airspeeds. On takeoff—the moment of greatest stress—a Rolls-Royce turbofan engine delivers the equivalent power of 228 Ferrari V-8 engines. In the time it takes to reach a cruising altitude of 35,000 feet, it gulps enough air to fill the *Graf Zeppelin*, more air than a normal man breathes in 20 years.

GRAPPLING WITH DRAG

By 1910 aviators had come to recognize the debilitating effects of drag—the retarding force opposing thrust. Spurred by fortunes in prize money, they strapped bigger and bigger engines to their rudimentary airplanes in search of a winning edge. But tripling the power of their whirling rotaries produced only marginal increases in speed. Their airframes had hit what seemed to be an impenetrable barrier of wind resistance. Here was drag of the most obvious type, and one that thrives on acceleration. Since this kind of drag increases at the square of the airspeed, doubling the velocity of their Blériots to a paltry 68 miles per hour meant a fourfold increase in the forces impeding them.

Moving bodies create drag. Joggers, cyclists and swimmers, boats, trucks and airplanes—all experience a resistance to forward motion through air or water. The bigger they are and the faster they go, the greater the retarding force. There are various ways to mitigate this effect, however. One is to streamline the shape of the moving body.

Streamlining reduces what scientists call form drag—the resistance caused by sharp angles and bulky protrusions. The art of streamlining is centuries old. Greek and Polynesian war canoes sped into battle with their slim prows raising barely a ripple. Today, bullet-shaped racing cars, Olympic bobsleds, even the wind-slicing helmets worn by sprint cyclists, slip through the air with little disturbance. Streamlining in aircraft design produces spectacular improvements. A well-shaped wing strut ten inches thick, and moving through the air at 200 miles per hour, produces less drag than a one-inch steel cable.

Another type of drag is skin friction, the resistance that develops in the thin layer of airstream that passes closest to the airplane. The smoother and more polished the craft's surface, the more easily the air flows past it, and the smaller the effect of skin friction. At low speeds, it plays a minor role in any case. But it jumps dramatically as velocity increases, and in high-speed jets a glassy-smooth surface becomes an essential part of overall design.

A third variety of drag applies only to aircraft. Whenever an airstream flows over a wing or other airfoil, creating lift, some of the energy is diverted to the rear. This has to do partly with the angle at which the wing strikes the airstream, and partly with forces of turbulence that come into play. In either case, induced drag, as it is called, tends to pull the plane backward. It is the price a pilot pays for lift.

Some induced drag is produced by the vortex that forms behind the wing's trailing edge. Another type of vortex also comes into play, and it occurs because air from the high-pressure area beneath the wing tends to leak around the wing's edges toward the low-pressure area above. Most of this seepage takes place at the wing tips, because the wings are slightly swept back. The result is a pair of powerful spiral vortices trailing from each wing tip.

Induced drag is greatest at low speeds and high angles of attack, such as during takeoff and landing. When an airliner lifts off the runway, it spews out massive amounts of wing-tip turbulence. Gliders and other slow-flying aircraft are partic-

THE POWER OF A WING-TIP VORTEX

Mini-tornados stream from an airplane's wing tips as eddies of high-pressure air curl up around them. A differing pattern of airflow above and below each wing—angled inward above, and inclined toward the tips below—helps impart the spiral motion. So strong are these trailing vortices that the wake of a large airplane can literally flip a smaller craft on its back.

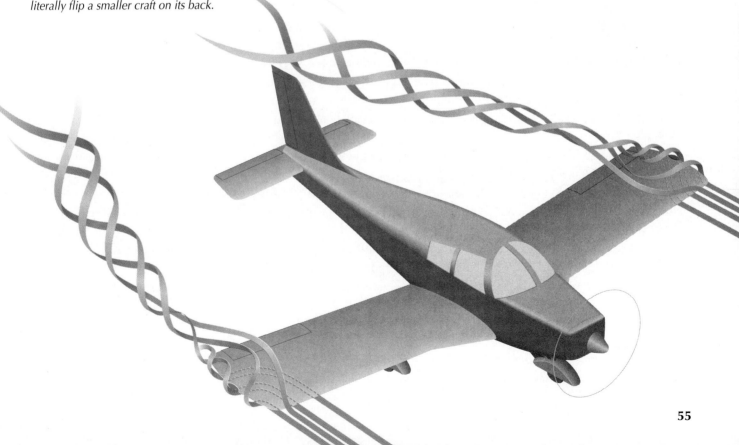

ularly vulnerable to drag of this type—one reason their wings tend to be long and slender. By extending the wings and tapering their tips, designers are able to minimize the vortex they create. Another technique is to add fin-like extensions, called winglets, which help reduce vortex drag. They stabilize airflow on the wing tips by decreasing the seepage from under the wing.

FORM FOLLOWS FUNCTION

When the Wright Brothers designed their first powered airplane they had only one concern—to fly. It mattered little that Orville's sand-hugging hop covered less distance than the wingspan of a Boeing 747 and that it could have been outpaced by a man on a fast bicycle. Speed, range, height and payload were all secondary considerations.

This is hardly the case in modern aviation. Today's aircraft are each designed with a specific function in mind. Commercial airliners must carry heavy loads of passengers long distances in reasonable comfort. Jet fighters must be fast, stunt planes highly maneuverable. As a result, designers must compromise, trading one quality of performance for another they deem more desirable.

In nature, the compromises have evolved over time. The shape of a bird's wing is closely linked to the life it leads—and faster is not always better. The broad, rounded wings of a grouse or pheasant allow for quick, furious takeoffs and short flights through heavy forest foliage. Ducks, geese and falcons fly fast over long distances, using slow, steady wingbeats; their wing tips are pointed to minimize drag. Seabirds such as gulls and petrels have long, narrow wings for gliding on steady ocean winds, while hawks and vultures have broad wings with slotted tips that allow them to ride the thermal updrafts generated over land.

In aviation, designers looking for speed must generally sacrifice range; maneuverability is achieved at the expense of miles per hour. But ingenious advances in technology allow a number of aircraft to have it both ways. For example, to minimize drag while cruising near the speed of sound, the 450-ton Boeing 747 has relatively small, swept-back wings. Its problems come during takeoff and landing, when this same wing configuration does not easily provide enough lift to raise the 400-passenger payload off a normal-sized runway. The solution is to temporarily alter the wings' shape by extending their surfaces with huge flaps at the leading and trailing edges.

The dart-shaped Concorde—designed for pure speed, but with a sharp reduction in payload—carries only 125 passengers. Its supersonic cruising speeds dictate a narrow body, and thin, delta-shaped wings that cannot be fitted with lift-generating wing flaps. It must land and take off at such high speeds that only a few of the world's runways are lengthy enough to accommodate it.

The ultimate compromise is the "swing-wing," used in the Navy's Grumman F-14 Tomcat. It has sharp, narrow wings that can be extended to increase lift and stability during low-speed takeoffs and landings. Then, as it soars into supersonic flight, the wings swing back into a delta shape.

Speed versus range, payload traded for speed or distance—such are the constraints that confront every airplane designer. They are resolved by judiciously modifying the shape and power of the airplane, and bringing the forces of lift, weight, thrust and drag into the desired balance.

In general, birds that fly rapidly have pointed wings to minimize drag. When pursuing prey, the falcon can sweep back its wings in a high-speed dive, attaining speeds of over 200 miles per hour. Like the falcon, the swing-wings of the supersonic F-14 Tomcat are extended for maximum lift at takeoff and landing, but are swept back to reduce drag at supersonic speeds. The computer-controlled, variable-sweep wings give the bulky F-14 surprising flexibility and mobility.

Long, narrow wings give the albatross a large sail surface in proportion to its weight, enabling it to glide for miles on a steady ocean wind. Likewise, the long, slender wings and lightweight, streamlined fuselage of the Voyager and modern sailplanes provide them with a generous amount of lift at little expense in drag.

Like a pheasant taking to the air with an upward leap, an RAF Harrier jet literally jumps from the ground. Broad, rounded wings, for maximum lift at low speeds, characterize vertical-takeoff-and-landing (VTOL) aircraft such as the Harrier. The pilot can change the direction of engine thrust by rotating the exhaust nozzles down to make the plane rise straight up like a helicopter, or by turning them to the rear for conventional flight.

From Blueprint to B-747

A worker installs floor panels inside the shell of the fuselage. The spars and ribs of the fuselage are visible overhead.

From a 90-foot perch, a crane operator lowers the completed rear section of the fuselage onto an adjustable cradle. The cradle supports the weight of the section and enables workers to align it precisely with the center section of the plane.

Jumbo jet—the word is synonymous with Boeing's 747 line of commercial airliners. More than 700 of them are in service around the world today, and by the year 2000, at least three billion people will have flown on them. The newest model—the 747-400—incorporates the latest advances in the fields of digital avionics, aerodynamics, and structural design—all intended to increase the plane's range while decreasing its fuel consumption and operating costs.

Assembly takes place at a sprawling 1,000-acre plant in Everett, Washington. The plant itself is 2,000 feet wide and 1,600 feet long, and contains 291 million cubic feet of work area; it rises 11 stories high and covers the space of 47 American football fields. The power required to operate and maintain the building is enough to light more than 32,000 average homes.

The complex assembly sequence begins with the construction of the wings, which span 211 feet from tip to tip—almost twice the distance of the Wright Brothers' first flight at Kitty Hawk. The framework of each wing—its spars and ribs—is made of lithium-aluminum alloys and covered with sheets

Towering 63 feet above the floor, the height of the tail is equivalent to a six-story building. A fuel tank in the horizontal stabilizer adds an additional 350 nautical miles to the plane's range.

A technician guides an engine onto its strut. Including the cowling, it has a diameter of 8 1/2 feet. Each of the four high-bypass turbofan engines delivers up to 58,000 pounds of thrust.

A technician inspects a six-foot-high winglet—the most noticeable feature of the new B-747-400's wings. Jutting out at a 29-degree angle, the winglets control the wing tip vortices—the swirling mass of air that comes off the ends of the wings when they are producing lift.

More than 10,000 people work around the clock to complete up to five 747-400s per month, on two production lines. The work area is so vast that Boeing supplies a fleet of over 200 bicycles for employee transport and railway spurs come right into the building.

A technician inspects a shock strut of a wing landing gear just prior to installation. Each of the four main landing gears—two on the body and one on each of the wings—has four wheels. The impact of the 300-ton plane at landing is absorbed evenly by the shock struts, which act as pneumatic springs, and the 16 wheels, which are filled with nitrogen.

The pressure bulkhead rests on the floor prior to installation. This unit, together with a rear bulkhead, keeps the cabin pressure at a comfortable level at high altitudes.

Sitting behind an integrated glass cockpit, a two-person crew monitors computers that control the flight. Advanced digital technology has reduced the number of conventional instruments from 971 to 365. The airborne systems and flight status, displayed on six large colored cathode ray tubes (CRTs), reduce the pilot's workload and make him more of a systems manager than a stick-and-rudder pilot.

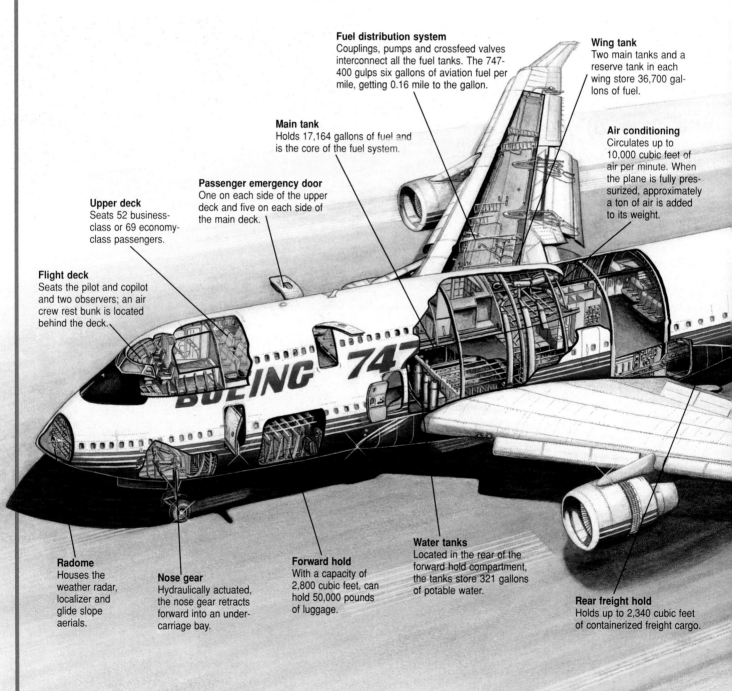

Fuel distribution system
Couplings, pumps and crossfeed valves interconnect all the fuel tanks. The 747-400 gulps six gallons of aviation fuel per mile, getting 0.16 mile to the gallon.

Wing tank
Two main tanks and a reserve tank in each wing store 36,700 gallons of fuel.

Main tank
Holds 17,164 gallons of fuel and is the core of the fuel system.

Air conditioning
Circulates up to 10,000 cubic feet of air per minute. When the plane is fully pressurized, approximately a ton of air is added to its weight.

Passenger emergency door
One on each side of the upper deck and five on each side of the main deck.

Upper deck
Seats 52 business-class or 69 economy-class passengers.

Flight deck
Seats the pilot and copilot and two observers; an air crew rest bunk is located behind the deck.

Radome
Houses the weather radar, localizer and glide slope aerials.

Nose gear
Hydraulically actuated, the nose gear retracts forward into an undercarriage bay.

Forward hold
With a capacity of 2,800 cubic feet, can hold 50,000 pounds of luggage.

Water tanks
Located in the rear of the forward hold compartment, the tanks store 321 gallons of potable water.

Rear freight hold
Holds up to 2,340 cubic feet of containerized freight cargo.

THE FINISHED PRODUCT

The newest member of the 747 fleet, the 747-400, is ready to join the work force of commercial aviation. Weighing about 850,000 pounds, the plane can carry 250,000 pounds of fuel, giving it a total nonstop range of more than 7,000 nautical miles—about one-third of the Earth's circumference.

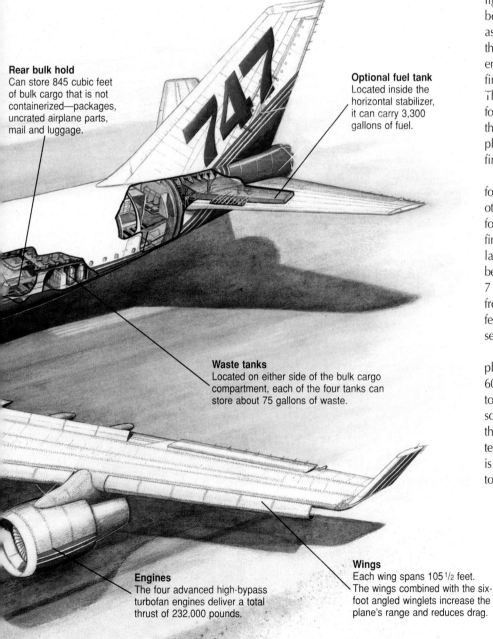

Rear bulk hold
Can store 845 cubic feet of bulk cargo that is not containerized—packages, uncrated airplane parts, mail and luggage.

Optional fuel tank
Located inside the horizontal stabilizer, it can carry 3,300 gallons of fuel.

Waste tanks
Located on either side of the bulk cargo compartment, each of the four tanks can store about 75 gallons of waste.

Engines
The four advanced high-bypass turbofan engines deliver a total thrust of 232,000 pounds.

Wings
Each wing spans 105 1/2 feet. The wings combined with the six-foot angled winglets increase the plane's range and reduces drag.

of honeycomb panels. Since the wings carry the weight of the plane, they are joined to a central wing unit and the plane is built around them.

The fuselage is assembled in three main sections. Its ribs, mainframes, and the one-half-inch outer skin are constructed of lightweight aluminum alloys. The center body section is lowered over the wing assembly by cranes and riveted to it. Next, the front and rear sections, assembled in other areas of the plant, are delivered to the final assembly area by an overhead crane. They are lowered onto adjustable cradles for precise alignment with the center section, then attached to it. With the body now complete, the horizontal stabilizer, tail cone and fin are attached to the rear section.

More than 100 miles of electrical wiring for the air conditioning, video, intercom and other flight systems are run through the 225-foot-long fuselage. Next, layers of woven, fire-resistant materials that insulate the fuselage are placed between the outer skin before the interior panels are installed. Only 7 1/2 inches of wall separates the passengers from the outside world. Finally the interior features are added—the stow bins, curtains, seats, toilets and galleys.

From the main site, the fully assembled plane goes to a paint hangar, where up to 600 pounds of paint (three coats) are applied to the exterior. Prior to its first flight, personnel check and recheck all systems. It is then flown through a series of rigorous flight tests. The plane is not delivered; once testing is completed, airlines must send a flight crew to pick it up.

MANEUVERING

Each year in August a swarm of pilots, airplane designers, backyard builders and flight buffs descends upon the quiet, rural city of Oshkosh, on the shores of Wisconsin's Lake Winnebago. The event is an eight-day celebration of aviation skill and technology known as the Annual International Fly-In Convention of the Experimental Aircraft Association. High above Oshkosh's Wittman Field, teams of stunt fliers like the one at right perform dizzying sequences of loops, spins, stalls, dives, barrel rolls, figure eights and other maneuvers that seem to defy all laws of gravity and aerodynamic sense.

Loudspeakers carry a stunt pilot's voice to the crowds on the ground. The words come in clipped, staccato bursts, taut with concentration and danger: "Nose down . . . full power . . . I'm leveling off . . . pulling up . . . harder!" Fifty feet above the runway, a tiny Pitts biplane snaps upward into a tight vertical spiral. At the top, as it breaks out of the maneuver, the plane flips momentarily onto its back, then rolls into a shallow dive. The forces of acceleration tug at the pilot's body with perhaps six times the pull of gravity, draining the blood from his brain.

None of this effort is visible from the ground. The plane sweeps and dives through its aerial ballet with the easy grace of an eagle cruising for its dinner, and the effect is a gasp of sheer wonder from the crowd. How can an aircraft corkscrew upward in this manner, or loop and roll, or even fly upside down, and not spin out of the air?

Pilots have been astonishing spectators with this puzzle since aviation's very first decade. Learning to guide an airplane's path was a basic step toward powered flight—as the Wright Brothers proved at Kitty Hawk *(page 36)*. The Wrights were soon followed by America's most flamboyant aerial showman, the designer Glenn Curtiss. On July 4, 1908 he flew a trim little biplane—the *June Bug*—through the six basic maneuvers of flight: takeoff, climb, turn, level flight, descent and landing, to win the prize offered by *Scientific American* to the first pilot to fly a measured distance of one kilometer. The air age had begun in earnest.

The Red Devils, an aerial stunt team, loop out of the clouds with gravity-defying grace. The team's three members, each flying his own home-built Pitts Special biplane, regularly perform at air shows across America.

Nature's Acrobat

No insect is a more accomplished aerialist than the common housefly. Performing loops, rolls, hairpin turns, flying backward, sideways, and vertically, it demonstrates flying techniques that involve extraordinary in-flight control.

During flight, the fly consumes prodigious amounts of oxygen, yet it has no lungs, nor does it breathe from the mouth or nostrils. Instead symmetrical rows of tiny air ducts run along its body. As the insect beats its wings, the muscles contract and force air out of the ducts; as the muscles relax, fresh air rushes in.

The power for flight comes from the insect's thorax, a box-like structure that houses the flight muscles. The wings are double-jointed, hinged to the thorax in such a way that they are free to move in any direction, the coupling system functioning like a ball-and-socket joint. Thus the insect rows through the air, beating its wings furiously in a figure-eight pattern like a sculling oar propelling a boat. To climb like a helicopter, it alters the angle of attack of its wings so that the current of air is directed downward rather than backward, just as the angle of attack of a helicopter's blades are altered in unison for vertical flight.

Unlike birds, which have a relatively weak upstroke, the fly produces equal propulsive power on the upward and downward strokes. To flap its wings about 200 times per second, the fly relies on a sophisticated mechanism in the thorax that boosts the speed of each wing beat. When the wings are midway through an upstroke or downstroke, the double-jointed wing couplings push against the elastic walls of the thorax and force it outward. As the wings continue through the stroke, the tension on the thorax is released. This forces the wings up or down with a snapping action, much like the action of a rubber band when it is being stretched and released.

Although most insects have two pairs of wings, flies and mosquitoes have given up the use of their hind wings. These have evolved into two small stumps, called halteres, that act like miniature gyroscopes. With the wings flapping, the halteres vibrate through their horizontal and vertical axes and send signals to sense organs at their bases. Any change in direction affects the normal undulations of the halteres and the fly is informed of its new flight course.

Its six legs are engineered so that the fly can land from almost any angle without slowing down. Instead it alights with a sharp jolt, each leg acting as a shock absorber. Unlike an airplane, the fly never runs forward after touchdown.

A multiflash photograph reveals how a fly lands upside down on a ceiling. Coming in for a landing, it approaches the ceiling at a 45-degree angle with its six legs fully extended. The two front legs touch down first, then it deftly cartwheels onto its other four legs to complete a perfect touchdown.

This sequence of photographs displays a housefly (Musca domestica) at takeoff. As its middle pair of legs lift off the twig (second photograph), a starter muscle automatically triggers the main flight muscles and the fly's wings begin flapping. The wings rotate once through a figure-eight pattern and the back legs lift off.

THE DYNAMICS OF CONTROL

The ability to make controlled maneuvers is essential to flight in any form—whether the fliers are birds, or stunt planes, or even humble insects *(page 66)*. There is no mystery to the process. In an airplane, a system of movable surfaces on the wings and tail allows the pilot to deflect the passing airflow, generating forces that pivot the plane in the desired direction.

From his seat in the cockpit, the pilot is within easy reach of all the necessary control devices. Each foot rests on a pedal, which connects to the rudder on the tail; his hand grips a control stick or wheel, allowing him to govern the elevators and ailerons. In most small airplanes the connection is mechanical—a basic "stick and string" arrangement as shown on the opposite page. Larger, high-speed craft require the extra muscle of hydraulics, which in some planes may even be activated by computers. But whatever the type of system, its operation is logical even to a non-pilot.

For example, a touch of pedal will angle the rudder to one side or the other—left pedal, left rudder, and vice versa. When the rudder swings over, it diverts the flow of air rushing past the tail fin. Two forces then come into play. One is a simple push, as the deflected airstream sets up an equal and opposite reaction; the tail is shoved to the side away from the deflection. At the same time, the rudder's new position changes the shape of the tail fin, making it a curve. The result is an airfoil, which in effect "lifts" the tail toward the low-pressure area on the outer side.

Together, these two forces move the tail; left rudder kicks the tail toward the right. This in turn pivots the airplane so that its nose swivels to the left. In much the same way, the helmsman on a boat controls his heading by using the rudder. If he angles the rudder to the left, the stern swings right, and the bow points left.

The same principle applies to the elevators on the tail's horizontal stabilizer. The pilot pulls back on the control stick—or, in larger planes, he pulls back the wheel—and the elevators tilt upward. This forces the tail down; the nose points up and the airplane climbs. By moving the stick from side to side the pilot manipulates the ailerons, one on each wing, which work in opposite directions. When the left wing dips the right one rises, and the plane begins to bank. It all seems like common sense.

Even so, aviation's pioneers took a while to arrive at it. The earliest known control system survives as drawings by Leonardo da Vinci, the great Italian painter and inventor of the 15th Century. Leonardo sketched a number of whimsical flying machines, including one in which a man rides a wooden framework strapped to a pair of bat-like wings. Since the pilot needs his hands to flap the wings, he must control his direction by other means. So Leonardo provided a harness, which the pilot wears on his head. A cable leads back from the harness to a dart-like tailpiece, which consists of crossed vertical and horizontal fins. By nodding or shaking his head, the pilot can wag the tail in any direction—presuming, of course, the craft can get airborne.

The Wrights fitted their *Flyer* with an elevator in front and a rudder behind, giving them both in-flight stability and basic control over altitude. But to bank the plane for turns, they used the wing-warping technique they had observed in turkey vultures: bending the wings' leading edges with wires attached to the wing struts. Ailerons would have accomplished the same job; indeed, a patent for them already

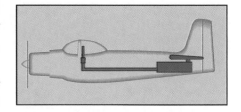

HYDRAULIC POWER

In a hydraulic system, manually operated cables lead to hydraulic activators, which do the work of moving the rudder, elevators and ailerons. A valve in each activator adjusts the pressure of hydraulic fluid, moving a piston connected to the appropriate control surface.

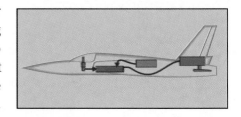

FLY-BY-WIRE

The most advanced control system uses electronic impulses to convey the pilot's commands. An on-board flight computer sends digital instructions to hydraulic activators, which in turn move the various control surfaces.

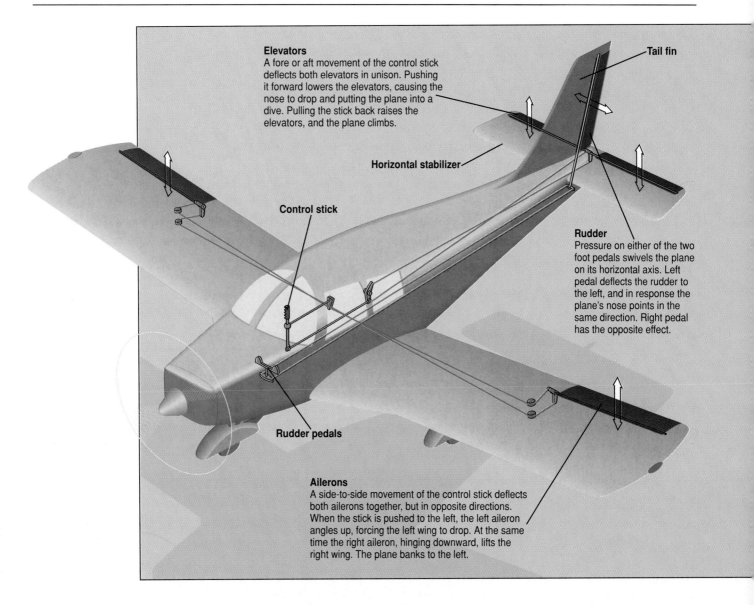

Elevators
A fore or aft movement of the control stick deflects both elevators in unison. Pushing it forward lowers the elevators, causing the nose to drop and putting the plane into a dive. Pulling the stick back raises the elevators, and the plane climbs.

Tail fin

Horizontal stabilizer

Control stick

Rudder
Pressure on either of the two foot pedals swivels the plane on its horizontal axis. Left pedal deflects the rudder to the left, and in response the plane's nose points in the same direction. Right pedal has the opposite effect.

Rudder pedals

Ailerons
A side-to-side movement of the control stick deflects both ailerons together, but in opposite directions. When the stick is pushed to the left, the left aileron angles up, forcing the left wing to drop. At the same time the right aileron, hinging downward, lifts the right wing. The plane banks to the left.

THE FLIGHT CONTROLS

The cockpit controls in this Piper Cherokee are linked to the external control surfaces through a system of cables, pulleys and levers: The rudder pedals connect with the rudder; the control stick with the elevators and ailerons. They allow a pilot to maneuver the plane in three directions—swinging its nose horizontally, pitching it up or down, or wagging its wings from side to side.

existed. But the device had languished in obscurity, and the Wrights did not know of it. What they did learn, by trial and error at Kitty Hawk, was something more basic: that neither the rudder alone, nor a dip of the wings, will cause an airplane to change direction. Only when the pilot uses both together can he put himself into a controllable turn. More than any other discovery, it was this that enabled the Wrights to fly.

Over the next few decades, the control of airplanes was considerably refined. By World War I most craft were equipped with ailerons. Then, in the 1920s, designers hit upon other devices to make the pilot's task easier. Trim tabs—miniature flaps on the trailing edges of the rudder and elevators—allowed pilots to adjust or "trim" their controls to a desired setting and leave them there. The tabs relieved pressure on the control surfaces, so that a pilot could remove his hands from the stick or wheel to perform other duties. The tabs also reduced the muscle power required to manipulate the controls.

The most recent advances in maneuvering are new systems for linking the cockpit controls with the external control surfaces. As airplane size and speed increase, so does the force of air against the flaps; eventually it becomes impossible to manipulate the flaps manually. Most planes today have hydraulic systems, much like an automobile's power steering. In some, the pilot's manual commands are amplified by means of hydraulic boosters, while in others hydraulics provide all the muscle. Should the main hydraulics fail, a backup system takes over. If that also goes, the pilot can still manually wrestle the plane into obedience. There is a price to pay, however; just as power steering dulls a driver's sense of the road, so hydraulic controls deprive the pilot of his "feel" for the air. To restore it, designers usually build in a system of springs or other simulators that convey a sense of the force pushing on the flaps.

The latest wrinkle is electronic. A small control column on the cockpit console—it resembles nothing so much as a computer-game joystick—sends electronic signals to one of several computers that control the appropriate flaps. The new Airbus A320 jetliner employs this "fly-by-wire" technique, as do such supersonic military craft as the Stealth bomber and the experimental X-29 jet fighter. The system has one main drawback. Strong electrical disturbances—lightning, for example, or even a powerful radar signal—can interfere with a plane's electronics. Eventually manufacturers may switch from wires to fiber-optic cables, which substitute pulses of light for electronic commands.

DEFYING GRAVITY

In flying any airplane, whether it be a commercial jetliner or the Piper Cherokee depicted in computer graphics on these pages, the pilot follows the same basic

TAKING OFF

With the plane's nose pointing into the wind, the pilot opens the throttle and the plane rolls forward, gathering speed. At the point where lift almost equals the plane's weight (about 65 miles per hour for a Piper Cherokee), the pilot raises the elevators. The tail drops, the nose tilts skyward, and the plane lifts off the tarmac. The steepness of its climb generates a heavy load of drag, slowing the plane down. To compensate, the pilot lowers the elevators. Leveling off, the plane builds up speed, allowing the pilot to raise the elevators once again for the climb to cruising altitude.

procedures, and uses his controls in virtually the same manner. Before taking off, he checks the weather, plots his route, and confirms his flight plan with the control tower. Once in the cockpit, he makes a thorough check of all systems. Are his flight instruments in order? Do his flight controls function as they should? Are his fuel tanks topped off?

On a large airliner like the Boeing 747, where the systems are more complex, the flight preparations are far more extensive. More than an hour is consumed by weather briefing, flight planning, assessing the fuel load and calculating such factors as weight and balance. Then, strapped in their seats, the captain and first officer systematically run through a checklist of almost a hundred items—electrics, hydraulics, pneumatics, pressurization, controls, instruments, radios, navigational equipment. "External power?" the first officer queries. The captain scans his power system voltmeter and advisory light, which show that the correct current is feeding the airliner's electrical system. "On and checked," he replies.

A dull moan announces the start of the first engine, quickly followed by the second. The elaborate checkout of the systems continues as the airliner taxis toward the runway for takeoff. "Starting systems?" . . . "Off." "Hydraulic panel and pumps?" . . . "Checked and on." "Generators?" . . . "Checked and parallel." As the plane approaches the takeoff point, the captain extends auxiliary flaps on the trailing edge of each wing, and a set of leading edge slats. With these moves, the airliner's streamlined, high-speed wings are reshaped into deeply curved, low-speed airfoils—essential for generating sufficient lift to overcome the jetliner's weight during takeoff.

Cleared for takeoff, the captain advances the four power levers; the engines roar and the plane begins to accelerate down the runway. As it picks up speed, the lift on its wings increases, and the plane feels eager to fly. When the airliner reaches 180 miles per hour, the captain eases back on the control wheel, angling the elevators upward into the airflow. In response, the plane's tail is forced slowly down and the nosewheel rises off the ground. This steepens the angle at which the wings attack the air. To overcome the increased curve, the air rushing over the wings speeds up still more, generating greater lift. Finally the upward force is sufficient to overcome the jetliner's weight, and the 390-ton plane rises majestically from the runway.

As soon as the plane is airborne, the captain momentarily levels off, with a resulting increase in forward momentum. In effect, he is giving himself the running start he needs to continue climbing. Watching his attitude indicator, which monitors the wings' angle of attack, he resumes his climb. He retracts the landing gear, reducing drag and gaining a further boost in acceleration. As the plane moves faster, its wing flaps and slats are no longer required, and they too are retracted. Pegging the airspeed at 270 miles per hour, the captain settles the plane into a smoothly rising climb and engages the autopilot.

ON THE STRAIGHT AND LEVEL

The airliner will make most of its journey at a cruising altitude of 31,000 feet, and the captain has already punched this information into his autopilot. About twenty minutes into the flight, as the plane nears the desired level, an amber light flashes. "Approaching Flight Level 310," the first officer announces. The next step is automatic. The autopilot lowers the nose of the plane into the cruising attitude and the airspeed increases. The airliner settles into a comfortable cruising speed of about 620 miles per hour.

CRUISING

Moving ahead at a constant altitude, a pilot must coordinate his speed with the angle of attack. To maintain sufficient lift at slow speeds, he raises the elevators, angling the wings upward into the airflow. When pushing forward at a faster clip—and thus generating more lift—he can level off to horizontal. At maximum speed, he must deflect his elevators and move in a slightly nose-down posture (opposite page); otherwise he will find himself gaining altitude. Cutting his throttle and again leveling off, the pilot finds his most efficient cruising attitude.

Outside the plane, invisible aerodynamic forces are at work. During acceleration, when thrust was greater than drag, the throttles were set to generate full engine power. The elevators were tilted upward to keep the plane climbing. Now, the autopilot signals minute nose-down corrections, lowering the elevators and edging back the throttles. Seeing that his electronic assistant has accurately pinpointed height, speed and outbound track, the captain sets the autopilot to cruise—a maneuver called "letting George do it."

With the aircraft stabilized in straight and level flight, the four aerodynamic forces are in perfect balance. Thrust equals drag, and lift is just sufficient to compensate for the airplane's weight. Should any force change, there will be a corresponding alteration in the plane's flight path. Thus, if the captain throttles back, decreasing thrust, lift will also diminish and the plane will start to descend. To maintain altitude he can open his throttle again. Or he can raise the elevators—increasing lift by pitching the plane's nose slightly upward so that the wings meet the air at a higher angle of attack. With the plane in a new equilibrium, it will move forward on a level path, but at decreased speed.

In practice, the automatic pilot will take care of most such adjustments. The occasional burble of turbulence as the plane passes through flecks of high-level cloud thus requires no action on the flight deck. Also, the plane's inherent stability tends to correct for minor disturbances in pitch and roll.

A routine fuel check shows that the engines are gulping almost 100 pounds of fuel per minute. In an hour's time, the airliner's weight will have decreased by nearly three tons. Less weight gives the captain lift to spare, and he can use it to fly more efficiently. As fuel is consumed, he dips the plane's nose slightly, lowering the angle of attack. Wing drag diminishes, and this translates into a bonus in thrust. The captain can then either fly ahead at increased speed, or he can cut back on power to save fuel.

BANKING AND TURNING

Glancing at his en-route navigation chart, the captain orders a course change: "Turn left onto 330 degrees." The first officer responds by gently swinging the control wheel to the left, putting the airliner into a banking turn. The left-wing aileron flips up into the airflow, generating a downward force that dips the wing. At the same time, the right wing rises as its aileron moves down, enhancing the wing's airfoil shape and increasing its lift. With the plane now banked at a 30-degree angle to the left, the turn begins. In much the same way, on a smaller scale, a bicycle rider can change his path by leaning to the left or right.

The force that pulls the airliner into the turn is a portion of its total lift. With both wings tilted, some of their lifting power is directed sideways toward the center of the turn—in this case, to the left. Thus the plane assumes its curving flight path. But other forces come into play, which tend to disrupt the maneuver. With its weight unbalanced to the left, the plane tries to sideslip. At the same time, with less lift pulling upward, the plane begins to lose altitude. The captain compensates by increasing the power, and by nudging up the elevators.

Another force causes the nose to swing toward the outside of the turn. This tendency, known as yaw, results from the extra lift on the rising right wing. By the laws of aerodynamics, more lift means more drag; this slows the right wing, holding it back from the turn. To bring the nose into line, the captain calls upon the rudder, the control device for correcting yaw. A touch of left rudder points the nose to the left, into the turn.

Used by itself, the rudder has little effect in changing an airplane's course. The nose will pivot, to be sure, but the plane continues forward in a kind of aerial skid—much like an automobile spinning on an icy road. Only when a rudder

MAKING A TURN

To turn smoothly without sideslipping, the pilot moves the rudder and ailerons simultaneously, banking the plane and swinging its nose in the direction of the turn. In this right turn, he uses right rudder and moves his stick to the right, raising the right-wing aileron while lowering the left. To complete the maneuver, he moves rudder and ailerons in the opposite direction to level the wings.

movement is combined with a banking maneuver do the forces act in harmony to turn the plane. But the amount of rudder pressure must be just right. Too little, and the yaw persists. Too much, and the plane can plunge into a disastrous spiraling dive from which it is difficult to recover.

With all three controls working together, the airliner swings toward its new course. The forces acting upon it are in perfect equilibrium. The inward lift that causes the turn is exactly matched by an outwardly directed centrifugal force. Vertical lift is equal to the airplane's weight; thrust balances drag. The result is a turn so smooth that no one in the plane can feel it. A pendulum would hang perpendicular to the plane's tilted floor, and a cup of coffee on a passenger's seat tray would not spill a drop.

As with the jetliner, so the Piper Cherokee shown below requires the same judicious application of ailerons, elevators and rudder to execute a smoothly controlled turn. There is one important difference, however. The Cherokee, light in weight and built for maneuverability, can snap through its turn in a matter of seconds. But a Boeing 747 has the unwieldy momentum of an airborne supertanker. It requires more than a minute and a mile or so of airspace to swing through an arc of 180 degrees.

THE THREAT OF STALL

As the passenger jet approaches a radio beacon en route, Air Traffic Control advises of congestion ahead and instructs the captain to cut his airspeed. He pulls back on the power levers, and the needle on his airspeed indicator drops. But the reduction in airflow also results in a loss of lift, and to maintain altitude he must increase his angle of attack by slanting up the nose. The aerodynamic forces rearrange them-

selves: Lift still compensates for weight, but a greater portion of it comes from the steeper angle of attack. The plane moves forward at a slower pace.

Cruising along at this high-pitched attitude is called "mushing," and there is a limit to how far a pilot can push it. At constant airspeed, raising the angle of attack from four degrees to eight degrees, for example, will about double the lift force. But when the angle of attack reaches about 14 degrees above horizontal, the airflow over the wings becomes disturbed and lift is destroyed. Eddies of turbulence break up the flow, and the difference in pressure above and below the wing begins to dissipate. By about 18 degrees above horizontal, the destruction of lift is complete.

PULLING OUT OF A STALL

Climbing too steeply to sustain the smooth flow of air over his wings, the pilot stalls. Raising the elevators to bring up the nose would only make matters worse. Instead, the pilot takes advantage of the dive by lowering the elevator and pushing the throttle. Powering forward at the lowered angle of attack restores the forces of lift, and the pilot regains command.

The plane goes into a stall: It begins to drop, nose down, with alarming rapidity. Most stalls are to some extent self-correcting. As the nose pitches down, the angle of attack declines, speed picks up and lift is restored. The pilot can assist by pushing the throttle forward and lowering the elevators—both lift-enhancing maneuvers. But this assumes that the plane has plenty of altitude, allowing the pilot to regain control before he crashes to the ground. All in all, the best strategy is to avoid stall in the first place.

Most commercial aircraft are equipped with stall warners. These devices, which are connected to the angle of attack indicator, sound bells, flash lights or emit loud beeps to alert the pilot to an impending stall. Should the angle of attack increase, the control column would begin to shake back and forth. Then, if none of these warnings were sufficient, an automatic stick pusher would literally thrust the control column forward, effectively changing a nose-up command into a nose-down maneuver.

TACTICS FOR LANDING

Coming in for a landing, the captain of the jetliner and the pilot of the Cherokee have the same goal: to touch down gently at the slowest possible speed. All kinds of factors can complicate the maneuver—wind and weather conditions, traffic patterns at the airfield, runway length, to name a few. Ideally, the pilot will land directly into the wind so that the onrushing air provides an extra bonus of lift and maneuverability. A crosswind makes his task more difficult, since it tends to slew the plane to one side or the other, called "crabbing." Constant adjustments of rudder, ailerons and elevators are needed to keep the nose of the plane pointed straight down the runway.

The Cherokee, light and slow-flying, can begin its approach with a minimum of preparation; its pilot need only alert the airfield's control tower and await instructions. The airliner crew must get ready for landing well in advance. Some 30 minutes or so before reaching the airport, the captain begins a slow, steady descent from cruising altitude. He radios Air Traffic Control and, if there are no other planes ahead of him, he is given clearance to land. At this point, he is still ten miles away from touchdown.

LANDING

The plane descends toward the runway at a steep angle, its engines turning slowly. Near the ground, the pilot raises the elevators slightly and revs the engines; the increased angle of attack and extra thrust help maintain lift. Less than 20 feet from the ground, power is reduced and the pilot keeps the plane airborne by slight increases in its angle of attack. With the wings almost at a stall, the plane seems to float momentarily above the ground. Then it softly touches down on the main wheels. The pilot quickly trims the elevators, then closes the throttle to cut speed even further. The nosewheel touches down, and a combination of reverse engine thrust and wheel brakes brings the plane to a standstill.

Disengaging the autopilot, the captain adjusts his course for the final approach. About seven miles from the runway he throttles back his engines, reducing airspeed; the nose dips and the plane enters a shallow dive. As the runway nears, the captain cuts the speed of the airliner way back, to around 160 miles per hour. To make up for the resulting loss of lift, he deflects the elevators, setting his wings at a higher angle of attack. He also reshapes the wings, extending their leading-edge slats and the flaps on the trailing edge—reversing, in effect, the procedure followed for takeoff. About 2,000 feet above the runway, the first officer lowers the landing gear, imposing an extra drag that helps settle the airplane into a steep, mushing descent.

Closing in on the runway, the captain notes the touchdown point, which rushes toward him at nearly 200 feet per second. He tilts up the nose until the plane is on the verge of stalling, and the plane seems to hover just above the tarmac. The captain activates a set of airbrakes, called spoilers, which emerge from each wing and create turbulence that destroys the remaining lift. The airliner touches down as the main landing gear makes contact.

With the plane firmly on the ground, a sudden roar announces that the captain has reversed the engines and applied full throttle. The plane quickly sheds its remaining speed, with only a touch of the wheel brakes—gingerly applied to prevent overheating—needed to maintain control. The captain moves his hand to the steering control that guides the nosewheel, ready to begin the slow taxi from the runway to the debarkation gate.

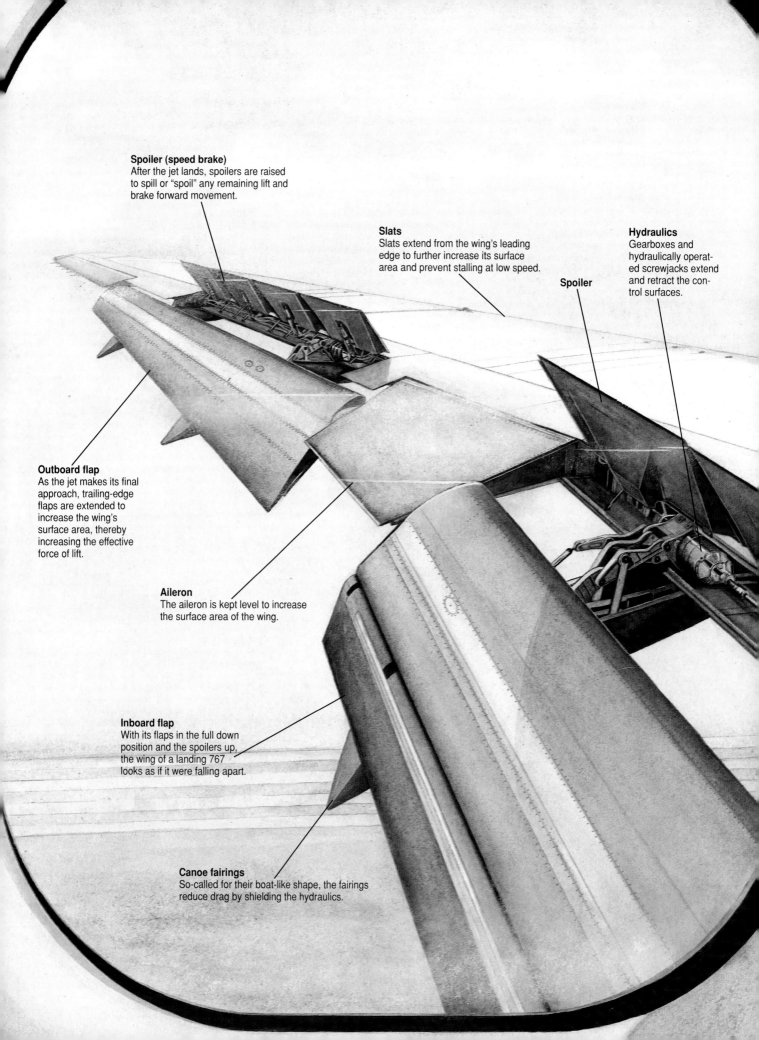

Spoiler (speed brake)
After the jet lands, spoilers are raised to spill or "spoil" any remaining lift and brake forward movement.

Slats
Slats extend from the wing's leading edge to further increase its surface area and prevent stalling at low speed.

Hydraulics
Gearboxes and hydraulically operated screwjacks extend and retract the control surfaces.

Spoiler

Outboard flap
As the jet makes its final approach, trailing-edge flaps are extended to increase the wing's surface area, thereby increasing the effective force of lift.

Aileron
The aileron is kept level to increase the surface area of the wing.

Inboard flap
With its flaps in the full down position and the spoilers up, the wing of a landing 767 looks as if it were falling apart.

Canoe fairings
So-called for their boat-like shape, the fairings reduce drag by shielding the hydraulics.

THE NAKED WING

Viewed through a passenger window at the moment of touchdown, the wing of this Boeing 767 is transformed into a slow-speed, high-lift device by means of flaps, slats and spoilers.

THE WING AT TOUCHDOWN

The most difficult period of any flight comes in the last few minutes before the wheels touch down. With the plane's speed slowed to a minimum, the pilot calls upon all his skill and concentration to maintain lift and avoid a disastrous stall. To help prevent it, airline designers have taken a cue from nature: Like a dove or an eagle, the 747 approaches the ground with its wings transformed.

When an eagle lands, it cocks its wings into the airflow and spreads its feathers, thus giving itself an extra degree of lift and control. The pilot of the 747, in extending the slats and flaps at the wings' leading and trailing edges, accomplishes the same goal. The wings are broadened so that they expose a larger undersurface to the air. At the same time they take on a more pronounced camber, which deflects the air sweeping over their upper surfaces into a speedier, more highly curving flow. As a result, air pressure increases below the wings, and decreases above them, for the needed boost in lift.

To a passenger with a window seat, the process may seem downright alarming. A wing before touchdown appears to self-destruct. The trailing edge flaps, mounted on guide tracks, slide downward and back in two, or often three, interconnected sections, activated by large hydraulic screwjacks. Spaces open up between each section, revealing gaps of naked sky. But this apparent disintegration is simply the mechanical equivalent of the eagle spreading its feathers. The openings actually augment the forces of lift by diverting additional airflow over the flaps' upper surfaces. Furthermore, the wing's total area increases by as much as 25 percent, virtually doubling the lifting power.

The final transformation occurs just before the airplane settles onto the tarmac. The spoilers, located directly ahead of the flaps, hinge upward like box lids and

What an airplane accomplishes with the help of mechanical devices, the bird does with natural grace and ease. Here an eagle inclines its body to catch the uplifting rush of oncoming air—thereby increasing its angle of attack—and extends its tail downward in a wide fan as an airbrake. At the same time, it spreads and twists its wingtip feathers to direct the airflow more effectively over the wings' upper surfaces. In the last few feet of flight, the eagle fans out its feathers to create a larger wing area, allowing it to float, parachute-like, to its perch.

push directly against the oncoming airstream. The sudden increase in air resistance kills the plane's speed and destroys its remaining lift. The spoilers thus serve the same function as the eagle's broad tail, which fans downward on landing to brake the momentum of descent.

MANEUVERING ON THE GROUND

When Louis Blériot made his pioneering flight across the English Channel in 1909, the mechanics of landing his craft were the least of his worries. His tiny monoplane weighed only 660 pounds, and he brought it back to earth on landing gear constructed out of bicycle forks, wire-spoke wheels and heavy rubber tires. Today's jumbo jets, touching down at speeds of up to 200 miles per hour, require far stur-

A telephoto freezes a Boeing 747-400 at the instant of touchdown. Descending at 170 miles per hour with its massive undercarriage lowered, the plane's 18 wheels slam into the runway with express-train force. The massive hydraulic shock absorbers take most of the impact. The 49-inch tires, inflated to 300 pounds per square inch, cushion the rest of the impact, and the airliner's carbon-disc brakes bring it to a gentle halt.

dier support. Their landing gear needs to be ultra-strong, relatively light in weight, and compact enough to retract neatly out of the way into the craft's slender wings during flight.

The main landing gear of a Boeing 747 is a near miracle of design ingenuity. It consists of four separate wheel trucks: one at the base of each wing, and two that retract into the plane's belly. Each truck carries four wheels rimmed with high-pressure, nonskid tires. Together they must support virtually double the 747's certified maximum takeoff weight of almost 900,000 pounds; as a fail-safe requirement, Boeing dictates that the jetliner must be able to land using only two of its main trucks. Two additional wheels, mounted on a truck extending from the nose, cradle the plane when it comes to a stop.

Toying with Aerodynamics

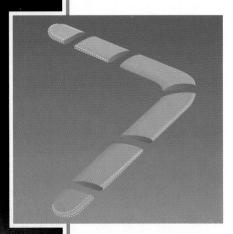

This cross section reveals the contoured top and flat bottom of the boomerang's two blades.

Captured by time-lapse photography, boomerangs circle the midnight sky, then float back to the ground like twirling maple seeds.

Humans have been tossing boomerangs since prehistoric times. The oldest example, a boomerang excavated from a cave in Poland, dates from 20,000 B.C. Neither is the flying disk a new invention. Natives of 18th-Century India threw metal rings as weapons. These aerodynamically complex devices have been transformed into intriguing toys by contemporary imaginations.

Scientists have used slide rules and computers in an effort to explain the flight of the boomerang. It works by a combination of lift, gyroscopic forces and Newton's Laws of Motion. When a boomerang is launched, a combination of linear and rotational velocities gives it forward motion, stability, aerodynamic lift (as well as drag), and gyroscopic precession *(page 100)*. Usually fashioned in the traditional V shape, boomerangs can take the shape of every letter of the alphabet except the letter "I." Once airborne, the boomerang becomes a set of rotating wings; the airfoil shape of the two arms *(inset)* creates lift like that of a helicopter's blades.

Gyroscopic precessional forces cause the boomerang to circle. In flight, the upper blade rotates into the wind, while the bottom blade rotates away from the wind, so that the top blade generates much more lift. This force causes a gradual rolling torque to the left (in the case of a boomerang thrown with the right hand). But instead of the boomerang actually rolling to the left, gyroscopic precession converts that torque into a left turn, like a "no hands" turn on a bicycle. Hence the circular flight path.

Unequal forces, created by the shape and twist of the blades, shift the center of lift slightly forward of the center of rotation. This results in a different kind of precessional force that causes the boomerang to gradually tilt over or "lay down" in a horizontal position midway through its flight. Finally the boomerang hovers, while still spinning, and drops into the hands of the thrower.

Recently, a mechanical engineer from California designed a flying ring with the help of NASA aerodynamicists and computers. Called the *Aerobie*, it has been hurled about a fifth of a mile—farther than any other man-made object without benefit of a power supply or external force like the wind. But although a flying ring may be thrown farther than a boomerang, it has to be retrieved.

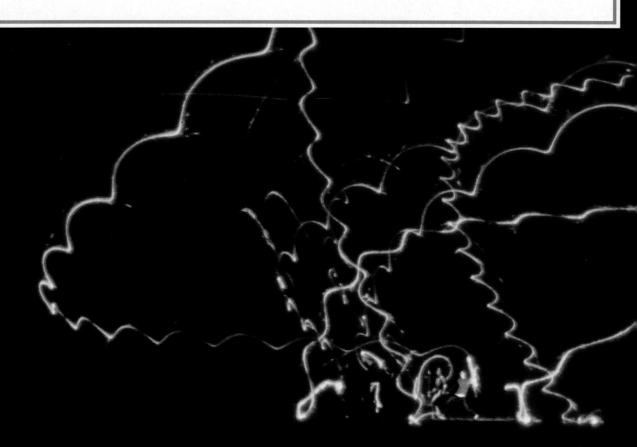

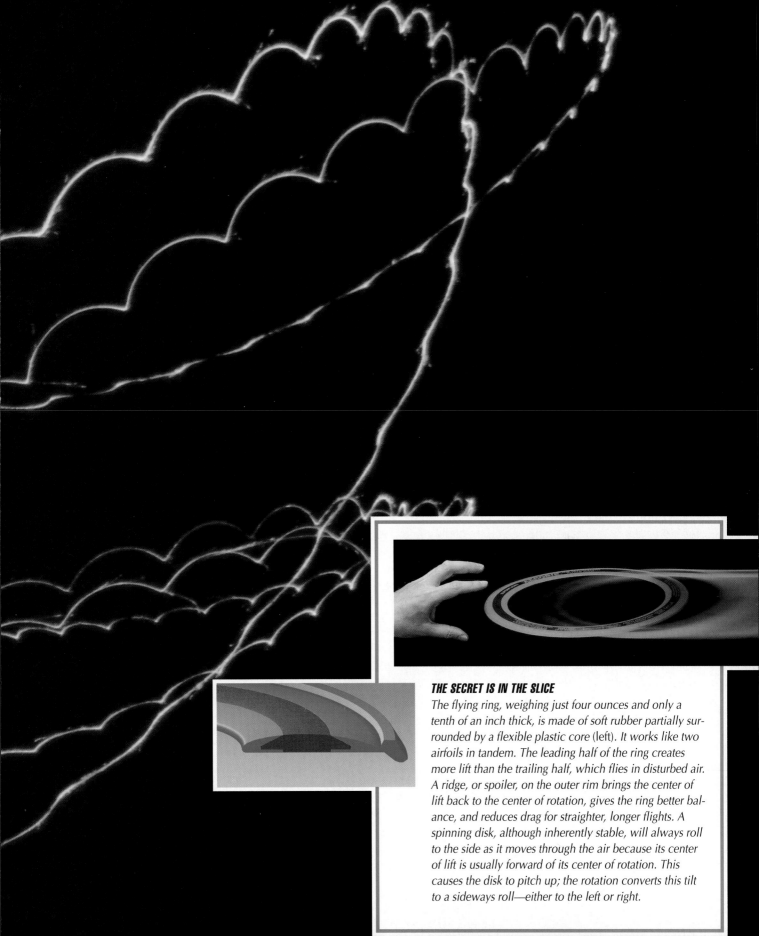

THE SECRET IS IN THE SLICE

The flying ring, weighing just four ounces and only a tenth of an inch thick, is made of soft rubber partially surrounded by a flexible plastic core (left). It works like two airfoils in tandem. The leading half of the ring creates more lift than the trailing half, which flies in disturbed air. A ridge, or spoiler, on the outer rim brings the center of lift back to the center of rotation, gives the ring better balance, and reduces drag for straighter, longer flights. A spinning disk, although inherently stable, will always roll to the side as it moves through the air because its center of lift is usually forward of its center of rotation. This causes the disk to pitch up; the rotation converts this tilt to a sideways roll—either to the left or right.

Most of the impact on landing is dissipated by heavy-duty shock absorbers attached to each wheel truck. In earlier aircraft, the shocks were generally heavy steel springs, but in the 747 and most other modern aircraft they consist of hydraulic cylinders filled with oil and either air or nitrogen. The trucks are hinged to powerful, hydraulically activated steel support beams that pivot down from their wing or fuselage housings on command from the cockpit. On the ground, the pilot rides three stories above the runway. Said one 747 captain, "It's like sitting at my attic window and trying to drive my house down the road."

Besides the forces of impact, each landing gear must also absorb deceleration loads of up to 64,000 pounds that occur when the pilot applies his wheel brakes. Assisted by reverse thrust, a 747 can screech to a stop in as little as 2,500 feet—a performance that absorbs as much kinetic energy as a million automobile brakes. Hydraulics again come into play, pushing multiple brake discs against each wheel drum. So intense is the friction that in some models the wheel assembly includes cooling fans to disperse the heat; otherwise the discs would soon glow a bright cherry red and cause the tires to burst. The plane's latest version, the 747-400, solves the heat problem with plastics: carbon-fiber discs that run cooler, last longer and save a valuable 1,800 pounds total over the earlier steel brakes.

Like all heavy aircraft, the 747 also incorporates an antiskid device in each wheel. An automatic brake-release mechanism balances the brake pressure against the wheel's speed of rotation, ready to ease pressure the moment a wheel threatens to lock. The pilot can stamp down hard on the brake pedal without fear of swerving off the runway, even when the surface is icy. Even the tire treads are specially designed for abrupt stops in less than ideal conditions. Since puddles of rainwater on the runway tend to make the wheels skate across the surface without gripping the pavement, the tires are engineered to forcibly pump the water aside. Some tires also have side ridges that act as spray deflectors, which keep heavy splashes from hitting the flaps or other vulnerable control surfaces.

HUMMINGBIRDS AND HELICOPTERS

No airplane will ever match the maneuverability of aviation's mechanical magic carpet—the helicopter. Commenting on its versatile performance, the inventor of the first practical model, Igor Sikorsky, said: "I didn't anticipate all the many uses helicopters find today. But I was sure an aircraft that could fly like a hummingbird would be immensely useful."

Indeed it is. For like a hummingbird, and also most insects *(page 66)*, the chopper can take off from a standstill, fly forward or backward, move straight up or down, dance to either side, or hover in midair. And the comparison does not end there. Both bird and whirlybird owe their aerial agility to a capacity for precisely angling the pitch of their airfoils—the wings of the hummingbird, and the rotor blades on the helicopter.

In a sense, the hummingbird is an evolutionary compromise between more conventional avians and the insects, whose nectar-sipping habits it shares. Whereas other birds flex their wings freely at shoulder, elbow and wrist, the hummingbird is equipped with long feathered paddles that are almost all hand. In function, they resemble the rigid fans of a butterfly or a honeybee. Virtually all movement comes from the hummingbird's shoulder, a joint of remarkable flexibility that allows the

WHIRLING WINGS

A hummingbird hovers motionless in midair, its body slanted steeply upward so that the arc of its wingbeat is almost horizontal. As the wings sweep forward on the downstroke, and upward on the backstroke, they bear a close resemblance to helicopter blades—though with the wingtips tracing a horizontal figure eight. A steady blast of air is forced downward by both strokes, producing continuous lift.

wing to twist from rightside up to upside down through almost 180 degrees. Powered by flight muscles that account for up to 30 percent of the hummingbird's total weight, they flap at a rate of 50 to 70 times per second—many times faster than any robin or wren.

As they flip through the air, the hummingbird's wings generate lift during both the upward and downward beats; no other bird can do this. During hovering flight, the wings move in a figure-eight pattern, eggbeater fashion, each one sweeping horizontally through one-third of a circle. On the downstroke they are positioned normally, their upper surfaces angled skyward and their leading edges cutting the air ahead. But on recovery they swivel, moving up and back with their leading edges facing to the rear. The whir of these oscillating motions resembles nothing so much as the spin of a helicopter's rotor—which also produces lift throughout its entire revolution.

To shift direction—which it can do in the blink of an eye—the hummingbird simply alters the plane in which its wings operate. To back off from a flower, it plies both wings behind its body. Darting forward, it beats its wings more vertically, giving itself a forward lift that can propel it ahead at up to 50 miles per hour. Additional control comes from the tail, which the hummingbird can spread, close, cup or angle in the propwash of its whirring wings. There is literally no direction in which it cannot maneuver. Banking into a sudden turn, it can even fly upside down for a brief moment.

The only creatures that rival the hummingbird in aerobatic prowess are the insects, which learned all the tricks some 100 million years earlier. Their wings vibrate at even higher frequencies than do the hummingbirds'—for bees, 200 to 300 beats per second on average, and an astonishing 1,000 beats per second for certain midges. The earliest known flying insect is the dragonfly, which dates from a swampy, sultry era of time some 345 million years ago known as the Carboniferous period. Some dragonfly species grew to enormous size, with wingspans of three feet. None of these giants survive, but their modern-day relatives can hover and dart just as ably as any hummingbird—or any helicopter. Indeed, with its long tapering body and its four slender, rotor-like wings, the dragonfly could have been the model for one of Sikorsky's first military choppers, the R-5, and its civil counterpart, the S-51. When the S-51 was manufactured under license in England, in 1946, the Royal Air Force dubbed it—what else?—the Dragonfly.

All appearances to the contrary, a helicopter is not simply an aircraft with a large propeller on its roof. Its rotor is in fact a set of wings that, powered by a piston or turbine engine, provide lift by spinning. Each blade is an airfoil, which connects to a central drive shaft through an elaborate hinge mechanism. Other mechanical devices, known as swashplates and control rods, allow the pilot to adjust the pitch of the blades, and also to tilt the entire rotor assembly in any direction. Like the hummingbird's swiveling shoulder joint, these mechanisms, which together comprise the helicopter's rotor head, provide the key to the helicopter's omni-directional maneuverability.

Three cockpit controls replace the standard stick and rudder pedals of a conventional airplane. A floor-mounted collective pitch lever near the pilot's left hand varies the pitch of the blades. With an increase in pitch each blade swivels so that its leading edge cuts the air at a higher angle of attack, for an overall increase in lift; the helicopter rises. For an extra upward boost, the pilot advances his throttle, which is incorporated directly into the collective pitch lever. Because helicopter flight involves the constant use of both hands, the throttle is a twist-grip on the lever's handle, similar to that on a motorcycle.

The second control arm is the cyclic pitch control, which is located in the familiar stick position. This governs the helicopter's horizontal movements. Should the pilot wish to move sideways to the left, he pushes the control stick in that direction. The command is transmitted to the rotor head, which tilts the whirring blades, like an enormous spinning pie plate, slightly to the left. The lift they generate becomes divided, as on a banking airplane. The upward force keeps the helicopter airborne, while the sideways force pulls it smartly to the left.

Other aerodynamic forces complicate the business of helicopter maneuvering. The strongest is a tendency of the spinning rotor to swivel the craft's fuselage in the direction opposite to the spin. This powerful torque, which results from the principle that any action calls forth an equal reaction, was a severe problem for early helicopter designers, since it made their craft virtually unflyable. They addressed it a number of ingenious ways. Some designs included two rotors, one mounted above the other and spinning in opposite directions. One particular model had a pair of short, fixed wings with the rotors rising from each tip. Large cargo-carrying helicopters still use paired rotors—one in front above the cockpit, the other spinning above the tail. But most passenger helicopters now carry a small

Blades
Lightweight, hollow fiberglass blades exhibit the classic airfoil shape. From rotor shaft to tip, a typical blade measures 37 feet.

INSIDE A ROTOR HEAD

On a typical helicopter rotor head, the blades are attached to a central rotor shaft powered by the engine. Pitch control rods leading from the upper swashplate attach to the leading edge of each blade. A rotating scissors device turns the swashplate as the blades revolve. The lower swashplate, linked to the cockpit, can be raised, lowered or tilted. These movements, conveyed to the upper swashplate, control the pitch and angle of the blades.

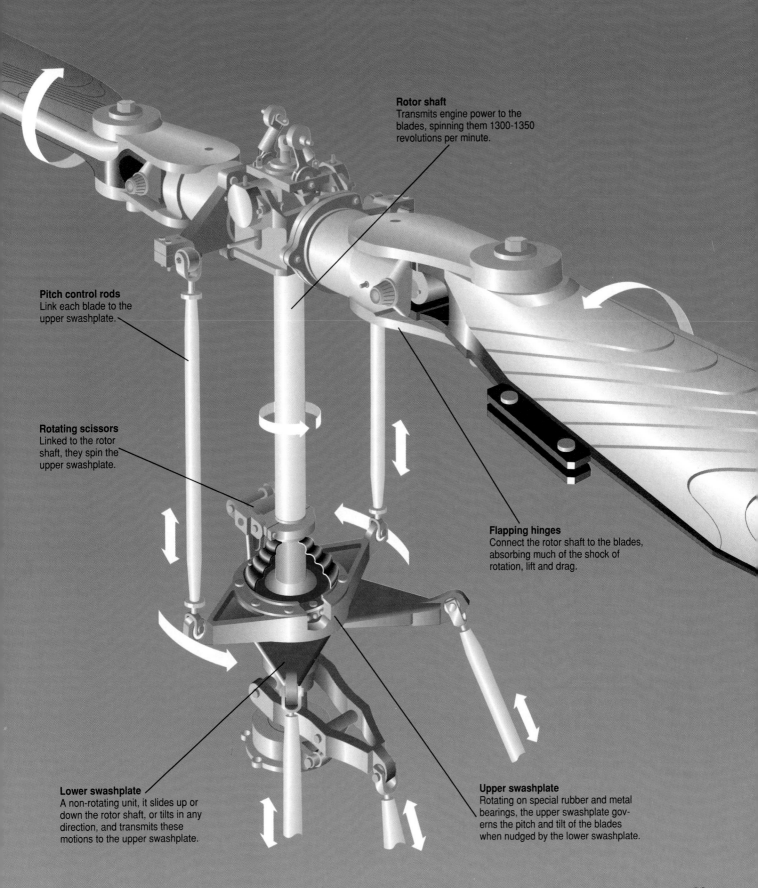

Rotor shaft
Transmits engine power to the blades, spinning them 1300-1350 revolutions per minute.

Pitch control rods
Link each blade to the upper swashplate.

Rotating scissors
Linked to the rotor shaft, they spin the upper swashplate.

Flapping hinges
Connect the rotor shaft to the blades, absorbing much of the shock of rotation, lift and drag.

Lower swashplate
A non-rotating unit, it slides up or down the rotor shaft, or tilts in any direction, and transmits these motions to the upper swashplate.

Upper swashplate
Rotating on special rubber and metal bearings, the upper swashplate governs the pitch and tilt of the blades when nudged by the lower swashplate.

tail rotor, mounted sideways like a propeller, which pushes the helicopter into line. It is governed by the third set of cockpit controls—the rudder pedals. These adjust the thrust of the tail rotor, allowing the pilot to steady his course, or steer to left or right during forward flight.

Another force takes effect whenever the pilot tilts the rotor assembly for horizontal movement. In the same way as a banking airplane develops more lift on its upper wing than on its lower one, the angled rotor blades are subjected to unequal pressures as they spin. In straight-ahead flight, with the assembly tilted forward, lift is diminished momentarily on the front-facing blade. Then, as the blade arcs backward, the lift upon it increases. It is this differential that pulls the helicopter forward; but it also puts a severe strain on the individual blade attachments. Thus the flapping hinges that connect the blades to the rotor shaft, which allow the blades to wiggle up and down, dissipating some of the force.

The pilot, one hand on the collective pitch lever, the other on the cyclic pitch control, his feet on the rudder pedals, can dart about the sky with almost total freedom, but the maneuvers take considerable skill. The hardest is hovering, in which he must pull just enough lift to balance the helicopter's weight, apply precisely the right amount of pedal to counter torque, and also hold the machine level by adjusting the rotor's tilt. It demands concentration, as one dyed-in-the-wool jet jockey found in his helicopter conversion course. "It's like trying to pat your head and rub your tummy while balancing on a bowling ball," he complained.

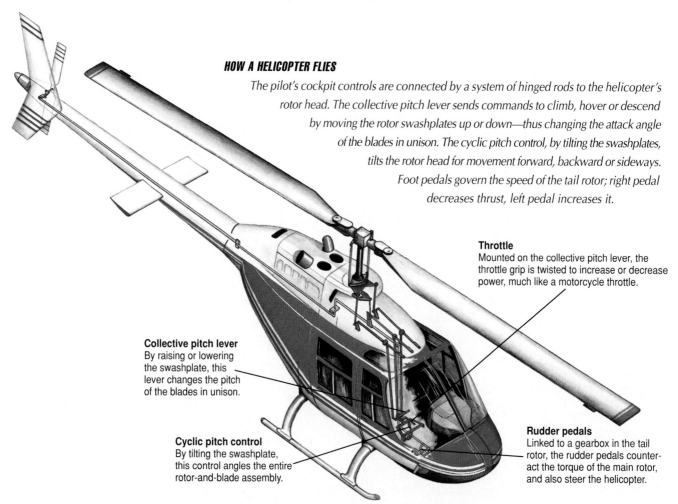

HOW A HELICOPTER FLIES

The pilot's cockpit controls are connected by a system of hinged rods to the helicopter's rotor head. The collective pitch lever sends commands to climb, hover or descend by moving the rotor swashplates up or down—thus changing the attack angle of the blades in unison. The cyclic pitch control, by tilting the swashplates, tilts the rotor head for movement forward, backward or sideways. Foot pedals govern the speed of the tail rotor; right pedal decreases thrust, left pedal increases it.

Throttle
Mounted on the collective pitch lever, the throttle grip is twisted to increase or decrease power, much like a motorcycle throttle.

Collective pitch lever
By raising or lowering the swashplate, this lever changes the pitch of the blades in unison.

Cyclic pitch control
By tilting the swashplate, this control angles the entire rotor-and-blade assembly.

Rudder pedals
Linked to a gearbox in the tail rotor, the rudder pedals counteract the torque of the main rotor, and also steer the helicopter.

Hovering

With the cyclic pitch control in neutral, the swashplate and rotor blades are level and the helicopter moves neither forward nor backward. The swashplate is raised to change the pitch, or angle of attack, of the blades in unison, producing lift to equal the weight of the helicopter and keep it motionless in the air.

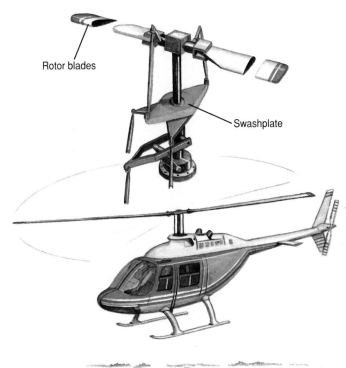

Rotor blades

Swashplate

Forward flight

When the pilot pushes the cyclic pitch control forward, he causes the swashplate and blades to tilt down at the front. Lift becomes a divided force. It increases over the back of the rotor to support the helicopter's weight, and decreases at the front of the helicopter to provide forward thrust.

Vertical flight

To climb, the pilot pushes the collective pitch lever forward. This raises the swashplate and steepens the blades' angle of attack to the point where lift exceeds the weight of the helicopter. To descend, he reverses the procedure.

Backward flight

The pilot pulls the cyclic pitch control backward, tilting the swashplate and blades toward the back of the helicopter. With the force of lift divided—some directed up, some to the rear—the helicopter retreats.

The Amazing Dragonfly

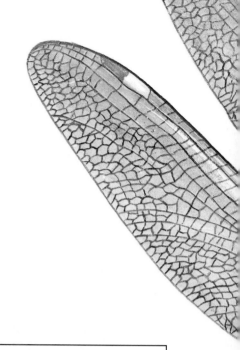

Dragonflies, which have been around for more than 250 million years, may provide clues that could revolutionize aircraft design. Studies show that dragonflies use "unsteady aerodynamics," a mode of flying radically different from the flight of airplanes and birds.

Dragonfly wings churn up the air to create whirling vortices of low pressure that the insect uses to provide lift. By contrast, airplanes and birds rely on the steady flow of air over the upper and lower surfaces of their wings. For them, turbulence can be deadly. Taking a lesson from this remarkable flier, designers may someday imitate it by fitting aircraft with wings that would put unsteady airflows to good use.

The wings of an airplane, they suggest, could be fitted with devices that would redirect turbulence back onto the wings, decreasing pressure and increasing lift; this energy is now wasted. One idea is a flap or "fence" that would automatically flip up near the leading edge of a wing when the aircraft goes into an unexpected stall. This roughness would create vortices that increase lift, possibly averting a crash.

Another possibility is to install a small extra wing near the front of an airplane. This "canard" would mimic a dragonfly by sharply shifting its angle of attack to generate vortices as needed. The larger rear wing of the airplane could then coast along on these aerial whirlpools.

Four wings; four muscles
In the upper cutaway of a dragonfly's thorax a pair of direct flight muscles (dark) has contracted, resulting in the upward movement of the wings. In the lower diagram the second set of muscles has contracted; the wings flap downward. In this way each of the four wings can beat independently, making the insect highly maneuverable.

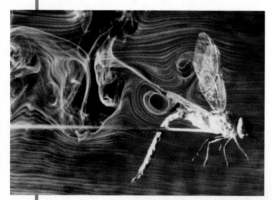

Articulated tail aids in steering and balance

With the help of miniature wind tunnels, stroboscopic photography and unwitting insect subjects, scientists are investigating the aerodynamics of turbulence. Here a dragonfly, tethered to a small wire, shows off its ability to generate lift by producing unsteady airflow. When the insect's front wings flap down, they stir up a small whirlwind which, when it passes over the back wings, generates enough lift to support up to 20 times the dragonfly's weight.

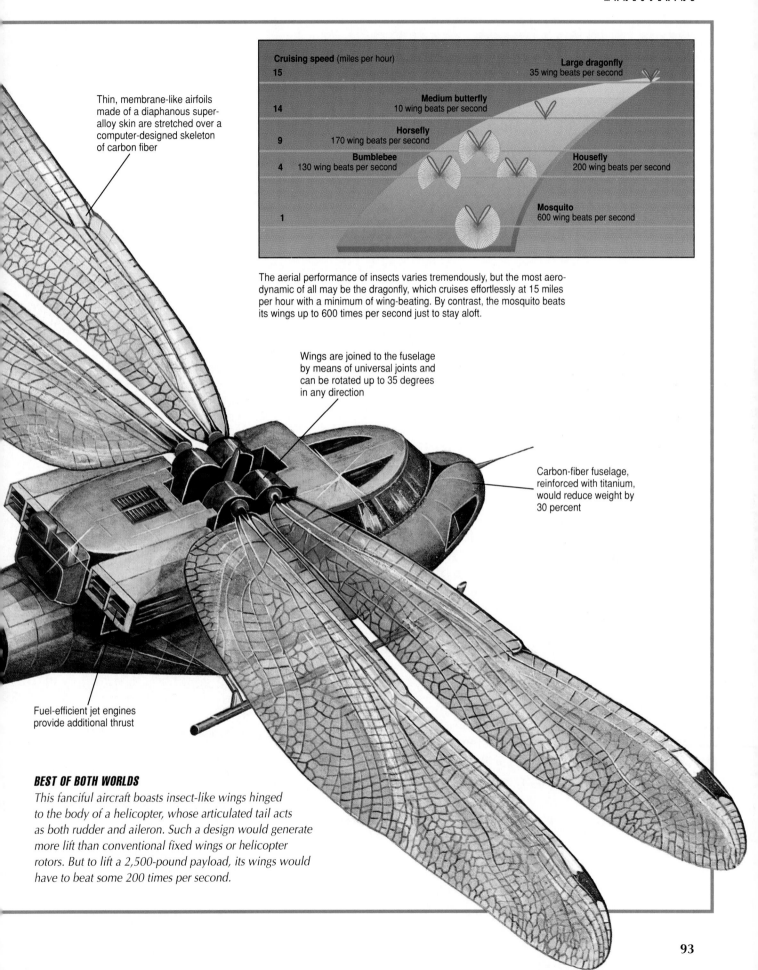

Cruising speed (miles per hour)		**Large dragonfly**
15		35 wing beats per second
14	**Medium butterfly** 10 wing beats per second	
9	**Horsefly** 170 wing beats per second	
4	**Bumblebee** 130 wing beats per second	**Housefly** 200 wing beats per second
1		**Mosquito** 600 wing beats per second

Thin, membrane-like airfoils made of a diaphanous super-alloy skin are stretched over a computer-designed skeleton of carbon fiber

The aerial performance of insects varies tremendously, but the most aero-dynamic of all may be the dragonfly, which cruises effortlessly at 15 miles per hour with a minimum of wing-beating. By contrast, the mosquito beats its wings up to 600 times per second just to stay aloft.

Wings are joined to the fuselage by means of universal joints and can be rotated up to 35 degrees in any direction

Carbon-fiber fuselage, reinforced with titanium, would reduce weight by 30 percent

Fuel-efficient jet engines provide additional thrust

BEST OF BOTH WORLDS

This fanciful aircraft boasts insect-like wings hinged to the body of a helicopter, whose articulated tail acts as both rudder and aileron. Such a design would generate more lift than conventional fixed wings or helicopter rotors. But to lift a 2,500-pound payload, its wings would have to beat some 200 times per second.

THE CROWDED SKIES

No creature on Earth maneuvers in more hazardous circumstances than an ordinary bat. The only mammal capable of true flight, bats congregate in huge colonies, sleeping by day in caves or tree branches, then emerging at sundown by the tens of thousands to feed and frolic under cover of darkness. Some species leave their dens in close-packed swarms, as densely concentrated as raindrops in a thunderstorm, moving out along well-established flight corridors. Then they start foraging, in the blackest of moonless nights, some browsing on fruits, others flapping after insects so small as to be barely visible even in bright sunlight. The feeding is virtually continuous; a bat must consume from one-third up to half its body weight each night just to keep going. How does the bat do it?

The secret is a short-range navigation system as sophisticated as any in nature. Each bat emits a succession of shrill clicks, about 20 or 30 times every second, that beam out ahead and bounce back from any object they strike. The bat's sensitive, oversize ears pick up the echo. This aerial sonar allows the bat to zero in blind on the tiniest gnat or mosquito—and also to avoid colliding with trees, cave walls or other bats.

The clicks from a bat have such a high sonic frequency—up to 2,000 vibrations per second—that they are largely inaudible to human ears. But the energy level of each click is intense. At a distance of two inches from the bat's head, the acoustic pressure has been measured at 100 decibels, roughly the same as a jackhammer. As the bat flies, it alters the frequency and the length of each burst. Some bats broadcast simultaneously on two or more wavelengths. The more rapid the vibrations, the smaller the object the bat can detect. A mouse-eared bat, for example, can detect a steel wire only one millimeter thick from a distance of two yards. Then, as the bat moves in closer, its emissions become shorter and more rapid—with the brief intervals between each vibration allowing the echo to return. The time delay from a hungry bat to a mosquito six inches from its mouth is about one-thousandth of a second.

Borne on wings nearly as thin as a plastic bag, a multitude of bats emerges from a Texas cave, avoiding midair collision by means of aerial sonar.

The Migration of Monarchs

Warmed by the Sun, monarch butterflies fill the skies of central Mexico, searching for water and flower nectar. As the days lengthen, they begin to mate; then they depart, flying at speeds from 10 to 30 miles per hour on their northward journey.

Birds are not the only masters of long-distance migration: Fragile monarch butterflies cover more than 1,500 miles on their annual odyssey.

Since monarchs feed and lay their eggs only on the milkweed plant, one theory of their migration suggests that the butterflies originated in Mexico, where over half of the world's 100 species of milkweed are found. After the last Ice Age, the monarchs spread northward as the Earth warmed, following the milkweed. In returning to Mexico each winter, they are simply going home.

Scientists believe that the start of the migration is triggered by changes in the number of hours of daylight, the angle of light from the Sun and nighttime temperatures. Responding to these signals in the spring and fall, the monarch butterflies become restless and depart.

How do these insects find their way to the same summer and winter sites each year? Scientists theorize that, like some migrating birds, monarchs have a built-in compass that tells them their latitude and allows them to read the angle of inclination of the Earth's lines of magnetic force. En route, they feed voraciously on milkweed flowers, each butterfly adding about 100 milligrams of body fat to its delicate frame.

One of the largest wintering sites is a secluded enclave of the Sierra Madre mountains in Mexico. At an elevation of 9,000 feet, the average temperature hovers around the freezing mark. The monarchs, in a semidormant state, cling to the branches and trunks of 100-foot fir trees, the weight of their numbers literally bending the boughs.

In early March, the butterflies descend in swarms to search for water—a signal that they are ready to begin their northward trip to feed and lay their eggs. Although the butterflies depart as huge flocks, by the time they reach Texas they are flying solo or in small groups. Some of the original migrants survive the return trip, but it is usually the second or third generations, bred en route, that arrive at the northernmost regions in the summer.

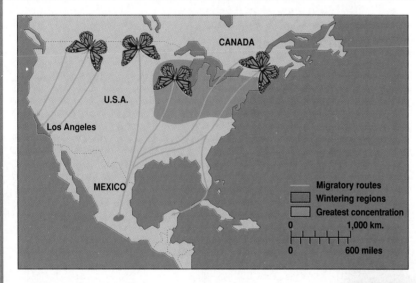

CANADA

U.S.A.

Los Angeles

MEXICO

Migratory routes
Wintering regions
Greatest concentration

0   1,000 km.

0   600 miles

After years of unsuccessful attempts at developing a way to tag monarchs, the solution came in the use of pressure adhesive labels, similar to the price tags found on glass bottles at the supermarket. At a hundredth of a gram, the weight of the tag is about one-fortieth the weight of the butterfly.

Flight plan of the monarchs
Western monarchs winter on the coast of California from San Francisco to Santa Barbara, returning in February to California's central valley and the cooler Pacific Northwest and central Rockies. Their more-numerous eastern relatives travel south in the fall from the Great Lakes region into south-central Texas and across the Gulf of Mexico to the Sierra Madre in Mexico. Some of the eastern monarchs fly down the Florida peninsula and may cross to Cuba, Mexico's Yucatan Peninsula and Central America.

Artificial horizon
This instrument *(top)*, also called the attitude indicator, represents the airplane's attitude in relation to the Earth. An artificial horizon is kept horizontal by the gyroscope to which it is mounted. A fixed bar represents the plane's wings. As the plane climbs or dives, the bar rises above or dips below the horizon line. When the plane banks, the bar banks accordingly.

Vertical speed indicator
The vertical speed indicator *(right)*, shows the rate at which the airplane is climbing or descending (as distinct from airspeed). The instrument case contains an airtight compartment housing a diaphragm. The case and the diaphragm compartment each connect to the plane's static pressure system, and can thus register two different pressures. In level flight, the two pressures are equal and the needle points to zero. As the plane ascends, the diaphragm pressure drops faster than the case pressure and the relative difference translates into the aircraft's rate of climb.

Heading indicator
The airplane symbol on the dial *(bottom)*, reflects the actual direction of the aircraft. Should the pilot change course, the compass card underneath will show the new heading. While the compass appears to rotate, its motion is only apparent. It is the airplane that turns and the compass that remains steady in relation to the Earth.

Other natural fliers are endowed with wonderfully accurate systems for long-range navigation. A homing pigeon, set free in unfamiliar territory as far as 600 miles from its home base, returns without hesitation in a single day's journey. Other birds migrate for weeks to far-distant feeding or mating grounds, often traveling thousands of miles over featureless desert or ocean. Arctic terns are the champion long-distance fliers. The annual round-trip flight from their summertime courtship area in the far north to their winter home on the South Polar icecap and back roughly equals the Earth's circumference. To find favorable winds, they may fly at the cold, oxygen-starved altitude of 21,000 feet.

At present there is no adequate explanation for how migratory birds are able to find their way. Some scientists suggest that they atune their flight paths to the positions of the Sun and stars, calibrating daily and seasonal changes by a mysterious internal chronograph. Other researchers believe that the birds may be guided by the Earth's magnetic or gravitational fields, by means of a genetic compass passed from one generation to the next.

BASIC INFORMATION
Whether the aircraft is a single-engine Cessna or the Boeing 747 shown above, six primary flight instruments form a basic T configuration on the flight panel.

Airspeed indicator

This gauge *(left)*, is connected to a pressure-sensing Pitot-static tube and a flexible diaphragm within the instrument case. The diaphragm compares the difference between the static air pressure on the tube and the rush of oncoming air, called ram pressure. An internal lever system translates this difference in pressure into air speed, expressed in nautical miles per hour, or knots.

Turn-and-bank indicator

This instrument *(middle)*, sometimes called the needle and ball, guides a pilot through banking turns. When the needle is centered, the plane is flying straight ahead. A deflection to either side shows that the plane is turning in that direction. The ball, in a sealed glass tube filled with kerosene, works on the same principle as a carpenter's level. When the ball is centered during a turn, the plane is in balance. But if the ball rolls in the direction of the turn, it shows that the plane is side-slipping. If the ball moves away from the turn, the plane is yawing.

Altimeter

The altimeter *(right)*, is in effect a barometer that senses the decrease in air pressure that accompanies an increase in altitude. A needle linked to the mechanism indicates the height of the plane relative to the sea or the ground over which it is flying.

To match these navigational feats, the pilots of airplanes make use of complex gauges, sensors and other flight instruments that permit them to fly in virtually all weather, and at all hours of day or night. Some planes are equipped with radar, the electronic equivalent of the bats' sonic detection system. And all pilots rely on a worldwide communications network that sends signals from the ground to guide them through today's crowded skies.

THE FLIGHT DECK

Flying blind is probably the proudest skill of pilots. It means being able to maneuver in clouds or fog when nothing can be seen from the cockpit windows but grey vapor—no horizon, no ground, no sunlight, no stars. Birds, who ought to know everything there is to know about flying, cannot fly blind. Over half a century ago, an army pilot blindfolded a pigeon and threw it out of an airplane. The pigeon tried all sorts of maneuvers and then went into a spiral dive. It simply gave up. It let itself fall, tilting its wings to brake its descent. In short, it bailed out.

A pilot can fly blind, however, with the aid of artificial senses—the instruments on the panel in front of him. He must learn to put complete faith in them. Under no circumstances must he react to a bodily sensation, however strong, particularly if it disagrees with what the instruments tell him.

The human body can sense almost all the motions a plane makes. Slip or slide, climb or descent, speed-up or slow-down—all these travel through an experienced pilot's nervous system like an electric shock. What he feels corresponds exactly with what his instruments report. But there is one maneuver his body cannot detect: a banked turn. That is because the combination of bank and turn keeps the plane in perfect balance. No matter how extreme the maneuver, the bank kills the feel of the turn, and the turn erases the sensation of the bank. The turn indicator on the instrument panel shows what the plane is doing, but the pilot's instincts tell him he is flying straight and level.

This paradox explains why early aviators became utterly confused when flying through clouds or fog. Unable to see the ground or the horizon, a pilot might veer to one side without realizing it. A banking plane tends to dip its nose and pick up speed, motions the pilot would quickly sense. To correct them, he would pull back on the stick—with often fatal results. The plane's nose would be forced more tightly into the turn, the bank would steepen, speed would increase even more, and the pilot would plummet in a screaming spiral spin.

It did not take many such mishaps for aviators to realize that without special instruments to guide them, controlled flight in murky conditions was next to impossible. Today, commercial aircraft can fly at night or above the clouds, at speeds

THE WORLD AS A SPINNING TOP

First observed by Isaac Newton in the 18th Century, gyroscopic forces have amazed man for hundreds of years. The gyroscope is one of the oldest mechanical devices; in fact, the Earth itself is a gyroscope, having been thrown off by the Sun and set spinning on its axis.

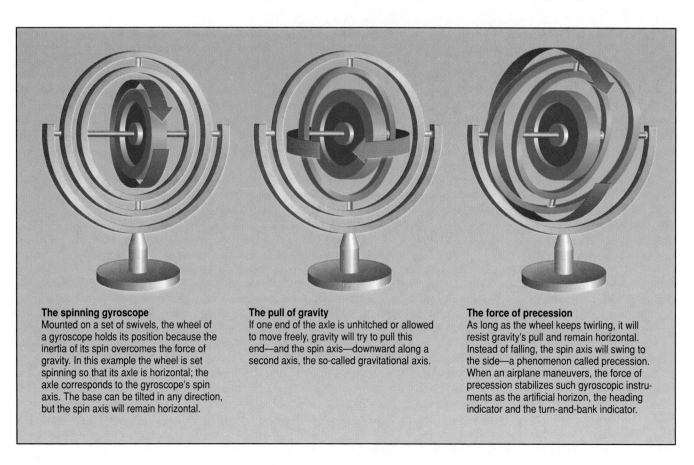

The spinning gyroscope
Mounted on a set of swivels, the wheel of a gyroscope holds its position because the inertia of its spin overcomes the force of gravity. In this example the wheel is set spinning so that its axle is horizontal; the axle corresponds to the gyroscope's spin axis. The base can be tilted in any direction, but the spin axis will remain horizontal.

The pull of gravity
If one end of the axle is unhitched or allowed to move freely, gravity will try to pull this end—and the spin axis—downward along a second axis, the so-called gravitational axis.

The force of precession
As long as the wheel keeps twirling, it will resist gravity's pull and remain horizontal. Instead of falling, the spin axis will swing to the side—a phenomenon called precession. When an airplane maneuvers, the force of precession stabilizes such gyroscopic instruments as the artificial horizon, the heading indicator and the turn-and-bank indicator.

PRESSURE, SPEED AND ALTITUDE

At the heart of flight instruments that measure altitude, speed, and rate of climb is a Pitot-static tube. A small, open-ended pipe installed on the nose or wing, the tube measures the difference between two pressure values. The airspeed indicator at right compares the static pressure of the atmosphere against the impact, or ram pressure, of the oncoming airstream.

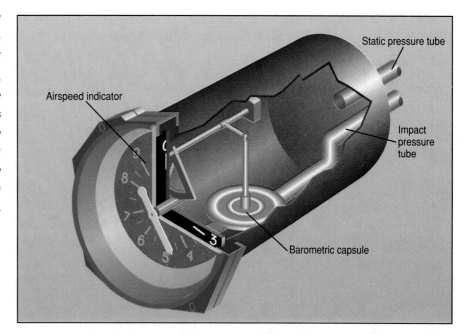

Static pressure tube

Airspeed indicator

Impact pressure tube

Barometric capsule

exceeding 600 miles an hour, often never within sight of the ground from takeoff to landing. The pilot flies entirely by his artificial senses, relying on the battery of alert and precise instruments in front of him. In the sweep of aviation's progress, this is perhaps the biggest accomplishment since the Wright Brothers.

The modern flight deck is a bewildering array of dials, gauges, switches, levers and buttons. The instruments fall into three broad categories. Systems instruments provide detailed information about the aircraft itself—its engines, fuel system, landing gear, flaps and control surfaces, and much else besides. Navigation instruments and equipment allow the pilot to establish his course and monitor his position. The third group—and the most important one for flying blind—depicts the moment-to-moment relationship between the aircraft and the ground below.

In virtually all planes, the "blind flying panel" contains six primary flight instruments: an altimeter, an airspeed indicator, a vertical airspeed indicator that shows the rapidity of ascent or descent, a heading indicator that tells where the nose of the plane is pointed, a turn-and-bank indicator, and an attitude indicator that serves in effect as an artificial horizon. The instruments can be displayed as cockpit dials or digital displays on a composite instrument screen.

Mounted either on the airplane's nose or near the leading edge of a wing, a so-called Pitot-static tube measures two kinds of air pressure and compares the difference. As the plane moves forward, the tube's front end catches the oncoming airflow and registers its force; this information, relayed to the instrument panel, will give the pilot his airspeed. At the same time, holes in the side of the tube measure the local atmospheric pressure—or, more precisely, the pressure the air exerts on the tube if both were motionless, or static. The static pressure serves as a benchmark for calculating the oncoming air pressure. It also provides the data for two other pilot instruments. Since air pressure naturally diminishes with height above sea level, the Pitot-static tube can determine the plane's altitude. And, when the altitude changes during a climb or a dive, the rate of change appears on the dial of the vertical airspeed indicator.

A few pre-flight adjustments are needed to ensure the instruments' accuracy. Since the prevailing atmospheric pressure varies with local weather conditions, this figure must be noted and the gauges adjusted. A correction must also be made for the air temperature. Cold air exerts more pressure than hot, and differences in local thermometer levels can skew altitude readings by hundreds of feet.

The other primary flight instruments—the heading indicator, the artificial horizon and the turn-and-bank indicator—rely on the marvelous ability of gyroscopes to hold their positions no matter what. At one time, gyroscopes belonged to a branch of mechanics of interest mostly to children: A spinning top does not topple over and a coin rolling across the floor stays upright on its edge. The Earth itself is a gyroscope, steadily spinning on its axis in space. Thanks to the inertia of its rotation, a gyroscope resists any effort to change its bearings. Mounted on swivels behind a plane's control panel, it is a center of stability, maintaining its level-headed attitude through the sharpest turns, rolls, dives and climbs. It becomes in effect a miniature, on-board representative of the stable Earth below.

The aviation genius who first put gyroscopes into the air was Lawrence Sperry, son of the founder of the Sperry Gyroscope Company and the American pioneer of blind flying. He designed a turn indicator that allowed pilots to navigate through heavy, low-lying clouds, make banked turns, and come out flying straight and level. His Sperry Artificial Horizon, developed in 1929, made visual reference to the

MULTI-TIERED SKYWAYS
Aerial superhighways crisscross the sky in regions of controlled airspace, where all planes are subject to compulsory air traffic control. Low Level Airspace extends to about 18,000 feet above sea level, where one set of traffic rules prevails. Another set governs High Level Airspace, which starts at 18,000 feet and is the domain of jets.

real horizon unnecessary. Although refined, it is still in use. The instrument's dial, connected to a gyroscope spinning at several thousand revolutions per minute, shows a line representing the horizon and a bar that mimics the position of the airplane's wings. When the plane banks, so does the wing bar. When it climbs or dives, the wing bar moves up or down in relation to the horizon line.

Another Sperry innovation was the Directional Gyro, a kind of non-magnetic compass that shows the plane's heading. On the dial's face is the image of a miniature airplane, superimposed above a freely-moving compass card. While on the ground, the pilot adjusts the card to agree with his conventional on-board magnetic compass; a spinning gyroscope locks in this information. In flight, the gyroscope holds its course, even when the pilot veers left or right. The card appears to swing, and the nose of the miniature plane points to the new heading.

Gyroscopes also play a critical part in two sophisticated computer systems. One is the autopilot, which takes over the routine control corrections needed for stable flight. A pair of gyroscopes, one spinning horizontally and the other vertically, detect any deviation from a set heading. A change in the plane's attitude or course activates small motors that make adjustments to the various control surfaces.

The other device is a navigational tool called the internal guidance system, or IGS. At its heart is a set of motion detectors, or accelerometers, that track the slightest changes in momentum. Should the plane accelerate in any direction—up or down, fore or aft, left or right—the accelerometers will respond. The role of the gyroscopes is to steady these devices while they take their readings. A continuous stream of data feeds from the accelerometers into the IGS computer, which uses it to calculate the plane's position. Even in the densest fogs and blackest nights, a pilot guided by an IGS always knows exactly where he is.

HIGHWAYS OF THE SKY

With so many airplanes in the sky today, a problem pilots face is keeping out of each other's way. Thus, a widespread system of invisible aerial highways and superhighways runs through congested areas to make this easier. Pilots travel along numbered air routes, or corridors, in a three-dimensional traffic network. A corridor is about nine miles wide and, at lower altitudes, 1,000 feet above or below any other; at levels beyond 29,000 feet, the separation expands to 2,000 feet. As a pilot flies along his designated corridor, he follows well-established rules of the air, just as an automobile driver respects local traffic codes. Interlocking radio signals, beamed from ground stations, direct the traffic and mark the intersections: his signposts. The transmissions are marked on charts that the pilot uses to track his position—the way a motorist takes his cue from road maps and road signs.

The system is a far cry from the techniques of aviation's early decades. Back in the 1920s, even airmail pilots tackled navigation the hard way: eye contact with the ground at any price. With little more than a clock, a compass and a map, they flew through good weather and bad, keeping a close lookout for landmarks below. They would try to slip under dense clouds and fog, often hedgehopping just above the treetops and using railroad tracks as guides. In those heroic days, bonfires marked the hills along the mail routes and helped serve as checkpoints. On clear nights, the pilots would fly by the stars, using an aeronautical version of the mariner's sextant and a chronometer to determine their positions.

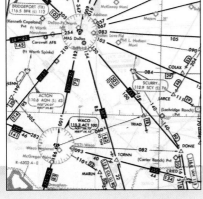

A PILOT'S SKY MAP

To an experienced pilot, this jumble of lines, numbers and codings indicates the normal traffic patterns between airports. Ground stations are designated as circles; they provide the pilot with vital navigation information en route.

Small private planes still fly at low altitudes and only under conditions of acceptable visibility, when the pilots can eyeball the ground and any other traffic in the vicinity. They operate under a convention of basic traffic ordinances known as visual flight rules (VFR). Most larger aircraft, and commercial jetliners in particular, now travel "on instruments" at all times, their pilots visually oblivious to the ground below. They observe a set of internationally-enforced instrument flight rules (IFR) and follow radio instructions from air traffic controllers along their route.

Air traffic controllers are the ground-based traffic cops of the airways. From the moment a jetliner pulls away from the terminal at one airport until it taxis to a stop at another, it is kept a safe distance from other planes by men and women on the ground. Every airplane, wherever in the world it may be, is in radio hailing distance of one or more controllers, who monitor the skies in their vicinity by means of radar. A microwave transmitter at the control center sends out radio signals on frequencies above those used in broadcasting, sweeping them across the sky in a 360-degree arc. The signals bounce off any airplanes they hit and return to the center; the radar echo from an airplane 200 miles away speeds back in only one-five-hundredth of a second. Large commercial and military aircraft carry a transponder in their bellies; triggered by the ground signal, it transmits a characteristic return signal that identifies the particular aircraft. The reflected signals, picked up by the center's antennae, show up as a blip on the traffic controller's computerized radar screen. The controller uses the tagged information to identify the plane, along with its precise location, altitude and ground speed.

A bird's-eye view of New York's John F. Kennedy Airport belies the frenzied activity behind the scenes—cleaning, inspecting, servicing, refueling, loading and unloading an average 400 airplanes per day; and boarding and disembarking some 80,000 passengers. It is the responsibility of ground controllers to make sure that the planes taxiing to and from the gates do not collide with each other, or with any of the dozens of service vehicles buzzing about.

At many airports, trained falcons such as this one keep runways clear of pigeons and gulls. Otherwise, the birds might be sucked into aircraft turbines or propellers, causing expensive damage and possibly engine failure. The falcons make few kills; their job is to scare the intruders away.

Like a line of cars waiting for the light to change, commercial jets queue up to await final takeoff clearance from the airport control tower. Once each plane is airborne and passes beyond local airspace, the controller will hand it off to the next control center along its route.

In heavily trafficked sectors, a controller must use all his skills and experience to convert the multitude of two-dimensional images on his screen into a three-dimensional picture of aircraft activity. And he must constantly think in a vital fourth dimension: that of time. For all of this air traffic is moving at different speeds, and he must respond quickly to divert any planes that are on collision courses. He alerts the pilots by radio and orders the necessary changes in their flying speed, altitude and direction.

Virtually every move that a large commercial jet makes, from the time it takes off until it lands at its destination, is dictated by air traffic controllers. The controllers are "pilots on the ground," and each segment of the flight comes under the supervision of one of four groups of them. The first group, the surface or ground controllers, directs the taxiing of planes to and from their proper terminals and gates. Local airport controllers govern takeoffs and landings; the bubble of space under their jurisdiction may reach upward as high as 8,000 feet and cover a diameter of more than 20 miles. Farther out, approach and departure controllers take responsibility for aircraft flying within 25 to 60 miles of the airport. As the plane speeds on its journey, its supervision is handed over to a series of en route controllers along its flight corridor.

In the United States, this system comprises 20 en route control centers dotted about the countryside, more than 400 airport towers and some 15,000 controllers. When air traffic is light, each controller may be responsible for only one or two aircraft. But during peak hours at busy hubs, like Chicago's O'Hare or Los Angeles International Airport, the number can increase to 15. Keeping so many planes on smooth courses requires cool nerves and split-second timing. One controller, asked what kind of people are good at his work, replied: "Not the ones who have trouble making up their mind."

FROM DEPARTURE GATE TO TAKEOFF

For a typical commercial flight, traffic control begins at least half an hour before takeoff, when the pilot files his flight plan with the departure flight planning office. The plan includes the aircraft's call sign, the type of aircraft, the requested route and altitude, the proposed departure time, the estimated flying time from takeoff to arrival, the amount of fuel on board and the type of navigation equipment being used. The flight planning office then alerts the airport's traffic controller, the depar-

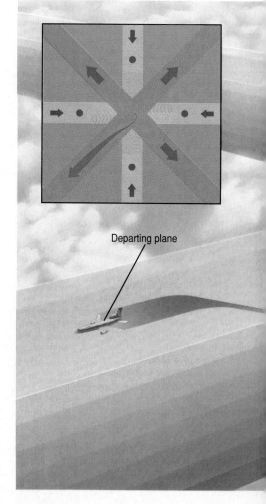

Departing plane

BEACONS AND BEDPOSTS

After takeoff, a commercial jetliner climbs along a pre-assigned access route to its cruising altitude, where it will level off. The pilot is guided by a set of four radio beacons, each situated about 40 to 45 miles from the control tower. Viewed from above, the beacons surrounding an airport resemble a four-poster bed (inset). Incoming planes fly directly over a bedpost as they descend to the runway; departing planes generally fly a course between two beacons as they ascend to their respective flight levels. Departure controllers track the plane until it attains its proper level, then hand off radar surveillance to the next controller en route.

TRACKING THE PLANE FROM THE GROUND

A transponder—standard equipment on all commercial airlines—transmits the plane's identification code, altitude, course and speed. This information appears as a data block of numbers and letters on the controller's radar screen.

ture controller and the en route controllers in the sectors that the plane will be flying through.

After the flight plan is approved, the pilot radios the airport control tower and receives his first instructions from the clearance-delivery controller, part of ground control. These instructions include the plane's takeoff position, the radio frequency it will use on departure, its transponder code, and any changes in the flight plan, such as delays caused by congestion. The pilot then pulls away from the gate and taxis to his designated runway. A runway number painted on the tarmac indicates the direction the plane will head as it takes off. For example, runway number nine means that the plane will point in a compass heading of 90 degrees, or due east. If two runways are laid out alongside each other, one is designated "right" and the other "left."

When the plane reaches the end of its runway and is number one for takeoff, a local airport controller takes over from ground control. From his post in the tower, the controller has a 360-degree, bird's-eye view of the scene below and the sky above; only when inclement weather obscures the runway does he resort to radar. But whatever the conditions, he knows where the plane is at all times. With a final check of the runway, he clears the flight for takeoff. The engines roar to full power and the jet accelerates down the runway.

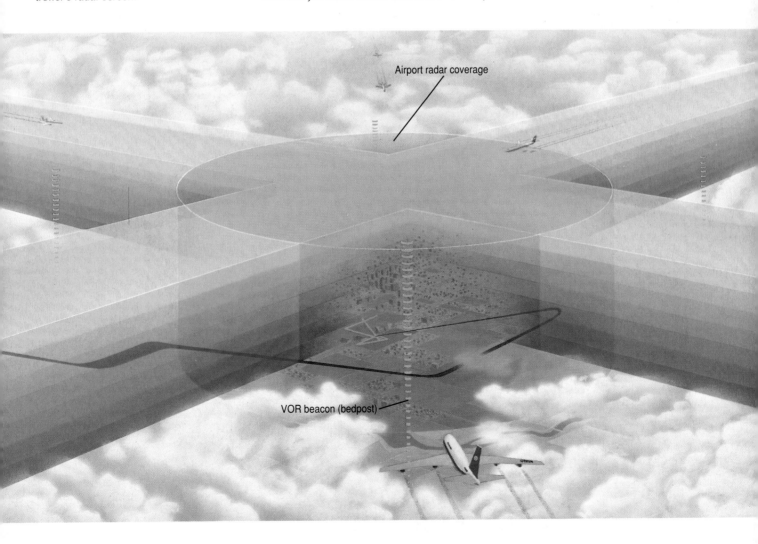

Airport radar coverage

VOR beacon (bedpost)

As the plane reaches takeoff speed, the pilot pulls back on the control column and points the nose skyward. Once airborne, responsibility for the aircraft passes from the airport tower to a departure controller, who should now be able to spot the plane as a coded blip on his radar screen. Receiving radio confirmation of this handoff, the pilot heads out on course. Within five to ten minutes, responsibility for the flight passes from the departure controller to the first in a succession of en route controllers.

TRANSATLANTIC RELAY

A plane crossing the Atlantic moves along a variable succession of corridors that are selected every 12 hours from an internationally published track list. The choice depends primarily on weather and wind conditions en route. One critical factor is the position of the jet stream. A migrating current of fast-moving westerly winds at great altitude, the jet stream can add or subtract as much as one hundred miles per hour from an aircraft's overall speed, depending on whether the plane is riding with it or bucking it. The flyways are identified alphabetically: Track Alpha, Track Bravo, Track Charlie, and so on. Over the Atlantic, they are 60 nautical miles wide and the same distance apart; the aircraft traversing them must be separated by a minimum of ten minutes in time and 2,000 feet in altitude.

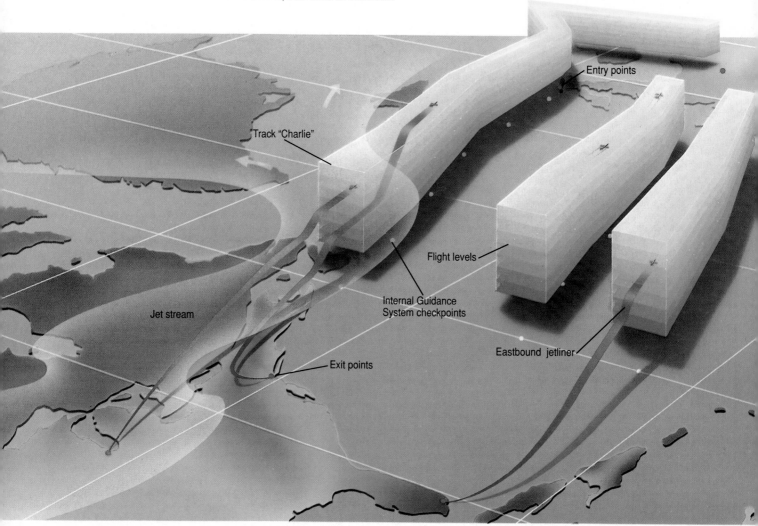

This particular flight from New York's John F. Kennedy Airport is destined for Rhein-Main Airport in Frankfurt, West Germany. When filing the flight plan before takeoff, the pilot requested a preferred corridor—Track Charlie, as it happens—along with a desired airspeed of Mach .85 and a projected altitude of 35,000 feet. The plan approved, he notes down the latitude and longitude coordinates of the geographic checkpoints he will be passing. As he reaches each one, he will report his altitude and position to the nearest traffic control center—which will also be monitoring the plane's progress on its radar screens.

The initial portion of the journey is over land. As the plane flies from checkpoint to checkpoint, it is guided by a series of VOR radio beacons, or Very High Frequency Omnidirectional Range beacons, that broadcast at static-free wavelengths just above normal FM household radio signals. As the plane travels, it is methodically handed off from one controller to the next. In North American air space, the control centers are typically 300 to 400 miles apart; a flight from New York to Frankfurt passes through four en route centers between the two airports. Each radar handoff represents a critical step along the way. Until a control center establishes contact with the next center on the plane's route and the handoff is accomplished, the plane is not allowed to proceed. If necessary, the pilot adjusts his flight plan on instructions from the ground.

The relay continues until the flight nears Gander, Newfoundland. By this time it has reached its designated cruising altitude of 35,000 feet. Gander Oceanic Control reconfirms the flight plan, or calls for an adjustment based on weather or traffic. The plane moves out over the North Atlantic. The pilot now activates the on-board Internal Guidance System (IGS). He has already punched in the appropriate transatlantic coordinates back at the departure gate, in anticipation of the flight's transoceanic leg. The IGS computer takes charge, noting each motion the plane makes and comparing it with earlier data to calculate the ongoing position. Should wind or weather jog the plane off course, the IGS, together with the autopilot, will make instantaneous corrections.

The IGS checkpoints are located at intervals of 10 degrees longitude; the plane flies unerringly to each one. At each point, the pilot radios his position to the appropriate control center. For the first third of the ocean passage he reports to Gander. His next reports go to Reykjavik, Iceland. Finally, as he nears Europe, he communicates with Shanwick Ocean Control in Ireland. Shanwick governs his movements until the flight makes its land entry over Cork, where it comes under the control of London Center. From this position on, the flight follows standard traffic procedures for Europe. Fortunately for American pilots, English is the international language of the air, and there is no difficulty in communicating with controllers in Frankfurt for the final approach and landing.

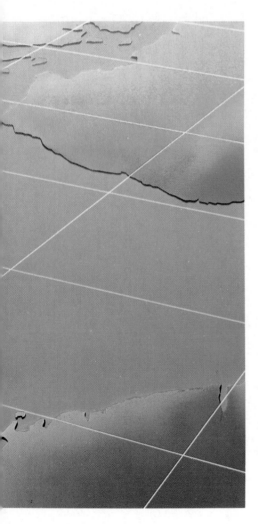

RIDING THE JET STREAM

Flying on Track Charlie, one of many east-west transatlantic air corridors, a commercial jetliner enters the jet stream, about 100 miles wide and some two miles from top to bottom. With the westerly stream at his back, a pilot can cut his travel time from North America to Europe by as much as one hour.

AVOIDING MIDAIR COLLISION

With the dramatic growth in commercial aviation, midair collisions present a very real peril; the definition of a near miss, meanwhile, varies with altitude. And they are not entirely modern problems. The first documented midair collision occurred over Milan, Italy, in 1910, during an air show. A Frenchman, piloting a monoplane, plowed into a biplane flown by an Englishman at a then breakneck speed of 50 miles per hour. Both survived the crash.

Since then, given the rapidity of growth in the number of air miles flown each year, it is remarkable that such disasters do not occur more frequently. It is statistically safer to fly in an airplane than to travel by automobile. But near misses have more than doubled in the past decade, and when collisions do happen, the heavy toll in human life can be devastating.

On the bright, sunny morning of September 25, 1978, Pacific Southwest Airlines Flight 182 was nearing Lindbergh Field at San Diego, California, at the end of its regular commuter run from Sacramento. The 66-ton Boeing 727 carried 128 passengers and seven crew, including flight attendants. The pilot, heading due east, was awaiting instructions from approach control to make the 180-degree turn that would put him on a direct path to his designated runway, number 27. At the same time, a two-seater Cessna 172 on a pilot-training exercise had just skimmed the same runway in a simulated landing maneuver. It too was flying east, and rapidly gaining altitude. Despite the day's brightness, visibility for both pilots was partly obscured. The Boeing pilot, flying level, could not see beneath him. The pilot of

the Cessna had a blind spot overhead, where the plane's high-placed wing cut off his line of sight. Both had the sun in their eyes. And so they collided.

The Boeing hooked the Cessna with its nose wheel, and the two planes plunged to Earth. The horrified traffic controller, unable to divert either plane in time, watched the falling wreckage on his radar sceen. "An aluminum shower," is how he described it. The chunks of flaming metal fell into a residential area, setting fire to ten houses. The death toll: all 137 occupants of both planes and 13 San Diego suburbanites. It was the worst air tragedy in North American aviation history.

The air fatality count over Europe has been as equally heartrending. For example, in September 1976, a British Airways Trident collided with a Yugoslav Air DC-9 in the skies near Zagreb; all 176 people aboard the planes perished. To head off such air disasters, traffic controllers direct every effort to keeping planes a safe distance apart.

Flight rules internationally have been tightened, and many commercial jets are equipped with a sophisticated, computerized radar that serves as a backup to the traffic controllers on the ground in their efforts at separation. Called the Traffic Alert and Collision Avoidance System (TCAS), it reinforces the "see and avoid" capability of the crew on the aircraft. A TCAS, which includes an advanced type of transponder, can track as many as 24 other transponder-equipped aircraft within a five-mile radius, evaluate collision potential, and recommend the proper evasive action for almost any flight circumstance. If a convergence seems imminent, it warns the pilot at least 40 to 45 seconds beforehand.

A TCAS-equipped aircraft is in effect surrounded by three concentric envelopes of radar-monitored airspace. The envelopes are defined by flight time, so that the extent of each one depends on the combined speeds of the aircraft involved. The space immediately around the aircraft is the collision area. Next, there is a warning area that represents 20 to 25 seconds distance in closing time between the collision area and an intruding plane. The largest envelope, the caution area, expands outward to mark the critical 45 seconds of response time.

These seconds are enough time to avoid aerial catastrophe. When an intruder enters a pilot's caution area, his radar immediately picks it up and displays its relative bearing and altitude on the screen. Should the intruder move into the warning area, the TCAS issues a series of flashes and beeps, along with specific instructions for averting a collision. Traffic controllers, monitoring the two planes from their posts on the ground, make the final decision as to which plane should change its flying altitude or course.

THE PROBLEM OF WEATHER

Fog, snow and rain are unavoidable conditions that pilots are trained to fly through. All restrict visibility and, at the very least, they add to the difficulties of traffic control and cause expensive delays in flight schedules. One of the worst weather problems is freezing rain, which may fall even in summertime at high altitudes. Consequently, every jetliner is equipped with deicers, electric heating coils embedded in the windscreen, and deicing fluid systems that protect the leading edges of the wings. An ice deposit of just half an inch in thickness on the leading edge of a wing can reduce its lifting power by as much as 50 percent and increase drag by an equal amount.

INTRUDER ALERT

The two airplanes in the diagram have diverged from their assigned air corridors and are flying on a collision course. Warned by his TCAS, the pilot picks up the intruder on his radar screen (inset). The screen shows the pilot's position as a jet symbol surrounded by a ring of dots. The intruder can be seen near the top of the screen. A blue diamond would indicate no threat; an amber circle represents a moderate threat; and a red square would mean an immediate threat that calls for quick evasive action. In this case, a synthesized voice alerts the crew to ascend to a higher altitude, while the other craft maintains its course.

One weather phenomenon that pilots attempt to avoid at all costs is a thunderstorm. The danger of a thunderstorm to them comes from three potential killers—high turbulence, hail and lightning.

All thunderstorms begin as unthreatening cumulus clouds. But an atmospheric imbalance can quickly transform these harmless-looking puffs into gigantic storm factories capable of producing every type of hazardous weather condition known to aviation. Their turbulence can bounce a plane like an ice cube in a cocktail shaker, severely stressing wings and fuselage. The hail often associated with them has been known to pummel down in chunks the size of golf balls. Even smaller particles can shatter a plane's windows, dent its wings, and dislodge the cover that protects its sensitive radar antenna.

A lightning strike can wreak absolute havoc. Packing enough electricity to light up ten city blocks, the average bolt can generate temperatures of 18,000° F., or twice the surface temperature of the Sun. It can punch holes in a plane's aluminum skin, damage engines, and destroy communication and navigation equipment. Lightning has been known to weld the hinges of ailerons, and momentarily blind pilots with its intense flash. And while it seldom harms anyone in the cabin of an aircraft, on occasion it has sheared off sections of the wings and tails from both light planes and heavy transports.

A pilot's first line of protection is the weather report, and whenever possible he plots his course to avoid likely storm areas. In flight, he constantly monitors

A SLICE OF THE SKY

A massive and menacing storm looms ahead of a commercial airliner. Using on-board weather radar, the pilot compares the instrument's cross-sectional profile of the storm to the view from his windscreen. Following the directions of the weather radar, he turns left to a compass heading of 100 degrees, while the storm moves slowly past him from left to right.

the sky for the anvil-shaped cumulus clouds that indicate thunderheads and he gives them a wide berth. The storms build rapidly and they typically cover many miles of airspace. Lightning can occur anywhere in their vicinity, striking from cloud to ground, cloud to cloud, or at any large passing object. Some discharges have streaked as far as 90 miles from their source. And an airplane, with its sharp, flat metallic surfaces, makes an ideal lightning rod.

All commercial airliners carry on-board weather radar to help them stay out of harm's way. Heavy precipitation sends back a strong radar echo, which shows up on a computerized, color-coded radar screen. Rapidly shifting colors indicate fluctuations in rainfall and possible turbulence or hail. The more advanced radars can pinpoint areas where lightning strokes are concentrated.

Radar is not infallible, however. Since its beams reflect only off moisture, it cannot detect fair-weather turbulence, or even hail, unless the ice crystals are coated with rainwater (nor can it distinguish between rain and wet hail). Furthermore, an airplane's most hazardous moments come not in the center of a storm nor at its height. Studies of lightning strikes show that a plane is more likely to be hit when a storm is dissipating, and when turbulence is low and rainfall light. Under these conditions, the air may still carry a potent charge of static electricity, which is triggered only when the plane flies through it. All things considered, a pilot's best strategy is to skirt the storm area entirely and radio the nearest ground controller for permission to change course.

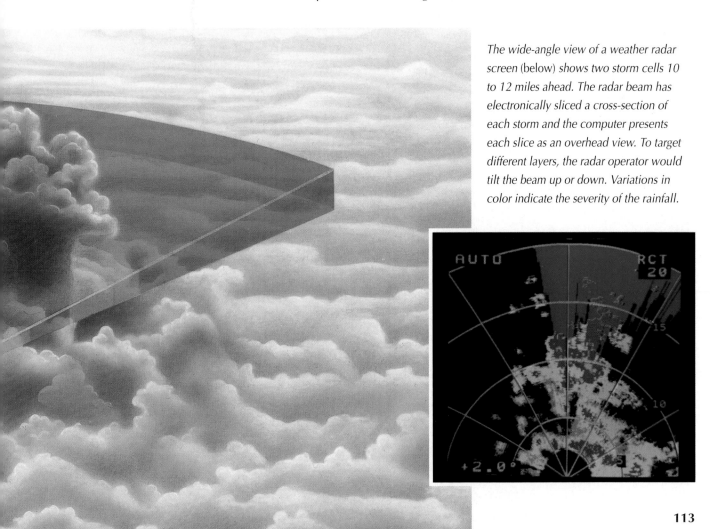

The wide-angle view of a weather radar screen (below) shows two storm cells 10 to 12 miles ahead. The radar beam has electronically sliced a cross-section of each storm and the computer presents each slice as an overhead view. To target different layers, the radar operator would tilt the beam up or down. Variations in color indicate the severity of the rainfall.

THE UPSIDE-DOWN TORNADO OF WIND SHEAR

"There are two critical points in every aerial flight—its beginning and its end." So wrote Alexander Graham Bell, inventor of the telephone, in 1906. Bell knew what he was talking about. Having turned his attention to aviation, he founded the Aerial Experiment Association, which was largely responsible for inventing the aileron. And what was true then is still true today.

As the jetliner begins its descent from 30,000 feet, the captain announces that things might get rough. The sky darkens suddenly as the plane enters the clouds and rain pelts across the window. Strong winds buffet the plane. Moments from touchdown, the engines roar to full power; a wing drops sharply, then rises. The plane breaks out of the rain, the ground rushes past and the plane is on the runway.

Many air travelers have white-knuckled their way through tense moments like these, relying on the skill of the professionals in the cockpit to deal with them. Maneuvering on or close to the ground—during takeoff or landing—is the most challenging aspect of flying. It accounts for nearly two-thirds of all fatal accidents since 1959. Fog, snow, haze and rain have contributed to more than half of these crashes. Recently another enemy has been pinpointed: low-level wind shear, or any sudden change in wind speed or direction.

Wind shear is produced by microbursts of violently moving air, sometimes referred to as upside-down tornadoes. Microbursts form in rain clouds when evaporating raindrops create a pocket of cool, heavy air that falls in a powerful downdraft. When the air hits the ground, it fans out like the contents of a bucket of water spreading across the pavement. Microbursts occur at low altitudes and are usually only one-half to two miles in diameter. They can build to maximum wind speeds in as little as two minutes, and sometimes last only five minutes. But they pack a mean wallop. Fifty were once detected near Chicago's O'Hare Airport in a span of forty-two days.

When an airplane enters a microburst, headwinds increase the speed of air flowing over the wings, giving the plane additional lift. Then, in rapid succession, the plane encounters a downdraft in the center of the microburst, followed by tailwinds that drive it toward the ground. A pilot's first reaction is to reduce power, as a motorist going downhill eases pressure on the accelerator. But as the headwind decreases, additional power is necessary. Apply full thrust? Yes, immediately, but the situation is complicated by the characteristics of a jet airplane. For all its virtues, the turbine engine does not react as fast as the old-fashioned piston engine. The plane, usually at 70 percent of full power during this stage of the approach, takes several seconds to generate maximum thrust. In that brief interval, it might lose too much altitude to pull out of its fall.

Wind shear's most severe tantrums are rare, short-lived and difficult to predict with accuracy. Following the dramatic crash in 1975 of Eastern Flight 66 at John F. Kennedy Airport in New York City (left), the Federal Aviation Agency declared war on wind shear. In cooperation with the major carriers, the agency embarked on a four-year program that included the development of ground-based devices to detect wind shear and airborne equipment to help pilots combat it.

To deal with these infrequent but significant threats to air safety, commercial pilots are trained to maneuver out of microbursts, assisted by on-board detection systems that compare the plane's speed with that of the wind. Large discrepancies

DIARY OF A DISASTER: JUNE 24, 1975

3:48 P.M.
Air traffic controllers at New York City's John F. Kennedy Airport note showers and lightning over the approach course of Runway 22 Left. Unseen are powerful downdrafts, called microbursts, that are the breeders of wind shear. Allegheny Flight 858 enters the first microburst, which apparently is weaker than those to come. Headwinds give the plane extra lift and the pilot reduces power. After passing through the microburst's central downdraft, the plane encounters lift-robbing tailwinds. To compensate, the pilot opens the throttle for maximum thrust and lands safely.

3:58 P.M.
Eastern Flight 902 suffers a drastic air-speed loss as it enters the backside of a second microburst. Reacting quickly, the pilot applies maximum power in order to abandon his approach. Even at full throttle and with the nose tipped up, the jet continues to sink—from 400 feet to 100 feet. Finally, a mere 60 feet above the ground, the aircraft levels off. The jetliner banks left and flies its shaken passengers to Newark Airport.

4:05 P.M.
At 500 feet, Eastern Flight 66 enters a rain shower and an extraordinarily strong third microburst. The captain calls for "takeoff thrust"—maximum power. But little more than a second later the jet crashes short of the runway, killing 113 people in the worst single-plane disaster in U.S. commercial aviation history to that date.

signal potential wind shear conditions. On the ground, airports maintain Low-Level Wind Shear Alert Systems, which have recently been upgraded by doubling the number of detectors.

CLEARED TO LAND

As an airliner nears its final destination at a typically congested big-city airport, the captain receives instructions from the local tower to enter a holding pattern—a tiered sector of the sky clearly defined by a radio signal beamed straight up from the ground. This aerial ladder is a traffic management system that keeps airplanes circling safely until a runway is free for landing. Each airplane enters the stack at the top—perhaps as high as 23,000 feet—and begins flying around an oval race-track nine miles long and four miles wide. Below it, at thousand-foot intervals, other aircraft circle the signal beacon, awaiting their turn to land. When the airplane at the bottom of the ladder is cleared to make its final approach, the plane above it takes its place. Each aircraft moves down one rung, with a new arrival entering the top position in the holding pattern.

When approach control clears a plane for landing, the pilot begins his intermediate approach, guided by a surveillance radar system. At about 3,000 feet, he then begins his final approach to the runway. Even if the runway is obscured by clouds or fog, the pilot can land the plane as long as he makes visual contact with the ground at 200 feet, and visibility on the ground is no less than a half mile. If not, the pilot must abort the landing unless he is specially certified to make a fully automatic approach.

Another set of radio beams—called the Instrument Landing System (ILS)—guides the pilot on his final approach. One is the glide-slope beam that indicates the plane's angle of descent. The second, intersecting the glide-slope beam vertically, is the localizer beam, which projects a pair of signals to the left and right of the runway centerline. Flying between them, the plane heads straight for the runway.

Inside the cockpit, the captain lowers the landing gear at about 2,000 feet and keeps his eyes on an ILS instrument with intersecting crossbars. These vertical and horizontal lines indicate the plane's relative position to the runway: left or right of the centerline and high or low on the glide path. When the lines are centered, the pilot knows his plane is descending toward the center of the runway at an angle of 2.5 degrees. Two vertical radio marker beams tell the pilot his distance from the start of the runway. The first marker is about five miles out, the second half a mile from the runway. At the second marker, the plane is about 200 feet up. Twenty seconds later it touches down, and the pilot reverses the engine's thrust and applies the brakes. Following the ground controller's directions, he taxies to the terminal building and gate.

The Instrument Landing System may ultimately be superseded by newer devices such as the Microwave Landing System (MLS). Unlike the single, fan-shaped glide path of the ILS, the MLS provides multiple entry points at varying distances from the runway to accommodate aircraft with different approach speeds. In this way, light aircraft can keep clear of dangerous vortices generated by larger, heavier planes. Given the continuing boom in air travel, increased capacity will be an important feature of tomorrow's airports; MLS may make it easier to schedule more takeoffs and landings in safety.

INSTRUMENT LANDING SYSTEM

The ILS is the current standard system of approach throughout the world. It consists of two ground installations that transmit narrow signal beams near the touchdown point on a runway. The glide-slope beam provides the proper angle of approach and a localizer beam provides direction. The localizer beam intersects the glide-slope beam vertically to define a glide path that leads the pilot straight to the centerline of the runway at an angle of between two and six degrees. His distance from the runway is indicated by two beacons called fan markers, one at about five miles and the other at one mile from touchdown point.

MICROWAVE LANDING SYSTEM

This latest advance in airport technology may eventually replace the ILS system. Two MLS ground stations, analogous to those of the ILS, transmit scanning beams that fan out in an arc 15 miles wide and to a height of 20,000 feet. This expanded flight channel allows access to several approach paths so pilots can avoid noise-sensitive areas—as well as other aircraft.

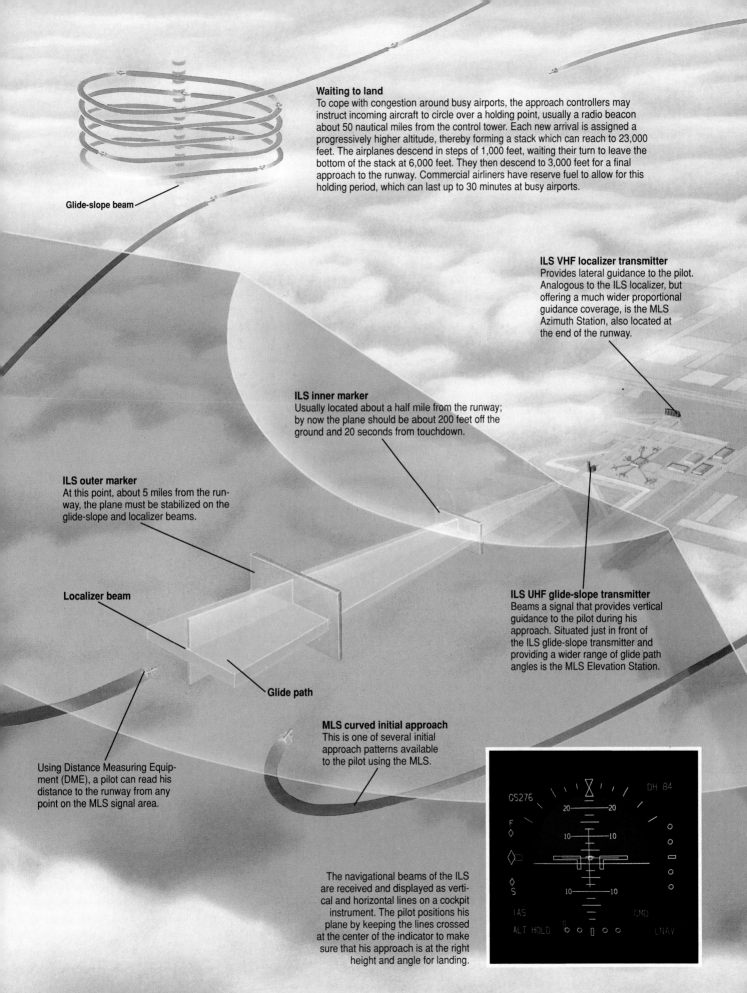

Glide-slope beam

Waiting to land
To cope with congestion around busy airports, the approach controllers may instruct incoming aircraft to circle over a holding point, usually a radio beacon about 50 nautical miles from the control tower. Each new arrival is assigned a progressively higher altitude, thereby forming a stack which can reach to 23,000 feet. The airplanes descend in steps of 1,000 feet, waiting their turn to leave the bottom of the stack at 6,000 feet. They then descend to 3,000 feet for a final approach to the runway. Commercial airliners have reserve fuel to allow for this holding period, which can last up to 30 minutes at busy airports.

ILS VHF localizer transmitter
Provides lateral guidance to the pilot. Analogous to the ILS localizer, but offering a much wider proportional guidance coverage, is the MLS Azimuth Station, also located at the end of the runway.

ILS inner marker
Usually located about a half mile from the runway; by now the plane should be about 200 feet off the ground and 20 seconds from touchdown.

ILS outer marker
At this point, about 5 miles from the runway, the plane must be stabilized on the glide-slope and localizer beams.

Localizer beam

ILS UHF glide-slope transmitter
Beams a signal that provides vertical guidance to the pilot during his approach. Situated just in front of the ILS glide-slope transmitter and providing a wider range of glide path angles is the MLS Elevation Station.

Glide path

Using Distance Measuring Equipment (DME), a pilot can read his distance to the runway from any point on the MLS signal area.

MLS curved initial approach
This is one of several initial approach patterns available to the pilot using the MLS.

The navigational beams of the ILS are received and displayed as vertical and horizontal lines on a cockpit instrument. The pilot positions his plane by keeping the lines crossed at the center of the indicator to make sure that his approach is at the right height and angle for landing.

An Airport at Sea

An aircraft carrier—a giant floating airbase—is one of the most formidable integrated weapon systems ever devised. It has been said that the act of maneuvering an aircraft carrier in the open seas is like steering the island of Manhattan from atop the Empire State Building.

The angled flight deck of a typical carrier is more than 1,000 feet long, 250 feet wide, and covers four-and-a-half acres. Planes are launched and retrieved every 45 seconds and in just under 500 feet, a tenth of the distance required by land-based planes. The activity on the flight deck can be so intense that it has been compared to O'Hare and Washington airports at peak hours squeezed into the size of four basketball courts.

The carrier is base to about 85 planes, each with a specific offensive or defensive role. There are all-weather fighters like the F-14 Tomcat, surveillance planes with all-seeing radar, tankers for midair refueling, both short- and long-range attack bombers, prowlers equipped with electronic jamming equipment to confuse the enemy, helicopters designed to detect and track enemy submarines or conduct search-and-rescue missions, and transport aircraft big enough to deliver entire jet engines.

For quick identification and maximum safety on the deck, the 2,000 men assigned to the carrier wear color-coded jerseys. The "air boss" and his assistant, the "mini-boss," wear yellow shirts and preside over the entire operation. Their commanding view is 140 feet above the sea in a glass-and-steel bubble called the "pri-fly" or primary flight control.

Because of the deafening noise level generated by incoming and outgoing planes, all flight-deck crew wear ear protection, and communications are coordinated by the use of hand signals. During night operations, the same signals are made with lighted wands. To further ensure the safety and efficiency of flight operations, all aircraft movement is replicated in the flight-deck control room with a scale model of the flight deck and miniature planes.

Like a shot out of a cannon, a 35-ton F-14 Tomcat blasts off a carrier's deck. In order to reach takeoff speed in such a restricted space, planes are hooked to a steam catapult system that generates over 70 million foot-pounds of energy. Even if the pilot locked his brakes, the sheer force of the thrust would hurl the plane over the bow at 130 to 150 miles per hour.

An F-14 drops like a brick out of the sky and slams onto the flight deck, its massive trailhook snagging the three-wire arresting cable, bringing the jet to a jolting halt just 30 feet from the edge of the ship. At the instant of touchdown, the pilot shoves the throttle full forward in case he misses the arresting gear. But once the tailhook snags the cable, the plane is halted in about two seconds.

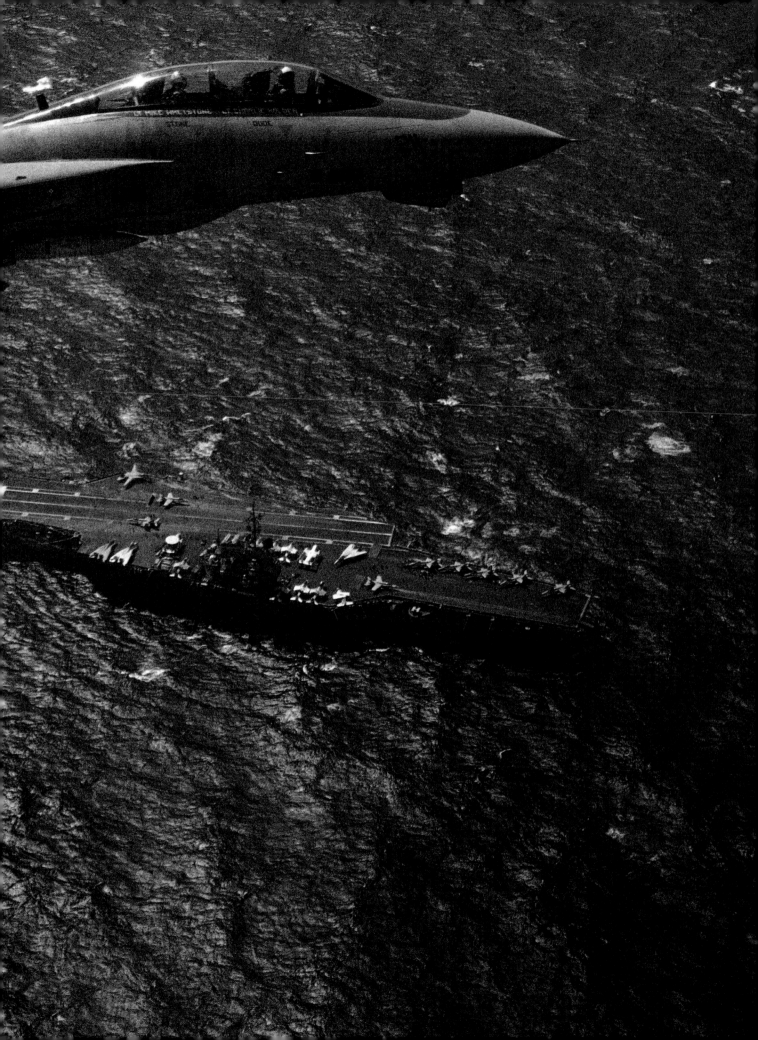

QUEST FOR SPEED

t is a rare and spectacular sight, visible only in humid weather and usually only to other high-flying pilots: An F-14 jet breaches the sound barrier like an explosion bursting from one universe into another. The barrier—a normally invisible wall marking the speed at which the compressed air piling up ahead of the plane creates a sudden, massive increase in drag—reveals itself for the wink of an eye and seems to stretch with the impact of the jet's nose cone. A vaporous radiance appears; from the front of the plane it looks like a halo but from the side it presents the illusion of a battery of sharp spikes darting off the wings and fuselage. The apparent barrier tears open, repairs itself in an instant, and is gone.

Ever since the first powered flight in 1903, designers have tinkered with airfoils, engines, fuel mixtures, materials, even the number of wings to coax additional speed from their craft. Toward the end of World War II, American pilots were pushing their P-38 Lightnings to 600 miles per hour in steep, powered dives. But at those speeds buffeting became so violent that some machines literally disintegrated: Wings and tails broke away and several pilots lost their lives. The problem lay not so much in attaining the speed of sound—about 760 miles per hour at sea level—but in overcoming the severe punishment the aircraft received as it approached that speed.

In 1944, even before the end of the war, the U.S. Air Force commissioned Bell Aircraft to design a rocket-powered experimental aircraft, called the Bell X-1 —Experimental Sonic-1—that would withstand the intense pounding. At that time the only way to test such an experimental plane was to hazard flying it. A 24-year-old fighter ace, Captain Charles E. "Chuck" Yeager, was chosen as the test pilot. On October 14, 1947, the slim, bullet-shaped X-1, painted bright orange for high visibility, was launched at 20,000 feet from the bomber belly of its B-29 mother ship and headed toward an altitude of 43,000 feet. Yeager was warned not to "push the envelope" beyond 96 percent of the speed of sound unless absolutely certain that he could safely do so. As the X-1, powered by a rocket engine capable of

In this rare photograph, snapped by the pilot of a second fighter, a shock wave appears as a circle of vapor around an F-14 Tomcat as it breaches the sound barrier.

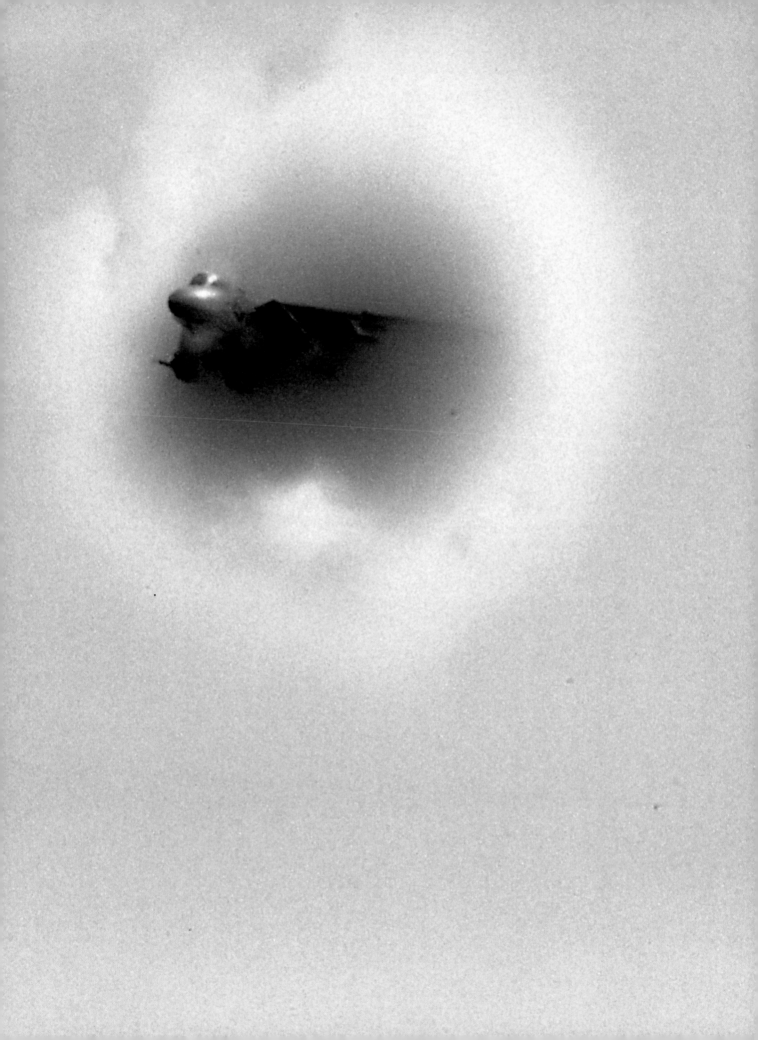

producing 6,000 pounds of thrust, climbed and accelerated, Yeager kept his eyes glued to a new device called a Machmeter, which recorded the aircraft's approach to the speed of sound.

When the meter rose above Mach .90, the machine began to shake alarmingly, nearly causing Yeager to lose control. Then, quite unexpectedly, the buffeting stopped. The Mach needle rose to Mach .965—then tipped off the scale for 20 seconds. On the ground, a tracking van reported a sound like the rumble of distant thunder: the first sonic boom made by an airplane. The data later showed that the X-1 had reached Mach 1.07—an event that Yeager later described disappointedly as a "poke through Jell-O."

BUILT FOR SPEED

By trial and error as much as by studied research, aeronautical engineers mapped the far side of the sound barrier and learned the properties of supersonic flight.

As an object—an aircraft, a bird, even a glider—moves forward through the air at any speed, it disturbs the air molecules through which it passes. They form alternating high- and low-pressure waves that radiate from the object like ripples of water around a moving boat. The pressure waves set up a vibration that to human ears becomes sound, but which in many cases is far too weak to be heard. These pressure waves always travel at the speed of sound regardless of the speed of the object.

However, the speed of sound itself is not constant but varies according to the temperature of the air. The warmer the air, the faster sound travels. At sea level in moderate temperatures, the speed of sound is about 760 miles per hour; at an altitude of 40,000 feet, where the temperature averages -60° F., its velocity drops to 660 miles per hour. To avoid ambiguity, scientists use a Mach number, named

SPEED RANGES

The shock waves generated by an aircraft traveling near, at, or beyond the speed of sound have radically changed the shape of aviation. To control turbulence at high speeds, designers sweep the wings back, so that both wings and the sonic shock waves they produce fit inside the shock cone streaming from the nose of the plane.

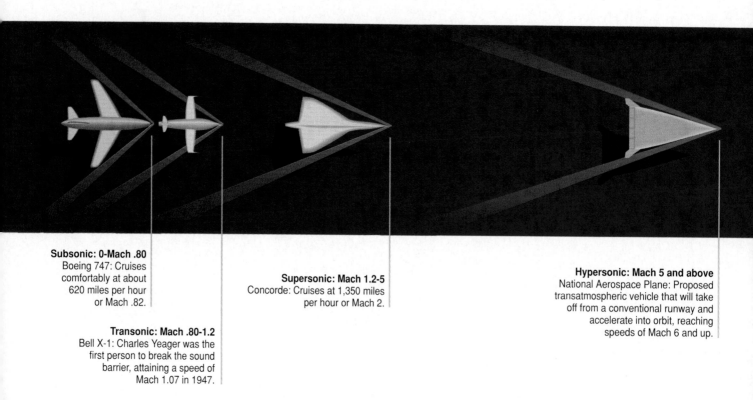

Subsonic: 0-Mach .80
Boeing 747: Cruises comfortably at about 620 miles per hour or Mach .82.

Transonic: Mach .80-1.2
Bell X-1: Charles Yeager was the first person to break the sound barrier, attaining a speed of Mach 1.07 in 1947.

Supersonic: Mach 1.2-5
Concorde: Cruises at 1,350 miles per hour or Mach 2.

Hypersonic: Mach 5 and above
National Aerospace Plane: Proposed transatmospheric vehicle that will take off from a conventional runway and accelerate into orbit, reaching speeds of Mach 6 and up.

for the noted Austrian physicist, Ernst Mach, who helped pioneer the study of sound. A plane's Mach number is equivalent to its speed divided by the speed of sound at the plane's altitude. Thus a plane traveling at the speed of sound—whether 760 miles per hour at sea level or 660 miles per hour at a high altitude—is said to be traveling at Mach 1, regardless of the temperature of the surrounding air.

The speed of an airplane dictates the reaction of the pressure waves to its passage. There are four main speed ranges, as shown on the chart at left: subsonic, transonic, supersonic and hypersonic. When a plane is flying at subsonic speeds—below Mach .80—the pressure waves have time to escape from the aircraft since they are traveling away from it at the speed of sound. In effect, the air particles ahead of the plane receive advance warning of its arrival from these changes in pressure and start to move out of the way. As a result, the air begins to adjust before the craft arrives, so as not to disturb its path.

As the plane approaches transonic speed—Mach .80 to Mach 1.2—the pressure waves do not have time to move out of the way of the oncoming plane since it is traveling along with them; the waves compress and the air becomes far more dense. When the plane meets the compacted air, it hits with a jolt and a series of shock waves builds up perpendicular to the direction of flight. The first shock wave attaches itself to the center of the wing's upper surface as the airflow there reaches Mach 1. As the plane's speed increases, the air under the wing also reaches Mach 1 and a second shock wave forms. Concurrently, the wave above the wing increases in size and strength.

The aerodynamic effects of these shock waves interfere with the thin layer of air that hugs the wing's surface, called the boundary layer, and result in a reduction in lift. This interference also hampers the effectiveness of control surfaces, and causes an increase in drag because the airflow behind the shock waves is turbulent. The turbulence induces shock stall, a condition that can subject conventional aircraft to violent buffeting.

As the plane accelerates to supersonic speed, the shock waves above and below the wing move back and join together at the trailing edge of the wing. Ahead of the wing, pressure is still building up and, as the plane reaches Mach 1, a new shock wave attaches itself to the wing's leading edge. Beyond the speed of sound, these waves bend back over the plane in the shape of a cone, with the nose of the plane at the cone's tip. This Mach cone trails behind like the bow wave of a ship, except that it is three-dimensional.

To solve the shock stall problem at transonic speed, aircraft designers looked at various ways of reducing the formation of shock waves. Their solutions took the shape of the DC-10, the 707 and the 747—all subsonic planes designed to cruise comfortably at Mach .80. The trick was to sweep back and "file down" the wings. In flight, air passes over these V-shaped wings at an angle, which delays the onset of shock waves until a much higher flying speed is attained. Making the wings thinner, with sharper leading edges, enabled them to slice cleanly through the compacted air.

Next was the problem of drag and turbulence beyond the speed of sound. Designers soon realized that a supersonic aircraft must fit neatly inside the Mach cone it produced. Since airflow at such speeds prefers sharp edges, the straight-

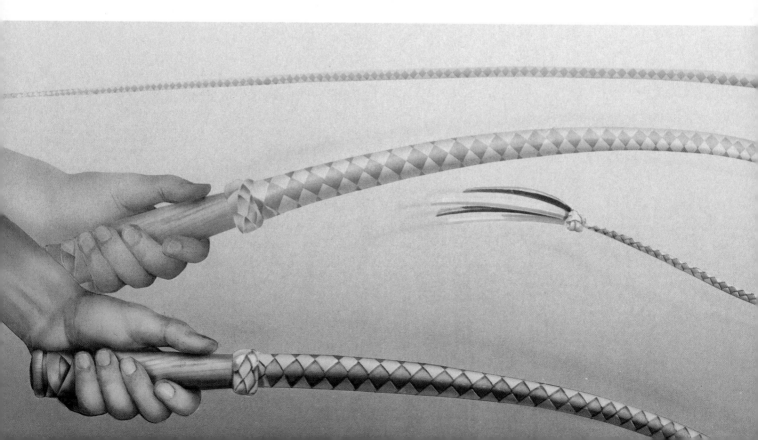

winged aircraft of the World War II era were redesigned with wasp-waist fuselages, needle noses and knife-edged wings. The faster the plane, the more such streamlining it required. Thus the Lockheed SR-71 Blackbird, the culmination of the great advances made in aviation since the days of the biplanes, was given a slender fuselage and stubby delta wings that blended smoothly into chines—horizontal surfaces attached to the fuselage to increase lift. In 1974 a Blackbird set a transatlantic speed record of one hour, 54 minutes and 56 seconds for the 3,470-mile route between New York and London—less than 1/17 the time it had taken Charles Lindbergh to fly to Paris 45 years earlier. The fleet has since been retired but still holds more than a dozen speed and distance records.

Heat, too, became a problem for aircraft reaching beyond the sound barrier, where the skin temperature of an airplane can exceed 500° F. and whirling turbine blades glow at 1700° F. New products had to be created or improvised: titanium skins that could withstand inferno-like heat; corrugated panels that would expand without warping; even gold-plated electrical connections that retained their conductivity at high temperatures.

In this classic photo by Harold Edgerton, a strobe light flashes for a third of a millionth of a second to freeze a bullet as it explodes through an apple at 1,800 miles per hour—Mach 2.38. The bullet's aerodynamic shape inspired early designers of supersonic aircraft, since ballistics experts agreed that the projectile's shape appeared to give it good stability at velocities above Mach 1.

SURPRISING SUPERSONICS

The roll of thunder, the sharp report of a high-caliber bullet, even the crack of a ringmaster's whip—all are sonic booms. They are generated by shock waves—the result of sudden increases in air pressure.

Nature's version of the sonic boom—probably the first ever heard by humans—is the thunderclap. It begins when a large, positive electrical charge builds up in the chilly upper layers of a thunderhead. Negative charges gather below and in turn set up an attraction with positive charges on the ground. At first no cloud-to-ground discharge of lightning can occur, because the air acts as an insulator. Eventually, though, a bolt zigzags to the ground at about 60 miles per second, opening up an ionized channel through the air—in effect, acting as a wire that the electricity can follow. The main bolt that follows discharges about 100 million volts of electricity, enough to light up the houses on ten city blocks, and heats the air in its path to more than 60,000° F. The heated air expands at supersonic speed, creating a cylindrical shock wave that is heard as a violent rumble. Although aeronautical engineers have succeeded in designing planes that can withstand the stress of high-speed flight, they have not eliminated its troublesome byproduct: the sonic boom.

The shock wave or cone that trails behind a supersonic plane eventually dissipates above it but also touches the ground below, some time after the plane has passed. The human eardrum vibrates in response to this dramatic change in air pressure; the vibrations are carried to the brain as electrical signals and interpreted as sound. The shock wave is heard as a loud, sudden bang—the sonic boom. It lasts for only a fraction of a second, but because the change in pressure is so sudden, the sound can be startling. Anyone within the shock wave's path will experience the sonic boom, a person in another plane flying at subsonic speed, a balloonist or an observer on the ground who has seen the plane pass by seconds before the sound is heard. Concorde's signature is a distinct double sonic boom—one from the nose and one from the tail. Only the fortunate passengers for whom the sonic booms are created do not hear them, since the shock waves travel away from them.

THE CRACK OF A WHIP

As a 12-foot bullwhip snaps, kinetic energy travels from the handle along the length of the whip to its tip, which gains speed and momentum as the whip unwinds. As the energy reaches the tip threads, called the "popper," the threads are momentarily accelerated to about 1,400 feet per second, or 900 miles per hour—Mach 1.26—and produce a shock wave that in turn produces a miniature sonic boom.

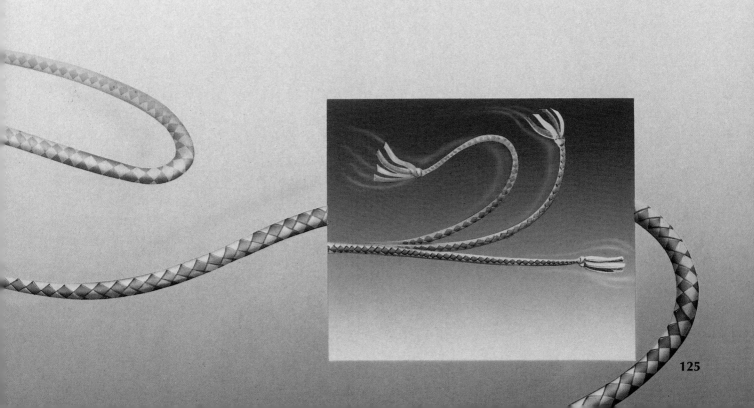

Flight Log: Concorde

Every line, curve and twist of the supersonic Concorde's 84-foot wingspan and 204-foot-long fuselage represents a precise marriage of form and function. The jet's wing, an aerodynamic compromise between the requirements of high- and low-speed flight, is swept back in a double delta compound curve known as an ogive.

Slung beneath it in two box-like structures are four Rolls-Royce turbojet engines, capable of producing 152,000 pounds of thrust. Thirty-four thousand gallons of fuel are stored in wing tanks. During flight, a system of pumps and valves shifts part of the fuel's weight to auxiliary tanks located fore and aft to compensate for changes in the plane's center of gravity.

At its cruising altitude of 50,000 feet, the Concorde's aluminum skin reaches a temperature of 262° F., and the fuselage stretches four inches, causing the windows to shift slightly. Looking forward from the flight deck, the pilot sees the horizon 300 miles away; the view is 270 degrees, spanning 250,000 square miles. Inside the Concorde's rapier-slim fuselage, up to 144 passengers are seated four abreast, where they can sip champagne and eat caviar while watching the jet's speed climb to Mach 2 on small screens called Machmeters.

0:00
TAKEOFF
Its nose tilted 5 degrees downward for better visibility, the Concorde lifts off the runway at New York's Kennedy Airport at 250 miles per hour.

0:15
SUBSONIC FLIGHT
Climbing 3,000 feet per minute, the Concorde temporarily levels off at 29,000 feet. Until the aircraft is clear of land, it will fly at 650 miles per hour—Mach .95.

0:20
ACCELERATION
Moving through the air, the Concorde produces sound waves that ripple outward like concentric spheres. At subsonic speeds, these waves move ahead of as well as behind the plane.

0:23
BREAKING THE BUBBLE OF SOUND
As the Concorde climbs to 50,000 feet and accelerates to the speed of sound (685 miles per hour at 30,000 feet), it catches up with its own pressure waves. The result: These disturbances pile up into a shock wave that is perpendicular to the line of flight.

During takeoff, the Concorde's nose must be dropped 5 degrees for the pilot to see the runway *(above)*. Once airborne, the nose is aligned aerodynamically with the fuselage in preparation for supersonic flight *(below)*. Because of its steep angle of attack, the Concorde's nose must be dropped 12.5 degrees during landing.

0:50
SUPERSONIC CRUISING

Beyond Mach 1, the pressure waves fold back over the Concorde in the shape of a cone. Where this cone touches the Earth's surface, the sudden change in pressure causes a sonic boom. At its cruising speed of Mach 2 (1,350 miles per hour), the Concorde covers one mile in less than three seconds.

2:58
SLOWING DOWN

Some 28,000 feet over Bristol, about 100 miles west of London, the Concorde slows to Mach .95 to avoid producing a loud sonic boom over southern England.

3:25
LANDING

Three-and-a-half hours and 3,500 miles from New York, the Concorde touches down at Heathrow Airport at 185 miles per hour.

For the time being, sonic booms are an environmental price for supersonic flight. Since the impact of the boom varies with altitude and aircraft design, measures can be taken to minimize the effects. A supersonic craft flying at an altitude of more than 40,000 feet creates a tolerable boom; the same plane traveling below 10,000 feet—too low to allow the shock waves to dissipate—causes an intense boom that may even shatter windows. Further streamlining of supersonic craft also helps to reduce the formation of shock waves, lessening the boom to a dull rumble.

MAN VERSUS SPEED

The final obstacle in the conquest of speed is not the brilliance of technology, but the vulnerabilities of the human body. In the whole history of aviation, only three men have "gone supersonic" without benefit of an aircraft. In separate incidents, two pilots have bailed out of planes traveling at supersonic speeds and lived to tell about it. A third set a world record for the longest delayed parachute drop and went supersonic in the process. In 1960, Captain Kittinger of the USAF stepped out of a balloon at 102,800 feet and dropped 16 miles before deploying his chute. For an instant—in rarefied air at 90,000 feet—he was plummeting earthward at a speed of 825 miles per hour, or Mach 1.25.

Wearing an anti-G suit and strapped tightly into the ejection seat of an F-15 Eagle (below), *a pilot's G-force tolerance is increased by about 1.5 Gs, raising his overall limit to about 8 Gs for 10 seconds during a high-speed maneuver.*

Unvarying speed, no matter how high, creates little stress on the human body, but sudden acceleration, deceleration or change of direction exerts a powerful, even violent force. Such a force is expressed in terms of the equivalent force exerted by the Earth's gravity at sea level (1 "G," as scientists say) and is experienced primarily as a change in body weight.

At 1 G, body weight is normal. At 2 Gs, effective body weight is doubled, and blood has difficulty reaching the head and eyes, causing cloudy vision. At 3 Gs, the legs are too heavy to lift, and peripheral vision starts to fade. By 6 Gs, arms cannot be raised above the head, and vision deteriorates from "gray-out" to blackout. A flier in excellent physical condition can "fight the G," warding off unconsciousness for the few seconds necessary to complete a maneuver by tensing the muscles of his legs and torso in an attempt to hold the blood in his head. But at 8 Gs, the blood circulation in the brain is so dangerously reduced that a pilot without a so-called anti-gravity suit will lose consciousness.

The anti-G suit *(opposite)* acts much like a corset. Bladders in the suit, located over the calves, thighs and abdomen, automatically fill with air as G-force increases, and exert a counter-pressure on the legs and abdomen, forcing some of the blood back up to the head and eyes. Reclining seats also reduce the vertical distance blood has to travel. But the suit's value is limited: It affords only about 1.5 Gs of protection at high G-forces.

The effects of the G-forces produced by a sharp turn in a speeding jet fighter have been described this way by a pilot: "An invisible force slams my helmeted head back, punching the air from my lungs and crushing my body. My internal organs are beginning to become unglued, separating layer by layer like reluctant Velcro, and I am suddenly acutely aware of the fluids in my body. I turn all my energies to keeping myself from coming apart at the seams. I have been rendered to jelly." The tighter the turn and the higher the speed, the greater the force. Even at only 450 miles per hour, a turn with a radius of one-half mile produces 5 Gs.

Military jets such as the F-16 are crammed with computers that perform multiple high-speed tasks and permit their pilots to maneuver at supersonic speeds and paralyzing G-forces. The cockpits resemble TV control rooms, with computer-generated displays replacing mechanical dials. Fighter pilots rely on these on-board computers to constantly report on the jet's speed, angle of attack and other flight attitudes, as well as to receive guidance from external navigation sources. Backed up by multiple computers that constantly monitor one another, the system senses any deviation from desired flight conditions and responds by sending out as many as 40 electronic instructions per second to the control surfaces—far more than any human pilot could carry out. The system can even override a command that is potentially dangerous to the plane or pilot, such as a turn that would generate excessive G-force.

Another feature of supersonic jets is a computer-generated cockpit display, aptly called the heads-up display, or HUD *(left)*. Virtually everything a pilot needs to know to fly and fight—his speed, heading, his weapons systems—is continually updated by computer and projected onto a transparent glass screen mounted at eye level. The saving of those precious few seconds previously needed to glance down at the instrument panel can mean the difference between success or failure, life or death, in a bomb run or dogfight.

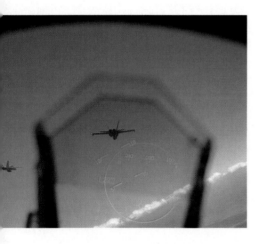

An F-15 fighter pilot keeps a watchful eye on both the "enemy" and his instrument panel (above) using a holographic heads-up display (HUD). The transparent screen, mounted at eye level, projects flight information directly to his line of sight.

FLY-BY-WIRE FIGHTER

To cope with supersonic speeds and the tremendous G-forces they exert, combat fighters like the F-15 Eagle below are constructed with an exotic superalloy skin stretched over a rigid airframe, which may be carved from a solid block of metal. The nose is crammed with as many as four flight computers, and the wings are essentially huge fuel tanks. The pilot does not directly "fly" the plane in a conventional sense; instead he "commands" the flight computers to execute the kinds of maneuvers illustrated below.

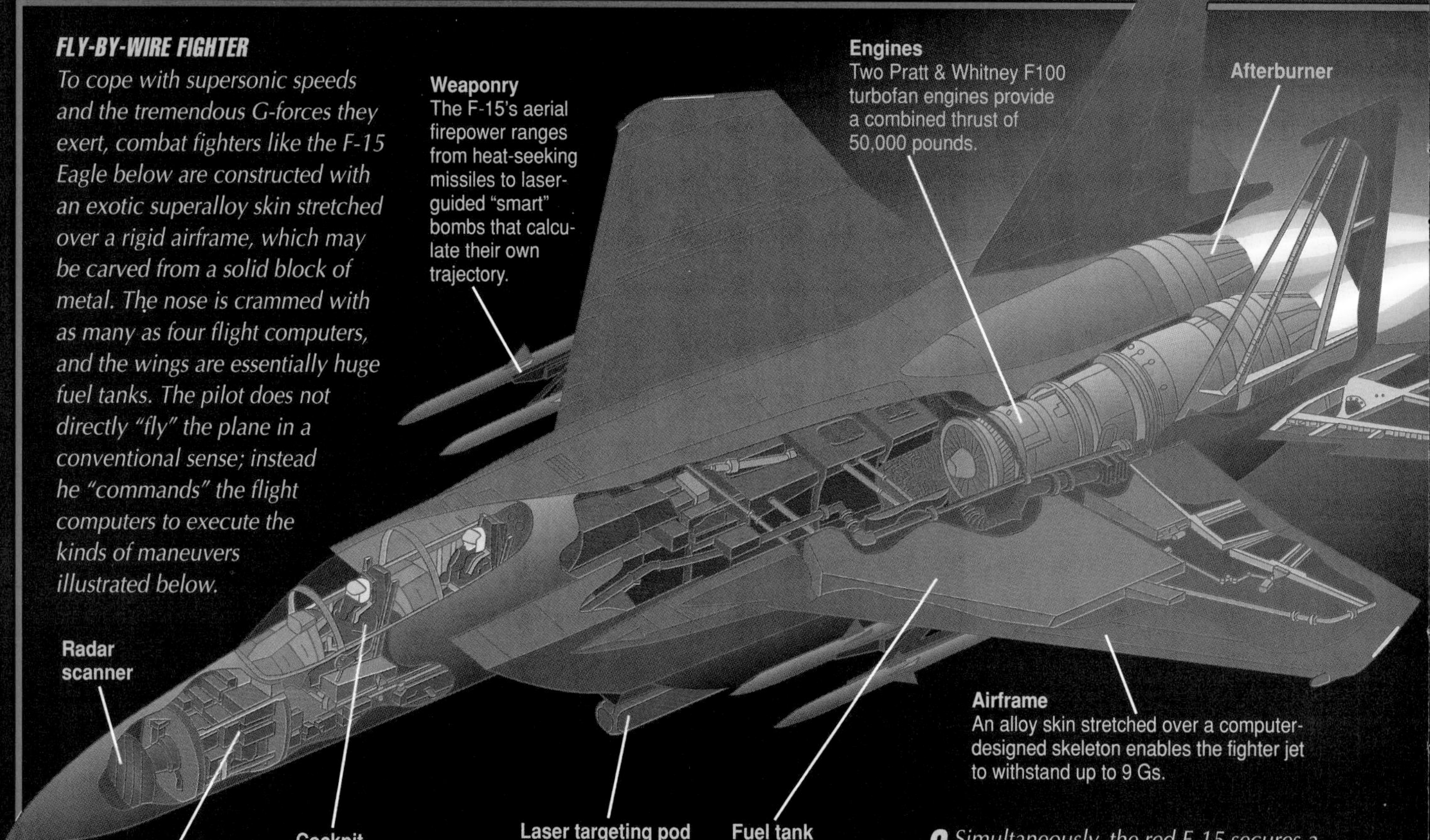

Radar scanner

Electronics bay
Advanced navigation and weapons systems include terrain-following radar and infrared night vision.

Weaponry
The F-15's aerial firepower ranges from heat-seeking missiles to laser-guided "smart" bombs that calculate their own trajectory.

Cockpit
Two crewmen are required to fly an F-15 in combat. The pilot handles the flight controls and navigation equipment while a weapon-systems officer monitors the complex radar and weaponry.

Laser targeting pod
Capable of quickly detecting long-range targets under all-weather conditions.

Fuel tank
The F-15 has a range of about 2,500 miles without refueling.

Engines
Two Pratt & Whitney F100 turbofan engines provide a combined thrust of 50,000 pounds.

Afterburner

Airframe
An alloy skin stretched over a computer-designed skeleton enables the fighter jet to withstand up to 9 Gs.

6 Simultaneously, the red F-15 secures a good lock and "squeezes off" one short-range, heat-seeking missile. As the bogeys become tiny, visible specks to the Eagle pilots, the missile destroys one enemy aircraft.

Dogfight at Mach 1

1 An Airborne Warning and Control System (AWACS) aircraft 100 miles away directs three F-15 Eagles (above) to intercept two bogeys—unidentified hostile aircraft—20 miles distant and closing in at the speed of sound (upper right). Traveling in a chain, the Eagles accelerate to Mach 1 just as they are detected on enemy radar.

The Korean Conflict of the early 1950s was the first jet air war. For the next four decades, new technologies and maneuvering techniques were put to the test in the skies above Vietnam, Israel and the Persian Gulf. Distances between fighters increased from several hundred yards to several miles in compensation for the much-faster speed and larger turning radius of the new aircraft.

The Vietnam War saw the advent of the air-to-air (AAM) missile. The dramatic increase in range of the AAM over the gun rendered traditional fighting wing formations obsolete.

Shown in numbered steps on these pages is a simplified version of what a dogfight of the future might resemble. Because of the distances and speeds involved, it is as much a battle of computers as of fighter pilots.

4 By the time the two formations are five miles apart, the bogeys know they have missed and lock onto the second (orange) Eagle.

5 As his radar receiver sounds, the orange Eagle climbs hard, dumps chaff, turns on his music and "snap-rolls" around the missile, which barely misses him.

7 The remaining bogey has little chance against the first F-15, which has rolled back into firing range and launched a heat-seeking missile to end the dogfight.

2 Thirty seconds later and 10 miles closer, one bogey "locks" onto the lead (yellow) Eagle and fires a radar-homing missile.

3 Alerted by his radar receiver, the yellow F-15 breaks hard right and dumps "chaff"—a cloud of metalized fiberglass that forms a huge radar image. He also turns on his "music"—electronic scrambling designed to deceive the enemy's radar. The missile loses its target and veers off.

The Leading Edge

An electronics specialist makes final adjustments to QN's delicate autopilot—the "brain" of the unusual model—in preparation for a test flight.

Bridging time, a replica of the prehistoric flying reptile Quetzalcoatlus northropi *swoops over Racetrack Dry Lake in Death Valley, California.*

In 1986, aviation aeronautical engineer Dr. Paul MacCready and his team of scientists gathered on a dry lakebed in Death Valley, California, for the debut of a curious flying machine. The fur-covered, electronics-filled contraption was a remarkably lifelike replica of *Quetzalcoatlus northropi*, the reptilian pterodactyl of 100 million years ago.

If MacCready and his team managed to fly "QN," as they called their contrivance, it would be more than just a bit of high-tech whimsy; never before had science achieved the flapping flight of birds—or pterodactyls. The replica was a self-contained, fly-by-wire model incorporating the latest advances in alloys and plastics, computers and robotics. Its electric-motor "muscles" were powered by six pounds of batteries and directed by an electronic "brain" that received instructions from a controller on the ground, an on-board gyroscope and a wind vane incorporated in its neck. The 18-foot wings, made of a carbon-fiber skeleton covered with latex, could flap at varying speeds, move backward and forward to control lift, and twist or "warp" to counter the effect of side winds.

QN was towed into the sky on a winch-powered sled. Once airborne, the platform was jettisoned and floated earthward on a tiny parachute. On the ground, the controller flicked a switch to activate QN's autopilot and 13 flight motors. Five hundred feet overhead, QN slowly wagged its wings and, as MacCready and his team held their breath and squinted skyward, soared and flapped for three glorious minutes before landing.

The modern sport of hang gliding was born of science and nurtured in serendipity. One of the earliest pioneers was the great German engineer Otto Lilienthal, shown sailing from a hill near Berlin in the late 1890s. Lilienthal believed that gliding, not flapping, would be the basis for powered flight, and to prove his contention he built and maneuvered around in hundreds of wood-and-fabric gliders until a tragic accident ended his investigations. Some of the flights were very dashing, though sport was the farthest thing from Lilienthal's mind.

More than half a century later, another engineer, NASA's Dr. Francis Rogallo, carried the idea of hang gliding a large step forward—again in the cause of science. Rogallo designed a remote-controlled kite that could be deployed to bring space capsules safely back to earth once they had reentered the atmosphere. Wind-tunnel testing confirmed that the delta-shaped wing possessed both aerodynamic stability and remarkable lifting powers. In the end, NASA found other ways to bring home its astronauts and satellites. But Dr. Rogallo's elegant and inexpensive wing, meant for space travel, has now made its mark as the basic design for the burgeoning sport of hang gliding.

Since the 1970s perhaps a million people worldwide have taken up this least expensive and most accessible form of flight. A typical flight begins atop a high ridge with strong prevailing winds. The pilot wears a shoulder-to-knee harness, which is in turn hitched to the frame of the glider. Facing into the wind, the pilot runs forward until the "sail" fills with air and lifts him skyward. To control the glider, he holds onto the crossbar and shifts his weight from side to side. To slow down and land, he pushes the bar forward.

An experienced pilot, with a keen eye for rising thermals, may climb to a height of 10,000 feet and stay aloft for ten minutes or more. Otto Lilienthal, who achieved glides of about 1,000 feet, would have been amazed—and delighted.

Otto Lilienthal tests one of his cotton-and-willow-rod gliders. Although he appears to be out of control, he has just bent his knees to adjust the glider's center of gravity.

A hang glider rides a thermal across the slopes of Mount Hood in Oregon. Unlike Lilienthal's manned kites, modern hang gliders are equipped with a harness and crossbar that allow the pilot to steer by shifting his weight.

Wars and the threat of war have given impetus to great leaps in aviation technology, sometimes at a pace too fast for the success of a design.

Jack Northrop, a high-school graduate who taught himself engineering before founding the firm that now holds his name, postulated that the most graceful and efficient shape for an aircraft was a flying wing. Among other benefits, the absence of a fuselage would reduce both drag and structural weight. Starting in World War II, a succession of promising designs led ultimately to the first flight in 1947 of the Northrop YB-49, a flying-wing bomber powered by eight turbojet engines *(inset)*. The design, although at least 20 years ahead of its time, was dogged by stability, control and mechanical problems. These—and, some sources say, political chicanery—led to the cancellation of the government contracts.

It was a bitter disappointment for Jack Northrop, but before his death in 1981, his dream came alive again in designs for the ultra-secret B-2 bomber. The B-2's unusual shape combines large interior volume—to house weapons, fuel and avionics—with a radar-avoiding profile. Ironically, the 172-foot wingspan of Northrop's new high-tech warbird is exactly the same as its untimely forebear, the flying wing.

The ill-fated Northrop YB-49 was the first tailless aircraft. In 1949, the 172-foot-wide "flying wing" set a world record by remaining airborne for more than nine hours without refueling.

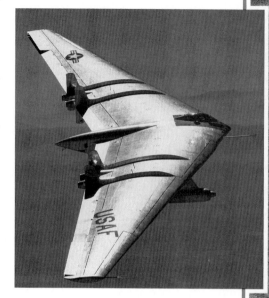

The bat-winged B-2 bomber flies over Palmdale, California, the home of the Northrop plant where it was built.

In this artist's rendition a high-speed civil transport (HSCT) cruises over Florida at more than three times the speed of sound.

Despite its age, the Concorde remains an inspiring design. Lifting off the runway at 250 miles per hour, this aircraft will reach the speed of sound in 23 minutes.

Passengers on board the Concorde are pampered for four hours as they speed across the Atlantic at Mach 2. Only military pilots and astronauts have flown faster—but without the creature comforts of leather seats and champagne. The Concorde fleet is aging, however, and designers are developing commercial airplanes that can fly even faster, farther and higher.

Fly-by-wire technology, in which on-board computers interpret the pilot's commands to adjust speed, direction and altitude, is replacing mechanical control systems. Lightweight plastics, carbon fiber and other space-age materials are already appearing in wings and fuselages. Early in the next century, jet fighters may even be flown in combat via remote control by pilots on the ground.

Armed with new technologies, companies such as Boeing and McDonnell Douglas are actively pursuing the idea of high-speed civil transports (HSCTs) like the design at left. But flight above Mach 3 poses new challenges. Since HSCTs would cruise above 60,000 feet, their engines must emit less nitrogen oxide than conventional jets so as not to harm the ozone layer. To be commercially viable, the new aircraft would require a range of 7,000 miles, twice that of the Concorde. Designers must also find ways of limiting sonic boom in order to win permission for supersonic flight over land.

There is no question, given the rapid march of science, that these problems will be solved—and that airlines may one day in the not-too-distant future carry passengers between New York and Tokyo in just over three hours.

Index

Numerals in *italics* indicate an illustration of the subject mentioned.

PICTURE CREDITS

Credits are read from left to right, from top to bottom by semicolons.

Front cover: Bruce Dale © National Geographic Society.

6 AllSport/Vandystadt/West Light; Tom & Pat Leeson; Alain Guillou. 7 C.J. Heatley III (2); Courtesy Rockwell International. 14, 15 Alain Guillou. 18, 19 Dave Becker/Viewfinders (5). 22 Gene Stein/West Light. 23 David Lawrence/The Stock Market. 26 Julie Ades/Visions. 30, 31 C.J. Heatley III. 32, 33 Stephen Dalton/Photo Researchers. 36 EMI Pathé News Library. 37 © S.C. Johnson & Son/Smithsonian Institution, courtesy IMAX Systems Corporation. 40, 41 Mark Greenberg/Visions (2). 44 NASA/Science Source/Photo Researchers. 45 NASA/Langley. 53 Charles O'Rear/West Light. 57 C.J. Heatley III; Barry Griffiths/Photo Researchers; Mark Greenberg/Visions; M.P. Kahl/DRK Photo; C.J. Heatley III; Roy Morsch/Bruce Coleman Inc. 58, 59 Courtesy Boeing Commercial Airplane Co. (3); Gabe Palmer/The Stock Market. 60, 61 Courtesy Boeing Commercial Airplane Co. (2); Philip C. Jackson (2). 62 Courtesy Boeing Commercial Airplane Co. 64, 65 James Sugar/Black Star. 66, 67 Stephen Dalton/Photo Researchers. 81 Tom & Pat Leeson. 82, 83 Courtesy Nikon Corporation. 84, 85 Peter Thomann/Stern/Black Star; Courtesy Superflight Inc. 92 Dr. Marvin Luttges/BioServe Space Technologies (2). 94, 95 Stephen Krasemann/DRK Photo. 96, 97 Ron Austing/Photo Researchers; Jeff Foott. 104 Chris Sorenson/The Stock Market; Ron Watts/First Light. 105 Chris Sorenson/The Stock Market. 107 Bob Klein, courtesy Unisys Corporation. 110 Courtesy Honeywell Inc., Sperry Flight Systems Group. 113 Courtesy Honeywell Inc., Sperry Flights Systems Group. 117 Allen Green/Science Source/Photo Researchers. 118, 119 C.J. Heatley III (3). 120, 121 C.J. Heatley III. 124 Dr. Harold J. Edgerton/Palm Press Inc. 128 Courtesy Boeing Military Airplanes. 129 George Hall. 132, 133 © S.C. Johnson & Son/Smithsonian Institution, courtesy National Air & Space Museum; © S.C. Johnson & Son/Smithsonian Institution, courtesy IMAX Systems Corporation. 134, 135 Bill Ross/West Light; National Air & Space Museum/Smithsonian Institution. 136, 137 Courtesy Northrop Corporation (2). 138, 139 Courtesy McDonnell Douglas Corporation; Yan Lukas/Photo Researchers.

ILLUSTRATION CREDITS

Credits are read from left to right, from top to bottom by semicolons.

8, 13 Robert Monté. 16 Steve Louis. 20, 21 Ronald Durepos. 27 Luc Normandin. 28, 29 Luc Normandin. 34, 35 Josée Morin. 52 Gilles Beauchemin. 62, 63 Gilles Beauchemin. 80 Gilles Beauchemin. 87 Josée Morin. 90, 91 Luc Beauchemin. 92, 93 Gilles Beauchemin. 98, 99 Luc Normandin. 102, 103 Guy Charette. 106, 113 Guy Charette. 114 Jean-Claude Gagnon. 116, 117 Guy Charette. 122, 123 Robert Monté. 124 Steve Louis. 126, 127 Pyrate Communications Inc.

ACKNOWLEDGMENTS

The editors wish to thank the following:
Alan Adler, Superflight, Inc., Palo Alto, CA; Delise Alison, Redpath Museum, McGill University, Montreal, Que.; Mark Allen, Mark Allen Productions, Oyster Bay, NY; Dr. Allison Andors, American Museum of Natural History, New York, NY; Ted Bailey, General Electric Aircraft Engines, Cincinnati, OH; Barkley E. Bates, Voyager Airways, North Bay, Ont.; Dr. Craig Bohren, Department of Meteorology, The Pennsylvania State University, University Park, PA; Christiane Brisson, Public Affairs, Air Canada, Dorval, Que.; Dr. Jim Britten, Department of Chemistry, McGill University, Montreal, Que.; British Airways, Toronto, Ont.; John Burton, Experimental Aircraft Association, Oshkosh, WI; Steve Buss, Experimental Aircraft Association, Oshkosh, WI; Gary Butcher, Porsche Aviation Products Incorporated, Reno, NV; Thomas R. Cole, Boeing Commercial Airplane Group, Seattle, WA; Martyn Cowley, Aerovironment Inc., Monrovia, CA; Mark Drela, Massachusetts Institute of Technology, Cambridge, MA; Abdul Elkeesh, St. Hubert Airport, St. Hubert, Que.; Carolyn M. Fennell, Orlando International Airport, Orlando, FL; Festival de Mongolfières du Haut-Richelieu, Quebec; Alan E. George, ILC Dover, Inc., Frederica, DE; John Gerike, Boeing Defense Space Group, Seattle, WA; Danielle Gerrard, Boeing Commercial Airplane Group, Seattle, WA; Dr. David L. Gibo, Department of Zoology, University of Toronto, Toronto, Ont.; Edward C. Gorman, Honeywell Air Transport Systems Division, Phoenix, AZ; Serge Guilbault, Montreal Area Control Center, Dorval, Que.; Richard Harrison, Boomerang Man, Monroe, LA; H. Keith Henry, NASA Langley Research Center, Hampton, VA; Dr. Frank H. Heppner, Department of Zoology, University of Rhode Island, Kingston, RI; Alain Jacques, Dorval Control Tower, Dorval, Que.; Linda Justo, Honeywell Inc., Business & Commuter Aviation Systems Division, Glendale, AZ; Charles Larocque, Bell Helicopter Textron Canada, St. Janvier, Que.; Michel LeBrun, Ecole Nationale d'Aérotechnique, St. Hubert, Que.; Pierre Legault, Bell Helicopter Textron Canada, St. Janvier, Que.; Captain K.D. Leney, British Airways (Concorde Division), Heathrow Airport, London, U.K.; Richard Leyes, Aeronautics Department, National Air and Space Museum, Smithsonian Institution, Washington, DC; Paul T. MacAlester, Hillsborough County Aviation Authority, Tampa, FL; Dr. Paul MacCready, Aerovironment Inc., Monrovia, CA; Vernon C. Maine, Kollsman, Merrimack, NH; J. Campbell Martin, NASA Langley Research Center, Hampton, VA; Judy Mills, Institute for Aerospace Studies, Downsview, Ont.; Bruce Nesbitt, Bell Helicopter Textron Canada, St. Janvier, Que.; Carol Petrachenko, NASA Langley Research Center, Hampton, VA; Pratt & Whitney Canada, Longueuil, Que.; Elizabeth V. Reese, Boeing Commercial Airplane Group, Seattle, WA; Richard F. Saler, The Goodyear Tire & Rubber Company, Akron, OH; Richard Shoemaker, Montreal, Que.; The Intrepid Sea-Air-Space Museum, New York, NY; Anna C. Urband, Navy Office of Information, Washington, DC; Captain Daniel P. Whalen, USN, Department of the Navy (Air Warfare), Washington, DC.

The following persons also assisted in the preparation of this book:
Dominique Gagné, Shirley Grynspan, Stanley D. Harrison, Carolyn Jackson, Brian Parsons, Shirley Sylvain.

This book was designed on Apple Macintosh® computers, using QuarkXPress® in conjunction with CopyFlow® and a Linotronic® 300R for page layout and composition; Stratavision®, Adobe Illustrator 88® and Adobe Photoshop® were used as illustration programs.

Time-Life Books Inc. offers a wide range of fine recordings, including a *Rock 'n' Roll Era* series.
For subscription information, call 1-800-621-TIME, or write
TIME-LIFE MUSIC, Time & Life Building, Chicago, Illinois 60611.

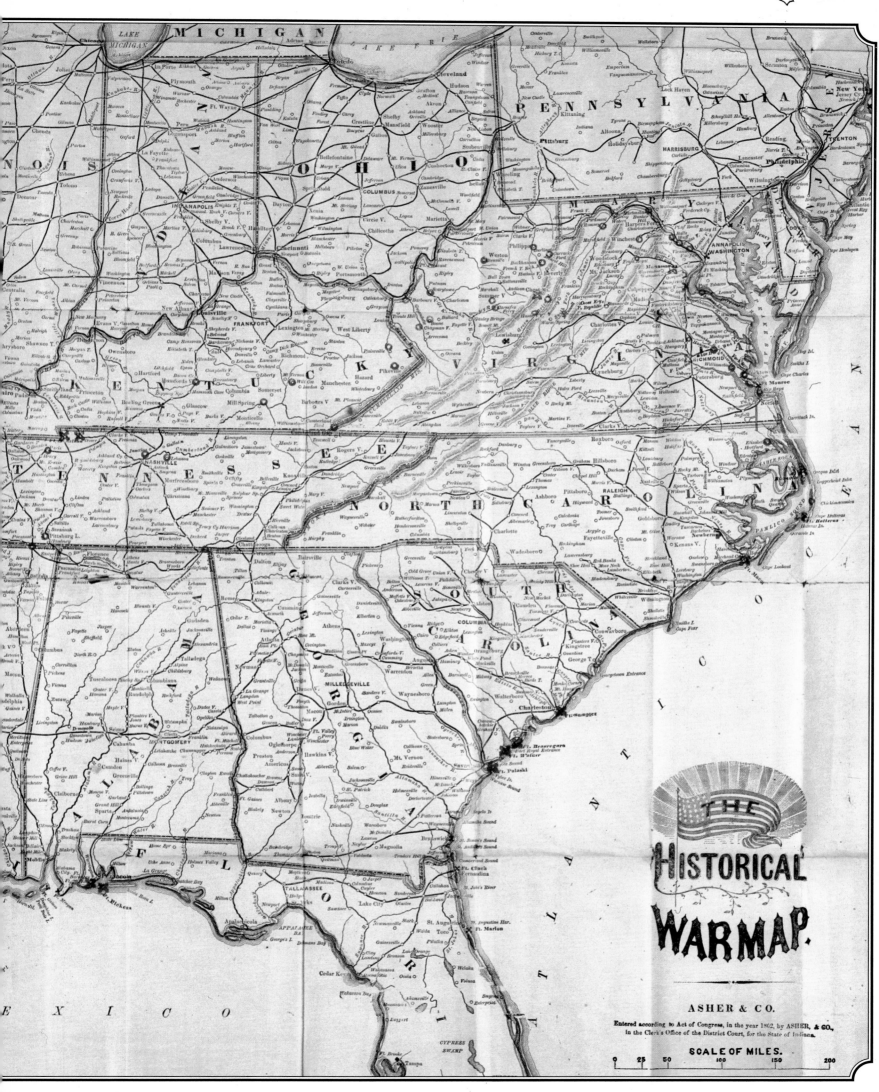

THE
Historical
WARMAP.

ASHER & CO.

Entered according to Act of Congress, in the year 1862, by ASHER, & CO.,
in the Clerk's Office of the District Court, for the State of Indiana.

SCALE OF MILES.

0   25   50   100   150   200

REBELS & YANKEES

THE BATTLEFIELDS
OF THE CIVIL WAR

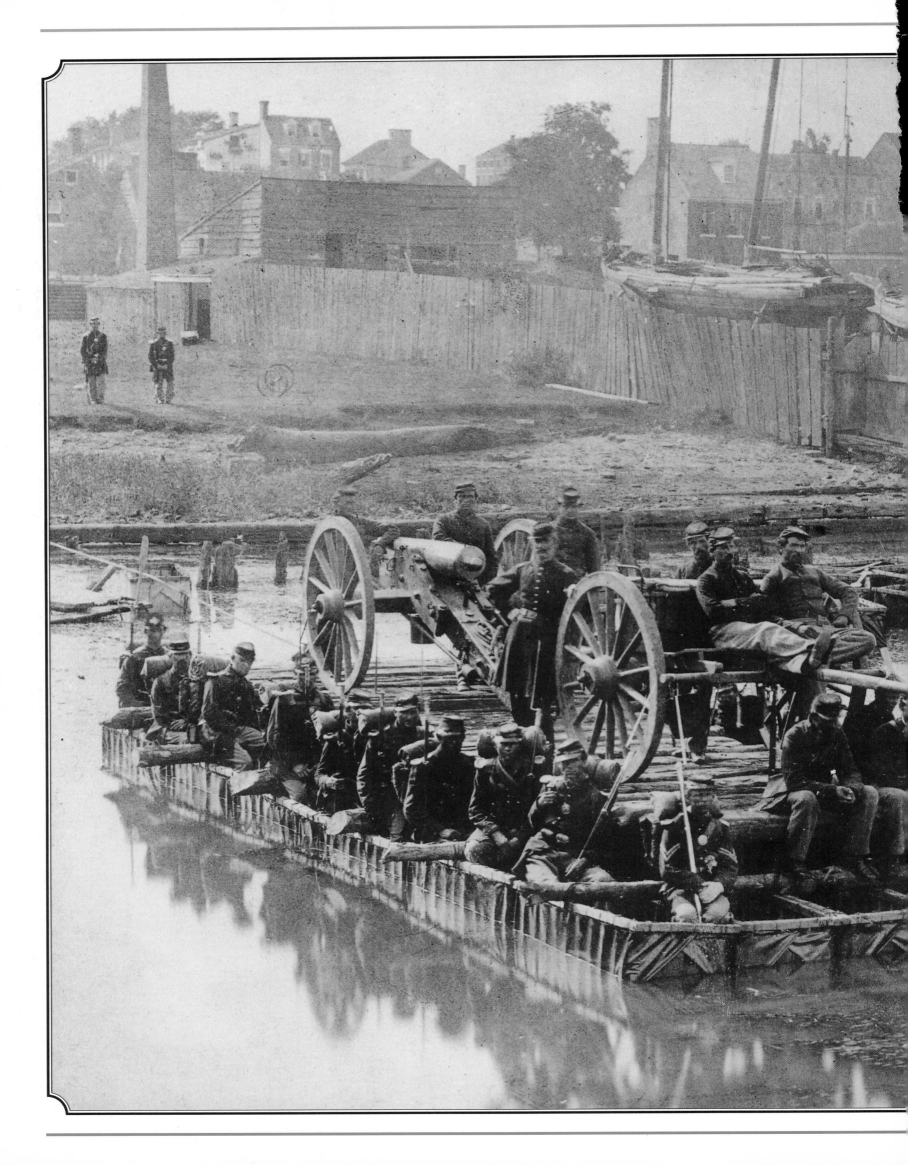

REBELS & YANKEES
THE
BATTLEFIELDS
OF THE CIVIL WAR

The bloody conflict of North against South told through the stories of its great battles.
Illustrated with collections of some of the rarest Civil War historical artifacts

WILLIAM C. DAVIS
TECHNICAL ADVISOR: RUSS A. PRITCHARD

SMITHMARK

A SALAMANDER BOOK

This edition published in 1991 by
SMITHMARK Publishers Inc., 112 Madison Avenue,
New York, N.Y. 10016.

SMITHMARK books are available for bulk purchase
for sales promotion and premium use. For details
write or telephone the Manager of Special Sales,
SMITHMARK Publishers Inc., 112 Madison Avenue,
New York, N.Y. 10016. (212)532-6600

© Salamander Books Ltd 1991

ISBN 0 8317 0702 X

CREDITS

Editor: Tony Hall   **Designer:** Mark Holt
Indexer: David Linton
Color artwork: Jeff Burn © Salamander Books Ltd
Map artwork: Derek Bunce © Salamander Books Ltd
Line artwork: Kevin Jones Associates
© Salamander Books Ltd
Color photography: Don Eiler, Richmond, Virginia
© Salamander Books Ltd
Filmset: SX Composing Ltd, England
Color reproduction: Bantam Litho Ltd, England
Printed in Italy

ACKNOWLEDGEMENTS

Without the help of the individuals and organizations
listed below, this record of Civil War militaria could
not have been assembled.

INSTITUTIONAL COLLECTIONS

Chester County Historical Society, West Chester, Pa
Roland H. Woodward, Executive Director
Margaret Bleeker Blades, Curator

The Civil War Library and Museum, Philadelphia, Pa
Russ A. Pritchard, Director
Constance Elaine Williams, Administrative Assistant

Gettysburg Museum of The Civil War, Gettysburg, Pa
Lawrence Eckert, Curator

Milwaukee Public Museum, Milwaukee, WI
H. Michael Madaus, Assistant Curator of History

The Museum of The Confederacy, Richmond, Va
Howard O. Hendricks, Curator of Collections
Malinda S. Collier, Registrar
Rebecca Ansell Rose, Curatorial Assistant
Corrine P. Hudgins, Photographic Assistant
Guy R. Swanson, Curator of Manuscripts and Archives
Eleanor G. Lewis, Curatorial Intern
Harry G. Gaydos, Curatorial Intern

The Union League of Philadelphia, Philadelphia, Pa

**U.S. Army Ordnance Museum, Aberdeen Proving
Ground, Md**
William F. Atwater, Ph.D., Director
Roger A. Godin, Curator
Armando E. Framarini, Museum Specialist
Lesley M. LeRoy, Administrative Assistant

U.S. Army Quartermaster Museum, Fort Lee, Va
Phillip M. Cavanaugh, Director
Luther Hanson, Museum Technician

Virginia Historical Society, Richmond, Va
Linda Leazer, Assistant Curator/Registrar, Museum
Services

West Point Museum, West Point, N.Y.
Michael E. Moss, Director
Robert W. Fisch, Curator of Arms
Michael J. McAfee, Curator of Uniforms and Military
History
Walter J. Nock, Museum Specialist Conservator

PRIVATE COLLECTIONS

William A. LePard, Ardmore, Pa
John G. Griffiths, Fredericksburg, Va
William L. Leigh III, Chantilly, Va
Wendell Lang, Tarrytown, N.Y.
William Smith, Boonsboro, Md
Don Troiani, Southbury, Conn

CONTENTS

AUTHOR

William C. Davis graduated as a Master of Arts in
history from Sonoma State College, Rohnet Park,
California in 1969. Since then he has written over
twenty-seven books on the Civil War period. Two of
his books, *Breckinridge: Statesman, Soldier, Symbol,*
and *Battle of Bull Run,* have received Pulitzer prize
nominations. He has recently completed a new
biography of Jefferson Davis.
Other consultancy work includes that for the Eastern
National Parks & Monument Association, the U.S.
Intelligence Historical Society, and the U.S.S. *Monitor*
Project for the N.O.A.A. William is the author of the
highly successful companion volumes in this *Rebels &
Yankees* series: *The Fighting Men of The Civil War,*
and *Commanders of the Civil War.*

TECHNICAL ADVISOR

Russ A. Pritchard, born 1940 in Memphis, Tennessee.
Graduate of The Choate School, Wallingford, Ct and
Washington and Lee University, Lexington, VA.
Director of the Civil War Library and Museum,
Philadelphia since 1976. Fellow of the Company of
Military Historians, Life Member of the National Rifle
Association and member of the Board of the National
Firearms Museum of the National Rifle Association.
Consultant for The Museum of the Confederacy,
Richmond. For thirty years a collector of arms and
accoutrements made within the Confederacy. Russ
has been the advisor to all of the books in this *Rebels
& Yankees* series.

INTRODUCTION

Throughout history men have revered and memorialized the places where their armies have met and fought. Among the earliest monuments surviving from antiquity are obelisks and stellae commemorating a victory by Assyrians or Egyptians, Macedonians or Romans. Arches of triumph, temples, inscriptions, and more, have proclaimed the glories of combat at Cannae and Marathon and a thousand other battlefields.

More often than not, they were built by victors, proud of their triumphs and boasting to the known world of their own greatness. But behind this vainglorying lay a deeper message, one understood throughout time by every people. And it is that no human experience is more personal or traumatic than a battle. When men in their legions go to war, it is as an army; when they go into battle, each soldier goes on his own, acutely aware that *his* may be one of the lives lost. Like nothing else, battle brings men face to face with themselves and what is best and worst in them individually, and collectively as a people. Thus it is that battlefields the world over have become sacred ground, and nowhere more than in that divided America that went to war with itself in 1861 and covered its young landscape in its own blood.

After almost a century of agitation, over issues that went back practically to the very foundation of the original colonies, it seemed inevitable that the descendants of those colonies, the states, should come to blows, and that the American landscape that had brought the settlers there in the beginning, that nurtured them and helped to define who they were, should in the end become their battleground, and for all too many, their burial place. In the seventeenth century, Europeans went west to become colonists. In the eighteenth century colonists went to war to wrest their freedom as Americans. In the nineteenth century Americans went mad, and went to war with themselves.

Of course, their's was a bloody landscape long before red, white, and blue split into Blue and Gray. Even before the coming of the white man, the native Americans battled with one another over now unremembered causes on long-forgotten fields. When the Europeans arrived, they only brought new ways and weapons to very old human failings, and the fighting went on as before, and the land continued to bleed. First they fought the Indian, and then one another – English, French, Spanish, and Dutch. Then came revolution, and finally those places where they met to settle the issues that reason could not solve started to assume symbolic, spiritual, even mythic importance. Lexington and Concord, Bunker Hill, Brandywine, Saratoga, and at last

Yorktown, became battlefields that unlike those that went before, lived in imagination and remembrance. In the decades to come, more would follow. New Orleans, The Alamo. The list, sadly, went on.

But no one expected what the 1860s would add to the tally, either for size or sheer numbers. When North went to war with South, an avalanche of armies were loosed that almost crushed the continent beneath their weight. Before they were done, this still-rural land would soak up the blood of more than 2,000 battles of varying size, from tiny skirmishes of a few dozen, to battles of 200,000 and more, the biggest conflicts ever seen in this half of the globe.

Most of those were unknown even at the time, and quickly forgotten by all but those who fought in them. Even more melted into the shadows of history as the old veterans and their generation died away. But a few, an epochal few, remained, and remain still, so traumatic that they cannot be escaped; so symbolic that they will not be forgotten. More than a century and a quarter after the fact, the world's generals – the great and the would-be – still study Stonewall Jackson at Chancellorsville. Ending his days after being one of the great military commanders of all time, Dwight D. Eisenhower chose to live amid the fields and monuments of Gettysburg. No American can ever forget that even D-Day in 1944 did not spill as much American blood as the single day at

In a heady moment for the new Confederates, victors inspect some of the damage done on the parapet of Fort Sumter just days after its surrender. In their euphoria they thought that this was all the war they would see.

Antietam in 1862. It was the most tragic of all wars, one between father and son, friend and friend, literally setting brother against brother. It was the American Iliad and Odyssey, a New World War and Peace, whose battlefields became sacred ground – not just spots on the map, but places in the heart.

No one expected either the trauma or the blood, much less the battlefields, when it all started to unfold. After all, there had been decades of talk and more talk, with always some blustering, some posturing, some threatening, and then some compromising. It may never have left everyone satisfied entirely, but at least it left them with some alternative to violence in their minds and hearts. Even when 1860 brought the final collapse, when the election of Abraham Lincoln left the slave states feeling so threatened that secession and the resultant challenge to the Constitution seemed to be the only recourse, most still hoped for a peaceful resolution. When state after state voted to withdraw from the Union, hopeful – some said naive – Southerners expected to be allowed to go in peace. When departed states banded together to form the new Confederate States of America in February 1861, they felt confident that their united front would be their defense and that the North would have to treat with them as independent. Even when local militia groups answered drum and trumpet calls to muster on the village greens, and when those companies joined others to form regiments, and when regiments started concentrating at Washington and Richmond and other cities to form small armies, still many said there would be no war. The boastful spoke of drinking all the blood shed in any conflict to come out of secession, so confident were they.

There seemed to be some hope in negotiations. Within days of the adoption of a constitution by the new Confederacy in February 1861, its commissioners in Washington started to talk with the authorities there, the only view being a final and eternal separation of the states. The old president James Buchanan, weak and vacillating, backed away and did his best to postpone the crisis until he had left office and the problem would fall to his successor. When Lincoln took power, however, he made it clear from the first that his oath to preserve the Constitution left him no choice or alternative course but to restore the Union and Federal authority in all of the states. Still he wanted to talk, to stall, to delay, always in the hope that time would calm rising passions. Yet even before he took power, there was already a festering sore that could not be relieved. For all of the Federal property siezed without resistance or violence throughout the seceded states, there were a few spots where men in Federal blue still held on. Fort Monroe, Virginia and Fort Pickens, Florida; both were secure in Union hands, with no fear of violent seizure. As for the Confederates, these places, while an irritant, did not pose a real challenge.

But there was another, right in the center of Charleston harbor, South Carolina, a slap in the face to the Confederacy right under the eyes of the people of the city called the seedbed of secession: Fort Sumter. Only a few score Yankees, in an unfinished fortress, with guns not yet entirely mounted, living in unfinished barracks, sat there in their blue like a splinter beneath a nail, a painful irritant that would not go away and that could not be ignored. Finally, abandoning negotiation that could never have yielded a satisfactory

Above: Fort Moultrie, abandoned by the Federals in December 1860, quickly became one of the Southern mainstays in the ring of batteries surrounding Fort Sumter. A shot furnace appears at left.

Below: A number of Charleston's dignitaries stand inside the parade ground of the captured Fort Sumter, savoring their moment of triumph and surveying the armaments captured in their first victory of the war.

Captain Abner Doubleday and Private, U.S. Artillery

There were some men, North and South, who quailled at firing the first shots of the war. Captain Abner Doubleday was not such a man, and when Confederates opened fire on Fort Sumter on April 12, 1861, he was only too delighted to fire the Union's first shot. He would rise to become a general, though he did not enjoy a distinguished career.

Doubleday appears here as an officer in the U. S. Artillery, as betokened by the red trouser stripe and the color of his shoulder strap. The crossed cannon insignia on the kepi of the private next to him also denotes the artillery. They stand inside one of Sumter's casemates, ready to fire on the Rebel Ironclad Battery, where old Edmund Ruffin himself fired one of the war's first rounds. Doubleday's uniform would not change markedly when he left the Regular Army and accepted his volunteer commission.

solution for either side in any case, the Confederates acted. On April 12, 1861, they turned Fort Sumter into the first battleground of the Civil War.

Early in the pre-dawn hours a series of last-minute negotiations passed back and forth between General P.G.T. Beauregard's emissaries and the Federals in the fort. Attempts to force a capitulation or evacuation short of hostilities came to nothing, however, and at 4:30 a.m. a signal shell from a James Island battery became the first shot of the Civil War. At once a ring of flame erupted around Charleston Harbor as battery after battery opened fire. For some time the Federals held their fire, then began a desultory answering fire from their few working guns. When the Yankees did shoot back, the Confederates cheered, not wanting their victory to be too easy. But soon Sumter's wooden barracks were blazing, and though not a man was harmed, Major Robert Anderson, in command of the garrison, decided to yield when he saw no hope of being reinforced.

As such, it was neither much of a battle not much of a field. Sumter could have been taken by a corporal's guard sooner or later, by starvation if by no other means. As for the site, it was not land or field at all, but a mountain of brick and mortar erected on an artificial island of rubble, most of it, ironically, granite from New England brought to the South.

But it was enough to start the war. Within hours of the surrender of the Sumter garrison on April 13, the war drums beat North and South. From towns great and small the regiments marched forth. Washington for a time found itself cut off from the rest of the North, with Virginians facing it across the Potomac and rebellious Marylanders at its back. When relief regiments marched through Baltimore to get to the capital, a riot claimed lives, and the first blood. Then, with Washington relieved and Maryland and its rebellious citizens more or less subdued, the Yankees began to put their feelers out south of the Potomac. There were skirmishes then. Men began dying in combat. Then in June came what they called a battle at Big Bethel, though by later standards it would be little more than a minor encounter, and soon forgotten.

But all the while, the armies kept growing. A time would come, soon enough, when they would meet, not just a few hundred to a side, but in their tens of thousands, to decide issues far greater than Baltimore or Big Bethel or even Fort Sumter could settle. More than 2,000 battlefields awaited their making, and among them waited places that were to become immortal. They already had their names. They needed only the unfolding of events. The earth and the blood were waiting to meet.

Above: Among the numerous innovative weapons used against Fort Sumter was this "floating battery", literally a barge with cannon mounted, covered by a heavy timber shelf and railroad iron cladding.

Below: The interior of the ironclad battery shows the beams forming its roof, the cramped interior, and one of its four heavy smoothbores. Some of the war's first shots came from this gun.

Edmund Ruffin, Palmetto Guard, Co.I., 2nd South Carolina Volunteers, C.S.A.

Old Edmund Ruffin led the pack among secessionists, even though he came from moderate Virginia. When secession came he could not wait to be at the center of it all, and went to Charleston, where the members of the local Palmetto Guard happily made him a member of their outfit. He cut a striking figure in the units black uniform, really nothing more than a civilian broadcloth suit with white military belt and cross-belt. The closest thing to a regulation item is his hat, with its distinctive ''P G'' surrounded by a wreath.

When reformed as Company I of the 2nd South Carolina, a more standard style of dress was adopted.

The Palmetto Guard drew duty manning the so-called ''Ironclad Battery'' facing Fort Sumter, and it fell to Ruffin to fire the battery's first shot. He also fired one of the war's last shots, committing suicide after the Confederate surrender.

CHAPTER ONE

FIRST MANASSAS (FIRST BULL RUN)

JULY 21, 1861

In the wake of Fort Sumter, both nations rushed to prepare for war, especially after Lincoln issued a call for 75,000 volunteers to put down the "rebellion". That act alone helped propel the wavering Southern states like Virginia, North Carolina, and Tennessee into the new Confederacy.

A peculiar self-confidence infused both sides as their citizens-turned-soldiers began drilling for the first time. In the South, men convinced themselves that the Yankees would never fight. Having been beaten at Sumter, they would not have the stomach for real battle. Southern physical, moral, and spiritual superiority had preordained success for the new nation.

Meanwhile, north of Mason and Dixon's line, men marching for the Union felt a keen mixture of resolution over avenging the insult to the Stars and Stripes, and at the same time an absolute conviction that the Confederates were nothing but blustering, posturing, comic-opera soldiers, and that with one swift campaign the rebellion would be destroyed.

Yet the war that began with such impact at Charleston, proceeded only by fits and starts during the three months that followed. There were only a handful of small engagements, mostly of a tentative nature, and even when the first "battle" finally took place at Big Bethel, Virginia, in June, it was little more than a skirmish by later standards. Still the Rebels won that day, and confidently expected to repeat their performance when and if the Yankees dared live up to their threat to invade Virginia and march "on to Richmond".

GENERAL IRVIN MCDOWELL won only one battle in his career, and that was with a watermelon. Certainly he was brave enough. A graduate of the United States Military Academy at West Point, he served with distinction in the 1846-8 war with Mexico, and won promotion and favorable mention in reports. After the Battle of Buena Vista he took a permanent place on General Winfield Scott's staff for the balance of the war, and became so close to Scott personally that he continued in the general-in-chief's military family for more than a decade afterward. Yet they shared little in common. Scott was outgoing, in the main popular with his subordinates, a man with an agile mind that helped compensate for his excessive pride and occasional petulance when challenged.

One thing McDowell did share with his mentor, however, was a Homeric appetite. He was, in fact, an unquestioned glutton. One of his own later staff members recalled that "at dinner he was such a Gargantuan feeder and so absorbed in the dishes before him that he had but little time for conversation." That may have been just as well, for McDowell made a poor social companion at best. Consistently unable to remember names and faces, he possessed little charm or personality. He lost his temper too easily, listened only carelessly to others, and often sat lost within his own thoughts, meeting almost everyone with what one of his closest supporters had to confess was a "rough indifference." It did not help that he capped his substantial obesity by wearing a ridiculous bowl-shaped straw hat that gave him more the appearance of an incongruously over-weight Chinese coolie than the general of the first army of the Union. Perhaps his seeming distraction and aloofness stemmed from his constant concern for his next meal. He touched neither wine nor spirits but "fairly gobbled the larger part of every dish within reach", recalled his adjutant Captain James B. Fry. And when done with that, he launched a culinary campaign against an entire watermelon, which he defeated and con-

It was all supposed to be a lark. Boys like these New Yorkers expected only ninety days' enlistment, a few weeks of training, and then a speedy campaign that would see them in Richmond before the summer was out.

sumed single-handedly, afterward pronouncing his vanquished foe to be "monstrous fine."

Yet however much he may have looked the comic opera general, McDowell was not a fool. He enjoyed powerful backing by politicians from his native Ohio, and this, plus the favor of Scott, saw him elevated overnight from a mere major to a brigadier general. Having spent his whole career in staff work, a man who never led so much as a corporal's guard in battle would now take over the task of building and leading to victory the largest field army ever yet assembled on American soil. There was a perverse sort of logic to it, conceivable only in the context of the burgeoning Civil War itself, which imposed a logical perversity upon almost every act of that tragically cursed generation.[1]

In the weeks immediately following the fall of Fort Sumter, Northern authorities belatedly commenced the task of raising volunteers in order to expand its pitiably small 16,000-man Regular Army. The new Confederacy, by contrast, had been the recipient of newly raised and equipped volunteer regiments from the seceded states even before the convention in Montgomery, Alabama, adopted its constitution. Lincoln issued a call for 75,000 volunteers from the states remaining in the Union, and even while the youth of the North flocked to the recruiting stations, Washington wrestled with the question of who should take command.

One thing was immediately clear. None of Lincoln's existing few generals would do for this active field command. Scott and John E. Wool were each over 70. Edwin V. Sumner was approaching 64, and the only other field grade general, David E. Twiggs, had gone over to the Confederates. It required a younger, more vigorous man to take over molding the new army being formed in and around Washington, especially since it was this army, and not others being raised in Ohio and the West, that everyone

Above: No one in American army history had gone so quickly from obscurity to instant high position. Irvin McDowell, one moment a major, found himself a brigadier the next, and charged by Washington with the greatest task asked of an American commander to date.

Below: Though taken in 1862, this image of the ruins of Manassas Junction still shows what it was that made the otherwise unnoticeable spot so vital. it was through this vital rail link that the Rebels were enabled to make their historic concentration of forces.

expected to put down the rebellion and win the war. Washington lay but a scant 100 miles from Richmond, Virginia, which in May became the Confederate capital. Thus North and South alike took it for granted that the fate of America as one nation or two would be decided on the rolling, iron-red soil of the Old Dominion. The man who commanded Lincoln's army in Washington would be expected to march into Virginia, meet and defeat the Rebels in his front, and march on to Richmond, and triumph.

No wonder, then, that even before McDowell received his new command, there were other men south of the Potomac, themselves new generals and amateurs at the business of command, hurriedly preparing to meet his inevitable advance. Jefferson Davis faced the same problem confronted by Lincoln when it came to command of his armies, but at least he could meet the issue with a few advantages. While Old Twiggs was too aged to be considered for command of the Confederates in Virginia, Davis did have at hand Joseph E. Johnston, himself a Virginian, and a battle-seasoned veteran of Mexico. He left the United States Army with the staff rank of brigadier which, by Davis' promotion policy, automatically entitled him to a generalcy. Johnston was an obvious choice to command in Virginia. Furthermore, in Beauregard, Davis had the war's first authentic popular hero. Even though a child could have taken Fort Sumter, still Beauregard's success demanded that he be made a high-ranking general as well, and with the scene of operations quickly shifting to Virginia, good sense dictated that the Creole be sent there as well.

Happily, Davis required not one but two armies in Virginia. Geography dictated such. One, the principal force, needed to position itself between Richmond and Washington, and as close to the Potomac as possible in order to protect northern Virginia. But barely 100 miles to the west there was another Virginia, the Shenandoah

Valley. Mountains made it, the Alleghenies on the west, and the less imposing but still significant Blue Ridge on the east. The Valley ran roughly northeast to southwest, from the Potomac down deep into the Virginia heartland. It presented a natural pathway of invasion to the Federals, for by marching south into the Valley, they could cross the Blue Ridge at one of the few gaps available, and emerge deep in the rear of any Confederates in northern Virginia. Thus Davis had to hold the Shenandoah, a task he gave to Johnston, while eventually he assigned Beauregard command of the newly designated Army of the Potomac being formed in and around Manassas Junction, about 25 miles southwest of Washington.

There was something special about Manassas. From the junction, the Manassas Gap Railroad led straight west, through Manassas Gap, to the Shenandoah, providing the northernmost east-west link between the two sides of the Blue Ridge. Immediately Davis and his generals saw the opportunity this offered. By maintaining control of that gap and by using the railroad, they could with relative speed shift troops from one side of the mountains to the other, allowing a concentration of forces back and forth to meet any threat. Only be meeting and pinning down both Johnston and Beauregard simultaneously, or by seizing the gap or breaking the railroad, could the Yankees prevent such a tactic from being employed. To the Confederates it presented a chance to hold Virginia with a minimum of its precious regiments. Moreover, since Davis and his generals hoped to more than merely defend in a coming engagement, it gave them a weapon with which they might be able to combine two numerically inferior Rebel armies in order to meet and overwhelm a Union force that neither of them could otherwise hope to meet successfully.

Since the map looked just the same to men in Washington, Lincoln and Scott saw the same possibility. Consequently, their strategy for the advance into Virginia called from the first for simultaneous movements. McDowell was to move against Manassas Junction, while another, smaller, volunteer force commanded by General Robert Patterson, was to cross the Potomac and take Harpers Ferry at the northern end of the Shenandoah, and then move south, keeping Johnston fully engaged and, if possible, driving him away from the Manassas Gap line.

First they had to raise their army. The young men of the North responded with heartening alacrity to the call for volunteers. Throughout May and June the regiments poured into Washington. Even as they arrived, the newly appointed

Captain, Co. G, 11th Virginia Infantry, C.S.A., Lynchburg Home Guard

A captain in Company G, 11th Virginia Infantry, the Lynchburg Home Guard, wears the regulation gray frock coat, trousers, and forage cap, specified for his regiment. The black trim on the trouser seams and cuffs and collar differs from the blue used on many other Virginia regimental uniforms.

His sword belt has the handsome two-piece Virginia State seal belt plate, supporting the scabbard for his Model 1850 foot officer's sword. An unusual feature are the brass shoulder scales with bullion epaulettes.

The enlisted men behind him might wear gray or blue, as both colors were to be seen in this regiment. Some also wore black jackets beneath their white webbing cross-belts, and gray trousers. Virginia was second only to North Carolina in its near-approach to fully equipping and outfitting its regiments, according to government regulations.

McDowell moved across the Potomac to occupy Arlington and Alexandria, giving Lincoln a foothold in Virginia and somewhat lessening the psychological trauma of having Confederate flags flying just across the river from the Capital. Meanwhile the new general built his army. From New York, Massachusetts, Ohio, Pennsylvania, Michigan, Rhode Island, Maine, Connecticut, and more, came the raw young regiments. By the middle of July there were more than 30,000 men under arms in and around Washington, most of them poorly trained, their enthusiasm expected to make up for their lack of preparation.

At least McDowell managed to put them under leaders who did have some experience. He divided his army into five divisions, commanded respectively by Daniel Tyler, David Hunter, Samuel P. Heintzelman, Theodore Runyan, and Dixon S. Miles. Tyler enjoyed an excellent reputation in the Old Army. Heintzelman was a veteran of the Mexican and Indian conflicts with a record for bravery. Hunter had spent forty years in uniform, though seeing no action. In the 1830s one of his closest Army friends had been Lieutenant Jefferson Davis. Runyan was a cypher, a militia general from New Jersey with no more experience than his small division. Miles had more experience than all the rest in real action, but also brought an unfortunate reputation for losing battles with the bottle.[2]

Among the brigade commanders in those divisions some soon-to-be distinguished names could be found, including future army commanders William T. Sherman, Ambrose Burnside, and Oliver O. Howard. But for now they were all almost uniformly men unheard-of outside the small confines of the old Regular Army, and many of them had already returned to civilian life when the war broke out. None of them was prepared for the demands of this war. Not one had ever led more than a company of 70 or 80 men in action before. Now they were called upon to lead thousands.

Most inexperienced of all, of course, was the man who had to lead them all, Irvin McDowell. Lincoln and Scott placed almost constant pressure upon him for a plan of campaign, even as his army was still forming. Repeatedly McDowell protested that his men were untried, untrained, "green" as he put it. Scott only replied, "you are green, it is true; but they are green, also; you are all green alike."[3]

McDowell insisted that Manassas provided the key to northern Virginia, as indeed it did. To drive Beauregard away and sieze the junction, he proposed to drive first for Centreville, about four miles northeast of Manassas. If successful that far, his move would force the Rebels back to their next natural line of defense, a stream two miles above Manassas, and running roughly northwest to south east. McDowell knew that it would only be crossable at a stone bridge on the Warrenton Turnpike and at a handful of fords, all of which he could expect to be heavily fortified and guarded. Instead of attacking there, he proposed merely to keep Beauregard occupied with demonstrations in his front, while he pushed a column far down the stream, crossing below the Confederate right flank. Once across, he could then drive straight west toward Manassas, putting himself in Beauregard's rear, cutting his rail communications both with Johnston in the Shenandoah and with Richmond as well. The Rebels would have no choice but to abandon their positions with little or no fight, and McDowell would

Above: Almost from the first, Brigadier General Daniel Tyler did not get along with McDowell. Their relationship soured even more when Tyler turned an intended reconnaissance at Blackburn's Ford into a substantial engagement that gave away Federal intentions.

Below: The full uniforms, the weapons on display, the blanket roll over the shoulder, all testify that these men of the 8th New York State Militia are ready to march, as they pose for a Brady camera on July 16, 1861, barely hours before marching for Bull Run.

find the road to the enemy capital open to him.

Considering the new general's inexperience and the pressures under which he worked, it was an excellent plan, and Lincoln and Scott accepted it without amendment. If successful, McDowell could win the campaign more by strategy than fighting, and have a fair chance of ending the war with a quick capture of Richmond. Inexperience aside, a few things stood in his way. He had almost no cavalry, which would severely hamper his ability to gather intelligence during the campaign, as well as denying him the role the mounted arm would play in guarding his flanks in action. Incredibly, though the region through which he would march was within a good day's ride of Washington, the War Department possessed no reliable maps of northern Virginia, and McDowell would be entirely dependent upon local guides whose trustworthiness he would rightly suspect. Moreover, the success of the whole plan lay inextricably entwined with Patterson's ability to keep Johnston occupied in the Valley. Most of all, he would have to overcome the greatest obstacle of all in getting his divisions where he wanted, when he wanted, across a sluggish but largely impassable little stream called Bull Run.

The Army of Northeastern Virginia marched out of its camps along the Potomac on July 16, unaware that the time and direction of its movement was such an ill-kept secret that word came to Beauregard almost at the same time that the order reached the Yankee troops. The Creole had been building his army just as frantically as his adversary, and against obstacles nearly as great.[4]

Beauregard took command in and around Bull Run on June 1, to find only 6,000 soldiers to hand, and a great deal of work needed to defend the several fords and the bridge over the stream.

Immediately Richmond began forwarding new volunteer regiments to him as soon as they became available, but the infant bureaucracy proved woefully inadequate to the needs of supply, and the general found his men ill-equipped and inadequately fed and clothed. By late June, however, he could see his army grown to the point that he could give it some formal organization, and he had in hand a number of officers, many with Mexican and Indian War experience, to whom he could turn for command.

The First Brigade, South Carolinians, went to Milledge L. Bonham. The Second Brigade went to the newly commissioned Brigadier General Richard S. Ewell. David R. Jones was to lead the Third Brigade. Colonel George Terrett led the Fourth Brigade for only two weeks before being superseded by another new brigadier, James Longstreet. The Fifth Brigade belonged to the hapless Colonel Philip Cocke. It was he who first anticipated the defensive possibilities of northern Virginia, and it was he who began the building of the army that Beauregard came to command. Yet the credit would elude him. His health ravaged, he would take his own life before the year was out. The Sixth Brigade went to the stuttering new Virginia colonel Jubal A. Early. By June Beauregard could count some 15,000 men in his ranks, arranged in a cumbersome organization that would better have seen all those brigades formed into two or three divisions, as McDowell had done. Then Beauregard would only have to issue orders to two or three subordinates, instead of to all six brigade commanders, a situation that more than doubled the paperwork necessary, and with it the possibility for confusion.[5]

At least he had the benefit of intimate familiarity with the countryside. During the weeks after taking command, he rode repeatedly over the hills, woods, and fields surrounding both sides of Bull Run. From the first, he saw that the stream itself was his best line of defense, and though he constantly bombarded Richmond with grandiose and impractical schemes for launching an offensive against McDowell, the odds from the first were that he would be the one attacked rather than the one attacking. This made Bull Run his best friend. Its banks were too steep for marching infantry and horse-drawn artillery and supply trains to negotiate, and its waters too deep to cross except at the fords. There Beauregard concentrated his defenses, expecting to counter McDowell's superior force by well placed batteries and breastworks.

At the far right of the Confederate line sat Union Mills Ford which, though Beauregard did not know it, was the one where McDowell actually hoped to cross his army. Beauregard feared for little there, in fact, since it required the enemy to make the longest march possible to reach Manassas. A mile and one-half upstream, however, sat McLean's Ford, a mile above it lay Blackburn's Ford, and less than a mile above that came Mitchell's Ford. All three were closer to the center of Beauregard's line, and could be threatened by the enemy. Happily they sat on a convex bend of the stream, so that all three lay within easy reach of a centrally placed force of two or three brigades, allowing Beauregard to use a minimum of his small command to defend along the interior line. A minor ford of no probable use to either side lay upstream of Mitchell's, but then came Ball's Ford, to which two good roads from Centreville led, and then above it sat Lewis' Ford. Not much farther along came the stone bridge on the turnpike from Centreville to Warrenton. It offered the best and easiest crossing of Bull Run available, served by the best road through that countryside. Obviously McDowell would make some attempt to take it if he advanced against the left half of Beauregard's line. Happily, after the road crossed the bridge, it was flanked on either side by hills, the gentler Matthews Hill on the north, and the more substantial Henry Hill on the south. The latter had an excellent view of the bridge, the road, and of the crossroad that led off northward to the ford at Sudley Springs, a mile and one-half northward, and the last of the crossings. By placing men on Henry Hill, Beauregard could impede passage toward Manassas from both the ford and the bridge.

Beauregard may have been a pompous blowhard at times, but he had an excellent eye for terrain, and looking at the positions along Bull Run he instantly recognized that he could defend all of the major crossing points with just two concentrations of his forces, one around Henry Hill and the stone bridge, and the other around McLean's Ford. But he was not to be satisfied with defense. In spite of having had more than one proposed offensive refused by President Davis, Beauregard moved several of his brigades in advance of Bull Run, and even beyond Centreville. With only 18,000 men in his army now, he implored Davis to order Johnston to come and join him in a concentration against McDowell, as if Johnston did not have a foe of his own to contend with in the Valley. And whereas Beauregard somehow believed that Johnston had more than

Below: This 1862 image shows the dense banks of Bull Run, near the Orange & Alexandria Rail Road crossing, at the southeastern edge of the battlefield. Though it does not appear deep, the run was only crossable at a few fords and the Stone Bridge.

22,000 men with him in the Shenandoah, in fact barely half that number faced Patterson. Finally, only the arrival of a message on July 16 that McDowell would march that afternoon brought Beauregard up against hard reality. Heavily outnumbered, he had no choice but to withdraw toward Bull Run.[6]

The Federals gave him plenty of time. McDowell's advance was a halting affair. The fact was, no one knew anything about moving tens of thousands of soldiers in their long, serpentine columns, over farm lanes and country roads. Then, too, the Yankees suffered from an exaggerated fear of hidden batteries and other deadly obstructions before them, when in fact they met

but little resistance from Confederate outposts. Still after the regiments left their camps around 2 p.m. on July 16, they covered barely more than ten miles before stopping and bivouacking for the night. The next day's march proved more stirring, as the Rebels before them evacuated Fairfax Court House without a fight, and only the fatigue of his men from a full day's march in the heat, dust, and humidity, prevented McDowell from ordering them on to Centreville that evening. Instead he gave Tyler orders to attack Centreville the next morning, and then set out himself to reconnoiter toward Union Mills and his hoped-for crossing place that would put him behind Beauregard's positions and thus win the battle.

What the general found upset all his plans at the last minute. The country leading toward Union Mills proved to be all but impassable for a large army. He would have to abandon his original intention. Thinking quickly, he simply reversed his plan and decided to send his main power in a flank march around the enemy's other flank by crossing at Sudley Ford, meanwhile continuing with his original concept of a feinted show of force at the fords in Beauregard's center in hopes of deceiving him into believing that an attack would be made there. Thus McDowell amended his orders. While Heintzelman and Hunter's divisions marched to be ready to turn northwest toward Sudley, Miles would occupy

Confederate National Flags

1 Confederate National Flag, First Pattern. Though never actually authorized by law, the "Stars and Bars" was adopted by the Provisional Congress of Confederate States on March 4, 1861 (in time to coincide with Lincoln's inauguration). The flag first flew above the capitol of Montgomery, Alabama, where the Congress was then sitting

2 Confederate National Flag, Second Pattern. The "Stainless Banner" was the first national Confederate flag adopted by law, under Senate Bill No. 132 of May 1, 1863. It was first used officially at the funeral of Stonewall Jackson where it covered the general's coffin. This particular example was the headquarters flag of General Jubal A. Early

3 Confederate National Flag, Third Pattern. From the very moment of its adoption, critics of the Second Pattern flag had said that it looked too much like a flag of truce. Criticism was especially strong from some officers of the Confederate Navy. The Confederate Congress consequently decided on this third and last pattern which was adopted officially on March 4, 1865

Artifacts courtesy of: The Museum of the Confederacy, Richmond, Va

Centreville as a reserve after Tyler had taken it, while Tyler himself was now to advance beyond the town toward Bull Run. He was to make a show of force to confuse Beauregard, but McDowell made his wishes emphatically clear: "Do not bring on an engagement."[7]

That same evening Beauregard frantically tried to augment his forces, calling to Richmond for more reinforcements, including the Hampton Legion from South Carolina, and pleading once again for Johnston to be ordered to him. With all of his forces withdrawn behind Bull Run at last, his line now stretched almost six miles, from Ewell at Union Mills, to Jones at McLean's Ford, Longstreet and Bonham at Blackburn's and Mitchell's, Cocke watching Island, Ball's, and Lewis' Fords, and a newly-arrived half-brigade of two small regiments at the stone bridge. Still expecting that McDowell would advance against his center at Mitchell's Ford, Beauregard placed half of his cavalry and almost half of his artillery in support of Bonham. Thus he stood arrayed at dawn on July 18. Then came news that should have been cheering. Richmond had ordered Johnston to move east of the Blue Ridge to join him. Beauregard met the telegram with disgust, throwing it down with the exclamation that "It is too late." Johnston could never arrive in time to keep them from being overwhelmed by the enemy's superior numbers. Just the same, one of his aides volunteered to ride to the west toward Johnston's last known position at Winchester, more than fifty miles distant. Beauregard almost thought it a fool's errand, but allowed Colonel Alexander Chisolm to go anyway. At almost the same time came word that Tyler was advancing through Centreville toward Bull Run.[8]

Tyler himself rode down toward Blackburn's Ford to learn what he could. He saw little other than enemy pickets and a few cannon, the balance of Beauregard's men being concealed. He should have known better, but despite his many years in the army, Tyler had never seen actual combat. Believing his eyes, he concluded that he could cross Bull Run at that point, almost

Federal Uniforms and Headgear

1 Frock coat of a drum major, 18th New York National Guard
2 Heavily decorated baldric (cross belt) with shield and drum stick emblem of drum major, 18th New York National Guard
3 Trousers of drum major, 18th New York National Guard
4 Shako with plate, chin strap and cockade. Shako is of the 7th Regiment, New York

National Guard
5 Frock coat of Lieutenant Colonel W.H. Armstrong, 129th Pennsylvania Volunteer Infantry. The coat carries his rank insignia and corps badge
6 Frock coat of 1st Sergeant Harlan Cobb, U.S. Engineers
7 Cobb's Hardee hat with plume, with insignia of company and branch of service
8 Cobb's trousers

9 Swallow tail coat of the Citizens Volunteers of Syracuse (N.Y.) 51st New York State Militia
10 Breeches for 9.
11 Modified Hardee hat worn by John M. Mitchell, Co. F, 79th Illinois Vol. Inf. Note bullet hole in crown from Liberty Gap, Tennessee, June 25, 1863
12 Model 1851 Albert style hat, 50th Regiment New York National Guard

Artifacts courtesy of: West Point Museum, West Point, N.Y.

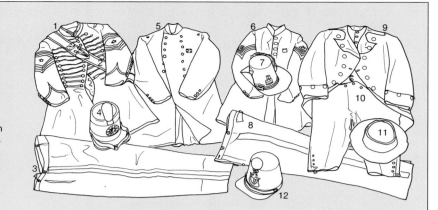

unopposed, brush aside the few Rebels he saw, and march directly for Manassas. At once he opened a tentative artillery fire, and then sent his skirmishers forward, men from Massachusetts dressed in gray militia uniforms. For the first time, though not the last along Bull Run, men found themselves confused by uniform colors. "Who *are* you?" the Yankees called out when they encountered gray-clad Confederate pickets. "Who are *you*?" came the confused reply. The confusion ended when a lieutenant from the Bay State replied, "Massachusetts men." An instant later a Rebel volley rang out, the lieutenant lay dead, and the unplanned and unwanted Battle of Blackburn's Ford began.[9]

It was a muddled affair. Following almost an hour of skirmishing, Tyler decided to send in more regiments, despite McDowell's order not to bring on a general engagement. Even though aides told him that he had done enough, that he had felt out the foe's strength and himself made a show of force, Tyler would not be dissuaded. He sent a whole brigade off toward the ford. Unfortunately, when finally he decided that he had gone too far and that the enemy was in too greater strength in front of him to push his way through to Manassas, he found his brigade too hotly engaged to extricate it right away. Some regiments broke and ran under the heavy Rebel fire, others stood their ground, pinned down by Confederate fire and unsupported. Others, out of the line of fire and left standing idle, broke ranks to pick blackberries. Only well into the late afternoon, when McDowell himself arrived and peremptorily ordered Tyler to disengage, did the so-called Battle of Blackburn's Ford conclude.

By standards soon to be established elsewhere along Bull Run, it was nothing more than a skirmish, Tyler's losses being a mere 83 killed, wounded, and missing. But it proved clearly to be a repulse, leaving Tyler's men dejected as they tramped back to Centreville for the night. By contrast, the Rebels, commanded chiefly by Longstreet and Early, lost only 70, most of them wounded. The only substantive damage that might have come out of the affair was when a Federal artillery projectile flew way over the

Above: The sleepy village of Centreville was first host to Beauregard's Confederates, expecting to hold it at all costs. Then McDowell advanced, and it was inundated by the flotsam of a beaten army.

Confederates and struck a house far in the rear, just when Beauregard and his staff were about to sit for a late midday meal, "very near destroying some of us", wrote an aide, "and our dinner spoiled."[10]

If Beauregard felt aggrieved over the loss of his lunch, still he experienced nothing but elation at the sight of Tyler's columns retreating toward Centreville that evening. And surely that was as nothing compared to his emotion the next morning when, at 6 a.m., Colonel Chisolm leaped from a panting horse and ran into his headquarters. He had found Johnston. After an exhausting 33-mile ride, Chisolm had reached Piedmont Station on the Manassas Gap Railroad, and there found the forward elements of Johnston's army. That general had achieved a signal success against the hopelessly timid and inept Patterson. After more than a month of feint and maneuver, Johnston issued orders to his men early on July 18 to prepare their rations and pack their baggage. Silently they simply marched away from the unsuspecting foe, moving first to Winchester, then turning eastward toward the Blue Ridge. Only when his army was well along in the day's march did Johnston issue an order informing them that they were on their way to join Beauregard at Bull Run.

Johnston himself rode ahead of the army, arriving first at Piedmont Station that evening. When he met Chisolm, he sent the colonel back into the night with the word that Johnston would start boarding his brigades as soon as they reached the station, and that his army should begin to arrive at Manassas Junction late on July 19 or the next morning. If McDowell did not initiate the battle before then, the concentration would be completed. Literally at the eleventh hour, it appeared that the Confederates might be able to make a revolutionary new use of modern transportation technology, and perhaps wrest a victory from what appeared until that moment to

be almost certain defeat. At the very least, Johnston's command merged with Beauregard's would give them almost equal odds.[11]

The first of Johnston's brigades arrived late in the afternoon on July 19. They were all Virginians, led by the oddest Virginian of them all, Brigadier General Thomas J. Jackson – hypochondriac, religious fanatic, disciplinary martinet, and, some thought, lunatic. He sucked on lemons, eschewed pepper, seldom laughed, and when he did merely opened his mouth without emitting a sound. In former days at Lexington's Virginia Military Institute, one aggrieved student had challenged him to a duel, and another tried to assassinate him. His men did not love him – yet.

Portions of Colonel Francis Bartow's Second Brigade arrived next, shortly after dawn July 20. A much-liked Georgian, Bartow had no real military experience, but represented the trend for prominent civilians who raised regiments of volunteers being given commissions to command. Soon after noon, another train arrived, this one bearing most of the Third Brigade, commanded by Brigadier General Bernard E. Bee of South Carolina, a career officer of high regard. Once more the train sped off to the west, this time picking up the Fourth Brigade, normally commanded by Colonel Arnold Elzey. However, Brigadier General Edmund Kirby Smith of Florida, commanding the Fifth Brigade, feared that the coming battle would be over before his command could reach Bull Run, and so he accompanied Elzey's brigade, and being senior, took command. Johnston, meanwhile, sent his artillery and his cavalry, commanded by Colonel James E.B. Stuart, overland toward Bull Run. Both arrived on July 20, to find that Richmond had also augmented Beauregard's command from other sources, sending scattered regiments and the Hampton Legion, which only arrived at 2 a.m. on July 21. When Elzey and Smith arrived later that morning, Johnston and Beauregard combined would total about 35,000, compared to the nearly 37,000 under McDowell's overall command. It was a wonderfully near thing, this concentration. Indeed, when the train bearing the Fourth Brigade and its two commanders finally chugged within earshot of

Manassas Junction, Smith and Elzey could already hear the sound firing in the distance toward Bull Run. The battle had begun.[12]

With a delicious irony, army commanders on both sides of Bull Run on July 20 planned almost exactly the same battle for the morrow. McDowell continued in his determination to send his main force off to his right, crossing at Sudley Ford, and then sweeping down on the Confederate left. This would uncover the stone bridge, allowing more Federals to cross in mass and then drive directly along the Warrenton turnpike to Gainesville, on the Manassas Gap Railroad. Unaware that Johnston was even then arriving at Manassas, McDowell believed that cutting the rail line at Gainesville would prevent troops from the Shenandoah from reinforcing Beauregard. That done, he could then turn on the Confederate army and drive it back from Manassas Junction itself. Some of McDowell's subordinates argued their belief that Johnston had already arrived. After all, they could hear the whistles of the locomotives arriving from Piedmont station, but their commander refused to countenance such thoughts. He had been assured that Patterson still had Johnston pinned down in the Valley.

In fact, of course, thanks to being senior in grade, Joseph E. Johnston was even then assuming command of all the Confederates massed on the other side of Bull Run. Arriving so late, and with no personal knowledge of the ground, he readily accepted a plan of attack presented to him by Beauregard, even though its planning was faulty, and the wording of the battle orders themselves was positively incomprehensible. It placed nine of their combined brigades on the right half of the 6-mile long line, from Mitchell's Ford down to Union Mills, but only one and one-half brigades were to hold the 3 miles to the left, including the stone bridge, reflecting Beauregard's unshakable conviction that the enemy would try to strike his center again as at Blackburn's Ford. But he intended to launch an attack himself, striking across at the enemy center with two brigades driving straight toward Centreville, while the bulk of his command swept around the Federal left to take his flank and cut him off from retreat to Washington. It was almost a mirror image of McDowell's strategy, which meant that the advantage, if any, could well go to the general who struck first.[13]

That was McDowell. At 2 a.m., July 21, while the Confederates still slept, he put his columns in motion. Miles would remain at Centreville. Hunter was to take the lead and find his way to Sudley Ford, cross, and push the foe back to uncover the next crossing near the stone bridge. This done, Heintzelman was to cross here. Then they would continue until they uncovered the stone bridge, where Tyler would cross after making demonstrations intended to pin down Rebels on the enemy left. The night march proved to be dreadful. Tyler slowed everyone as he moved forward, and did not reach the slope leading down to the stone bridge and Bull Run until 5 a.m. or later. Sporadic rifle and artillery fire broke out from either side of the stream, but the man commanding the spare brigade and one-half on the Confederate side soon guessed that this was only a demonstration, and that Tyler would not attack. For nearly two hours he remained unsure of the foe's intentions until a message reached him revealing that Yankees had been seen crossing Bull Run at Sudley's Ford. "Look out on your left", it said, "you are turned."

Above: Near here, at Sudley Ford, Hunter's division spearheaded the first Yankee crossing of Bull Run. Here, too, the routed Federals streamed back across in their panicked flight to Centreville.

Below: Arguably one of the ugliest generals serving in the Civil War, and certainly among the most unpopular amongst his own side, David Hunter had the misfortune to be among the first wounded in the battle, and on his birthday, at that.

That was Hunter. His had been a miserable march in the darkness, his men often having to hack their way through a virtual forest of trees and undergrowth on the so-called "road" they traveled to Sudley's. Worse, a local guide took Hunter on a "short cut" that in fact added three miles to the route. Consequently, it was 9 a.m. or later before the leading elements of Burnside's brigade finally soaked their cuffs in Bull Run and crossed unopposed to the Confederate side. Quickly they turned to their left and rushed toward the Warrenton road, barely more than a mile away, with nothing between them and a clear route to the railroad except a regiment from South Carolina, a battalion from Louisiana, and one incredible man.

Colonel Nathan G. Evans brought West Point and Old Army experience to his command of this half-brigade, but more to the point, he brought instincts as a rough-and-tumble brawler. His staff thought him the most accomplished braggart in the Confederacy. There was no question of his being among the most intemperate drinkers. He even kept a special orderly with him whose chief duty was to carry a small keg of whiskey and keep it near to hand at all times. But woe to the man who tangled with "Shanks" Evans, as his friends called him. Outflanked and outnumbered, and far from any supporting troops, him immediate instinct was to attack. When Tyler appeared in his front at the stone bridge, Evans quickly saw through the demonstration, and now with word of Hunter's approach, he immediately led most of his tiny command off to meet the threat, leaving a mere four companies concealed from view above the bridge to face Tyler's entire division, fewer than 400 men holding down almost 10,000.[14]

In the next couple of hours, Evans and his brave little command saved Beauregard and Johnston from disaster. Indeed, even as "Shanks" rushed his men and his keg toward Hunter, the Creole had almost entirely lost control of the battle even before it began. McDowell's siezing the initiative before him virtually negated his own offensive plans, and during the hours immediately after dawn and Tyler's appearance at the stone bridge, Beauregard repeatedly adopted and then abandoned a succession of extravagant new plans, sometimes so confused and contradictory that in one instance one of his orders literally directed Jones and his brigade to attack Ewell Johnston, meanwhile, showed a peculiar reluctance to interfere with his subordinate's conduct probably because of his own ignorance of the ground, as well as a fear of responsibility that he would manifest throughout the war ahead. Only after word of Hunter's crossing reached Beauregard did he finally abandon his own notion of attacking, and start rushing reinforcements toward Evans. He ordered Bee, Bartow, and Jackson to move without delay.[15]

Evans, of course, did not wait for help. He rushed his men to the crest of Matthews Hill, north of the Warrenton road, and there emplaced them on the forward crest. As soon as Hunter's leading regiments from Burnside's brigade emerged from the trees in the distance, Evans opened fire. Even though Hunter sent forward an advance to drive the Confederates away, Evans held his ground, soon deceiving the Yankees into thinking they faced far more than a few hundred Rebels and one resolute colonel. Before long, Hunter himself fell with a painful wound in his neck, the lead from an unknown Confederate being probably the first gift received on this, his birthday. Burnside assumed command of the

division, and soon believed that he faced two full brigades posted on Matthews Hill. He brought Hunter's other brigade, under Colonel Andrew Porter, up in support, but still Evans held them in check.

And then the cocky Confederate launched an attack of his own, sending the 1st Louisiana Battalion, later known as the Louisiana Tigers, straight toward the center of Burnside's line. They were the sweepings of New Orleans' gutters, men of a dozen nationalities, drunkards, brawlers, criminals, who waved their gruesome bowie knives in the air as they charged. Their numbers were too few to drive the Yankees back, but they stopped Burnside's advance in its track,

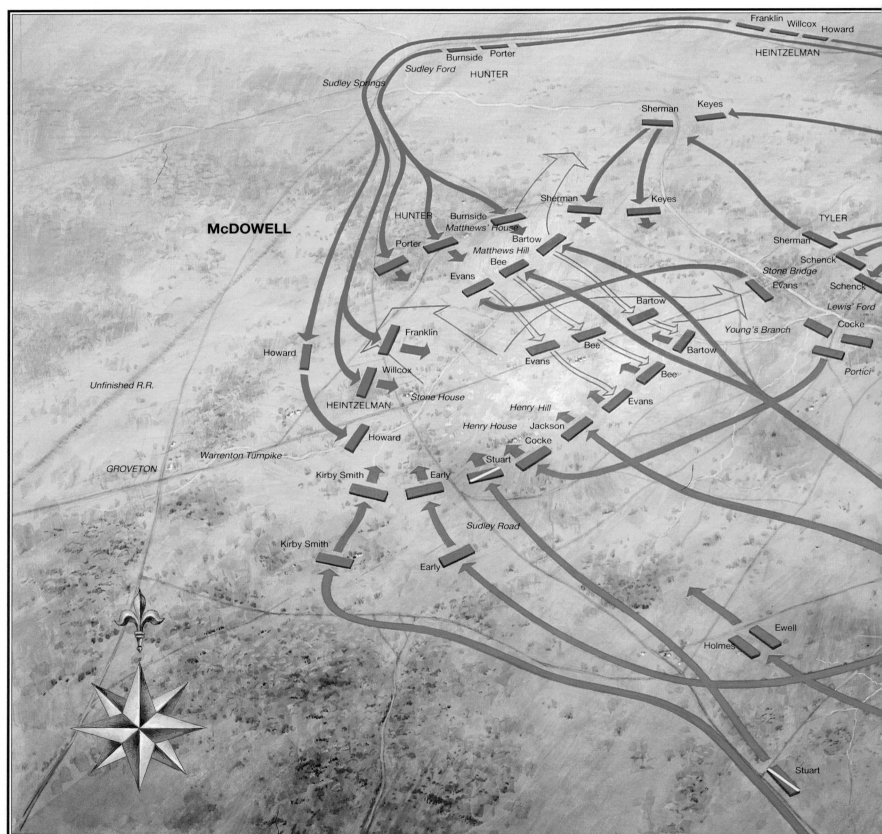

Battle of First Manassas (First Bull Run), July 21, 1861

No one had ever expected this lovely countryside to play host to warring armies. Moving out from Centreville, at upper right, McDowell's leading column under Hunter moved along the Warrenton Turnpike until turning off in a wide sweep upstream to cross Bull Run at Sudley Ford, being followed by Heintzelman. Meanwhile Tyler was to move straight along the turnpike to the Stone Bridge.

Beauregard, who was chiefly responsible for the Confederate dispositions, was taken much by surprise. Having expected an attack on his right, he had put the bulk of his forces in place to cover the lower fords, while his left at the Stone Bridge had nothing more than the half-brigade of Nathan Evans to cover it.

McDowell's plan worked wonderfully at first. Tyler did his job well, and Hunter crossed at Sudley without opposition. When the first Federal masses swept down from the ford, Evans was hard-pressed to hold them at Matthews' Hill, but hold them he did, even after Heintzelman's division started to come on the field. Meanwhile the Rebel high command reacted quickly and started sending reinforcements to the endangered left flank. Bee arrived, then Barlow and his brigade, and the Confederates delivered a stunning counterattack. Still, Yankee pressure forced them back from Matthews' Hill and across Young's Branch, toward Henry Hill. Now Jackson arrived, to become a "stone wall", and even more reinforcements finally stabilized the line, though charge and countercharge continued for two hours or more.

Steadily the lines grew longer, extending to the southwest as more and more units reached the scene. Tyler finally started to cross at the bridge and a ford just upstream, while the fight for Henry Hill became ever more desperate. Then accidents gave the battle to the Confederates. Howard's Yankee brigade came into line at the far Federal right and went into attack, just as new Rebel troops arrived at the right place and time to hit him in the flank. His brigade melted, and then when more Confederate regiments emerged on this crumbling Union flank, and Jeb Stuart's cavalry delivered a demoralizing charge, the Federal right simply fell apart, spreading panic all along the line. Running for their lives, the bluecoats raced back across the bridge and Sudley Ford, for Centreville, and eventually Washington.

McDowell lost 2,900 out of about 20,000 engaged: the Confederates 2,000 out of about 17,000 engaged. As well as the men, the North lost cannon and supplies, and the morale boost this first battle would have given them.

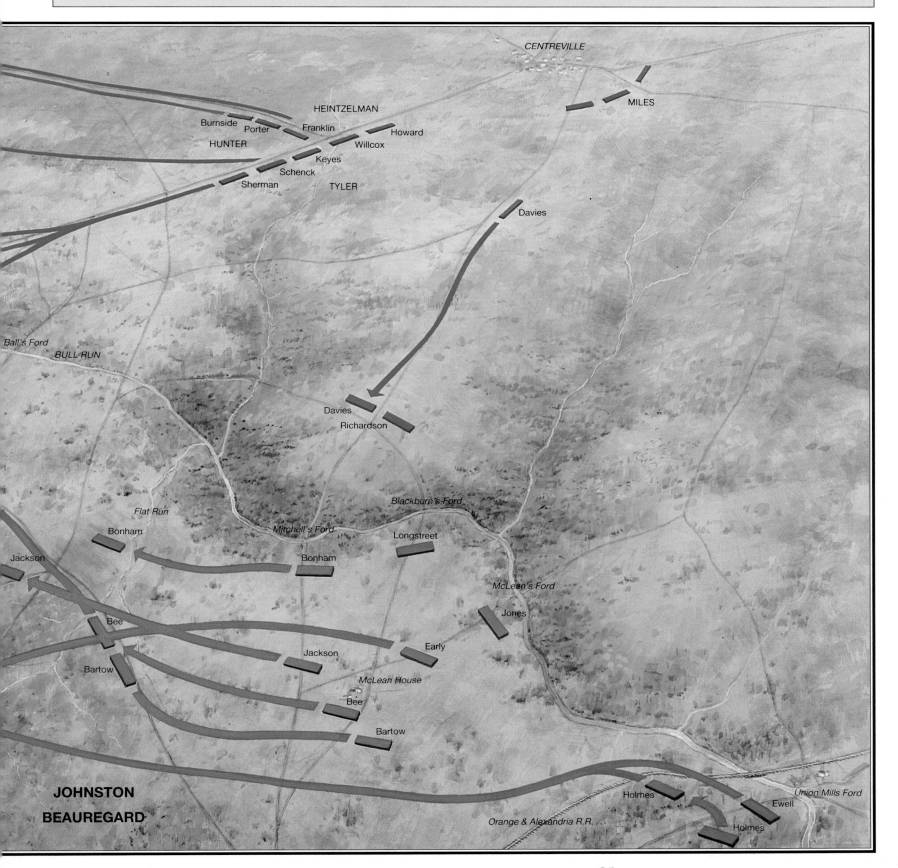

exchanging lives for time. Minutes later, as Evans was about to withdraw from the hill to avoid being overwhelmed by the lengthening enemy line that now extended far beyond both his flanks, he saw Bee's brigade rushing up from the rear, with Bartow not far behind. All their forces combined totaled not more than 4,500, barely half the number of Yankees now facing them, but Bee, now senior officer on the scene, immediately took his cue from Evans and prepared to attack once more. They must continue to gain time for Jackson and other reinforcements to arrive.

With Evans on the left, Bee in the center, and Bartow on the right, the Rebel line swept forward. Down the slope of the hill they sped, hurling themselves into Burnside's lines on the edge of a wood. Immediately the fighting became intense, though not hand to hand. Rather, the Confederates took cover in thickets or else simply stood out in the open, trading volleys with the Federals. None could say with certainty how long they stood the fire. Some said fifteen minutes, other thought it more like ninety. Whatever the time, its toll was terrible. First Evans was forced to withdraw, then Bartow, leaving Bee's men alone as the enemy stretched his lines and moved forward to overlap and envelop them. In some companies, more than a third of the men were hit. Officers, trying to inspire the men by example, exposed themselves heroically, and paid a dear cost. Finally Bee, too, had to withdraw to avoid being overwhelmed, and joined Bartow and Evans, already establishing a new defensive line on the south bank of Young's Branch, a stream several hundred yards in the rear, south of Matthews Hill, and roughly parallel to and in advance of the Warrenton pike. Within minutes the Yankees began to pursue.[16]

Suddenly, after a frustrating morning, everything seemed to be going McDowell's way. Despite losing Hunter early in the action, the Federal advance was going according to program, though behind schedule. Then, just as Bee's attack lost its force and the Confederates started to pull back, McDowell ordered Tyler to attack across Bull Run. Sherman had found an unguarded shallow spot above the bridge earlier that morning, and now he led his 3,400-man brigade across just in time to come up on the right flank of Bee's line and add impetus to the Confederate withdrawal. As Sherman moved his brigade to join with Burnside's and Heintzelman's, an elated McDowell rode along the lines waving his hat and joyfully shouting "Victory! Victory! The day is ours!" And so it appeared. Tyler was sending more men across after Sherman, Heintzelman's freshly arriving troops constantly streamed into the line, and Burnside's brigades, though badly mauled by the morning's fighting, still had strength. Half the Yankee army was across Bull Run and ready to hammer down Beauregard and Johnston's flank.[17]

The Confederate generals had not been idle all morning, yet neither had they really taken control of their destinies as yet. Beauregard continued to expect his brigades on his far right to cross the stream and move against Centreville, a threat that must certainly force McDowell to pull back his own attacking divisions. Only Beauregard had issued so many orders in the past 24 hours, and such conflicting ones, that he had forgotten what he said to whom. Some of his brigade commanders never even got their orders to advance. At last Johnston started to act decisively. After diplomatically suggesting re-

Below: General Joseph E. Johnston would prove to be one of the disappointments in the Southern command. Nevertheless, at First Manassas, after the surprise, he was the chief architect of victory.

peatedly that they should send everything available to their endangered left, he finally announced to Beauregard that "The battle is there. I am going." As Johnston rode off, Beauregard finally woke up to the state of affairs, perhaps because Johnston's own decision now signalled that he was assuming active command. Immediately the Creole began ordering most of

Below: The Stone House on the Warrenton Turnpike became a major landmark after the battle. While it did not figure prominently in the fighting, its wooden floors soon soaked up the blood from the wounded of both sides.

the rest of his unoccupied brigades to hasten to the left.[18]

By noon McDowell still appeared to hold the advantage. His line north of Young's Branch was poised, ready to continue its advance, and he still heavily outnumbered the Confederates, who had pulled back to the better ground on the northern slope of Henry Hill. More reinforcements sporadically reached them. Hampton arrived and helped stabilize the line when Bee and the others retreated south of Young's Branch. Then, even as the lines were reforming, he looked off to his left and saw a fresh brigade come into view on the crest of Henry Hill. It was Jackson. That morning he had not waited for orders from Beauregard. Hearing the increasingly heavy firing off to the left, he led his brigade in that direction on his own initiative. By 11:30 he watched his men tramp up the back slope of Henry Hill, but that was as far as he went. He had his men take cover by laying down, and resolved to await either further orders or the enemy's attack.

Bee, meanwhile, had retired with one of his regiments, the 4th Alabama. Though the musketry was hot, the Yankees had not resumed their advance as yet, so Bee rode over to see Jackson. "General, they are beating us back," he said. Jackson only replied that "we'll give them the bayonet." Apparently without further discussion, Bee rode to the 4th Alabama and asked if the men had it in them to follow him back into the fight. Jackson had said nothing about launching an assault, and made no preparations in his own command to do so. Either Bee interpreted the bayonet remark to mean that Jackson intended that they should charge, or else he felt disgusted that Jackson seemed inclined to stay put on the crest behind his bayonets, and therefore decided to attack again on his own.

In either case, before he led the Alabamians back toward Young's Branch and the Yankee line, Bee uttered some last words to his men. No one who heard them left an immediately contemporaneous record of what Bee said, but four days later a reporter with the army repeated what

someone present told him. "There is Jackson standing like a stone wall", cried Bee. "Let us determine to die here, and we will conquer. Follow me." Many have thought it a tribute to Jackson, yet at the time the Virginian was doing nothing, his men resting on their arms atop the hill in relative safety. A few believed otherwise, and a major on Jackson's own staff later claimed that Bee soon complained to him of Jackson standing still "like a stone wall" while his Alabamians and some of Bartow's Georgians went back into the fight.[19]

They did not conquer, as Bee had promised, but they did die there, being almost shattered by Yankee artillery. Bee himself took a mortal wound, and died within hours. Beauregard and Johnston reached the crest of Henry Hill just in time to see the survivors come streaming back up the forward slope. At once they agreed that the former would remain on the battle line to direct the placement of the troops, while the latter rode to the rear and, in his capacity as overall commander, saw to the coordination of the reinforcements rushing toward the battle-line, sending them where most needed. It was a good arrangement. Then, even as Johnston rode off, Beauregard sent Bartow and his remaining Georgians around to extend Jackson's left flank, and in moments a bullet killed the beloved colonel almost instantly.

In the next hour, more and more regiments and fragments of regiments came on the scene, and Beauregard used them to extend his flanks, both of which were still in danger of being overlapped. By 1 p.m. his line stood almost complete, with nine full regiments and bits of others in place, perhaps 8,000 men in all. He was still heavily outnumbered, but he had thirteen cannon on the field, too, and from their elevation on the hill they could do good work in repulsing a Yankee attack. They soon got their chance, for McDowell decided now to resume his offensive. Strangely, he chose to do it by sending two of his own batteries out in advance of his infantry, across Young's Branch, and up the forward slope

Below: A pre-war portrait of Professor Thomas J. Jackson of the Virginia Military Institute, gave little hint that this man was to become the mighty "Stonewall". They said his eyes burned with blue fire in battle.

of Henry Hill, almost under the guns of Jackson's infantry. It seemed an absurd move, exposing the artillerymen to a withering rifle fire, but McDowell no doubt trusted that the current and widespread fear of artillery fire would demoralize the Confederates before they could do damage to his cannoneers. He also promised to send forward infantry supports.

Below: The Robinson House sat northeast and to the right of the Henry House, and much of the fight raged around it as well, especially after Sherman crossed Bull Run and started to assault the Confederate right.

Nearly another hour passed before the eleven guns went forward, and the whole movement proved to be a shambles. The promised infantry support almost collapsed when it got just past the guns, and Jackson's line gave them a first volley. Then as the Yankees tried to rally, "Jeb" Stuart and his cavalry came on the field, by accident, at exactly the right time and place. They emerged from a wood beyond Beauregard's left, and one of the first things they saw was the exposed flank of the infantry support as it staggered under Jackson's volleys. Stuart immediately ordered a charge, and put much of the Federal infantry to rout. What remained was pushed back when the 33d Virginia of Jackson's brigade advanced without orders. The Yankees did not see where the Virginians came from. Further, this regiment wore blue uniforms, as many Confederate outfits did at the beginning of the war. Consequently, the commanders lost fatal time arguing among themselves about whether the advancing Virginians were the enemy, or another friendly support regiment. The Rebels' first volley settled the argument, but too late for the Federals to turn their guns against them. Almost all of the artillery horses were killed on the spot, and from the two batteries, only one cannon was successfully withdrawn. The Virginians swarmed over the rest and happily found them loaded, turning and firing them against the fleeing Yankees. Now Jackson ordered the rest of his brigade forward to pursue the retreat, and Beauregard sent other regiments into the advance as well, to capitalize on the enemy's repulse.[20]

From now until 4 p.m. or later, the battle raged on without a clear advantage to either side, McDowell benefitting from his superiority of numbers, and the Confederates from their good position atop Henry Hill. Charge followed upon charge. Colonel Wade Hampton went down with a wound, and the Federal cannon were retaken, then taken again, before they could be hauled out of the no-man's land between the armies. Both army commanders made the mistake of sending forward single regiments at a time, as if afraid to risk too much at a single assault. The result, of course, was that units were used up savagely by such piecemeal actions. Only Jackson made full brigade strength assaults, and they proved devastating. When McDowell sent his own men against Jackson in counter-assault, the Virginian fully earned his infant nickname of "Stonewall," and this time there could be no mistake that it was a tribute to the general and his men, who would soon thereafter be known to posterity as the Stonewall Brigade.

The continuous fighting on the hot July afternoon took a heavy toll on both sides, but as the Confederates stubbornly held their ground, and as more and more of McDowell's regiments were repulsed, the sight of Union soldiers retreating, and the disorganized flight of many of them back through the lines, exerted a demoralizing effect upon the fresh regiments still arriving, as well as on the men still in the lines. Ever so gradually it became McDowell's turn to lose control of the battle as, in seeming exasperation, he could think of little to do but keep sending forth more small assaults. Then Heintzelman fell with a painful, though not fatal, wound, and he had to leave the field. Just as he went to the rear, he met the fresh brigade of Colonel Oliver O. Howard, just arrived. At once McDowell ordered Howard off to the far right, to charge up Henry Hill immediately. The Confederates stopped first one of his regiments,

then another, and within minutes Howard's brigade was streaming back toward Young's Branch in near panic. Meanwhile, the Rebels had sufficiently mauled Sherman's brigade over on their own far right, that this portion of the line was now relatively secure. The arrival of more regiments sent forward by Johnston extended and stabilized the extreme left, and Jackson, of course, anchored the center.

Just as McDowell's strength waned, Beauregard's grew, and chance and coincidence now combined to give the Confederates even better news. Another train had arrived near Manassas Junction – Smith and Elzey at last. Still it took them three hours or more to march to the front.

Along the way, Smith encountered Johnston at his headquarters. Asking for orders, Smith was told simply to "Go where the fire is hottest." He led his men first toward Jackson's right, but then learned that it was the far left that needed support. Before he could move, Smith fell from his saddle with a Yankee bullet in his chest, and Elzey took command once more. He moved his brigade in a wide arc around to the left, moving in part through the cover of a wood in the left rear of the line. As a result, when he emerged, he discovered that he had gotten somewhat beyond the left flank, and in advance, and right there before him sat the exposed flank of Howard's already demoralized command.

He charged, and blasted into Howard with a force that carried him all the way to Young's Branch. Howard simply crumpled before him. Within minutes, Early's brigade, which had been marching almost all day, finally arrived and joined Elzey, along with some of Stuart's cavalry and a few field pieces. They all drove forward, and faced with these fresh brigades, Howard's men melted, running in panic for their lives, and spreading their demoralization to the other Federal units through which they passed. Seeing Early and Elzey advancing almost without resistance, Beauregard now ordered his whole line to charge the remainder of McDowell's line. The battle was won.[21]

Distinctive State Flags of the Confederacy

1 North Carolina State Flag, unit unknown. At a convention held at Raleigh, on May 20, 1861, North Carolina became the eleventh state to secede from the Union. Its state flag was adopted on June 22, 1861. This follows the pattern of the Texas flag in many respects, including the use of the lone star, a popular device among several states.

2 Virginia State Seal Flag, unit unknown. The Old Dominion, in convention, took its first vote to secede on April 17, 1861, its flag was adopted on April 20. The state seal was adopted during the Revolutionary War and features the figure of 'Liberty' and the Latin inscription, *Sic Semper Tyrannis*, meaning, "Ever Thus to Tyrants". Virginia state troops were first

called to the colors, to defend the state from the Federal authority, on the same day that the convention voted on secession.

3 Flag of the Florida Independent Blues, Company B, 3rd Regiment Florida Volunteer Infantry

4 South Carolina State Seal Flag of Company B, 5th Regiment South Carolina Volunteer Infantry

Artifacts courtesy of: The Museum of the Confederacy, Richmond, Va

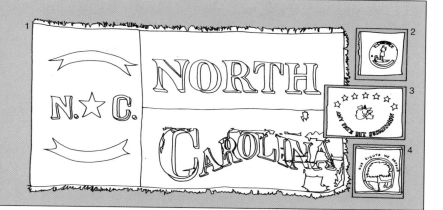

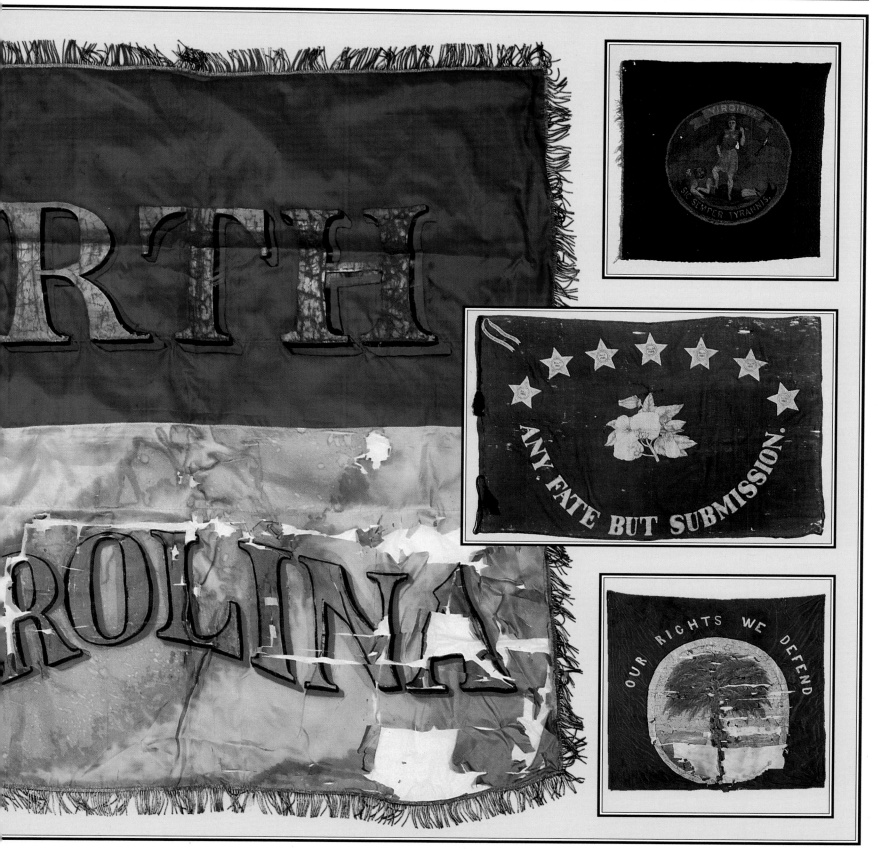

Private, 69th New York, U.S.A.

The fighting 69th New York achieved an enviable reputation during the progress of the war, and their exploits lived on in song and story for generations.

His uniform was very much the regulation Union blue, from his dark short jacket or tunic, to his sky blue trousers. This was largely an unornamented outfit, no stripes on the trousers, no piping or other color on the jacket, and all its leatherwork a simple black, from the belt to the shoulder strap, cartridge case and cap box, and even the black haversack. The only color about his uniform or personal equipment was the brass in his beltplate, strap badge, and the infantryman's horn insignia and unit number on his kepi.

The brightest thing about him may have been the burnished steel barrel of his Springfield percussion rifle and the polished blade of his bayonet, which, like most soldiers in this war, he rarely if ever used in combat.

In fact, McDowell's army already felt shaky before Howard arrived on the scene. Its commander, and most of his subordinates, were exhausted from their long night march, and then spent themselves in their ill-advised piecemeal assaults that sacrificed the initial advantage of numbers, all the while allowing Johnston and Beauregard to speed more reinforcements to Henry Hill to even the odds. Then when Howard retreated so spectacularly, the willpower of the rest simply gave way. Well before McDowell gave the order for a general retreat, the men had made the decision for themselves and a disorderly scramble for the rear commenced. McDowell did manage to get a few units to stand their ground as a rearguard to cover the withdrawal, but for the rest, there was little semblance of anything other than terror. "The retreat soon became a rout", McDowell would report, "and this soon degenerated still further into a panic." The Confederates sent Early and Stuart and a few other fresh units in pursuit, but they were for the most part too exhausted themselves to mount an effective chase. Only some of Beauregard's artillery made a significant contribution to the enemy rout when its shells upturned a wagon on the bridge over Cub Run, a stream some distance toward Centreville. The result further demoralized the Federals who needed to use that bridge to retreat, and all remaining semblance of order vanished. Intermixed now with the fleeing bluecoats were a good number of civilians from Washington, including several politicians, who had come along with the army to picnic during the battle and watch the fun as McDowell presumably walked right over the Rebel upstarts. They, too, scrambled for safety.[22]

It did not help that when McDowell finally reached Centreville that evening, he found Miles drunk. Immediately he relieved him of command, and personally assigned Miles' brigades and the few others not battered in the day's fight to establish a defensive line between Centreville and the victorious Confederates. But though Johnston and Beauregard tried to organize a substantial pursuit, urged on by President Davis who arrived on the field just as the fighting concluded, it simply could not be done. They had suffered at least 387 killed, and 1,582 wounded, with a few more missing, and all-told casualties probably reached 2,000, about twelve percent of the 17,000 or more who bore the brunt of the fighting. One regiment lost fully one-third of its numbers. But still they had the victory, and with it the spoils. McDowell's losses totaled 460 in killed, another 1,124 wounded, and a staggering 1,312 in missing, most of whom would soon be in Southern prisons. Moreover, in the wake of their hasty retreat, the Federals left cannon, wagons, supplies, and at least one Congressman.[23]

Following so soon after the victory at Fort Sumter, the Confederate triumph gave the Southern cause an indescribable boost of morale and confidence. "Joy ruled the hour", wrote a clerk in Richmond. Many believed that the war was already over, though more mature heads thought otherwise. The Rebels might have a great success to their credit, but those present knew how close a thing it had been most of the day, and that the Federals had fought every bit as well as the Confederates until circumstances turned against them. McDowell was no more personally responsible for the defeat than Johnston and Beauregard were for the victory. But for chance, they might

easily have reversed roles. Moreover, thinking men in the South knew that their foes were not just Northerners but Americans, for whom a lost battle had never been a lost war. The Yankees would come again, predicted one Richmonder, with "renewed preparations on a scale of greater magnitude than ever." On the other side of the lines, in McDowell's discouraged encampments, one Northern volunteer spoke for many when he wrote home that "I shall see the thing played out or die in the attempt."[24]

Above: It was over fields like these that the Battle of Bull Run raged. Simple quiet Virginia farmland became the war's first battlefield, and was never to be the same again.

Above: After the rout, when McDowell's army had gotten back to Washington and safety, men and officers posed for the camera almost proudly, relieved to be still alive.

This battery was not heavily engaged, staying on the north side of Bull Run during the fight. Called the Sherman Battery, it had struck fear into the hearts of many Confederates.

References

1 James B. Fry, *McDowell and Tyler in the Campaign of Bull Run* (New York, 1884), pp.7-9; Warren B. Hassler, *Commanders of the Army of the Potomac* (Baton Rouge, La., 1962), pp.3.

2 US War Department, *War of the Rebellion: Official Records of the Union and Confederate Armies* (Washington, 1880-1901), Series I, Volume 2, p.761; I, 51, Part 1, pp.411, 413-4 (hereinafter cited as *O.R.*)

3 US Committee on the Conduct of the War, *Report of the Joint Committee on the Conduct of the War* (Washington, 1863), Part 2, p.38.

4 *O.R.*, I, pt.2, pp.718-21.5.

5 *Ibid.*, pp.943-44.

6 *Ibid.*, pp.447-48; Alfred Roman, *The Military Operations of General Beauregard in the War Between the States* (New York, 1884), I, pp.77, 89; Rose Greenhow, *My Imprisonment and the First Year of Abolition Rule at Washington* (London, 1863), p.16.

7 *Report of the Joint Committee*, p.39.

8 *O.R.* I, 51, pt.2, p.177; A.R. Chisolm, Notes on Blackburn's Ford, Chisolm to George P. Smith, April 15, 1901, Alexander R. Chisolm Papers, New-York Historical Society.

9 Warren H. Cudworth, *History of the First Regiment Massachusetts Infantry* (Boston, 1866), p.42.

10 John C. Gregg to "Friend Heber," July 25, 1861, Peter Schmitt Collection, Western Michigan University, Kalamazoo; *O.R.*, I, 2, pp.306-307, 314; John L. Manning to his wife, July 18, 1861, Williams-Chesnut-Manning Papers, South Caroliniana Library, University of South Carolina, Columbia.

11 Chisolm, Notes on Blackburn's Ford.

12 William C. Davis, *Battle at Bull Run* (New York, 1977), pp.136-43.

13 *O.R.*, I, 2, pp.779-80; *Report of the Joint Committee*, p.207.

14 E.P. Alexander, *Military Memoirs of a Confederate* (New York, 1907), p.30; Thomas Pelot to Lalla Pelot, September 15, 1861, Lalla Pelot Papers, Duke University Library, Durham, North Carolina.

15 *O.R.*, I, 2, pp.489, 518.

16 *O.R.*, I, 2, pp.384, 390; Thomas Goldsby, "Report," M.J. Solomons Scrapbook, Duke University Library.

17 William Todd, *The Seventy-Ninth Highlanders* (Albany, N.Y., 1886), p.34.

18 Alexander, *Military Memoirs*, pp.32-34.

19 Charleston, *Mercury*, July 25, 1861.

20 *Report of the Joint Committee*, pp.169, 216; *O.R.*, I, 2, pp.394, 407, 495.

21 Bradley T. Johnson, "Memoir of the First Maryland Regiment," *Southern Historical Society Papers*, IX (1881), p.482.

22 *O.R.*, I, 2, p.320; Davis, *Bull Run*, p.239.

23 *O.R.*, I, 2, pp.568, 570, I, 51, pt. 1, pp.17-19; *Report of the Joint Committee*, p.41.

24 Davis, *Bull Run*, p.255.

CHAPTER TWO

SHILOH (PITTSBURG LANDING)

APRIL 6-7, 1862

The months following the Federal disaster at First Manassas amounted almost to a "phoney war" by later definitions, as almost nothing of consequence took place in the East for nearly nine months. But out in what they all called "the West", that vast area between the Appalachians and the Mississippi, there was a lot going on, and much of it exerted a material influence on the outcome of the war.

There were substantial – if confused and ill-managed – battles. At Wilson's Creek in Missouri both sides struggled for control of the Show-Me State in the fall of 1861, then in November newly commissioned Brigadier General U.S. Grant led his first little campaign at the other end of the state, leading to a fight at Belmont. In the months following, the Union cemented its hold on the upper Mississippi, and in late winter Grant moved against and took Forts Henry and Donelson, thus gaining control of the lower Tennessee and Cumberland Rivers. Other places with prosaic names like New Madrid and Island No. 10 fell to the Federals, all tending toward their progress down the Mississippi and into the heart of the Confederacy. Grant's taking of the forts, however, had been the most significant achievement. It severely disarrayed Confederate strategy for the whole region, forcing the Southern high command to respond if they had any hope of holding onto northern Mississippi, Tennessee, and any part of Kentucky. It called for great risks, and men willing to run great hazards for great gains.

WILLIAM T. SHERMAN produced probably more memorable quotations than any other general of the Civil War. "Hold the fort", he said, and if legend may be believed, perhaps he did years later declare that "war is hell." But certainly there is one memorable expression that he would have preferred everyone forget he ever uttered, and that was his confident assurance to his commander U.S. Grant on the morning of April 5, 1862, from his riverside position at Pittsburg Landing, Tennessee. "I have no doubt that nothing will occur today", he declared, and in words that would return to haunt him later, he added that, "I do not apprehend anything like an attack on our position." He was only half right. Nothing did happen that day. But the next morning he would be fighting for his life in the bloodiest battle ever yet seen in the hemisphere.[1]

An odd chain of circumstances brought a host of unusual men together that next April dawn. There were some who had trained for war much of their lives, yet never heard a shot fired. Others came with no training at all, yet considerable experience of fighting. As for the two men commanding the opposing armies, one had seemingly been a congenital failure at life before the war, and just a month before the coming battle had been removed from his command briefly for "neglect and inefficiency." Yet soon after the guns sounded he became one of his nation's first war heroes. The other brought with him a career of leaping from one achievement to the next, a reputation as probably the greatest soldier on the continent, yet from the moment of taking up arms in this contest he had disappointed almost everyone, and came to this field of action with demands for his removal ringing in his ears. The

Pittsburg Landing, Tennessee, shown shortly after the Battle of Shiloh, reveals nothing of the fury that swept right up to these river banks on April 6-7, 1862. The *Tigress*, at left, was U.S. Grant's headquarters boat in the campaign.

result of their meeting on this day and ground where Sherman expected nothing to happen decided the rest of their lives.

Direct, unsubtle, apparently guileless, Ulysses S. Grant presented the proverbial "face in the crowd." Other than his horsemanship, nothing about him before the war commanded notice or admiration. His lackluster performance at West Point, followed by a noteworthy but all-too-brief spate of glory in the Mexican War, repeated itself in a post-war peacetime service from which he resigned in 1854 under suspicion of more than the customary intemperance for frontier officers. He unwittingly summarized his own utter lack of sophistication when he reportedly commented on his musical tastes. He only knew two songs, he said. One was "Yankee Doodle" . . . and the other one wasn't. Only the desperation of Illinois' governor for experienced officers to train his new volunteer regiments in 1861, brought about Grant's return to uniform to organize and command the 21st Illinois. No one was more surprised than Grant when soon thereafter Washington created four new volunteer brigadiers for Illinois and, almost by default, one of the stars fell on his shoulder. Soon he attracted notice, however, with an unsuccessful attack on Belmont, Missouri, which, though it failed, still gained him wide notoriety in the Union. Soon thereafter he began urging that an attack be made on the center of the Confederates' long extended Western line, reaching from Cumberland Gap clear across Kentucky to Columbus on the Mississippi River.

Many in the North, including Grant's immediate superior, Major General Henry W. Halleck, and even President Lincoln himself, believed that the weak spot in that line lay where the Tennessee and Cumberland Rivers cut across it as they flowed north to south through western Kentucky and deep into Tennessee. It fell to Grant to prove them right, when he took Fort Henry on the Tennessee River and Fort Donelson on the Cumberland in February 1862. Doing so gave him access to both streams, providing pathways straight into the rear of the Rebel line. The foe had no choice but to abandon all of Kentucky and almost all of central and western Tennessee. It was a major triumph that, for a time, made Grant a hero eclipsing all others in the North. Indeed, so formidable was his popularity that when a jealous Halleck removed him from command just three weeks after Fort Donelson on erroneous charges, Lincoln himself intervened, and Halleck returned Grant to command of his army on March 17.

As vital as Forts Henry and Donelson were to keeping Grant the lifelong failure in command, so did their fall severely threaten the tenure of the man who came to this war riding such a crest of success and fame. General Albert Sidney Johnston won acclaim wherever he went, from an exemplary performance at West Point in the 1820s, to service in the Black Hawk War a decade later, before he resigned and went to Texas where he became a general and later secretary of war in the infant republic. Then came the Mexican War, in which he led a volunteer regiment with distinction, later a return to the Regular Army, and in 1855 he received a commission as colonel in command of the elite new 2nd US Cavalry, his subordinates including the likes of George H. Thomas and Robert E. Lee. By 1860 he was a brevet brigadier general in command of the Department of the Pacific, but when his adopted Texas

Above: Corinth, Mississippi, shown here later in 1862, was Sidney Johnston's staging point toward Pittsburg Landing. The Tishomingo Hotel beyond the tracks would become a landmark for thousands of soldiers.

Below: "Jeff Davis and the South" reads the colorful, if rustic, sign at this well-armed young Rebel's elbow. He and thousands more like him made up the Confederate army that fought the first great battle in the West.

seceded in 1861, he resigned his commission, at first resolved not to fight for either side.

Yet both sides wanted him, and in the end his Southern birth and sympathies won out. Friendship may have helped, too. Confederate President Jefferson Davis had been a cadet at the Military Academy at the same time as Johnston, though two years behind. For some reason Johnston, to whom all the young men looked up, adopted Davis as one of his friends. As a result, the younger cadet idolized his comrade with an unquestioning hero-worship so compelling that to the end of his life Davis could not or would not admit of any fault or failing in Johnston. Even when chosen president of the new Southern nation, Davis still stood in awe in his old friend,

and immediately sought to woo him into Confederate gray. Even without being able to communicate with Johnston, who was still on the west coast, Davis commissioned him a full grade general, in seniority second only to Adjutant General Samuel Cooper, who was too old to do active service. This meant that Johnston would be the senior field general in the entire Confederate Army, with everyone else – Joseph E. Johnston, Lee, Beauregard, and more – his subordinates. Yet still Davis heard nothing from Sidney Johnston, who was believed to be making his dangerous way clear across the continent. Early in September 1861, Davis lay ill in his sickroom on the second floor of the Executive Mansion in Richmond, his staff under orders not to disturb him. But one day he heard the muffled sounds of a caller downstairs, and a distinctly familiar boot tread in the entrance foyer. "That is Sidney Johnston's step", he cried joyfully. "Bring him up." Within days General Johnston was on his way west to take command of Department No. 2, embracing almost all of the Confederacy between the Appalachians and the Mississippi as well as parts of the territory further west.[2]

Then his troubles began. His command was too large, his resources too few, and his own apprehension of Yankee intentions let him down. When Grant moved in February, 1862, the forts on the rivers were unfinished and ill-prepared. Fort Henry fell without a fight, and though the struggle for Donelson was tough and sanguinary, still the outcome was almost never in doubt. Disasters elsewhere along the line in eastern Kentucky only added to Johnston's problems, and the loss of so much territory, especially the transportation and supply center at Nashville, was a severe emotional blow to the Confederacy. At once, many who had hailed Johnston as a genius before, now called for his censure, even replacement. Only the unwavering support of Davis kept him in command. Meanwhile, Beauregard, who continued to be the most popular Southern military hero, had made himself an increasing nuisance in Virginia after the victory at Bull Run. He fretted over his rank and started to feud with Davis over credit for the Manassas victory, all the while chaffing at being under Joseph E. Johnston's continued command. Even before the loss of the forts, Davis had reassigned Beauregard to serve as second-in-command to Sidney Johnston, thus ridding himself of a problem. After Henry and Donelson, Beauregard's prestige would actually work in favor of restoring confidence in Johnston and his army.

Even while retreating into northern Mississippi, Johnston himself was already planning a restoration of his own. There were two Yankee armies that posed threats to him. One was Grant's, and by far the smaller at about 30,000. The other was the Army of the Ohio, commanded by Major General Don C. Buell, numbered about 50,000. Originally posted in Kentucky, Buell moved forward to Nashville immediately after Johnston's withdrawal. Johnston immediately feared that the two armies would link to form one massive force against which he could hardly hope to contend, for even by the end of March after receiving heavy reinforcements, he still could not muster more than 45,000 of all arms. Indeed, this is exactly what Halleck hoped to do, but even after Washington officially expanded his command to include Buell, the Army of the Ohio's slow and recalcitrant general persistently dragged his feet and delayed. He feared that Johnston would try to get into his own rear if he moved too far from his base. Unspoken was the inner fear he almost certainly felt that if he joined with Grant, he would become merely a subordinate and that Grant, being senior major general by only a few days, would reap all future glory, of which some thought he had already taken more than his share.

Johnston's one hope of preventing any such combination (given his unawareness that Buell was his unwitting ally on this point) lay in meeting and defeating the separate forces in detail before they could merge. Once all of his scattered forces withdrew from Tennessee, he concentrated them in and around Corinth, Mississippi, so that by March 23 his army was nearly complete, and now he reorganized it into the newly designated Army of the Mississippi. It proved an odd and unbalanced organization, devised in fact by Beauregard, to whom Johnston delegated far too much responsibility, the same mistake made by another General Johnston during the Battle of Bull Run. Beauregard created four corps, the I Corps commanded by General Leonidas Polk.

Polk was everyone's mistake, from Davis on down. A West Pointer, he resigned immediately after graduation, and never did a day of active military service. Yet he, too, enjoyed an intimate friendship with Davis, having been a part of the Johnston clique at the Military Academy. He bore much of the responsibility for the failure to have the river forts properly completed, and for the rest of the war he would be a liability to a succes-

Captain, 11th Indiana Volunteer Infantry

The 11th Indiana Infantry had the distinction of attracting an inordinate amount of attention when it took the field in 1861. More than anything else, this was because of its uniform. It was a zouave pattern, but in a far from common gray. The jacket was trimmed in red, and obviously not intended for buttoning, since it had no buttons. Even though an infantry unit, the 11th's officers wore the red shoulder straps normally used by artillery, as does this captain.

A red sash and red piping on the trousers completed his attire, while on his head he wore a regulation kepi in gray, topped in red.

He wore buttoned gaiters to gather the cuffs of his baggy zouave pantaloons to his shoes or boots. Oddly enough, the 11th Indiana had two zouave uniforms during the war. This first one is excellently documented. The second version, issued at the end of 1861, is almost a complete mystery. The outfit was raised initially by Lew Wallace, later a general.

sion of commanders, especially the man leading the II Corps, General Braxton Bragg. He, at least, had battlefield experience from Mexico and elsewhere, though combined with one of the most irascible and unstable temperaments in the Old Army. Though not a favorite of Davis before the Civil War, he would become one very quickly. Inexplicably, while Polk's corps numbered 9,400, Bragg's totaled a staggering 16,200, and both seriously outnumbered the III Corps assigned to the best trained and skilled soldier of the lot, General William J. Hardee. He led barely more than 6,700, yet every professional soldier on the continent knew him at least by name thanks to his authorship of the favored drill manual then in

use by both sides, commonly called simply *Hardee's Tactics*. Finally Beauregard created a Reserve Corps numbering some 7,200, and give it first to General George B. Crittenden, who soon found himself relieved of command for drunkenness. In his place Johnston assigned Brigadier General John C. Breckinridge, a man with no battlefield experience at all. Yet he was a Kentuckian, and the Confederates needed to woo Kentucky into seceding and joining the Confederacy, and supporting Southern forces when again they entered the state, as Johnston confidently expected he would before long. Having been a senator and vice president of the United States just before the war, the youthful and immensely

popular Breckinridge brought political benefits that might outweigh his lack of experience.

All the while that Johnston and Beauregard reorganized and struggled to arm and equip the army, they watched Yankee movements in Tennessee, taking some heart from what appeared to be indecision and sloth. Grant sent occasional reconnaissances and probes out of his bases on the Tennessee at Savannah and Pittsburg Landing, but there was no further advance. Obviously he was waiting for Buell, and that general, though finally moving from Nashville, was doing so in such a dilatory fashion that it took him two weeks just to cross the Duck River, less than halfway between Nashville and Grant's divisions.

Union National and Regimental Flags

Federal regiments carried two colors; a national flag and a regimental flag. The safety of the colors was entrusted to the Color Guard of the Color Company. Inevitably, they were the focal point of hostile fire in battle

1 State national color purchased by officers of the 50th Pennsylvania Volunteer Infantry at their own expense, to replace their flag lost at the Battle of Second Manassas (Second Bull Run) August 29-30, 1862. This particular flag was later carried by the regiment during the Battle of Fredericksburg. The flag was later sent to Philadelphia to have battle honors painted on and so was not carried at Gettysburg. A new state color was issued to the regiment in fall, 1863,

and Colonel Hoffman who commanded the regiment sent this color home.

2 State Regimental Color presented to the 138th Pennsylvania Volunteer Infantry by the citizens of Bridgeport and Morristown, Pennsylvania, during Christmas 1864 when the regiment was stationed before Petersburg, during the siege of that city.

Artifacts courtesy of: The Civil War Library and Museum, Philadelphia, Pa

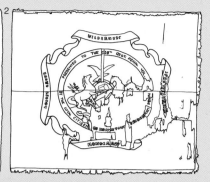

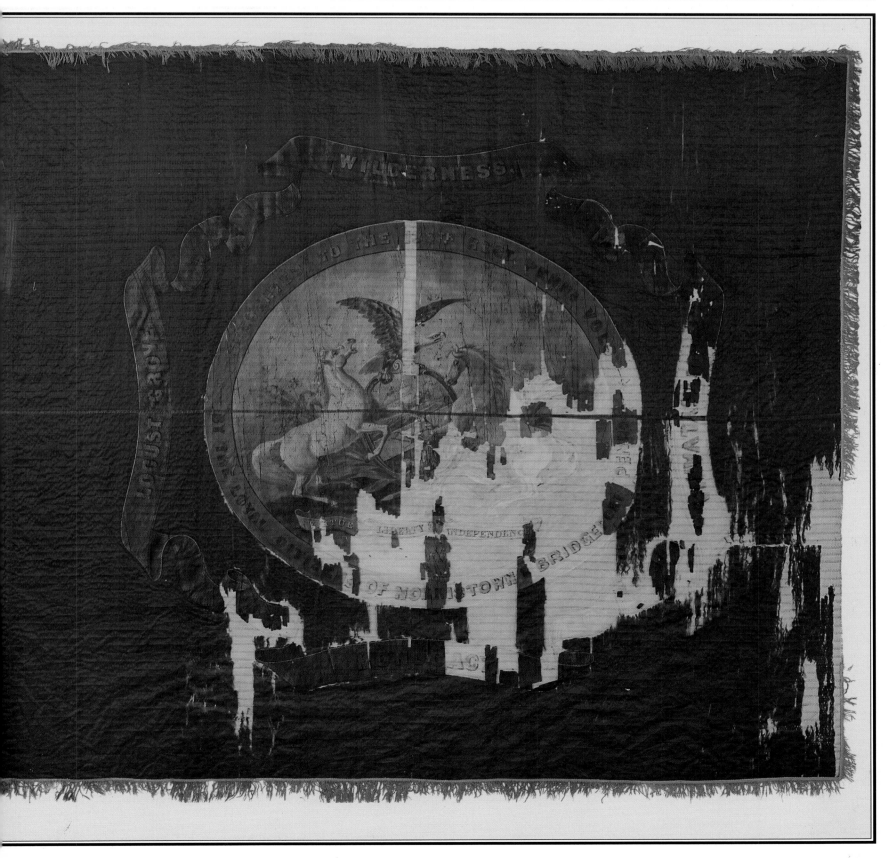

Above: Young Northern men in Illinois flock to their mustering point to go off to fight the war for the Union. These were the boys who held the Hornets' Nest and rallied, despite surprise to salvage a victory at Shiloh.

Within two or three days of his arrival at Corinth, Johnston decided to take advantage of the opportunity this presented by moving against Pittsburg Landing as soon as possible. On April 1 he put his troops on notice to prepare for active campaigning. The next day, receiving intelligence that Buell was across the Duck and moving toward Grant, Johnston resolved to delay no more. Orders went to the corps commanders to be ready to march at 3 a.m. on April 3. Pittsburg Landing lay a scant twenty-five miles from Corinth. Despite the recent heavy rains that had all rivers and streams swollen, and most roads turned into mires, the Confederates could hope to reach and strike the enemy in two or three days at the most.

Unfortunately it started to go wrong from the first. Confusion and badly worded orders resulted in the troops being a full twelve hours late in starting the march. Meanwhile, the peculiarly detached Johnston allowed Beauregard to devise the actual battle plan, as he had at Manassas, and just as happened there, the final drafts of the orders were complicated, contradictory, and ill-conceived. Units took wrong routes, others bogged down in the muddy tracks that passed for roads, and by nightfall on April 4 all of the corps were miles short of where Johnston expected them to be. Still that afternoon the commanding general issued attack orders for the next morning, but late that night a torrential storm blew through that soaked the army and the already dismal roads. All movement had to be postponed until the rain stopped. Johnston began to lose his customary calm, yet by late afternoon of April 5 not all of his troops were yet in position. The fear that Yankee scouts, encountered from time to time during the day, would have alerted Grant and ruined a hoped-for surprise attack tormented him. Moreover, some of his units, like Breckinridge and his Reserve Corps, had not been able to leave their original camps until that

Below: General Albert Sidney Johnston had the trust of Jefferson Davis and the entire South when he led his army to Shiloh. Yet his performance there, cut short by his death, left behind a host of questions about his skill.

morning, and had slogged more than twenty miles that day, arriving wet, muddy, and exhausted.

There could not have been a worse time for Beauregard and Bragg to lose their nerve. Late that afternoon, even while units were still arriving, those two generals concluded that the enemy must by now be aware of their presence. After all, Hardee's men had been in line since the previous evening a scant two miles from Federal camps around Pittsburg Landing. Incredibly, when Johnston joined the two generals, his second-in-command proposed that they abandon the offensive completely and return to Corinth. Their commander himself remained cool, supported by Polk and Breckinridge who expressed themselves ready to continue as planned. "Gentlemen, we shall attack at daylight tomorrow", he curtly announced, then left them.[3]

Almost within earshot of the uneasy Confederates as they tried to sleep that night, lay another army, incredibly still unsuspecting what dawn would bring. Five of Grant's six divisions made their camps here. It was Sherman who had selected the spot originally. Despite the loss at Bull Run, Sherman had performed well and won promotion to brigadier. Almost immediately Washington sent him west, where he soon commanded the Department of the Cumberland, only to lose the command when nervous exhaustion, seemingly erratic behavior, and newspaper sensationalism all combined to create a rumor that he had lost his mind. A much-needed rest at home restored his health, but his reputation as "crazy Sherman" would take longer to heal. Still, Grant instinctively saw something in him to trust, the beginning of one of the most successful commander-subordinate relationships in military history. Hoping to launch a speedy offensive against Johnston, Grant allowed Sherman to select Pittsburg Landing as a likely meeting place for the expected link-up with Buell, should he ever arrive. Meanwhile, it served as a good spot on the Tennessee to maintain supply and communications with their base at Cairo, Illinois.

Sherman had actually moved his own and the other divisions with him a mile or two inland from the landing, to the vicinity of Shiloh Church, and it was a sound move. The Tennessee here ran north-south for a stretch. Roughly parallel and about four miles to the west flowed Owl Creek. Thus the two streams afforded excellent protection to both of his flanks. In between Sherman posted his own division on the right, almost extending to the banks of the creek. On his left sat the division of Benjamin Prentiss, a new general who, like Breckinridge, owed his commission to political, not military, experience. His men were new and barely trained, and perhaps it was because of this, in part, that Sherman sandwiched them in the line between two of his own brigades on the right, and a third far to the left. Because of the wooded and often tangled terrain, and the route of the roads necessary for movement and communication, Prentiss and his two brigades stood at a slight angle in the Yankee line, and a bit in advance of the units on either side. Some distance behind Sherman camped the division commanded by another politico, Major General John McClernand, a ruthlessly ambitious wire-puller and consummate egotist already disliked by both Sherman and Grant. Off to McClernand's left, some distance in rear of Prentiss, sat the division of Brigadier General Stephen Hurlbut, yet another political appointee with no military experience. And completing the forces immediately present was Brigadier General W.H.L. Wallace's division. Characteristically, he, too, owed his stars to politics. Sherman had placed him far in the rear, and not far from Pittsburg Landing itself.

In fact, of the five generals commanding divisions here, only Sherman possessed either West Point training or practical experience in command, and even he never experienced action until Bull Run. Grant had one other division, commanded by yet another amateur, Brigadier General Lew Wallace, stationed a few miles down the Tennessee at Crump's Landing, and though he might have had good reason to feel a bit uneasy with such an overwhelming preponderance of amateurism in his high command, thanks to Lincoln's penchant for giving important positions to men of political influence, still Grant had confidence in Sherman and little or no apprehension that the position around Shiloh stood in any peril. As for his own plans, he was only waiting for Buell to arrive before launching a drive of his own against Corinth and Johnston. To be sure, reports of enemy activity on the roads south of Sherman did come in during the first days of April, but incredibly no one seems to have spotted more than Confederate outposts, or else no one credited that such contacts might presage a major enemy movement. Reportedly Lew Wallace received one report that Johnston's whole army was on the move, but if such news did reach him, when he passed it on to Grant it was not believed. Wallace himself later admitted that it was too incredible to believe that more than 40,000 Confederates could have gotten that close and not been spotted long before. And thus on the morning of April 5 Sherman could assure Grant that no danger existed of an attack, even as Hardee's men spent the entire day no more than two miles from Shiloh.[4]

It was shortly prior to dawn on April 6 when Federal cavalrymen out on patrol rather more than a mile south of Shiloh Church suddenly saw Confederate horsemen in the distance ahead of them. Shots were fired, and hastily they withdrew, but not before Mississippians from Hardee's corps exchanged volleys at long range with men of Prentiss' First Brigade. The battle had begun.

Johnston's plan for the battle came almost entirely from Beauregard, superseding one of his own that he abandoned sometime on April 5. Instead of advancing with his corps lined up abreast of one another, with Breckinridge held in reserve, Johnston would now launch his attack with the corps lined up one behind another along the entire front, a foolish plan that sacrificed all the advantages of the numbers Johnston had concentrated for the attack, just as the Creole's battle plan for Bull Run would have done. Only McDowell taking the initiative first prevented Beauregard's plans from going into motion in that battle, thereby probably saving the bombastic general from his own folly. Here at Shiloh, however, the initiative lay with the Rebels, and for the rest of the day they would pay for Sidney Johnston's unwise reliance upon Beauregard. Indeed, almost Johnston's only amendment to the plan was an order that the attacks should concentrate on the Federal left, and there was sound thinking in that. Johnston hoped thereby to drive Sherman away from the Tennessee, separating him from any hope of linking with the advancing Buell, while at the same time driving him back against the rain-swollen Owl Creek where he would be trapped and forced to yield.

The skirmishing continued for more than half an hour and grew in intensity, yet for the time being no one in the Federal camps around Shiloh Church seemed to appreciate what faced them. Many continued to believe that this was nothing more than a reconnaissance, albeit a large one. But then Prentiss' advance parties saw something that made their hearts stop. Suddenly from out of the wood in the distance they saw an endless line of gray and butternut clad soldiers, with a seeming sea of fluttering red and white flags waving above them. It was Hardee's corps, augmented by a brigade from Bragg, and nearly 9,000 strong. Moving like what Beauregard called "an Alpine avalanche", they surged forward shortly before 7:00 a.m. Some distance in their rear, General Johnston heard the sounds of increased firing,

Above: Transports on the Tennessee River tie up on the bank at Pittsburg Landing. It was command and control of the all-important rivers that led the armies of both sides to fight at Shiloh in the first place.

Below: William T. Sherman, shown here as a major general late in the war, was an obscure brigadier still living under a cloud of ugly rumour when he got his chance of command under Grant at Shiloh.

swallowed the last of his morning coffee, and mounted his steed "Fire Eater". Wheeling in his saddle before dashing off to the front, he confidantly told his staff that "Tonight we will water our horses in the Tennessee River."[5]

Hastily, even as realization of what faced him slowly dawned, Prentiss roused more and more of his units from their morning meals and rushed them toward the sounds of the firing, though he had not yet himself seen the enemy. In fact, subordinates actually gave what little management there was on the firing line, and made the actual dispositions of the troops that the general sent forward. With Hardee rapidly overlapping both flanks of the heavily outnumbered advance, the Yankees worked to set up a stiffer line of defense on a crest overlooking a wooded ravine several hundred yards behind the firing. By about 7:30

two regiments stood atop the crest when they saw swarming over the hilltop on the opposite side of the ravine seemingly uncountable thousands of the enemy. At once they fired the first volley, briefly halting the foe, and then the two sides commenced trading blast after blast at each other across the tree-tops between them.

A few Confederate units broke and ran at this, their first taste of fire, and the confusion they caused slowed Hardee's advance for precious minutes. But soon after 8 a.m. the Southerners recovered and pushed forward in an assault, bayonets glistening at the muzzles of their rifles. They almost overwhelmed the Yankees before Prentiss' men pulled back in some confusion, withdrawing right into their camps. Here the general hoped to establish a better defense, and fresh regiments awaited the coming of the foe. But then they saw Rebels advancing through the woods in their front, while they could hear the growing sound of advancing enemy fire on their right. Fire briefly stalled the Rebel juggernaut, but not for long. Making matters worse, some of the Yankees took their places in line woefully unprepared. In the rush that morning, the 15th Michigan actually marched off without any ammunition for its rifles, an oversight only discovered now that they stood in line with the enemy advancing. "We stood at order arms and looked at them as they shot", one enlisted man recalled. There was nothing else they could do.[6]

Still Prentiss' resistance did credit to his division, and they inflicted some grievous casualties on the foe, including Brigadier General Addley Gladden, leading Bragg's single brigade attached to Hardee. A Yankee artillery shell practically ripped off his left arm. He would die within a week. The loss of Gladden and the by now heavy fire coming from Prentiss succeeded in slowing Hardee's advance. At least an hour before the general sent word to Hurlbut, advising that this was no mere reconnaissance, and pleading for assistance, but none was yet forthcoming. Meanwhile, more and more of his inexperienced volunteers were either falling to Rebel bullets or losing their nerve and racing for Pittsburg Landing. A situation that started out bad was rapidly becoming critical. One of his brigades, the first to be engaged that morning, lay dispersed and battered, with only isolated pockets of men now holding out in and around their camps, and then a Rebel bullet killed its commander. The line collapsed and the remnant of his men abandoned their camps and raced for the rear. Prentiss once more ordered his remaining troops to fall back in hopes of linking with reinforcements, should they ever come. A renewed enemy attack put one of his regiments completely to rout. From his entire division the general now had barely two regiments that had retained any organization, and almost his entire command was in retreat toward the Tennessee. Johnston might indeed be drinking from the river before nightfall.

Over on Prentiss' right, Sherman, too, had been taken by surprise, though Hardee struck Prentiss first and the sounds of firing there gave the others at least some warning. Still as late as 7 a.m. some of his regiments remained around their breakfast fires, even though they could hear the sound of guns to their left. Uncertain of the extent of the Rebel advance, Sherman himself rode to his front to see what he could, narrowly missing death when an enemy skirmishing party suddenly appeared not fifty yards distant and sent a volley at him that killed men around him and slightly injured his hand. He had seen enough, and hastily rode back to rush forward reinforcements. The battle he thought so unlikely had come to find him.

Thanks to a better position, a little more warning, and at least some previous experience at Fort Donelson, Sherman's division reacted more quickly and successfully than Prentiss'. Despite the momentum of Hardee's attack in their front, these Federals managed first to slow him substantially, and then to stop him cold, sending some of his own regiments back in disorder and confusion. Moreover, Sherman outnumbered the single brigade of Rebels facing him, led by General Patrick R. Cleburne, and now held him in check as Cleburne impatiently awaited support from the second Confederate wave, Bragg's corps, more than a mile to the rear, and probably at least half an hour from arrival. The folly of Beauregard's battle plan was already becoming apparent as it launched the attack with its smallest corps, with real strength and support too far behind to capitalize on an advantage.

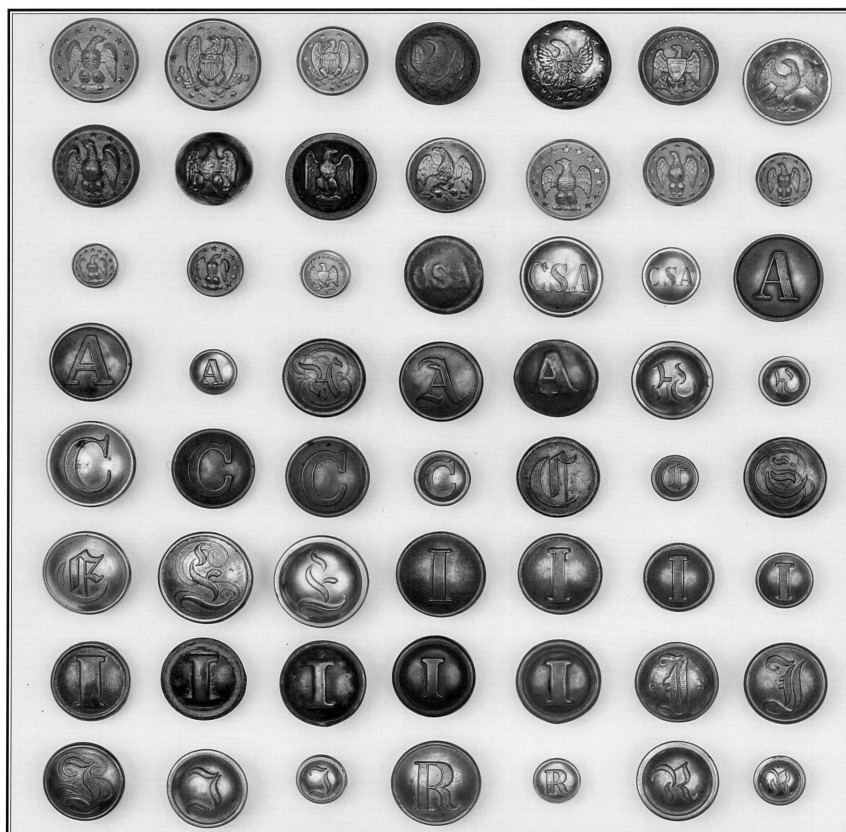

Confederate General Service, Naval and State Buttons

1–17   Staff officer's and officer's buttons. Most English imports
18–20   Enlisted man's coat and vest buttons
21–28   Artillery, various
29–35   Artillery, various
39–52   Artillery, various
53–56   Artillery, various
57–65   Naval officer's uniform buttons. English imports
66   Confederate Navy enlisted man's hard rubber coat button

55   Alabama Volunteer Corps coat button
68–69   Alabama state seal coat and cuff button
55   Arkansas officer's
55   Florida state seal
55   Florida Cherokee Rose
73–75   Georgia state seal coat and cuff buttons
76–77   Kentucky state seal
78–82   Louisiana state seal button variations
83–84   Maryland state seal coat and cuff buttons

55   Mississippi state seal button
86–91   Mississippi star buttons
92–95   N.C. state seal buttons
96–97   N.C. starburst variants
98–102   S.C. state seal buttons
103   Tennessee state seal button
104–106   Texas buttons
107–112   Virginia state seal buttons

Artifacts courtesy of: Virginia Historical Society, Richmond, Va

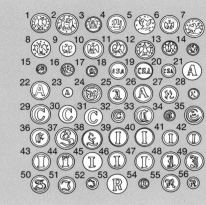

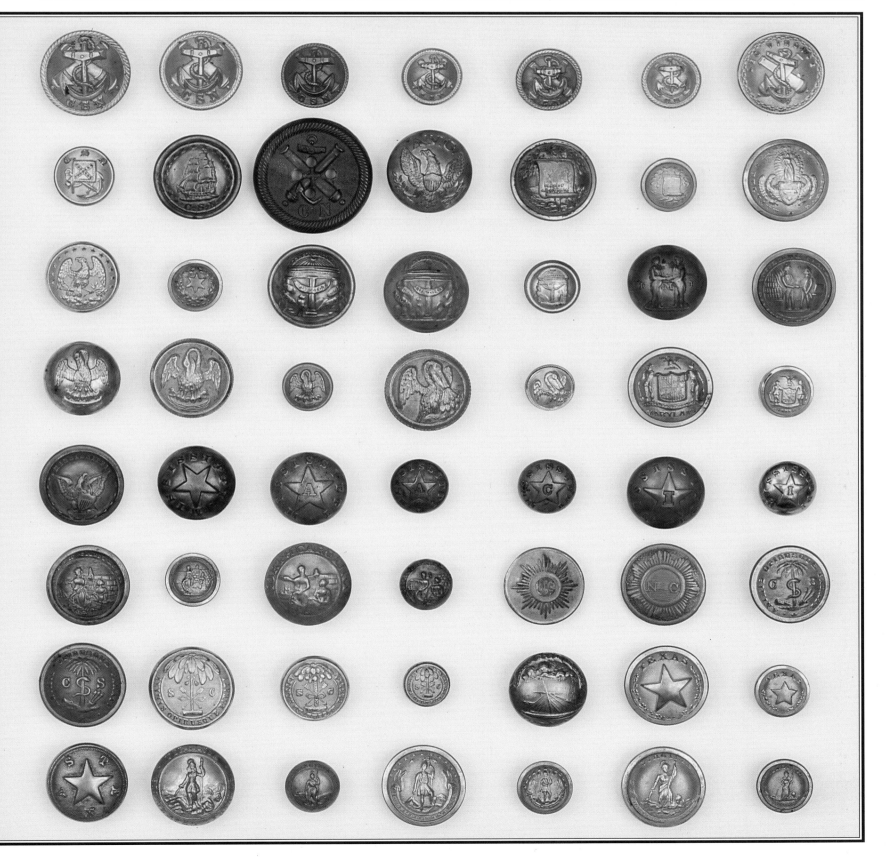

At this very moment, General Bragg was grumbling to himself about Beauregard's "faulty arrangements" and "blunderings". Having his corps spread out across a front several miles wide forced some of his units to march across country rather than over good roads, slowing them and disturbing the integrity of his line. As a result, when after considerable delay he finally got elements of his line into action to support, it was only a single brigade that reached the right spot, and it was not enough. Worse, much of the ground before them worked to Sherman's advantage, being so damp and marshy that progress through it proved agonizingly laborious.[7]

For the next two hours the intensity of the combat on Sherman's front steadily increased as more and more of Bragg's units finally reached the battleline. Gradually the Confederates extended their flanks on either side. Prentiss' withdrawal left Sherman's left terribly vulnerable, and soon some of Hardee's units turned to their own left to join with Bragg in driving against Sherman. Meanwhile, as several brigades struggled along the center of the lines, one of Bragg's units had a clear path to get around Sherman's right as well. Only its dilatory advance prevented an opportunity to get into the Yankee rear and perhaps overwhelm Sherman. Even then, as more and more of his units came into the fight only to be sent reeling in confusion from well-placed Rebel artillery fire and the incredible show of force, Sherman began to see that he could not control the battle in his front. His own performance was exemplary, however. Under the heaviest fire, with repeated near escapes from enemy bullets, he remained calm and inspiring, chewing a cigar and showing himself seemingly everywhere to try to enspirit the men.

It proved to be of no avail. Before very long his own camps lay in imminent danger of being overrun. First his left gave way entirely under the combined pressure of Hardee and Bragg, and this only further jeopardized his right, though it held on stubbornly, sustained by the fortitude of the men in the ranks, and the delays and mistakes plaguing the foe in that sector. It was all buying vital time, for Sherman had earlier sent urgent messages to McClernand, Hurlbut, and W.H.L. Wallace, advising them that this was no reconnaissance in their front, but a full-scale Confederate attack that endangered their position, Pittsburg Landing, and the whole army.

The first reinforcements to reach Sherman were Colonel Julius Raith's brigade from McClernand, though they arrived just in time to become a portion of the left flank that Hardee and Bragg chewed up and crumpled back. Shortly after 10:00 Sherman had been forced to withdraw from the area of Shiloh Church. An hour later the whole line was clearly a shambles. Sherman himself had two horses killed under him in the heavy fire, and was now riding a commandeered artillery horse, only to have it shot down within a few minutes. His men could hold no more. Rallying those whom he could inspire to stand firm, Sherman saw the rest of his division melt into the woods to the rear. Now it would be McClernand's turn.[8]

By now Grant had reached the scene and was trying to ascertain just what had happened and how bad the damage was. He first learned of the attack around 7:30 when a messenger arrived at his headquarters. Stepping outside, the general could hear for himself the sound of distant artillery, and though uncertain as yet of the extent of

Above: In a sentimental age, any excuse to romanticize was seized. Little Johnny Clem *may* have acted bravely in the battle; later legend turned him into "the drummer boy of Shiloh."

the engagement, he took it for granted that this was a serious attack. He sent word off to Buell to rush toward Pittsburg Landing. Then he boarded the transport *Tigress* and steamed upriver, stopping first at Crump's Landing to meet with Lew Wallace and order him to have his division ready to move in any direction ordered whenever Grant sent further instructions. Then the *Tigress* steamed off again, reaching Pittsburg Landing around 9 o'clock. As soon as he got there and spoke with men on the scene, he realized that the Confederates intended to contest his hold on the landing itself, and he immediately sent word back downriver to Wallace to come on right away, though he did not specify what road he should take from Crump's Landing.[9]

Immediately Grant rode forward to the scene of action to meet with Sherman and McClernand, and there he saw with his own eyes the deplorable state of affairs. He had been forewarned by the sight of hundreds of demoralized soldiers huddled under the sheltering banks of the Tennessee when he reached the landing. Now at the front he saw even worse. Never an excitable man in the face of the enemy, Granted stayed calm now, though in estimating Johnston's forces at 100,000 or more, and in sending repeated appeals to Buell to rush to join with him, he betrayed the anxiety he felt for his army. He knew he was in serious trouble.

Fortunately, a gathering pocket of stiff Federal resistance began to cluster in the vicinity of a peach orchard well to the left, where parts of General John McArthur's brigade of W.H.L. Wallace's division had come forward. Meanwhile, off to their left the brigade of Brigadier General David Stuart had come up. In fact, when first spotted, Stuart's command was mistaken for an attacking party of Yankees seeking to turn the Confederate right, for here, again, Beauregard's plan worked against them. Even without the disorganization caused by the rough terrain and the

morning's fighting, Hardee's, and then Bragg's lines simply could not stretch themselves sufficiently to cover all the ground between Owl Creek and the Tennessee. As a result, and thanks to the unfolding of the fighting on their left against Sherman, the Southerners' right lay "in the air", largely unprotected. As they advanced, they encountered Stuart in the distance, and mistaking his purely defensive posture as a threat, halted their advance in that sector. Word went back to Johnston, who now saw his plan to drive the enemy away from Pittsburg Landing seriously jeopardized. He was pushing the Yankees back, to be sure, but if he continued as he was, he would wind up driving them back to the landing instead of away from it. Now this appearance of Stuart only emphasized the need to shift the pressure of his attack from the left and center to his right. With Hardee and Bragg already fully committed, and with Polk's corps inadvertently wandering piecemeal into places in the line without directions, he had only Breckinridge's Reserve Corps available. At once Johnston ordered the Kentuckian to take his two brigades to their right. They must dislodge Stuart and any other resistance near that peach orchard, and do it quickly, so that the balance of the attack could be pressed as planned.[10]

Even before Breckinridge arrived, another brigade struck Stuart and put all but a few hundred of his men to flight, thus clearing the Yankee left flank and opening what could be a clear path along the Tennessee to the landing. But then the Confederates who routed Stuart failed to press forward; they were out of ammunition. This plus McArthur's timely arrival brought precious stability to the Yankee left at a crucial moment. Yet the few men that Stuart rallied, added to McArthur, posed no match for any serious renewal of the Rebel pressure. There could not have been a better time for Hurlbut and two of his brigades to come rushing forward at last. Hurlbut brought them straight forward, past McArthur, through the peach trees and slightly beyond. Here he posted one brigade facing south, and the other on its right facing slightly west. McArthur took position on Hurlbut's left, and somewhat behind him, while recently arrived units of W.H.L. Wallace's prolonged Hurlbut's right. He may not have intended it, but Hurlbut's two brigades formed a considerable salient projecting dangerously in advance of the rest of the newly established line. His exposed position soon became evident, and before long he pulled his men back to the orchard itself, and the cover of the trees and some fences. Attacking Confederates would have to come at them over open ground with little or no cover.

Confusion continued to hamper Johnston's efforts to press through this shaky portion of the Federal line. The two brigades sent to work here while he waited for Breckinridge to arrive seemed unable to capitalize upon any of the several advantages offered them. Worse, being the last corps in the advance to the fight, Breckinridge and his men were far in the rear, and could not reach the field much before noon, if then. Johnston chaffed at the delay and inactivity, all of which gave Hurlbut vital time to consolidate his position. Better yet, as men became accustomed to the sound of rifles and artillery, the panic left some of Prentiss' men who had fled earlier in the day. He rallied portions of four of his regiments, and these plus bits and pieces of other units almost added up to half a brigade, which he put in

Above: Gallant young Tennesseeans like these members of Captain A.M. Rutledge's battery posing for the camera in Nashville on July 4, 1861, were the heart and soul of Johnston's Confederate army.

Below: The martial pose was all the rage North and South, yet the Rebel boys seemed somehow more prone to contrast their innocent young faces with a small arsenal of weapons. The knives were never used.

position immediately to Hurlbut's right in a sunken road whose bank afforded excellent defenses.

No one on either side in the battle could have foretold that this peaceful orchard and rustic road were about to become the scene of the most intense fighting ever seen. Near about noon, with Breckinridge not yet on the scene and the rest of the Confederate corps hopelessly intermingled and confused, Bragg simply told Hardee to take command of the left, Polk the center, and he would direct affairs on the right. One of his first acts was to send a fresh brigade forward toward the sunken road, not knowing what awaited them. The first volleys from Prentiss were devastating, inaugurating what would be more than five hours of vicious combat. Immediately after that first repulse Grant himself arrived at the road, took in the vital importance of the position, and made it clear to Prentiss that the fate of the army could depend on his holding out "at all hazards". Prentiss would do his best.[11]

In a short while he did better than that. Aided by his position, and the fact that the curve of the road allowed for a concentration of fire, his men and their artillery shattered the Confederate attackers. By 3 p.m., at least four successive assaults against the position had suffered a bloody repulse, and perhaps already Rebs and Yanks alike thought they could hear in the constant buzzing of the bullets a distinct similarity to the sound of hornets. Ever after that small piece of ground so vital to Grant would be known as the Hornets' Nest.

By this time the advance elements of Breckinridge's corps finally arrived on the scene, just as broken by the rough terrain as the units that preceded them. Their commander immediately began sending them into the assaults aimed at pushing back Prentiss and Hurlbut, and he, too, found himself unable to break through. Johnston himself rode to the peach orchard early in the

afternoon to see firsthand the resistance that was stalling his advance, and one point a frustrated Breckinridge rode up to him with the complaint that he could not get one of his Tennessee regiments to fight. Governor Isham Harris of Tennessee happened to be with Johnston, heard the remark, and himself went to rally his native sons. That done, Breckinridge tried to get his Third Brigade to go forward in a bayonet charge personally ordered by Johnston. The men balked. They had seen too much death already in the Hornets' Nest.

Again frustrated, Breckinridge came to Johnston, and the commanding general returned with him. His presence and calm and collected manner soon fortified the wavering soldiers' resolve, and

with Johnston, Breckinridge, and Harris, in the lead, the brigade advanced with a yell at about 2 p.m. They almost devastated the defenders, pushing back Hurlbut's left dangerously, taking much of the ground near the peach orchard, and only stopping when they came up against stouter defenses at the northern edge of the orchard. The hail of fire had been terrific. At least two spent bullets struck Johnston, causing nothing more than bruises. A third hit the sole of his left boot, narrowly missing his foot. Johnston was rather more elated than concerned about the danger, and that plus the excitement of the assault that had definitely gained good ground probably made him entirely oblivious to a fourth rifle bullet, one that did not harmlessly bound off or tickle his foot. This one came from behind, quite possibly fired by a Confederate whose aim was wild in the confusion, and it buried itself in his right leg behind the knee, nearly severing the popliteal artery. Profuse bleeding commenced immediately, yet Johnston seemingly felt nothing, and the blood filled his boot rather than flowing out where someone might have seen it. Only when Johnston suddenly swooned, went pale, and started to fall from his saddle did Harris become alarmed.

"General, are you wounded?" he cried.

Johnston was just conscious enough to say, "Yes, and I fear seriously."

They may have been his last words. Harris quickly got him to the rear, but within scant minutes the general was completely unresponsive. As others gathered around, they got a bit of brandy down his throat, but still no one had discovered the wound. If they had, a simple tourniquet above the knee might have saved his life. Johnston had such a tourniquet in his personal effects, in fact. Instead, the bleeding continued, and by about 2:30 he was dead, the first army commander in American history to die in combat, and the only one the Confederacy would so lose. Even while his friends and staff cared for the body and tried to conceal the disaster from the trooops nearby, a messenger rode to notify Beauregard that he commanded the army now. The battle must go on.[12]

It took time to find Beauregard, and when Bragg and other Rebel commanders learned the news, at first they did not know what to do, causing a costly delay in pushing the advance and giving Grant more time to dig in. But it proved to be time gained at the expense of Prentiss and Hurlbut. Even before orders came from Beauregard, Breckinridge on the right and left, and other units in the center, moved on the Hornets' Nest once more, hammering it until well after 5 p.m. Both of the Federal flanks finally gave way, and by about 5:30 the brave defenders were literally surrounded. With Sherman having been driven back much earlier, W.H.L. Wallace, on Prentiss' right, was also in danger of being surrounded. Then a bullet passed through Wallace's head, killing him before he fell from his saddle. Minutes later isolated pockets of cut-off Federals began to surrender. Prentiss and Hurlbut had about 2,000 men left, facing several times their number encircling them. Finally those who could tried to escape the noose spreading around them. Prentiss could not, and was soon seen waving a white flag, though Hurlbut got away and was soon at Pittsburg Landing itself, where Grant assigned him the task of reforming the fugitives there for what might have to be a last ditch stand on the river bank.

Grant could not yet know it, but the succession of disasters during the day was actually working in his favor in a perverse way. The Confederates were exhausting themselves in their uncoordinated assaults, and many units were hopelessly scrambled. Moreover, in pushing the Federals back so rapidly in places, the Rebels had moved faster than their supplies of ammunition and water. Thus, when they passed through abandoned Federal camps, all too often they broke ranks to plunder and gobble the food still warming over the fires. Prentiss, though at terrible cost, had slowed them for more precious hours, and actually distracted Breckinridge later in the day from what, had the Kentuckian but known it,

was a clear open road to Pittsburg Landing. And the death of Johnston, of course, would be an enormous morale blow when it became known. As a result, the Southern momentum was considerably slowed, with nightfall approaching. Unfortunately, Lew Wallace had inadvertently taken the wrong road from Crump's Landing, and his delay seriously upset Grant, but good news came from Buell, whose long overland march from the Duck River was nearly done, and he expected to reach the banks opposite the landing that evening. Furthermore, the closer the Confederates came to the river, the more Grant could bring them under the fire of heavy guns aboard a number of gunboats anchored off shore. Certainly on

their face, the events of the day looked like a shambles for the surprised Yankees, but with every minute their prospects improved. If Grant could hold out until the next morning, Buell's arrival might not only save him, but allow for a counterattack against the exhausted foe.

Once the confusion of the surrender of the Hornets' Nest passed, the Confederate high commanders met to plan their next move. Bragg, Breckinridge, and Polk met to confer, agreeing that with a bit more than an hour of daylight left, they should press forward. Though Beauregard had been notified of command devolving upon him at least an hour and one-half earlier, no orders had yet been received from him.

Battle of Shiloh (Pittsburg Landing): the first day, April 6, 1862.

Ironically, the Battle of Shiloh on a map looks far more organized and orderly than it did on the ground, where even units as small as a company had trouble staying together.

The advance of Johnston and his four corps is seen at upper left. But very soon after the first elements struck Sherman's camps, their organization dissolved, as brigades intermingled. Still, despite good resistance, the Federals were forced back. While Grant frantically tried to

organize his surprised army, the Confederates pressed on until they came to Benjamin Prentiss and his command in and around the peach orchard at left center. Here the brave Prentiss and his men stopped the Rebel advance cold, and held it up for precious hours at what would ever after be called the Hornets' Nest. The fighting here became so severe, and the bluecoats so stubborn, that Johnston committed his Reserve Corps early in the battle, and then

himself took a mortal wound leading a Tennessee regiment into the fight.

Meanwhile, Sherman frantically tried to hold the Federal right against renewed attacks, while the Rebel pressure on Prentiss continued. At last the Hornets' Nest had to fall, and with it the Union left flank collapsed. The Federals had no choice but to fall back all along their shaky line, contracting their lines yet again as they took position under the cover of the

batteries that Grant massed on the bluff above the Tennessee River. Meanwhile, Lew Wallace finally began to approach from the far right, and Buell's marching divisions sped down from the north, at bottom right. The closing Confederates actually got within sight of the Tennessee, having pushed the Federals until their backs were to the river. But now Beauregard took command, and lost his determination, calling off the advance just as victory was

seemingly within grasp.

Only now did the Federals' situation begin to improve. The night of April 6-7 saw the bombardment of Rebel positions by Federal gunboats, as well as the arrival of substantial reinforcements under the command of Buell and Lew Wallace (bottom of the map and far right). Having ferried his men across the river, Buell launched an attack on the morning of the 7th, taking the already exhausted

Confederate troops by surprise. The battle was now renewed in earnest, as the Confederates tried to hold on to their positions in the face of increasingly successful assaults by fresh Union troops. By mid-afternoon, Beauregard ordered the withdraw back towards Corinth, detailing Breckinridge to cover the retreat.

The toll was terrible. Grant lost 13,000 out of his 40,000 engaged, and 10,700 out of 44,000 Rebels fell.

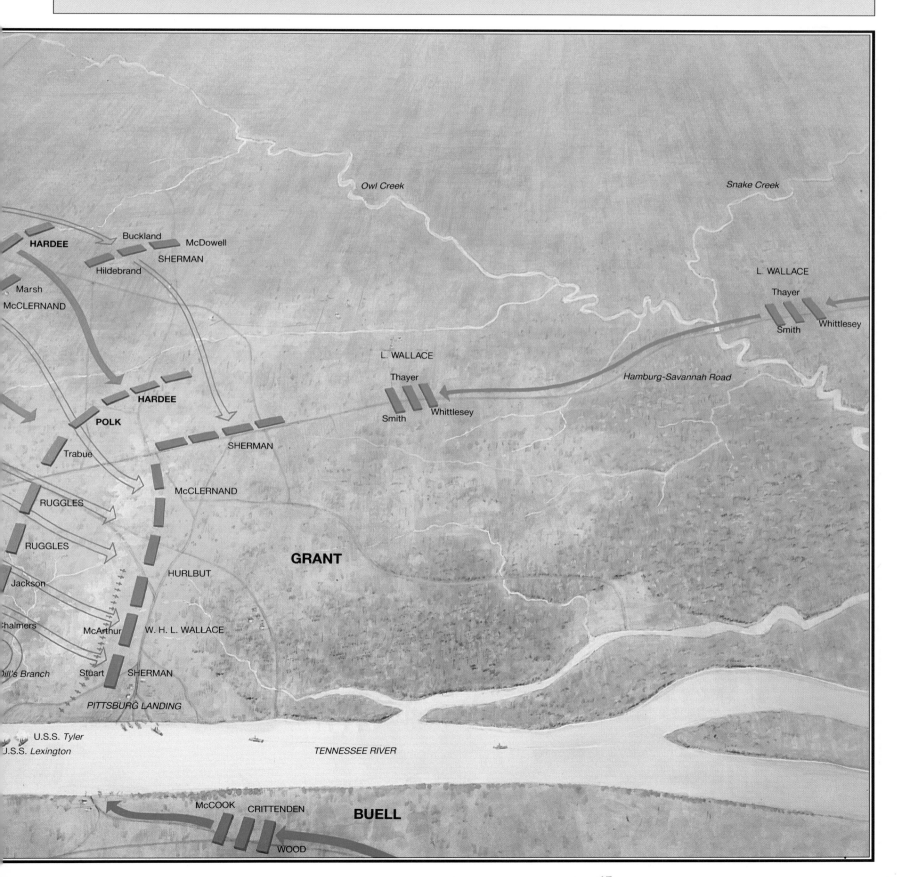

Consequently, these front-line generals started the advance once more on their own. Their exhausted men moved slowly, even though victory seemed surely in their grasp now. By 6 p.m. they came in sight of Grant's batteries and remaining infantry massed for the final defense of Pittsburg Landing on a crest actually overlooking the landing itself. The Federals could not afford to give way another inch.

The Confederates' final advance began sometime after 6 p.m. It proved to be tough going. On the right, along the river itself, the Rebels were checked by heavy fire and their advance stalled. On their left the charge progressed farther, only to be pinned down by murderous fire from Grant's massed batteries. Hurriedly reinforcements were rushed to support the advance and get it going again. "One more charge, my men, and we shall capture them all", shouted Bragg. But then Beauregard was heard from at last. Seemingly on the verge of victory, he ordered them to break off the attack and withdraw for the night.[13]

Beauregard had never had much nerve for this attack. Moreover, he suffered considerably at the moment from a respiratory infection that sapped his energy. Out of touch with the battle's events from his position far in the rear – to which Johnston had assigned him – he feared disaster in his own ranks thanks to the seemingly huge number of stragglers that he saw. He remembered from Bull Run how demoralized even the victor can be, and convinced now that they already had a victory, he decided that the men had done enough. He would rest them, let them eat and drink, and then finish off Grant in the morning. It was the biggest mistake in a career filled with bad military judgments. When Bragg got the word, he fumed. Breckinridge, on the verge of making what he believed would be a climactic assault, complained to his staff that "it is a mistake."

Indeed it was. Though badly hammered, Grant was only then starting to feel his strength. "Not beaten yet by a damn sight", he was heard mumbling to himself that evening. Ordering his

Union Hardee Hat and Branch Indicative Insignia

1 Pattern 1858 army hat
2 Pattern 1858 hat insignia for regiment of mounted riflemen
3 Pattern 1851 hat insignia for regiment of mounted riflemen
4 Pattern 1851 infantry enlisted man's stamped brass hat insignia
5 Pattern 1851 infantry officer's stamped brass insignia
6 Pattern 1851 infantry officer's embroidered hat

insignia. The numeral above the horn indicates the regiment
7 Pattern 1851 ordnance enlisted man's stamped brass insignia
8 Pattern 1851 engineer enlisted man's stamped brass insignia
9 Pattern 1851 artillery officer's stamped brass insignia
10 Pattern 1851 artillery enlisted man's stamped brass insignia

11 Pattern 1851 officer's embroidered eagle – all service insignia
12 Pattern 1851 cavalry enlisted man's stamped brass insignia
13 Cavalry officer's stamped brass insignia
14 Unofficial badge of the Cavalry Corps, Army of the Potomac, of stamped brass
15 Pattern 1851 cavalry officer's embroidered hat insignia

Artifacts courtesy of: West Point Museum, West Point, N.Y.

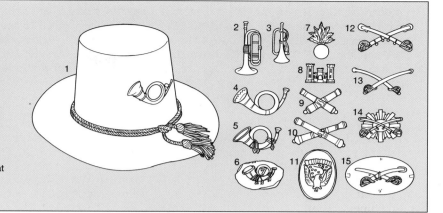

8th Texas Cavalry, "Terry's Texas Rangers", C.S.A.

One of the toughest mounted outfits in the South was made up of the rugged plainsmen of Texas, men for whom the Civil War was little more than the continuation of a struggle they had been waging for years anyhow, either with Indians, Mexicans, or Free-Staters. They went into battle under their distinctive "Wigfall" flag, named for a leading Texas politician, Louis T. Wigfall. "God Defend the Right", it proclaimed.

To help in that defense, the sergeant in the foregound carries a rare Dance revolver, a Confederate-made copy of the Colt .44 Dragoon. Indicative of the disdain with which most cavalrymen regarded the saber as a weapon, these two do not even carry them. They sit on Hope saddles, and carry the minimum of gear, adhering to the dictum to travel light and fast. Several men, like the one in the background, wore a silver star on their hats, symbolizing Texas.

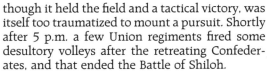

gunboats to fire steadily during the night to keep the enemy awake, Grant concentrated on getting Buell's men across the river as quickly as they arrived. By the next dawn he would have at least 17,000 of them in his lines, along with Lew Wallace's 7,500-man division, which finally arrived about an hour after the fighting on April 6 ceased. While Beauregard slept in Sherman's captured tent, full of self-satisfaction and confidence that the enemy was thoroughly beaten, Grant spent a miserable night in the rain, huddled under a tree, planning his own attack to start on the coming morning.[14]

In fact, Buell launched his own attack the next morning, apparently without discussing the matter with Grant, displaying his obvious disdain for the nondescript little man who still outranked him. When his mile-wide line advanced, Buell took the Rebels almost as much by surprise as they had taken Sherman the day before. Beauregard felt little confidence in his army, so battered and exhausted from the day before. Still he tried to stop Buell's advance with a charge by Breckinridge and Hardee. In response, Buell sent forward a counterattack that sent the Rebels reeling. Charge and answering charge rushed back and forth during the morning, but almost from the first the Federals had the upper hand, their men fresh and well supplied, while the weary Confederates found that many of their rifles would not fire from having been soaked by the rain during the night. Worse, their spirits were dampened. They went to sleep thinking that they had won the battle. Now they awoke to find a newly invigorated enemy in their front, and no end of the battle in sight. All along the line Beauregard could see the lack of enthusiasm in their faces, and the lethargy in their movements. By noon he seriously questioned whether he could ask more of them.

By 1 p.m. Beauregard came to a tentative decision, and ordered captured arms and munitions taken to the rear. He needed these spoils of war. Meanwhile he tried to rally his brigades to stop the enemy drive, and take the offensive once more for the victory he still thought within reach. They did manage a fair assault on the right of Grant's line, now held by Wallace and elements of Sherman and McClernand, but the enemy beat them back handily. Over on the Confederate center, Bragg ordered an assault that also failed to

push the enemy back, and then the arrival of fresh regiments forced the attackers back in confusion. Then a renewed threat arose back on the left, and only a counterattack led by Beauregard personally managed to stop it. Yet he knew that the check was temporary. Fresh Yankee troops had been coming into the fight all day. By 3:30 Beauregard could see more rushing into the enemy lines. At several places along the line, even his brave veterans were starting to break in the face of renewed Federal advances. He gave the order to withdraw. Detailing Breckinridge to act as rearguard and cover the retreat, Beauregard put his tired, battered army on the road back to Corinth. Grant realized that his own army,

Above: Major General Don Carlos Buell's timely arrival on April 6-7 may have saved Grant from a renewed battering. Slow, plodding, and contentious, poor Buell would never come that close to glory again.

Below: Arrayed amid their flags and their stands of arms, these resplendent young Missourians are typical of the Union volunteers who stood the test at Shiloh in the bloodiest battle of the West.

though it held the field and a tactical victory, was itself too traumatized to mount a pursuit. Shortly after 5 p.m. a few Union regiments fired some desultory volleys after the retreating Confederates, and that ended the Battle of Shiloh.

The cost to both sides had been terrible. Of just over 40,000 engaged on the first day, Grant lost more than 10,000 in casualties, and for the whole battle his killed, wounded and missing totaled more than 13,000. The Confederates lost 10,700 of 44,000 engaged. In the two-day fight, 3,500 Americans lost their lives in a battle that did little credit to the foresight of either of the commanders present. Grant and Sherman allowed themselves to be shamefully surprised, though they quickly lived it down in the general euphoria over what the Union perceived as a victory. Johnston and Beauregard never had a good battle plan, and the latter plainly never had much stomach for this fight, either from illness or a fear that all the credit for a victory would go to Johnston. Sidney Johnston himself squandered his life, and though his early death in battle would engender speculation over the fate of the Confederacy had he lived, there were those even then who questioned his management throughout his tenure in command. Davis, of course, would hear none of it, and spent the rest of his life defending and honoring his dead friend's memory. When Beauregard later seemed to claim credit for Johnston's achievements, and uttered veiled criticisms of the general, it fuelled a feud with Davis that lasted the rest of their lives, and seriously affected the course of command decisions in the Confederacy.

Therein lies the lasting significance of Shiloh. At the end of the campaign, the armies were where they began, only bloody and tired. But the Yankee high command emerged with laurels. Grant's role as hero was firmly cemented, and Sherman was rising along with him, each to be supported unfailingly by Lincoln. Johnston was dead, and Beauregard perilously close to being discredited, never to enjoy the confidence of his president again. Therein lay the beginnings of the internal and command problems that would plague this Confederate army until the end of the war, and seriously damage the cause itself. Shiloh would be the first and last major Rebel offensive in the region, while for Grant it was only the first step on the glory road to victory.

References

1 *O.R.*, I, 10, pt.2, p.94.
2 Charles Roland, *Albert Sidney Johnston, Soldier of Three Republics* (Austin, Texas, 1964), pp.252-60.
3 Wiley Sword, *Shiloh: Bloody April* (New York, 1974), pp.92-95; William Preston Johnston, *The Life of General Albert Sidney Johnston* (New York, 1878), p.568.
4 Lewis Wallace, *Lew Wallace, An Autobiography* (New York, 1906), pp.450-58; *O.R.*, I, 10, pt.2, p.93.
5 *O.R.*, I, 10, pt.1, p.386; Johnston, *Johnston*, p.582.
6 Sword, *Shiloh*, p.160.
7 Braxton Bragg to William Preston Johnston, December 16, 1874, Mason Baret Collection of Johnston Family Papers, Tulane University, New Orleans.
8 *O.R.*, I, 10, pt.2, p.404.
9 Wallace, *Autobiography*, p.461.
10 *O.R.*, I, 10, pt.1, p.404.
11 *O.R.*, I, 10, pt.1, p.278.
12 Johnston, *Johnston*, pp.614-17.
13 C.C. Buel and R.M. Johnston, eds., *Battles and Leaders of the Civil War* (New York, 1884-88), I, p.605.
14 Sword, *Shiloh*, p.368.

CHAPTER THREE

THE SEVEN DAYS'

————— JUNE 25 – JULY 1, 1862 —————

Following the Yankees' qualified victory at Shiloh, the Union took new heart. Later in the month New Orleans fell to Admiral David G. Farragut's fleet, and he started moving up the Mississippi toward Vicksburg. Out west of the great river the Federals also saw some brightening in what had once been a bleak picture of defeat.

But it was in Virginia that Lincoln and Washington most wanted a victory. Following the debacle at First Manassas, it seemed that nothing would happen. When the "Young Napoleon" George B. McClellan replaced McDowell and started rebuilding the newly dubbed Army of the Potomac, Lincoln and the North expected much of him. Yet the months of 1861 simply passed by, one by one, with no movement, and all the while Joseph E. Johnston and his army sat there around Manassas glaring at them. It was even worse in the spring of 1862. The Yankees sent three small armies to take the Shenandoah Valley from the enemy. Yet in a series of battles from late March to early June, General Thomas J. Jackson, newly nicknamed "Stonewall", defeated each in its turn and saved the Valley for the Confederacy.

This put all the more pressure on McClellan. "Little Mac" had to move. He must attack the Confederate capital.

ONE OF THE war's most painful ironies struck the Confederacy in the spring of 1862. In the course of all of American history only three army commanders would fall in battle, all in the Civil War. Two wore the gray. Each went down in this same spring. Both were named Johnston. Though the irony alone is more than enough to make the coincidence worthy of note, any comparison between the events ends there. Debate may continue to the end of time over the impact of Sidney Johnston's death at Shiloh. Though his time in command was so short, and his conduct of it so indecisive, that none may truly say how matters might have run had he lived, there is almost no doubt that those who followed him to command in the troubled Western theater were lesser men and leaders, one of them destined to be that other Johnston in time. But of the fall in battle of Joseph E. Johnston himself there is, and can be, no doubt. However unfortunate for him at the time, his wounding and removal from command was the greatest blessing bestowed upon his cause's struggle for survival. Literally, the shell fragment that struck him in the chest in the gathering twilight of May 31, 1862, changed the course of the war.

Johnston may have been the hero of First Manassas, but in the months that followed he revealed an increasing number of character and personality traits that clearly unfitted him for army command. He feuded with President Davis over matters of rank and seniority, fell under the influence of his subordinate Beauregard, became so suspicious of the Richmond government that he persistently refused to share his plans with the War Department, and often as not, in fact, had no plans. Instead, he revealed an indecisiveness and hesitation, an unwillingness to take responsibility, and a genuine obtuseness when it came to understanding the president's instructions that can only have been feigned in a man of his intelligence. As the winter of 1861-2 came to a close, Davis' confidence in him lay severely shaken, and relations between the two, though still polite, took on an increasing formality as each grew to suspect the other.

Onto this troubled tableau stepped yet another monumental ego destined to leave its mark on the war. Major General George B. McClellan owed his rank and position almost entirely to matters beyond his control or influence. Fortune always smiled on him. Gifted with unquestioned academic brilliance, he graduated second in his class at West Point when only nineteen. Better yet, he took his commission in 1846 just as the war with Mexico commenced, obtained a position on Winfield Scott's staff, and won plaudits and promotion in the land of the Montezumas. Nevertheless, he left the army in the 1850s, but when the war broke out the governor of Ohio quickly commissioned him a major general of state troops, and Washington followed suit with a formal commission and assignment to command the small army raised to invade western Virginia.

A magnificent image of one of McClellan's hundred or more regiments on parade on the Virginia Peninsula in 1862. These were the units he relied on to drive Union hopes for victory "on to Richmond."

A handful of minor battles and skirmishes cleared the vastly outnumbered Confederates from the region, and though the fights were really conducted by his subordinates, McClellan received – and took – the credit. Immediately thereafter, the defeat of McDowell alaong Bull Run left Washington and the nation virtually crying for McClellan to come east. The Union desperately needed a hero in the midst of its series of disasters in the summer of 1861, and McClellan was all Lincoln had. Three years later in the war, achievements such as McClellan's in western Virginia would scarcely have attracted national attention, and garnered him little more than a letter from Lincoln and perhaps the thanks of Congress. In the dark days after First Manassas, however, it got him McDowell's command and, not long thereafter, appointment as general-in-chief of all Union armies.

This was heady stuff for a man just turned thirty-four, and Lincoln knew it. He even asked McClellan straight out if it were not too much responsibility, but the man already revelling in being hailed as the "young Napoleon" only brimmed with supreme self-confidence. "I can do it all", he proclaimed.[1]

Certainly McClellan did much in the days after he took command. He took McDowell's dispirited army and trained and molded and augmented it into a virtually new entity – the Army of the Potomac. Relentlessly he worked to equip the men with the very best in uniforms and weapons, provided the best rations he could find, and struggled tirelessly to build their morale and self-confidence into a mirror of his own. In return they rewarded him with an almost reverential love and admiration. At the same time, he built a command structure with almost unquestioning loyalty to him at its foundation. In short, he worked almost a miracle that perhaps no other man in the Union could have achieved at that place and time.

Time, of course, was part of his secret. Following the debacle at Bull Run, the conflict east of the Appalachians entered a period that in later conflicts would be called a "phoney war". Through

the entire balance of 1861 and the winter that followed, virtually nothing of consequence happened. The timid and hesitating Johnston nitpicked and temporized, the frustrated Beauregard asked for assignment to the West, and McClellan took advantage of the months they gave him to form and perfect both his army and his plans. Indeed, that luxurious gift of time spoiled him, for he became so accustomed to having all the time he desired that he could never afterward be compelled to act quickly or until every laborious bit of preparation had been performed to his satisfaction.

The longer McClellan took to act with his grand new army, the more certain character flaws revealed themselves. His ego and arrogance proved positively offensive. He could not and would not stand for challenge or disagreement from anyone, and even characterized Lincoln as a "well-meaning baboon" and "an idiot". Very quickly he came to regard himself as a power unto himself, above the people, the army, even Congress and the president. Worse yet, while he looked upon his own presumed friends with contempt, he bestowed an exaggerated respect upon his foes. Whereas the Confederates in northern Virginia numbered barely more than 40,000 through the balance of 1861, McClellan happily accepted erroneous reports that magnified Johnston's numbers to 200,000 or more, even though accurate intelligence was coming in from his own officers.[2]

Belief in such exaggerated enemy strength justified McClellan in his own mind in using every bit of time possible to plan his inevitable offensive, and to his credit he planned well. In the indignation following First Manassas, the North wanted a speedy advance directly against Richmond. To capture the Rebel capital, they

reasoned, would end the rebellion. McClellan shared this delusion, though he quailed from meeting the enemy head-on in battle.

Instead, he finally submitted a plan that would end the conflict by strategy, with the least risk to his beloved Army of the Potomac, now approaching 120,000 strong. He would not march his army south, overland, to drive Johnston out of his defenses around Manassas. This would only force the enemy back on Richmond, with a succession of wide rivers in between, each of which offered a natural defensive barrier. McClellan proposed to board his army on a fleet of transports, steam down the Potomac into the Chesapeake, and then make for Fort Monroe, a massive casemated masonry fortification at the tip of the peninsula formed by the York and James Rivers. This fort was one little piece of Virginia that never fell to the Confederates in the days after succession. Unfortunately, thus far its value had been chiefly symbolic, for the two rivers were so wide here at their mouths that the fort's guns could not effectively interdict Rebel traffic. The James led directly upriver to Richmond, seventy-five miles northwest, while the York flowed roughly parallel and north of the James to West Point, about thirty miles east of the capital. From there the Pamunkey, a smaller but still navigable stream, flowed northwest to its origin directly north of Richmond. The several rivers together formed a peninsula not wider than fifteen miles on average, and in places such as Yorktown barely over five.

Confident that his supply line via the Chesapeake and Fort Monroe could not be threatened, McClellan proposed to drive up this peninsula over its fairly even ground, straight toward Richmond. The Confederates would be forced to pull all of their troops out of northern Virginia to meet him, thus relieving any threat to Washington, and abandoning their carefully prepared fortifications. Moving swiftly, McClellan could get the jump on them, would have only two tiny streams to cross, and could drive toward the capital before the enemy had time to erect proper defenses. Conversely, should Johnston move his army to

Lieutenant, Orr's Regiment, 1st Carolina Rifles, C.S.A.

The 1st South Carolina Rifles, raised in 1861 by prominent politician James L. Orr, was an unusually large regiment, over 1,500-strong in the summer of 1861. It served with the Army of Northern Virginia from Gaines' Mill all the way to Appomattox, when only a few hundred were left to surrender.

Theirs was a distinctive garb, matching coat and trousers of Confederate gray, trimmed with black at the cuffs and front seam and hem, with a black stripe running down the trouser legs. Among the other unusual features was a special vertical stripe running up from the cuffs to show the rank of officers like this lieutenant. No other Confederate regiments seems to have used this device, and several companies of the 1st South Carolina later adopted the more standard regulation sleeve braid and collar insignia in use by most other regiments.

the Peninsula in time to meet McClellan short of Richmond, then "Little Mac" would have had time to prepare his own defenses and force the Rebels to attack him on his own terms and ground of his choosing. Moreover, thanks to the narrowness of the Peninsula, the two rivers would protect his flanks, and as he advanced, either could be used as a direct line of supply more quickly and less laboriously than any overland avenue of communications. Though there were definite flaws in the plan, if executed quickly it offered considerable promise.

Unfortunately, "quickly" was a word foreign to McClellan. The initial movement went well. On March 17 McClellan began embarking his army,

and within two weeks had nearly 60,000 men and 100 cannon at Fort Monroe, with more arriving every day. All that lay in his front were 13,000 Confederates in the old Revolutionary War earthworks around Yorktown, twenty miles north. But immediately McClellan made excuses as to why he could not attack. Though he initially estimated enemy numbers at Yorktown accurately when he arrived on April 2, within only a few days he reported them grown to more than 100,000 – when they had not grown at all. Possessed of a four-to-one advantage, he now convinced himself that he was himself outnumbered. No attack could succeed, he said. Instead he must take Yorktown by siege. He lost a full

month, a month that gave Johnston all the time he needed to get most of his army to Yorktown, and Richmond vital time to prepare to defend the Peninsula and the capital. "No one but McClellan could have hesitated to attack", Johnston reported contemptuously two weeks into the siege.[3]

McClellan never did attack. When Johnston arrived, he, like McClellan, was siezed by timidity. He wanted to evacuate, but Davis insisted that he hold Yorktown as long as possible. Thus Johnston held out until May 3 when he finally ordered his men to put out of their works and retire during the night, convinced that McClellan was finally about to move. Davis was mortified at

Battleflags of the Army of Northern Virginia

1 Probably the flag of the 4th North Carolina Infantry Regiment, captured by Federal troops at the Battle of Chancellorsville. This flag is of the second bunting issue

2 Flag of the 7th Regiment Virginia Volunteer Infantry. During the Battle of Gettysburg, this regiment was part of Kemper's Brigade of Pickett's Division. The color was captured by Federal forces during the battle

3 Flag of the 9th Regiment Virginia Volunteer Infantry. This color was captured by Federal forces at the Battle of Five Forks, Virginia, April 1, 1865. Held by General Pickett, Five Forks was on the far right of the A.N.V. line at Petersburg. The position was taken by General Sheridan

4 Flag of Courtney's Virginia High Constabulary, captured at Spotsylvania Court House, May, 1864. This flag is of the third bunting issue

5 Flag of an unknown unit captured at Sayler's Creek, Virginia. April 6, 1865. This was the last major engagement fought between the Army of Northern Virginia and the Army of the Potomac. This flag is of the fourth bunting issue

Artifacts courtesy of: The Museum of the Confederacy, Richmond, Va

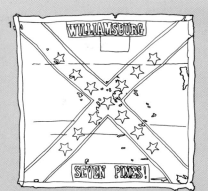

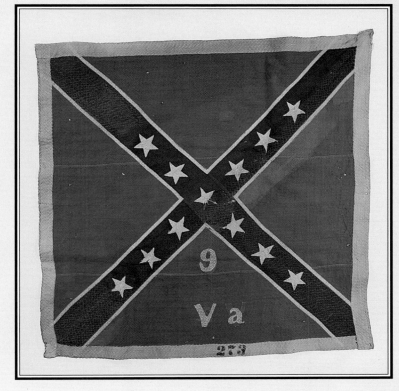

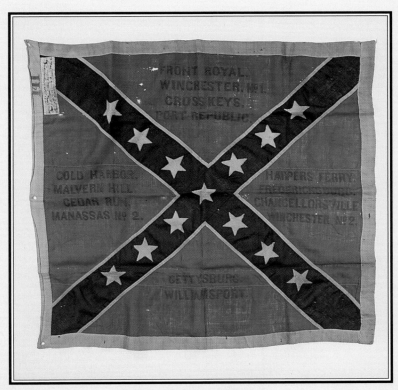

the abandonment of the position without a fight, and even more chagrined that Johnston pulled back all the way to the line of the Chickahominy River, one of the streams that crossed the Peninsula, and barely ten miles from Richmond. McClellan spent the entire month in leisurely pursuit, all the time telling Washington that enemy strength was growing and that he was in mortal danger of losing his army, which of course could not be his fault. When he came himself to the Chickahominy late in May, he expressed a resolution to die with his men, if need be.

It was all bombast, of course, but now, in fact, he did face real danger. Johnston's army had now grown to nearly 75,000, most of them emplaced

in the vicinity of Fair Oaks Station on the Richmond & York River Railroad, and the nearby village of Seven Pines. Worse, when McClellan arrived, he split his army by putting something less than half – about 40,000 men – on the south side of the Chickahominy. Month-long rains had so swollen the little stream that now it was crossable only by a few bridges, and thus "Little Mac's" left flank, south of the river and on the same side as Johnston's entire army, stood vulnerable. Then an especially heavy downpour on the night of May 30 so flooded the river that its waters carried away almost all of the bridges. If Johnston struck quickly, he could overwhelm nearly half of McClellan's army.

Unfortunately for the Confederates, Johnston conducted a dreadful battle at Fair Oaks or Seven Pines on May 31. Confused and disordered attacks, units marching in the wrong direction, and an overall lack of coordination showed that Johnston had not profited much by his experience at First Manassas. His own ordnance chief, Colonel E. Porter Alexander called the whole affair "phenomenally mismanaged."[4]

Yet one supreme benefit did come out of it, and that was the errant bit of a Yankee shell that knocked Johnston from his saddle and put him out of the war in Virginia for good. As his performance in the past several months had shown, and as he would demonstrate in future years after

Federal Uniforms of Enlisted Men

1 Zouave-pattern jacket of 95th Pennsylvania Volunteer Infantry, known as "Gosline's Zouaves"
2 Four-buttoned sack coat of 1st Sergeant of Artillery. The coat appears to lack any branch of service insignia
3 Distinctive shell jacket with yellow piping of the 15th Pennsylvania Cavalry, known as the

"Anderson Cavalry"
4 Distinctive shell jacket of light blue color of the Veteran Reserve Corps
5 State issue infantry jacket of 1863. This particular example is from the New York State Militia and National Guard
6 Sergeant's shell jacket of the 6th Pennsylvania Cavalry, known as "Rush's Lancers". Note the non-issue lining on

inside of the the jacket.
7–8 Shell jacket, vest and trousers of the 23rd Pennsylvania Infantry, known as "Birney's Zouaves". Note the distinctive zouave cut to the jacket, and the elaborate piping. The 23rd Pennsylvania fought at Antietam, where it was one of six regiments in Brig. Gen. John Cochrane's 3rd Brig. 1st Div, 4th Corps

Artifacts courtesy of: West Point Museum, West Point, N.Y.

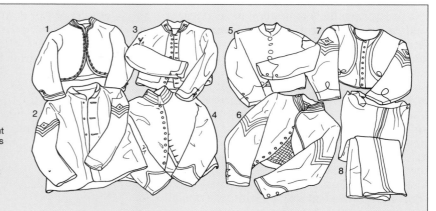

recovering from his wound, Joseph E. Johnston was not the man to command a field army, much to the detriment of his cause. But there was another, also a Virginian and near at hand, with whom the story would be different. President Davis was on the battlefield during the fighting, and soon came to the wounded Johnston's side, their feuding forgotten in the solicitude of one old soldier for another. Fortunately the dying sun was bringing the day's engagement to a close, and as Johnston was taken to Richmond, command of the army devolved upon the next senior officer, Major General Gustavus W. Smith. But later that night, as the president rode back into the capital, he informed his companion and chief military advisor that on the morrow, he was to take command of the Army of Northern Virginia, as it was to be styled. The officer to whom Davis spoke was General Robert E. Lee.

There were many in the Confederacy who did not think too highly of the fifty-two-year-old Lee. Though he had served without interruption in uniform since 1829 and won praise and promotion in Mexico as one of Scott's favorites, he spent almost his entire career in the engineers, and had never led soldiers in battle until 1859 when he commanded the contingent of marines who attacked and captured the radical abolitionist John Brown and his raiders at Harpers Ferry. Still, his seniority alone made him automatically one of the Confederacy's top ranking officers when secession came, placing him just junior to Sidney Johnston, and just ahead of Joseph E. Johnston (the source of the latter's feud with Davis over rank). In the war thus far his role had not been exemplary. He commanded unsuccessfully in western Virginia, where his soldiers derisively nicknamed him "Granny" Lee for what they took as caution on his part. Then he superintended the erection of coastal defenses in South Carolina, leading other critics to deride such work by calling him "Spades Lee" and the "Ace of Spades." South Carolina governor Francis W. Pickens actually warned friends that Lee was not "the man his reputation makes him." Rather, Pickens declared that "Lee is not with us at heart, or he is a common man, with good looks, and too cautious for practical Revolution."[5]

But Davis saw things in Lee that others did not, and in March 1862 he brought the general to Richmond to act as his military advisor and *de*

Above: McClellan's magnificent Army of the Potomac at Cumberland Landing, during the Peninsula Campaign, its camps stretching for miles in the distance. Against such legions, the Rebels seemed doomed.

facto general-in-chief, though Davis, being Davis, never allowed any truly important decisions to be made by anyone but himself. Unlike Johnston and Beauregard, Lee knew how to get along with the president, and accepted the subordinate role of the military to civil authority. Moreover, he never argued or complained, kept Davis freely and fully informed, and even at times flattered the president's ego a bit. Diplomacy behind the lines could count for almost as much as brilliance on the battlefield in this war, and Lee would prove to be effective at both.

Quite sensibly Lee did not immediately assume command. The battle was as yet undecided and Smith knew the ground better than Lee. But no sooner did the dawn come on June 1 than Smith collapsed, virtually overcome by nervous trauma over the responsibility given him. Thus Lee assumed command the next day after both sides had withdrawn from the indecisive fighting around Seven Pines. His first act was to set his army to work strengthening the fortifica-

tions around Richmond, setting off a new round of outcry that he would not be a fighting general. But to Davis Lee confided his intent to strengthen those works so that only a small portion of his army could man them effectively, while he would take the bulk of the Army of Northern Virginia, as he now called it, and drive McClellan from the Peninsula. Davis liked this kind of talk, and gave the general his head.

During the next three weeks Lee used his spades and his brains, while McClellan remained almost stationary, still convinced that his 105,000 or more troops on the Chickahominy line were heavily outnumbered. The Army of the Potomac was divided into five corps, of roughly equivalent size, and all led by Old Army professionals. Generals Erasmus D. Keyes, Samuel P. Heintzelman, and William B. Franklin were all veterans of First Manassas. The elderly Edwin V. Sumner was the oldest of the lot, while Fitz-John Porter was the youngest, just in his thirties, and a particularly loyal favorite with "Little Mac". None would demonstrate great talent in the war ahead, and some would resign in obscurity before its close, but still among them they commanded the finest army yet seen on the continent. Unfortunately, their commander was not inclined to use it. For fully two weeks after Seven Pines, McClellan sat in his works slowly planning his next set-piece battle. Typically, he proposed to fight it with his artillery, trusting to his guns alone to drive Lee back, and then he would close in upon Richmond and bombard it into submission, cleanly, bloodlessly, without risk to himself and his precious army.

Three problems would thwart him. One was his own sloth and timidity. Another was the fearfulness in Washington. Lincoln, terrified that the Rebels would take advantage of McClellan's isolation on the Peninsula to send a force overland against the capital, withheld most of the corps commanded by McDowell, keeping it to protect the Potomac line. McClellan did not really need those 25,000 or so withheld from him, but their absence gave him a perfect excuse to complain that Lincoln was to blame for the delay in his

Left: At Fair Oaks, or Seven Pines, the campaign did not start auspiciously for the Confederates when a shell fragment from a Yankee gun like one of these knocked Johnston from the saddle, and his command.

commanding general would use these young leaders in a complex and closely timed offensive. He would leave the two small divisions of Generals John Magruder and Benjamin Huger south of the Chickahominy, with only about 22,000 between them, to face 75,000 Yankees across the lines. The divisions of the two Hills and Longstreet would join with Jackson, almost 60,000 strong, to strike Porter on the north side of the river. It was a great gamble, for while Lee was overwhelming Porter, McClellan could overwhelm Magruder and Huger and have a clear road to Richmond.

Jackson was to sweep down on Porter's northern flank while the others crossed the Chickahominy, thus forcing the enemy back from Beaver Dam Creek. If timing slipped or if someone's part in the plan miscarried, however, it could all come to nought. The daring revealed two things about the man who had only exercised command of the Army of Northern Virginia for a few weeks. One was that he still had much to learn about managing large numbers of men, and that they could not be made to move as quickly and easily on open ground as over a map at headquarters. Complex coordinated attacks would prove frustratingly elusive of success throughout this war. The other thing revealed, however, was that "Granny" and "Spades" Lee was a thing of the past. This general, unlike Johnston, would fight without excuse and without delay.

Consultation with Jackson revealed that his Valley command could be in the vicinity and poised for action by June 25, and accordingly Lee set the next morning as the time for the attack to commence. However, other circumstances precipitated unexpected fighting on the 25th when a Yankee reconnaissance near Seven Pines escalated into a severe skirmish in the vicinity of Oak Grove. McClellan pushed two of Heintzelman's divisions forward to within four miles of Richmond in order to secure some ground that he wanted before launching his artillery attack on Lee's lines. He had learned by now of Jackson's advance, and was sending even more predictions of gloom and defeat to Washington, washing his hands of any responsibility. Yet McClellan had finally stopped delaying and had himself set June 26 as an attack date. Had he followed through and struck first, he might severely have disrupted Lee's plans, though there is little in "Little Mac's"

campaign, and that in the event of a Union defeat "the responsibility cannot be thrown on my shoulders."[6]

The third problem was Robert E. Lee. Part of the reason that Washington was in a panic was the brilliant Shenandoah Valley Campaign conducted that spring by Major General Thomas J. "Stonewall" Jackson. Having routed three separate small Yankee armies, Jackson and his three divisions, numbering about 17,000, were now in undisputed command of the Valley. Lee thought briefly of a holding action against McClellan while sending heavy reinforcements to Jackson to mount a counter-invasion of the North. However, the mighty Stonewall and his men were too exhausted from the rigors of their recent campaigning to sustain any such movement. But they could still fight. Instead, Lee finally decided to rush Jackson east from the Shenandoah to join the Army of Northern Virginia. McClellan was offering him a tantalizing opportunity, and Jackson could play an integral part in capitalizing upon it.

After Seven Pines, McClellan made only minor shifts in his positions. By the third week of June, the Chickahominy still cut his army almost in half. Shifting some of his units south of the stream, he left only Porter's V Corps on the northern side and somewhat in advance of the rest of the army, in an exposed position that McClellan felt warranted since he expected reinforcements from McDowell to be marching overland to link with Porter. Unfortunately, Lee's dashing cavalryman "Jeb" Stuart led a lightning swift raid around McClellan's army that discovered Porter's exposed position, and brought the news back to Lee. Porter's 30,000 men were emplaced along a tributary of the Chickahominy called Beaver Dam Creek, near Mechanicsville. They were separated from the rest of their army, and that alone made them an appetizing target. Better yet, however, McClellan was now succoring his army from a supply base at White House on the Pamunkey, almost due east of Porter. If Lee could push Porter aside, he could drive straight for White House. Denying McClellan that base would force the

Above: A military bridge over the Chickahominy. McClellan's decision to divide his forces on either side of the river exposed half his army to extreme danger when the river flooded, isolating its left flank.

Yankees either to withdraw down the Peninsula to the next best location for a new base – thus freeing the immediate pressure on Richmond – or else to fight to hold on to White House. To do the latter, McClellan would have to send his corps across the Chickahominy over the severely limited number of bridges, and right into the waiting guns of the Confederates, whom Lee presumed would be able to defeat them one by one.

It all depended upon his division commanders, many of them men he barely knew, and none of them truly experienced in command except for Jackson. General James Longstreet hardly participated in the fighting at First Manassas. A.P. Hill had only recently been promoted to major general after service at Yorktown and Seven Pines, but he was still new to divisional command. Major General Daniel Harvey Hill of North Carolina had fought one of the war's first small battles not far from here in June 1861, and again under Johnston on the Peninsula, yet he was still an unknown quantity to Lee. Nevertheless, the

Right: After the Battle of Seven Pines, this Union battery takes up position at Fort Richardson, an earthwork situated in front of Fair Oaks, between Nine Mile Road and the Richmond & York River railroad.

record to suggest that he could have capitalized on the advantage and actually gone ahead to threaten Richmond. As it was, when he learned about Jackson, he called off his June 26 attack, but then did nothing more than strengthen his White House guard and warn Porter to be vigilant. Then he sat back and let Lee run the rest of the campaign.[7]

Not surprisingly, when Lee attacked on June 26 nothing seemed to go as planned at first. His orders to Jackson were to swoop down on Porter's exposed flank at first light. When A.P. Hill heard the sound of Jackson's guns, he was to cross the Chickahominy, brushing aside a small Yankee outpost guarding the crossing, then drive through the enemy posted in Mechanicsville. Pushing them back would open the way for D.H. Hill and Longstreet to cross the river uncontested, form on A.P. Hill's left and right, and then the united Confederate line would push Porter before them and away from White House.

Unfortunately, there was no sound of Jackson's guns. It remains one of the most frustrating mysteries of the Civil War. A stickler for obeying orders to the letter, Jackson had been known to keep an entire column of troops waiting in line to march while he stared at his watch, unwilling to move one second before his instructions specified. Certainly he had his orders now, and they were specific. Yet when the sun rose, he did not move an act that defied adequate explanation. Enemy cavalry did face him in front, but for his three divisions, they could present nothing more than a nuisance. So, too, the trees and brush felled to impede his advance presented little real obstacle to seasoned warriors who could march twenty miles in a day and still be ready to fight. But perhaps, in the end, that was it. In the past three months they had simply marched and fought too much. They covered 400 miles on foot in their Shenandoah campaign. Now they were just arrived after an exhausting combined march and jolting train trip. Even then, with Stonewall to prod them, they would have moved. But he did not. As exhausted as the rest, the indomitable

Confederate Vandenburgh Volley Gun

This weapon was developed in 1860 by General Origen Vandenburgh of the New York State Militia. After failing to sell it in England, Vandeburgh despite his position, sold it to the South.

1 View of the muzzle of the Vandenburg Volley Gun, showing its multi barrels. In this particular example, the rifled barrels are of .50 caliber

2 Breech view of the Volley Gun, showing the handles of its screw breech mechanism. When tightened, the screw forced the breech holding the individually loaded rounds into an air-tight seal with the weapon's firing chamber. All the barrels were fired simultaneously by means of a percussion cap located in the center of the breech handles.

3 Side view of the brass, breech-loading Volley Gun, showing the large gun-sight on the top of the barrel, and the breech secured in position. This weapon was made by Robinson and Cottam, London, and is marked *85 No. 4*. It is only 36 inches long, but weighs 400 pounds. It was captured by Union cavalry near Salisbury, North Carolina in April, 1865

Artifact courtesy of: West Point Museum, West Point, N.Y.

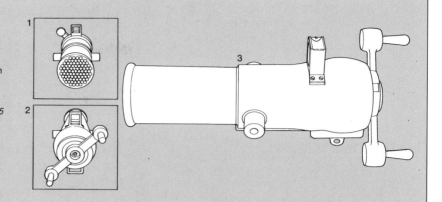

Jackson temporarily ceased to function. Disorientation, apathy, and sleepiness conspired to rob him of his accustomed iron discipline. He halted six miles short of where he was to have been the night before the attack, then stayed up praying through the night instead of resting himself, and finally started marching toward Porter later than ordered. Other than a late morning message to Lee that he was running late, Jackson communicated with no one, and by 5 p.m. was still almost three miles from the battleline. There he simply put his men into bivouac and himself went to sleep without his command having fired a single shot all day. Alas, it was not to be the end of his erratic performance on the Peninsula.

With Jackson seemingly off in limbo, it fell to A.P. Hill to carry the fight to the enemy. Vainly he waited through the morning to hear firing from Jackson. Finally by late afternoon he gave up waiting. It did not help that Lee had been of little or no help. He waited, too, throughout the morning and into the afternoon, seemingly frozen by his battleplan and unable or unwilling to act until Jackson was in place. Lee was still learning, and this was to be his first battle. It did not help that President Davis and other dignitaries were with him, waiting. A few months from now a more seasoned Lee would improvise, think on his feet and act, rather than allow the failure of one part of a plan to endanger all of it. But now he listened, and it was Hill who put the battle in motion by acting on his own, crossing the river, and driving toward Mechanicsville.

It was 3 p.m. or shortly thereafter that the firing finally commenced, as Hill's 16,000 advanced toward Mechanicsville. The Federals occupying the village yielded it with nothing more than minor skirmishing, but as Hill's brigades swept on through the town, they could see the Yankees posted in strong positions on the far banks of Beaver Dam Creek a mile ahead of them. Quickly the battle got out of anyone's control. Two of Hill's brigades struck at Porter's right flank, and one regiment actually got across Beaver Dam Creek, though to no purpose. The

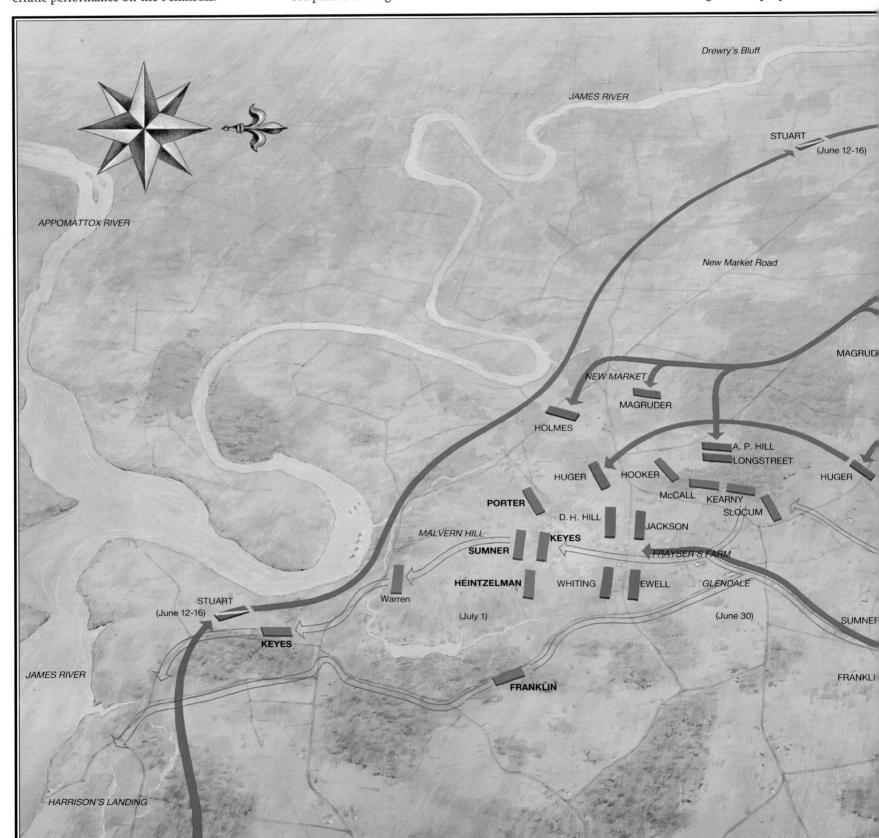

The Seven Days' Battles, June 25-July 1, 1862

Few campaigns producing major battles saw as many fights, over as much ground, as the Peninsular Campaign of 1862, and particularly The Seven Days'. It began at Fair Oaks, or Seven Pines, a few miles west of Richmond, at right center. The battle was not well managed by either side, and its only positive result was the elimination of Johnston and the emergence of Robert E. Lee.

Lee struck first at Beaver Dam Creek at the extreme right. Unfortunately, circumstances, including a tardy Stonewall Jackson, prevented Lee's plans from working. Undaunted, he pressed on, striking again at Gaines' Mill on June 27, at lower extreme right. Here Lee scored a definite victory, but still missed a chance to make it decisive. Thus he pushed on, and once again had a chance to catch the Yankees split in two at Savage's Station on June 29, at lower right center. Yet once again a crushing victory eluded him as Jackson once more failed in his vital part of the plan. Lee's frustration at not being able to get his army to work in the field the way it could on his maps began to tell on him, yet he held onto the offensive nevertheless.

The next day he struck again at Glendale, at left lower center. With his divisions set to come at McClellan from three directions at once, Lee hoped for the decisive victory that had eluded him for the past several days. But again execution did not equal planning, and McClellan slipped the trap to take position on Malvern Hill about two miles further south, and almost overlooking the horseshoe shaped bend of the James River at left.

Unfortunately, by now Lee was at a loss, worn out by campaigning, frustrated in every plan. In desperation, he resorted to a massive frontal assault against McClellan's positions, thinking them battered by his artillery. But again the Yankees mowed down his weary Rebels with dreadful cost, and McClellan withdrew the next day, unharrassed.

The prestige of Northern arms and that of McClellan himself had taken a serious blow, while in contrast the Confederacy had discovered in Robert E. Lee a general in whom it could fully place its trust and loyalty.

Holding onto his small piece of the Confederacy, McClellan and his largely undefeated army sat out the coming months. McClellan refused to budge, and despite orders from Washington, maintained vociferously that the proper line of advance should be through the Peninsula. This would have great bearing on the outcome – the Battle of Second Manassas (Second Bull Run), August 29-30, 1862.

The campaign cost him more than 10,000 of nearly 100,000 in his army. Lee suffered nearly 20,000 out of his 80,000.

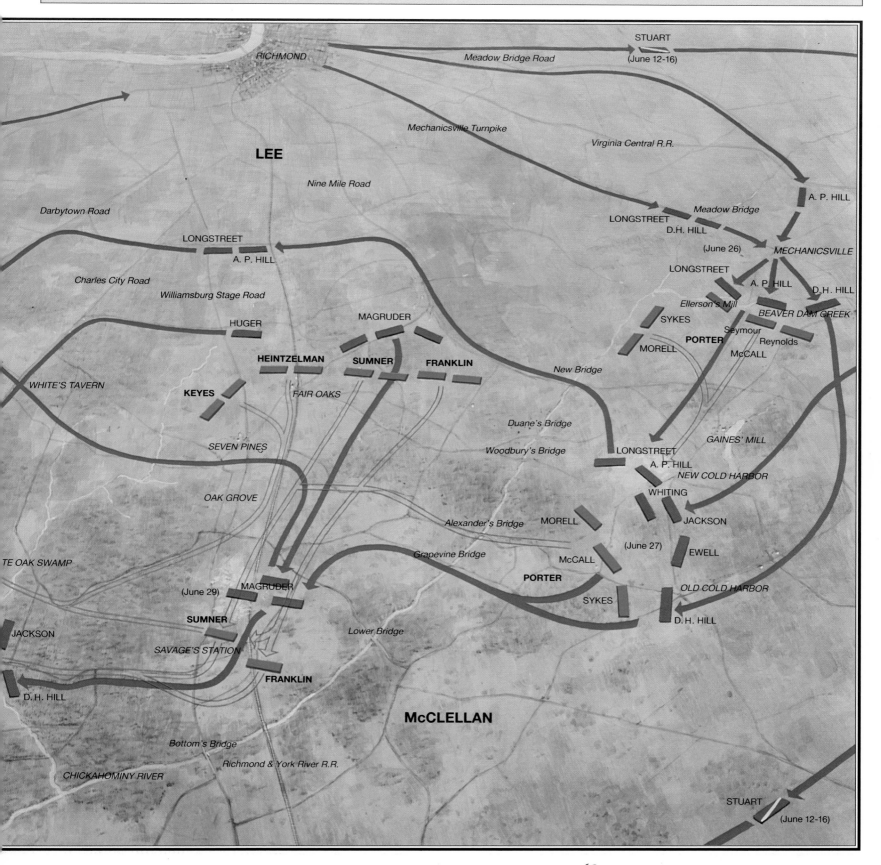

stream, though barely ten feet wide, had steep brushy banks that provided excellent defensive positions from which the Federals sent forth sheets of flame and lead. Another of Hill's brigades got enmeshed in brutal volleying at Ellerson's Mill on Porter's left center, and when a brigade from D.H. Hill finally got across the Chickahominy, it rushed to its support, only to be chewed up in like fashion.

Only nightfall put an end to a fight that had gone wrong for Lee from dawn to dark. Instead of forcing Porter away from the creek by maneuver, Lee had found himself fighting for it, and losing. There was no Jackson, no Longstreet, and no D.H. Hill. A.P. Hill had borne the fight all alone,

and suffered for it. He lost almost 1,400 killed and wounded, while inflicting not more than 360 casualties on the enemy. At the end of the day, Porter held his ground, and Lee had nothing to show for his first battle but his wounds.[8]

Yet Lee's losses proved to be only tactical, for later that night McClellan gave him a strategic victory by deciding to abandon the line of Beaver Dam Creek without further fight. Though Jackson had never moved against Porter's flank, the knowledge that Stonewall was near put a fright into "Little Mac". He recognized the magnitude now of the threat to his supply line north of the Chickahominy to White House, and that alone was enough to unnerve him. McClellan did not

collapse as had G.W. Smith or Jackson. But inwardly he accepted defeat, or, more accurately, he defeated himself. He decided that he would have to change his base of supply from the York to the James River. That in itself was not such a desperate move. But the York was the only way by which he could practically get his heavy siege artillery up to bombard Richmond. By shifting to the James, he tacitly admitted that he would not, and could not, pursue further his only announced plan of attack on the Confederate capital. Thus McClellan's offensive evaporated. He was still a dangerous foe if only for his strength of numbers, but in the mental test of wills between the commanders, Lee now held the initiative.

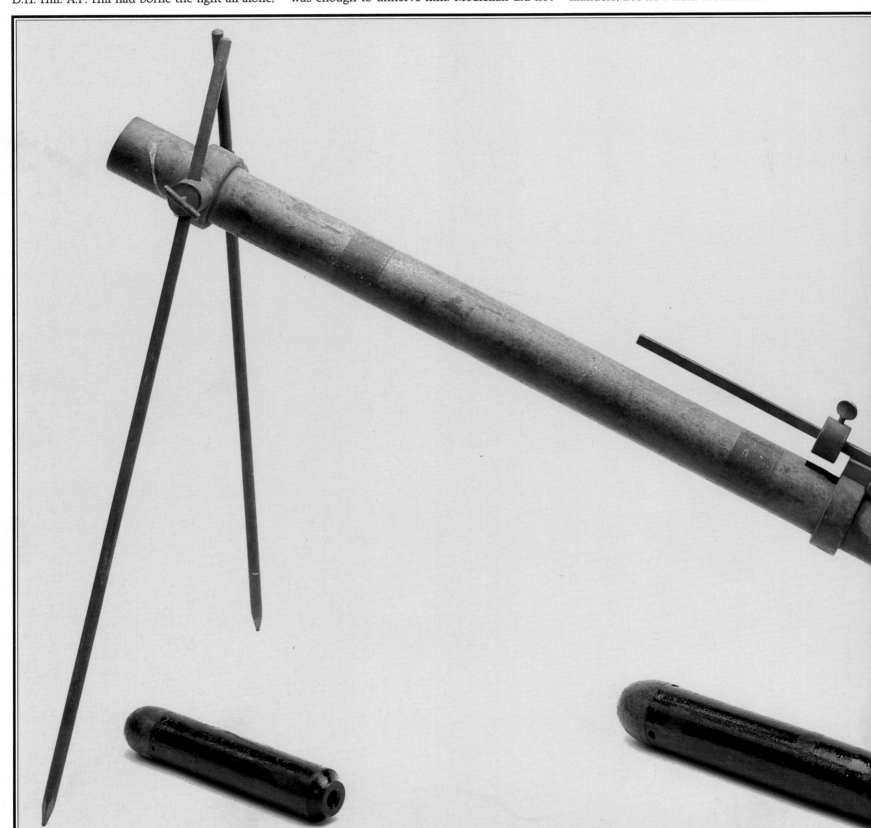

Union Rockets and Signal Pistols

1 Hale patent rocket launcher, 2.25 inch. This rocket system was invented by British civil engineer William Hale in 1844. This pattern was approved by Union naval ordnance expert, Captain, later Admiral John A. Dahlgren in 1847, when he introduced the system at the Washington Navy Yard. Although it was inaccurate, Hale rockets were used throughout the war in several calibers. The Hale rocket system superseded the Congreve rocket. This had been used by the British in the Napoleonic Wars and against Americans during the War of 1812. It was noisy, inaccurate and generally ineffective

2 and 3 Variant 2.25 inch rockets for Hale's rocket launcher. The projectile was placed in the open breech. Elevation was adjusted at the bipod using the sight on the top. Once launched, the rocket achieved stability in flight by means of a vane inside the exhaust nozzle, which caused the rocket to spin

4 U.S. Navy Model 1861 percussion signal pistol. Made at the U.S. Navy Yard, Washington

5 U.S. Army Model 1862 percussion signal pistol. Made by William Marston in New York

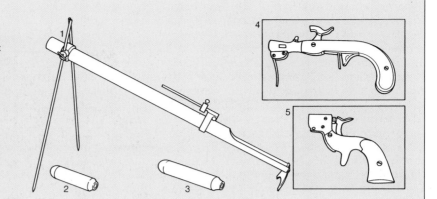

Artifacts courtesy of: West Point Museum. West Point, N.Y.: 1-3; Russ A. Pritchard Collection: 4,5

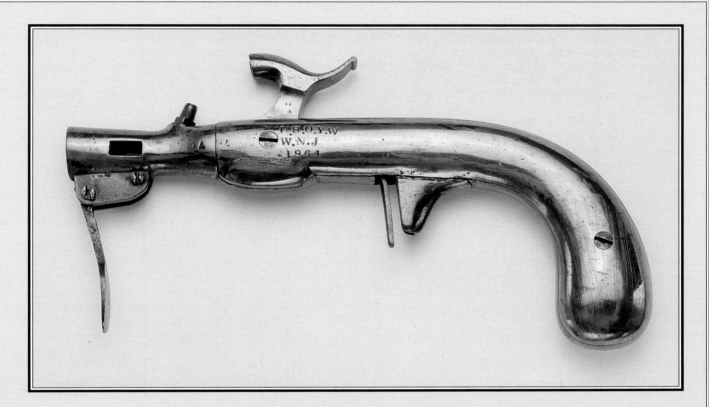

Porter began to pull out of the Beaver Dam Creek positions soon after 3 a.m. on June 27, falling back toward Boatswain's Swamp, four miles to the rear, near Gaines' Mill on Powhite Creek. There Porter formed with his left on the Chickahominy where the Powhite flowed into it and fronted by the swamp, and his line extending northward along a creek to Old Cold Harbor. His front stretched in a convex arc some two miles, giving him an advantage of interior lines, and with the Grapevine Bridge over the Chickahominy just in his rear, allowing access to McClellan on the other side. It was a strong position, but Lee, his mind like his army on the offensive, was not about to recoil from renewing the fight.

Apparently overlooking Jackson's lapse of the previous day, Lee once again planned an attack that depended largely upon the sleepy Stonewall. Hill was to follow Porter's line of withdrawal directly. Longstreet, now safely across the river, would move parallel to Hill, and between him and the Chickahominy. Jackson was to come in on Porter's right flank as he should have done the day before, while D.H. Hill was to march his division way around the left to come in on Jackson's flank and form the far left of the army, striking Porter's rear in or around Old Cold Harbor, thus cutting off his line of retreat to White House.

Once again, Lee infused too much subtlety into his plans for his commanders and his army at this early stage of their development. Longstreet was supposed to threaten Porter's left while the main attack came on the right. However, if Porter showed signs of giving way to Longstreet or of weakening his left to strengthen his right, then Longstreet was to attack in earnest. It was still too complex, and again Jackson failed to appear when and where ordered. His unaccountable lethargy persisted, and in a repeat of the day

before, it was A.P. Hill who bore the brunt of the battle with the Federals.

Hill sent his bloodied division forward around 3 p.m. The Confederates rushed past Gaines' Mill then wheeled slightly southward to strike the right center of Porter's arc. By then D.H. Hill had his division in position, but Jackson was late in arriving, and again seemed disoriented. It did not help that Porter's flank was some distance to the rear of where they all expected it to be. Instead of facing an exposed and vulnerable enemy right, they were looking squarely down Porter's guns. The failure to meet the expected seemed to confuse Jackson all the more and he simply stopped, uncertain what to do even though he could hear A.P. Hill once again in the maelstrom of battle less than a mile to his right.

Hill was fighting for his life near New Cold Harbor. As the day before, Porter held a good position behind a stream bank, while Hill's men had to advance against them across open, swampy ground. The Virginian's men had been in action almost since dawn, when they began their pursuit of Porter, and by now they were exhausted, at times actually having run to catch up with the retiring foe. When they approached Porter's main line, they came upon a rain of rifle and artillery fire. Porter had the luxury of several fresh brigades unblooded in the fight at Mechanicsville, while Hill's division was already badly winded. More than 25,000 Yankees were able to concentrate their venom on Hill's 13,000 before Jackson and the rest got into the fight, as indeed they finally did.

It turned out to be a jumbled, confused series of little local fights, as the swamp and the forests prevented any overall control by Hill. Brigades attacked and withdrew, attacked again, were broken up or pinned down, depending on the

inclinations of their commanders and the severity of the Yankee fire. After a time Porter actually gained the upper hand, pinning down almost the entire division, and by 4 p.m. there was danger of the Confederate attack breaking up entirely. Only then, still waiting for Jackson, did Lee take a personal hand, ordering Longstreet into the battle to support Hill's right. Then, without orders, one of Jackson's division commanders went to Hill's left on his own initiative, and though the situation continued to be desperate, at least the faltering Rebel line stiffened. Finally, around 6 p.m., Lee finally met with Jackson himself, and the balance of Stonewall's command came into line, with little thanks to him.

Having given up on his original plan of that morning, Lee now decided to resort to a general assault all along his line. It came around 7 p.m. and it gained good ground. Perceptibly Porter's resistance to the 50,000 or more arrayed against him slackened. Rebels ran to the creek in his front, waded across, then threw their rifles up the banks and clambered up after them. Lee sensed that a critical moment was at hand. He found Brigadier General John B. Hood, who had served with him in the prewar army, and now commanding a Texas brigade.

"Can you break his line?" asked the commanding general.

"I can try," replied Hood, and he led them down to the creek, across, and up into the teeth of the tired and dispirited defenders. An advance line of

Below: Among the toughest fighters in the whole Army of Northern Virginia were the men of Hood's Texas brigade, men like these rugged boys seen here out of the front line in winter quarters. They were arguably the best "shock" troops at Lee's command.

Yankees gave way, and Hood rushed on to the main line. It, too, broke in the ferocity of his charge, and thus he penetrated the very center of Porter's line. Within minutes the gallant Federal's line collapsed generally, though not in rout. Deliberately, and aided by reinforcements sent by McClellan, Porter withdrew to the bridges over the Chickahominy, and after nightfall took his command to the other side, leaving behind 6,800 men killed, wounded, or captured. Yet Lee paid for his victory with almost 9,000 casualties, the brunt of them, again, coming from A.P. Hill's valiant division.[9]

While the fighting of the past two days went on, McClellan kept 70,000 or more men virtually idle on the south side of the Chickahominy, virtually flimflammed by the activity of the enterprising Magruder into thinking that he actually faced superior numbers. Magruder marched his troops back and forth, lit extravagant campfires, made bold reconnaissances, and used every other artifice of deception to make the foe think that his 22,000 were many times their number. McClellan happily cooperated. Indeed, by the close of the Gaines' Mill fighting, between Magruder's flamboyant histrionics and Lee's ill-managed but relentless attacks, "Little Mac" was a thoroughly beaten man. Even then – or perhaps especially then – the extent of his petulant megalomania revealed itself. "I have lost this battle because my force was too small", he whined in a letter to Secretary of War Edwin Stanton. "I am not responsible for this". He protested that if he only had 10,000 fresh men he could still beat Lee, when the fact was that he had almost four full army corps, better than 60,000 "fresh men" who had done nothing at his disposal. "If I save this army now, I tell you plainly that I owe no thanks to you or to any other persons in Washington. You have done your best to sacrifice this army."[10]

In fact, of course, McClellan was doing a perfectly good job of that on his own. The next day saw no general engagement, as the Federals withdrew from the Chickahominy towards the James River. Through the day Lee remained uncertain of McClellan's movements, until Stuart brought him intelligence that the supply base at White House had been abandoned and everything not removable destroyed. Still, Franklin, Heintzelman, and Sumner remained in Magruder's and Huger's front, and this left Lee puzzled. He thought that cutting off the enemy from the base at White House would force McClellan either to retreat entirely, or else come north of the Chickahominy to attack him. Seemingly the Yankee was doing neither, and Lee could not foresee what McClellan was in fact trying to do – withdraw to the south to the James, near Harrison's Landing, there to erect a new base. "Little Mac" had abandoned any idea of an offensive, but as long as those three corps menaced Magruder, Huger, and Richmond, he could not be certain what course to pursue. Thus the Federals gained a day, vital time in what was now a race to escape the Rebels' crushing attacks.

The next morning, June 29, Lee discovered that McClellan had pulled out entirely from the works in front of Magruder and Huger. At least the mystery was gone now. McClellan was in retreat, and obviously toward the James. Lee had already formulated a plan for putting all nine of his divisions on the march, by a bewildering variety of roads – some barely more than tracks – to try to intercept McClellan before he could reach the James and the covering fire of Yankee

gunboats on the river. Once again, however, it was all simply too complex a movement even in the best of circumstances, and circumstances did not favor anything like this now on the Peninsula. Many of his divisions were tired. There were few good maps, the roads were a mire after the spring rains, and some commanders like Longstreet, Huger, and D.H. Hill were strangers to the landscape of Virginia.

Lee first caught up with elements of the Army of the Potomac by about 10 a.m., three miles south of the Chickahominy at Savage's Station on the Richmond & York River line. Yet again Jackson failed to take his position on the left flank at the proper time, and Magruder, who was to launch the attack, had to delay until around 3 p.m. Huger, who was to come up on Magruder's right was equally tardy. It did not help that there were few roadsigns, and Lee seems not to have thought of giving his division commanders local guides. Longstreet fared worst of all. Ordered to move to the far right, below Savage's Station, to cut off the Yankees' retreat to the south, Lee told him to move on the Darbytown Road. It took its name from a local farm whose occupants, thanks to queer local tradition, called themselves "Darby" when their actual name, and the sign on their fence that identified the road, was spelled "Enroughty." Longstreet, a South Carolinian unfamiliar with local customs, lost precious time trying to find the right road, and A.P. Hill, ordered to follow him, did the same.

Below: The aftermath of The Seven Days' was scenes like this one, as the dead lay scattered over much of the Peninsula. The level of fighting paled the previous experience of both armies, and gave a taste of what the next three years held for them.

As a result, it was Magruder alone, assisted by a few other units, who opened the battle, facing three divisions from Sumner's and Franklin's corps. Magruder was himself exhausted from the tension of the three previous days of facing a virtual army all by himself, and today he did not perform well. His assault was half-hearted, and almost ignored by some of McClellan's commanders, with the result that the Yankee retreat continued unimpeded and Lee lost, or so he believed, a real chance to cut the enemy off from the James.[11]

He tried again on June 30 near Glendale, this time directing seven divisions to concentrate against four posted by McClellan to act as rearguard while his supply trains and the bulk of the army continued their flight to the James. Again it all failed to work. Only Longstreet and the ever-combative A.P. Hill managed to get themselves to the scene of action, where they launched a series of vicious attacks through the afternoon that did push the Yankees back and inflicted serious casualties, but the Confederates themselves suffered some 3,500 or more without achieving Lee's aim. Incredibly, McClellan had abandoned the land entirely, leaving his corps to fend for themselves almost without orders while he went to the comfort of a gunboat on the James and began organizing his new supply base. His absence probably accounts for much of the Federal success in thwarting Lee's designs.

But Lee would not be deterred. His frustration showed itself early on July 1 when he exploded that "I cannot have my orders carried out!" Unmindful that his orders called for what was practically unattainable, he was unwilling to abandon his hope of crushing McClellan without one last effort. The tension of the last week clearly showed. Having tried clever strategy only to meet

15th Virginia Infantry, C.S.A.

These two soldiers probably started the war in other regiments, most likely the 33rd Virginia or the 179th Militia. The 15th was in part formed by amalgamating companies from these other regiments. They got a fairly regulation issue uniform of gray, with blue piping. The wounded field officer at left wears a two-piece Virginia state seal belt plate, and is supporting himself in part on a converted flintlock musket altered to use the Maynard tape primer system. The private helping him wears a more unusual white belt, and if he were in full uniform he might even have matching white cross-belts. His Virginia Manufactory musket is being held by his officer. That officer, if *he* were in full uniform, might very well have fringed epaulettes on his shoulders. The 15th Virginia was not destined to be one of the leading glory regiments of the Confederacy, but it did its full share in the struggle to keep Virginia inviolate from Federal invaders.

massed enemy cannon, and Lee funneled more isolated commands into the slaughter. Only nightfall mercifully ended perhaps Lee's worst tactical performance of the war. It had been a butcher's picnic. Nearly 5,500 of Lee's men were killed or wounded, as against Federal casualties of half that number. The acerbic General D.H. Hill, especially bitter over the bulk of the losses falling in his command, declared that Malvern Hill "was not war – it was murder."[12]

When night closed on that terrible field, even Lee finally gave up further attempts to stop McClellan. In seven days of almost constant fighting – soon the battles collectively would be called The Seven Days' – he had lost a fourth of his army in killed and wounded. He had not destroyed McClellan as he hoped, and on July 2 "Little Mac" completed his withdrawal to Harrison's Landing where Lee could not assail him. Still Lee had achieved much. The threat to Richmond was ended, at least for the moment. McClellan had suffered 10,000 casualties himself, and stood with his prestige in the North seriously damaged and the confidence of Lincoln deeply shaken. True, he was still on the Peninsula, but he would never mount an offensive there again.

Yet despite Lee's disappointment when he reported to Davis that he had not achieved all that he had hoped, he was the one person who could not see the greatest benefit to the Confederate States from The Seven Days'. It was the making of Robert E. Lee. Though he only won one actual battle, at Gaines' Mill, still he drove McClellan back and saved the capital against seemingly overwhelming odds. That eradicated forever all recollections of "Granny" Lee. It cemented his hold on the command of the army that he would subsequently lead to imperishable glory, an army that Joseph E. Johnston had seemed disinclined to lead at all. And there, in the end, lay the great benefaction of that shell that took Johnston out of the war for the next several months, though it was a tragically mixed blessing. Out of Johnston's blood, the South got Lee, who undoubtedly prolonged the war in the East far longer than Johnston or any other might have done. Yet out of that prolongation would come untold suffering, for The Seven Days' brought an end to the "phoney war," and the revelation of a whole new scale of vicious fighting, and the longer the Confederacy lasted in this new level of warfare, the more of its sons would mingle their blood with Johnston's. After The Seven Days', there was barely a man in the South who would not readily bleed for Robert E. Lee.

with repeated disappointment, he now resorted in desperation to an almost utter lack of finesse. The Federal rearguard had taken a position atop Malvern Hill, an imposing elevation where they massed eight divisions and more than 200 cannon. Looking at this position, Lee somehow deluded himself that the Federals could be driven off by a resolute attack. Longstreet seconded the idea, himself locating a position from which nearly 60 Confederate artillery pieces could bombard the Yankee lines. Most of his own divisions were in the vicinity now, including even Jackson for a change, and making his decision, Lee issued the oddest attack order of his career. His batteries might be able to punch a hole in the Yankee defenses, it said. If so, Brigadier General Lewis Armistead commanding one of Huger's brigades was in a good position to see it. He would charge "with a yell," said Lee, simply ordering his other commanders to "do the same."

Below: The Peninsula looked as if a hurricane had swept over it, the trees and brush mowed down as if with a scythe as the armies turned everything at hand into hasty defenses in their deadly dance across the landscape.

Above: Cold Harbor, which would be the scene of some of the war's bloodiest moments two years later, also played host to the grimmest realities of war in 1862. Months later the hastily buried dead were reinterred.

It was a disaster. Federal artillery fire quickly silenced the Confederate barrage, negating the conditions for Armistead's order to advance. But Lee forgot or neglected to cancel that order. Then, quite unexpectedly, Yankee sharpshooters advanced on their own directly toward Armistead, who repulsed them and pursued. Just at that moment Magruder came on the field, armed with Lee's order to follow Armistead's lead. Unaware of the utter failure of the artillery barrage, he sent word to Lee that Armistead appeared to be advancing successfully, and a desperate Lee thought he saw a chance for victory still, and ordered an attack. Magruder went forward just before 5 p.m. The Yankee artillery mowed them down like wheat. Then D.H. Hill arrived on the left. Hearing the firing and all the yelling, he obeyed Lee's terse order and launched his own brigades into the advance against Malvern Hill. Huger, too, sent his division up against those

References

1 Tyler Dennett, ed., *Lincoln and the Civil War in the Diaries and Letters of John Hay* (New York, 1939), p.33.
2 George B. McClellan, *McClellan's Own Story* (New York, 1887), pp.168-77.
3 *O.R.*, I, 11, pt.3, p.456.
4 Alexander, *Memoirs*, p.102.
5 Francis W. Pickens to Milledge L. Bonham, July 7, 1861, Milledge L. Bonham Papers, South Caroliniana Library, University of South Carolina, Columbia.
6 *O.R.*, I, 11, pt.1, p.51.
7 Clifford Dowdey, *The Seven Days* (Boston, 1964), pp.161-62.
8 *Ibid.*, pp.200-202.
9 *Ibid.*, pp.236-38.
10 *O.R.*, I, 11, pt.3, p.266.
11 Dowdey, *Seven Days*, p.283.
12 Buel and Johnson, *Battles and Leaders*, II, p.394.

_____ CHAPTER FOUR _____

ANTIETAM (SHARPSBURG)

SEPTEMBER 17, 1862

Never in the course of the war would there be a greater season of woe for the Union than the summer of 1862. McClellan's series of defeats in The Seven Days' left him virtually powerless on the Peninsula, yet he would not budge his army. Instead, Washington created a new Army of Virginia and placed at its head Major General John Pope, the victor in small engagements on the Mississippi that had helped secure Yankee control of the river in Tennessee and Missouri. Tactless and blustering, Pope alienated most of his officers from the start, leading one subordinate general to declare him not worth "a pinch of owl dung."

Still Pope advanced to meet the Confederates in northern Virginia, only to be beaten by three men. At Cedar Mountain on August 9, Stonewall Jackson met and defeated a large portion of the Yankee army in what would be Jackson's last battle in independent command. Then at the end of the month Pope met Lee on the old Manassas battleground, and once more the Federal army was driven from the field, though this time not in humiliating rout. And through it all, Pope's most potent enemy was McClellan, who refused to help him, and instead seemed to take actual delight in seeing him beaten. Meanwhile, in faraway Mississippi, bureaucratic problems were keeping Grant from following up the Shiloh advantage. Lincoln desperately needed a victory from someone, somewhere.

GEORGE B. MCCLELLAN showed seemingly remarkable staying power. Even after his failure on the Peninsula, even after his petulant and insubordinate communications with Washington, even after his snubs to President Lincoln himself, still "Little Mac" seemed able to dominate Yankee high command. He never stopped complaining over his peremptory orders to remove his army from the Peninsula and return with it to northern Virginia and the capital area. The true strategic line of advance was still via the York and James, he would argue; his hidden objections lay in being told to co-operate with a new commander, Major General John Pope, and his growing army. McClellan intentionally dragged his feet and raised every objection possible to impede carrying out his orders, and by late August 1862, as Pope was about to be thoroughly beaten by Lee at Second Manassas, McClellan was still inventing excuses to withhold two full corps that might have made the difference for the unlucky Pope.

Almost everyone saw through McClellan's behavior, yet he got away with it, especially after news of Pope's defeat reached Washington. It fulfilled McClellan's own prophecy of disaster if control of any of his troops were turned over to another, a prophecy he did his best to make happen. And thus after Pope's debacle, with the renewed hysteria over the threat of a Rebel army marching on Washington once more, McClellan came once more to the fore. Despite their anger at "Little Mac's" part in the disaster – Lincoln privately accused him of relishing in Pope's defeat – the administration asked him to take overall command of all forces once more.[1] It was a humbling moment for Lincoln; McClellan exulted in his triumph, though with customary false humility he proclaimed that "I only consent to take it for my country's sake."[2]

On September 2, 1862, McClellan once more resumed command of all Union forces in northern Virginia and Washington, receiving rousing cheers from the troops who never ceased loving him, even in the face of defeat. The euphoria could not last long, however, for the very next day General-in-Chief Halleck warned McClellan that with Pope's army cleared out of the Manassas area and Lee in the ascendant, they should expect the Rebels to capitalize on their recent success by invading Maryland and even Pennsylvania. McClellan must rush to be ready.

The waters of Antietam Creek flow past one of its bridges at Sharpsburg shortly after the great battle was fought. The scene of tranquility is deceptive considering that along these banks will pass the bloodiest day of the war.

Rushing, of course, was not in the Young Napoleon's makeup. Organizing, on the other hand, most certainly was, and if a battle could be won on paper, McClellan was the man for the job. He took the battered but not demoralized remnants of Pope's Army of Virginia and shuffled them back with his own command to make a revitalized Army of the Potomac. His I Corps, almost entirely New Yorkers and Pennsylvanians, he gave to Major General Joseph Hooker, who won the nickname "fighting Joe" in the newspapers during the Peninsula Campaign. Its three divisions were commanded by Generals Abner Doubleday, James B. Ricketts, and George G. Meade. Old "Bull" Sumner led the II Corps, three divisions under Israel Richardson, John Sedgwick, and William H. French. McClellan's favorite Fitz John Porter stayed at the head of the V Corps, with Generals George Morell, George Sykes, and Andrew Humphreys commanding the divisions. Franklin took the VI Corps, with Henry Slocum, William F. Smith, and Darius Couch at the head of his divisions, while another McClellan favorite, Burnside, led the IX Corps and its four divisions under Orlando Willcox, Samuel D. Sturgis, Isaac P. Rodman, and Eliakim Scammon. Old but valiant Joseph Mansfield completed the infantry complement with his XII Corps, the smallest at only two divisions under Alpheus Williams and George S. Greene. Each division in the army had its own artillery except for the XII Corps, which combined all its guns into a single separate command. McClellan also put together five brigades of cavalry as a separate mounted division, led by Brigadier General Alfred Pleasonton.

It was a magnificent army, once again well equipped, well fed, trained to perfection (for volunteers), and filled with high spirit and élan. The only question was whether or not McClellan would use it. No such quandary existed in anyone's mind about that other great force in the East, the Army of Northern Virginia. Lee could be counted on to use it, and after his performance at Second Manassas, to use it with daring and imagination. He finally abandoned the cumbersome organization by divisions that he inherited from Johnston. While not yet authorized by Davis and Congress to create formal corps, he did so informally by forming two corps-like commands. The larger went to Longstreet, the steady and

Above: One of McClellan's oldest and most trusted friends, General Ambrose Burnside looked every inch a soldier. Unfortunately, his brains never matched his looks, as he proved at Antietam.

dependable – if slow – South Carolinian. He inherited control of the divisions of Generals Lafayette McLaws, Richard H. Anderson, David R. Jones, John G. Walker, and the hard-hitting Hood, along with the independent brigade of the very independent "Shanks" Evans of Bull Run fame. Lee's second "corps" went to Jackson who, though apparently forgiven for his deplorable

Below: If possible, the Army of the Potomac was even more magnificent than when on the peninsula, and it loved "Little Mac". The legions in its camps were ready and anxious to fight for him.

performance on the Peninsula, still received the lesser command, just four divisions led by Generals Richard S. Ewell (who lost a leg at Second Manassas and was replaced by Alexander R. Lawton), A.P. Hill, John R. Jones, and D.H. Hill. Each division had its own artillery, and each corps an artillery reserve. As usual, Lee's gallant cavalryman Stuart led the division of horse.

Even if Lee had not been the pugnacious fighter that he proved to be, events in 1862 would have forced him to follow up his double gains over McClellan and Pope by taking the offensive. Domestically, the victories heartened the Southern people, who were just now realizing that this would not be the brief war their politicians had promised. The close call around Richmond had taken some of the pomposity out of their attitude, replacing it with a sense that they could still win their independence, but only by beating the Yankees and beating them again until the foe no longer had the stomach for fighting. Furthermore, having regained most of northern Virginia, they must move quickly to maintain their hold. Yet the ground just gained had been ravaged by the armies after a year of war, and could not sustain Lee's need for provisions. He must draw his rations elsewhere. Then there was the international dimension to consider. The South desperately needed European help in winning its independence, yet Britain and France were slow to lend their potentially decisive aid, each waiting to be sure it would be backing a winner. The victories of the spring and summer worked to this purpose, but it would be even more effective if the Confederacy demonstrated that it could take the war to the enemy on its own homefront.

All of these influences and more combined toward the inevitable logic of invading the North. There Lee could draw fresh supplies and make the Yankee civil population feel the hard hand of war. There he could draw the foe away from any advance back into Virginia. And a victory gained there must assuredly convince world powers that the Confederacy was a safe bet for formal recognition and military assistance. "The present seems to be the most propitious time since the commencement of the war for the Confederate Army to enter Maryland," Lee told Davis in outlining his invasion plan on September 3. He might even go on into Pennsylvania, thus posing threats to both Washington and Baltimore, and in such a position, Davis could then bargain from a position of great strength in proposing an end to the war in return for Southern independence. But this was not to be just a raid or a show campaign. "I went into Maryland to give battle," he would say later. He expected to meet McClellan somewhere on those Northern fields, and he expected to beat him.[3]

Lee confidently believed that McClellan's army was still demoralized and dispirited from its defeats of the last three months. If the Rebels struck now, "Little Mac" – given his sloth, which Lee took for granted – would not be able to respond to the invasion for three or four weeks, more than enough time for Lee to wreak havoc all the way to the Susquehanna River.

Never willing to waste a moment, Lee put his campaign in motion the very day that he notified Davis of his intention. Indeed, the precise details of the movement were still forming in his mind as his legions took the road northward to Leesburg. The next day they reached and began crossing the Potomac at White's Ford, twenty-five miles northwest of Washington, and already well

Privates, North Carolina Infantry, C.S.A.

North Carolina troops were among the best clothed of all Confederates, and enjoyed perhaps the greatest degree of uniformity of dress among their regiments. These "Tarheel" infantrymen are quite typical in their gray sack coats, reaching halfway down the thigh, with loose collars that were just as often worn turned down as standing up. On their shoulders they wore strips of cloth in colors denoting branch of service. Contrary to usual Confederate regulations calling for blue as the infantry color, these infantrymen were to wear black shoulder strips.

The trousers were of matching gray cloth, with black stripes down the seams. Some soldiers were equipped with single or double crossbelts, and while weaponry varied, most carried the 1842 musket.

Only because North Carolina could produce its own textiles and had ports for blockade-run items, could it arm and equip its men so effectively.

in McClellan's rear. It was a heady moment for Lee's ragged veterans. At last they would give the enemy to know how it felt to play host to a hostile army.

General Halleck had predicted exactly what Lee would do, and warned McClellan on the very day that Lee told Davis of his plans. Confirmation reached Washington the next day as witnesses reported seeing the Rebels crossing the river. Almost immediately Lincoln gave McClellan orders to put the army in the field, and "Little Mac" acted with unwonted alacrity, immediately putting the lie to Lee's prediction of a several week paralysis. The Army of the Potomac was in much better condition and morale than he realized. McClellan was still reorganizing the army, and refitting and equipping it. Now he worked feverishly, and by September 7 was actually ready to move in response to Lee.

Other than the alacrity with which it happened, this was exactly what Lee wanted. He could have moved into Maryland by the Shenandoah, masking his march and keeping the enemy in the dark longer. But he wanted to make McClellan pull out of the Washington defenses to have to chase him, and thus the wily chieftain virtually broadcast his line of march. But what he secretly intended was that once at Frederick, Maryland, with McClellan rushing to catch up, he would then turn west, cross the low range of the Catoctin, to South Mountain. It could be crossed only at a few gaps, most notably Turner's and Crampton's. By crossing over, then closing those gaps behind him, he would leave McClellan trapped on the eastern slope while Lee was free to rest his army, receive supplies sent up via the Shenandoah, and then move northward into Pennsylvania. When an exhausted and attenuated Yankee line finally caught up to him to fight, Lee could then force McClellan to fight on ground of Confederate choosing, and at a great disadvantage.

This is almost exactly how it worked. Lee entered Frederick on September 6, to stay for five days. There the men rested and replenished their

74

Confederate and Southern State Bonds

1 Six percent coupon bond issued under Act of Feb. 17, 1864. Printed in Columbia S.C.

2 Six percent coupon bond issued under Act of March 23, 1864. Central vignette is the Old Customs House, Richmond

3 Seven percent coupon bond issued under Act of Jan. 29, 1863. Trivalued in pounds sterling, French francs and 1000lbs of cotton; sold in Europe to raise funds

4 Alabama state eight percent stock certificate issued for military defense; Act of Jan. 29, 1861

5 Eight percent coupon bond issued under Act of Aug. 19, 1861. Printed in Richmond, Va

6 Georgia state seven percent coupon bond – Act of Dec. 11, 1861

7 Eight percent coupon bond – Act of Feb. 20, 1863

8 Seven percent coupon bond issued under Act of Feb. 20, 1863

9 Eight percent coupon bond issued under Act of Feb. 20, 1863

10 Eight percent coupon bond issued under Act of Aug. 19, 1861. Printed in South Carolina

11 Eight percent coupon bond issued under Act of Aug. 19, 1861

Artifacts courtesy of: The Civil War Library and Museum, Philadelphia, Pa

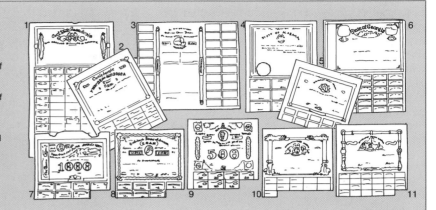

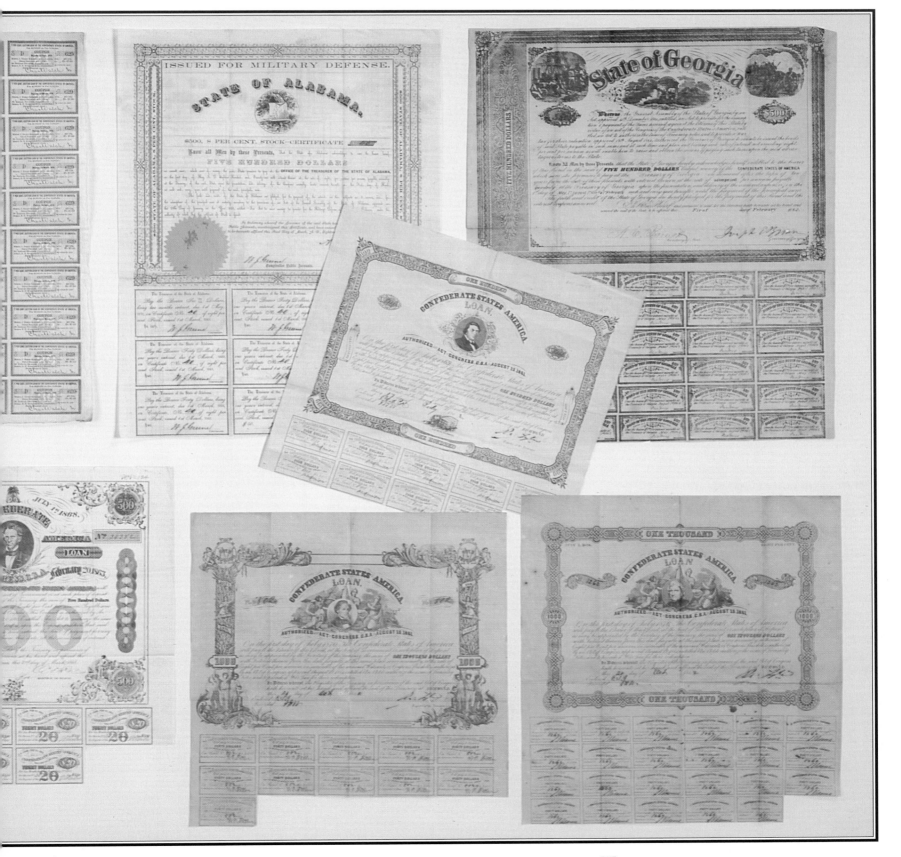

haversacks from Maryland's abundant fields. Lee watched with pleasure as the Army of the Potomac moved slowly toward him, being still more than twenty miles distant by September 9. But what did not please the Confederates was what did not happen at Harpers Ferry. Having effectively cut it and its garrison off from Washington, he had expected that its defenders would abandon the post without a fight, leaving its munitions to Lee. Furthermore, in the event of having to abandon the invasion, his natural line of retreat would be via Harpers Ferry into the Shenandoah. He could not leave that garrison, small though it was and commanded by the drunkard Dixon Miles of First Manassas infamy, in his rear. He decided now to take it since it would not leave of its own accord. He gave Jackson the job of taking nearly half the army, by three separate routes, to strike Harpers Ferry from all sides simultaneously. Jackson was given such overwhelming force that they expected to accomplish the task quickly, by September 12, after which Jackson would march north to meet Lee west of South Mountain to resume the campaign.

It was a brilliant, if desperately chancy, conception, but Lee had already divided his army in the face of the enemy before and emerged the victor. He only had 50,000 with him now, though, and if McClellan's army caught up with Longstreet before Jackson returned, the Rebels would be outnumbered three-to-one.

Fortunately, McClellan, though moving, moved slowly and thus gave Lee time. McClellan continued to organize his army even while it moved. He created three "wings." Franklin commanded the first, his own VI Corps and an additional division. Sumner took the center wing with the II and XII Corps. And Burnside led the right wing, his own IX plus the I Corps. Porter and his V Corps remained behind to guard Washington for the moment, but still McClellan has almost 85,000 men on the move. Even then, he believed that Lee had at least 100,000, and probably more.[4]

By September 11 McClellan was only fifteen miles from Frederick, now convinced that Lee actually had 150,000. Though reports of Jackson's movement toward Harpers Ferry reached him almost daily, he remained uncertain about Lee's overall intentions, knowing only with certainty that Lee had left Frederick. Many around him, including McClellan himself, seemed to think that Lee was already retreating back toward the Potomac. Thus when McClellan finally marched into Frederick on September 12-13, his own plan of campaign remained unformed. There could not have been a more propitious time for one of his aides to walk into his headquarters on the morning of the 13th to hand him a dispatch from General Williams, commanding the XII Corps. An hour or so earlier, a corporal in one of his Indiana regiments stumbled upon a bundle of three cigars lying in the grass near his camp. The wrapping turned out to be no ordinary paper, but an actual copy of Lee's order detailing the plan for taking Harpers Ferry, and his own intended movements. The news was stunning. McClellan held in his hand exact details on where every Rebel division was, was headed, and its timetable. Lee had divided his army into two weak halves, and armed with this information, McClellan could move to defeat him in detail with overwhelming force. "Now I know what to do", he exclaimed. "Here is a paper with which if I cannot whip Bobbie Lee, I will be willing to go home."[5]

Above: Sharpsburg was a small, sleepy village beside Antietam Creek when the armies came. But here, on the war's bloodiest single day, they left it a graveyard in their wake, with more than 2,000 dead.

But then McClellan proceeded to fritter away all the fruits of his good fortune. Instead of moving immediately, as he wired Lincoln he would do, he wasted nearly a full day before putting his army on the road again. He expected to reach South Mountain on September 14 and fight Lee the next day, while sending Franklin's wing to save Harpers Ferry. He still believed that the half of the army with Lee equalled or outnumbered his other two wings, but expressed himself more than willing to risk a fight.

Just then Lee was being severely inconvenienced by more than the lost order. The Harpers Ferry expedition fell seriously behind schedule. Even with the overwhelming numbers committed to the side operation, and the equally monumental incompetence of Dixon S. Miles, did not make it the easy walkover that Lee expected. The Rebels first approached on September 12, driving a garrison of Yankees from nearby Martinsburg before them. But it took time for all of the Confederates to come up, and lesser Yankee commanders, in spite of Miles, put up a better resistance than expected. The defense continued through the next two days, and only on the morning of September 15 did the beleagured garrison finally surrender. Jackson bagged 11,500 prisoners, thousands of rifles, more than 70 cannon, and tons of munitions and *materiel.* One of the few casualties on either side was old Miles himself, mortally wounded by one of the last artillery shots fired. Jackson promised to lead five of his divisions toward a junction with Lee later that same day, while A.P. Hill would remain behind to oversee getting prisoners and materials off to safety before leading his division north.

Lee should have been pleased, but if so he kept it to himself. He was starting to realize that he and his army faced the potential for serious trouble. He had nine divisions in his army, and had committed six of them to the Harpers Ferry operation. That left only two under Longstreet and one of D.H. Hill's as the balance of his command, and Longstreet had mistakenly taken his two all the way to Hagerstown, ten miles north of where Lee wanted them. His command was split into three isolated portions, and McClellan – while still agonizingly slow – was moving quicker than Lee anticipated. Indeed, the day before Harpers Ferry fell, McClellan had an opportunity to hit several separated portions of the Army of Northern Virginia at once. A man of real moral courage might have taken them all.

On September 14 McClellan put his army on the road out of Frederick toward's Turner's Gap, while Franklin's corps marched in desultory fashion a few miles to the south, on a parallel route to Crampton's Gap. When the main van approached Turner's, led by Jesse Reno, now commanding Burnside's IX Corps, there was nothing in front of them but D.H. Hill and two brigades. But those Rebels were determined to fight, for the gap meant everything to Lee just then. Hill and his gallant 2,000 fought like demons, often hand to hand, delaying McClellan's advance for more than two hours. Then the Federals blew their advantage by calling a lull in the fight to wait for the arrival of more supports. That gave Hill's other brigades time to reach the gap, with Longstreet hurrying to join him. In the end, when the Federal attack renewed, both the I and IX Corps were involved, with still a heavy numerical advantage over the Confederates. Hill and his supports managed to set up sufficient defenses that they held off the enemy advance until nightfall.

Seven miles to the south, Franklin made an equally tardy approach to Crampton's Gap, and it is well that he did. The gap had been virtually undefended until General Lafayette McLaws rushed his division from Harpers Ferry. Though Franklin still outnumbered McLaws four to one, he believed himself to be the underdog, and withheld his attack, giving the Confederates ample time to rush their reinforcements. Franklin did not attack in the end until well into the afternoon, and then only pushed it hesitatingly. Nevertheless, by nightfall he had forced the Rebels out of the gap and into the valley beyond. But there he stopped, unwilling – or too afraid – to follow up his success. He had gotten himself between two halves of Lee's army, giving McClellan a clear road to victory. Instead they all decided to wait until the morrow to plan their next move.

the past, he planned a Napoleonic massed cavalry attack on Lee's center should either of the flank assaults achieve success. McClellan seems utterly to have missed the lesson of the last year of warfare that the rifled shoulder arm had made cavalry charges obsolete. He also seemed unconcerned that his mounted men would still have to ride no more than four abreast across the Middle Bridge, right into the face of Rebel cannon, in such a charge. It was absolute foolishness. With the time and men at his disposal, even McClellan should have been able to produce something better.[6]

Making matters worse, he proceeded to tamper with his high command just the day before the fight. He broke up Burnside's wing, making Hooker independent, and returning Burnside to the command of the IX Corps, whose commander, Reno, had been killed at Turner's Gap. Now Hooker was to handle the attack on Lee's left, and Burnside the attempt to turn back the enemy right. Furthermore, Hooker was to be backed up by Sumner with the II and XII Corps, but they were all to be subject to Hooker's orders, effectively removing "Bull" Sumner from command. Franklin was to form the reserve, along with Porter's recently arrived V Corps and, of course, Pleasonton's cavalry. All combined, excluding non-combatant ranks, McClellan had about 75,000 fighting men at hand – five times what faced him across Antietam Creek.

Lee's dispositions by nightfall on September 16 revealed just how weak he was, and just how daring. Longstreet commanded his center and right, a long thin line with only the three divisions of Walker, Jones, and D.H. Hill right in the center along a sunken road. When Jackson and the first of his divisions arrived that afternoon, Lee placed them on his far left after McClellan's movement of Hooker across the Upper Bridge in broad daylight gave away his intent to strike Lee there first on the morrow. Jackson put Jones' division in a small forest known locally as the West Wood, just to the left of the Hagerstown road, with Lawton on the right behind a cornfield. Hood took position in Jones' rear in the wood alongside a Dunker Church. Even the addition of elements of Jackson's command from Harpers Ferry still only brought Lee up to 26,000 by nightfall. He desperately needed those absent divisions of McLaws, Anderson, and A.P. Hill.[7]

It was a restive night on both sides of Antietam Creek. Rarely in this war did two armies come this close without immediately launching into one another. Lee and McClellan stared at each other fully a day and a half, taking their time, making their dispositions. Thus their men knew when the battle was going to start, and that when it did commence, everything would be in place for each to do the maximum damage to the other. No one exactly predicted that the next dawn would inaugurate the bloodiest single day in American history, but most sensed that it would be something very definitely out of the ordinary.

They would argue over who fired the first shot, a pointless exercise, since there were probably dozens of "first" shots, all uncoordinated. The artillery started it sure enough, even while the morning fogs were still lifting. Within a few minutes, around 6 a.m. or later, Hooker began to move forward, driving straight along the line of the Hagerstown Turnpike toward the Dunker Church. Meade moved in first, heading toward the cornfield where he ran into a Virginia brigade that held him at bay despite repeated assaults.

Above: Antietam Creek could only be crossed at a few fords and at bridges like this one, destined to be famed as Burnside's Bridge after his abortively wasteful attempts to get across when he could have waded nearby.

Lee would not wait. Knowing the danger that faced him, he decided that night to abandon Turner's Gap and pull his army together. The next morning, with Jackson soon to be on the road, he ordered all of the remaining divisions to concentrate near Sharpsburg, seven miles northwest of Turner's Gap and eight miles due north of Harpers Ferry. It was a good location for a concentration. Sharpsburg sat astride a number of roads running to all points of the compass. The village itself lay barely a quarter mile west of Antietam Creek, a stream just wide and deep enough to discourage crossing anywhere but at three places. The Upper Bridge sat more than two miles northeast of town. More than a mile downstream, and directly opposite, was the Middle Bridge. Three-fourths of a mile further down lay the Rohrbach Bridge. There was one good ford just below the Upper Bridge, and two more well below the Rohrbach. Still, with only about 15,000 men with him when he selected Sharpsburg, Lee was wise to pick a place that offered only limited crossings. That alone would put McClellan at a disadvantage, and help even the disparity of odds. Furthermore, the ground just west of the creek offered a slight rise or ridge along the Hagerstown Turnpike, running north to south through the village, and directly parallel to the creek. Lee could position his thin ranks along this high ground, thus commanding the bridges and fords. Along the way a succession of woods, fences, "sunken" roads, and more, offered good defensive terrain.

It was clear that Lee came to Sharpsburg to defend. It was not what he had wanted for his campaign's climactic battle, but in the circumstances, with his army fragmented and heavily outnumbered, and with McClellan reacting faster than he had hoped, any thought of himself taking the offensive was totally impracticable. Of course Lee was always a dangerous opponent, and never more so than when in trouble. But for now he had taken risk enough in simply deciding to stay on Yankee soil. Moreover, his position offered serious vulnerabilities as well as strengths. The Potomac ran in an arc only a mile or so west and south of Sharpsburg, swinging eastward to the point, four miles below the village, where the Antietam flowed into it. The two streams made a peninsula of sorts, but one that was open at

the top. Lee could not hope to prevent McClellan from crossing the Antietam at every point upstream – he simply did not have enough men. The Yankees would get across somewhere on his left flank; that was a given. So Lee selected the best ground he could find, about a mile north of Sharpsburg, and there anchored his left, virtually abandoning the Upper Bridge to McClellan. But the gray chieftain's far greater vulnerability lay at the opposite end of his line. If the Yankees could get across the Rohrbach Bridge, or thereabouts, and force their way just 1,500 yards on the other side, they could take the road leading to Harpers Ferry, cutting off Jackson's marching divisions. Worse, if McClellan could press a bit farther to the Shepherdstown Road, he would cut Lee off from all avenues of retreat to the other side of the Potomac and safety. Take those roads, and McClellan would have Lee trapped with his back to an uncrossable river. Even a bad general could make of that an end to the Army of Northern Virginia.

Whether or not McClellan was that bad a general remained to be seen, but when he brought the advance of his own column up to the Antietam on September 15, he showed no signs of hurry. Instead, as he had all along, he proceeded to make Lee the gift of vital time. McClellan did nothing for the rest of that day, and spent all of September 16 studying and thinking, playing with the perfect alignment of his own corps, and methodically planning what was to be, in fact, his very first offensive battle.

Even McClellan could appreciate the weakness in Lee's position, while he continued to entirely misapprehend the weakness of the Rebel army. "Little Mac" planned to mass for an attack on Lee's left where he would not have to contest a crossing for the Antietam, while at the same time – more or less – moving across the bridge in front of the enemy right. Neither attack, however, was to be coordinated, and neither seemed aimed at the possible isolation of Lee from support or retreat. Instead, revealing his utter immersion in

Hooker soon had Doubleday and Ricketts with their divisions ready to go into the fight, elements of the latter's division going straight into the high standing corn stalks toward unseen Confederates. They were slaughtered. The brigade in the actual field lost a third of its number in the next few minutes and soon left the field – and the battle – for good.

What caused the fearful toll in that brigade was a lesson that no one seemed yet to have learned after more than a year of war. Generals were sending their regiments and brigades in one by one, instead of attacking in mass formation to take advantage of their numbers. Rickett's lead brigade went into the cornfield with no supports

at hand, and thus allowed the enemy – outnumbered overall – to concentrate their fire on the isolated unit. Then in the two brigades sent to support the advance, one brigade commander was wounded and the other turned and ran in terror, leaving them temporarily leaderless.

Soon Lawton sent a new brigade to support the by-now badly blooded defenders of the cornfield, launching a counterattack that regained some ground for a time. The nature of the fighting became brutally intense, though not yet hand to hand. Some regiments almost ceased to exist, the 12th Massachusetts taking 67 percent casualties in one hour. Finally the fighting for the cornfield settled as each side held to opposite fringes of the

acreage and kept firing at the other. The fury of the past hour had simply exhausted them. Meanwhile, Doubleday went forward toward Jones in the West Wood. Barely had fighting begun when Jones himself was put out of action, and General W. E. Starke had to take over from him, only to fall himself soon thereafter with mortal wounds. Before that happened, however, Doubleday's people ran into a seemingly solid wall of flame and lead as they hit the western edge of the cornfield and the West Woods just across the Hagerstown road. Only persistence gained them a grasp on the edge of the woods, a perch from which to launch further drives forward. Starke fell in an attack aimed at driving them out. Even as he was

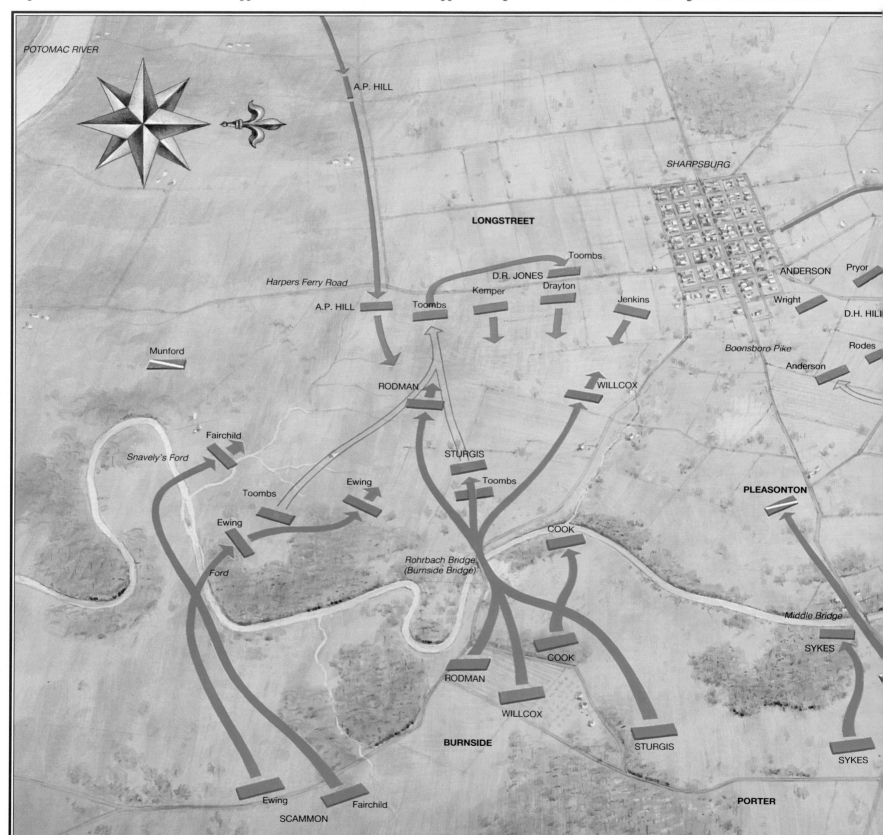

Battle of Antietam (Sharpsburg), September 17, 1862

Robert E. Lee's great daring almost got him into fatal trouble when his invasion of the North, and his splitting of his army in the face of McClellan, found him with only half his command in Maryland, with his back almost literally to the Potomac. McClellan uncharacteristically opened the battle with an offensive, but soon abandoned almost any direct control of the fighting, leaving it to his corps commanders on the field. The

fighting started when Hooker's corps at extreme right center, and Mansfield just below him, crossed the creek and pushed forward into the North Wood, the East Woods, and finally the terrible Cornfield.

The Yankees pushed forward to the Dunker Church before determined Rebels pushed them back. Meanwhile the II Corps under Sumner crossed the creek from bottom right center and rushed across the fields until it encountered D.H. Hill's

determined but desperately outnumbered Confederates in the Sunken Road. Here the fighting raged like a tornado, sucking every available unit into it for hours.

At the other end of the lines, where Lee had depleted his forces to reinforce his left, McClellan gave the enemy valuable time by doing nothing. Not until late morning did Burnside start trying to get across the creek, and then instead of looking to find the fording places available, he

persisted in sending his men across the Rohrbach Bridge. Here, at left center, Burnside squandered his men, brigade after brigade, while Rodman managed to get his people across at Snavely's Ford less than a mile downstream, and virtually unopposed. As a result, it was late afternoon before Burnside finally got across. Even then he was moving well toward Sharpsburg and Lee's rear, reaching the outskirts of the village when A.P. Hill

providentially arrived at the last moment from Harpers Ferry and hit the Federals a devastating counterblow that stopped Burnside.

Hill saved the Armyof Northern Virginia from almost certain destruction. Reacting quickly to his fortuitous delivery, Lee rushed reinforcements – especially artillery – into the line and succeeded in pushing Burnside back almost as far as the Rohrbach Bridge. There the battle ended at nightfall,

with both sides exhausted and unable to continue the slaughter. Lee, however, refused to use the cover of night to make sure of his escape and held the field for another day in defiance of McClellan. Here he stayed moving wounded and supplies until the nightof the 18th-19th when his battered army finally made its way out of Maryland.

Lee lost more than 10,000 casualties out of his 50,000. McClellan suffered 12,400 from his 85,000.

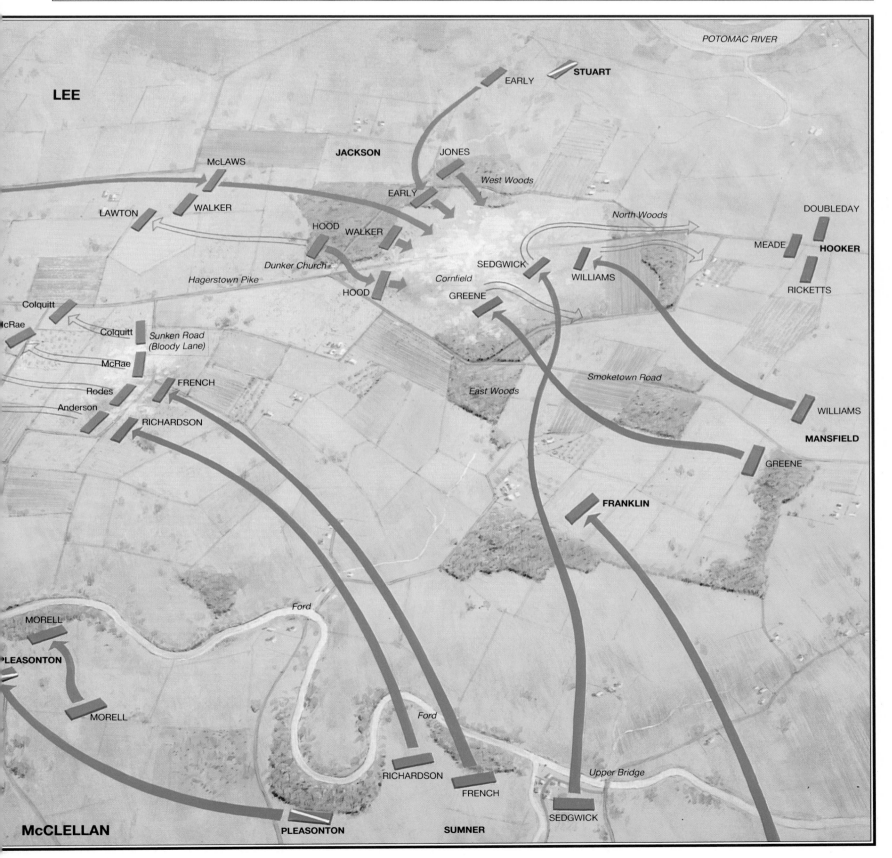

carried from the field, dying, he had the fleeting satisfaction of knowing that he had succeeded. He had stopped the Yankees cold in his front.

Yet the situation was desperate. Lawton had been so badly mauled in the cornfield that he could not hold on much longer. Jones' division, now being led by a colonel, still held its grasp on the West Woods, but the Federals could clearly overlap his left, and worse, the line was so weak at the cornfield now that his right could be turned too. Fortuitously, at this moment, just after 7 a.m., Hood brought his division screaming up from the rear. Barely halting to form ranks, they hammered into the Federals in the cornfield and the East Woods on its right and drove forward. Hooker, stunned by the ferocity of the attack, threw in the last of his corps reserves and finally called on Mansfield's XII Corps for reinforcements. The level of fighting in the next half hour became the most brutal yet seen on the continent. Men fell in rows where they stood, laid down by savage enemy volleys. Between the remaining fog and the growing pall of low-hanging battle smoke, few could see clearly any distance. Some even noticed a queer phenomenon. Fearful at first of going into battle, some men found that when it began, they lost their terror and, instead, were seized by a peculiar fearlessness and compulsion heralded by everything in sight briefly taking on a crimson hue. Literally, they "saw red".

Hood's men almost reclaimed the entire cornfield, only to have Meade's last brigade destroy them. One regiment lost 80 percent in only a few volleys. Finally, by about 7:30, Hood began to fall back, still stubbornly holding on to a piece of the East Woods. Both Jackson and Hooker were exhausted, beaten to tatters in ninety minutes. The Yankee lost a third of his numbers in killed and wounded; Jackson lost nearly as great a percentage, and Hood's division was so mauled that, when asked where it was, he sadly replied, "dead on the field."[8]

McClellan's battleplan – such as it was – already lay severely compromised with the halting of Hooker. This meant that Mansfield and his XII Corps must now go into the fight, not to capitalize on Hooker's success, but to redeem him from utter exhaustion. Mansfield himself led them forward at about 7:30, coming in on Hooker's left and advancing straight into the East Woods. It was a confusing movement, passing through the wood without benefit of good roads or reconnaissance. His right elements actually

swung in behind Hooker's battered left. Still, the fresh Yankees came on the field almost in time to make a difference. When Hood saw them coming he began to pull out of the cornfield, even though reinforcements were on their way to him from D.H. Hill. The fighting became particularly unmilitary, with nothing like formations and lines, but small clusters of men taking cover behind bushes, fences, trees, and rocks. It was not one battle, but thousands of small personal fights. One of those Confederates scored a big victory when a bullet from his rifle slammed into General Mansfield's chest, mortally wounding the old man. Alpheus Williams succeeded to the corps command at once, and he continued to try to funnel reinforcements into the fight. In the continuing maelstrom, the corporal who originally found Lee's lost order himself fell with a bad wound. Finally, after a North Carolina brigade fled in terror, a Yankee charge drove through the cornfield to come to hand-to-hand blows with the Rebels. Finally about 9 a.m. the weight of numbers told decisively, and the Confederates began pulling back, abandoning the West Woods, the cornfield, and the woods to its east. For three brutal hours they had done their duty and held Lee's left flank against overwhelming odds. They could hold no more.

Lee implored Hood to hold on, for off on the road to the south he could see the head of McLaws' column not more than a mile away. But before they could arrive, Hood and Hill had been pushed back clear to the Dunker Church, and slightly beyond, and only the fortunate intervention of another bullet, this one taking Hooker out of the battle with a wound, brought a temporary lull in their advance. Hooker left the field believing that the battle was almost won.[9]

At almost the same instant, old General Sumner began to bring his corps up for the fight. Unfortunately, having given only the vaguest orders the day before as evidence of his battleplan, McClellan now exercised almost no control at all over the fight once it commenced. He did not tell Sumner where to go, and left it almost entirely to the old man himself despite "Little Mac's" own loudly proclaimed belief that Sumner was little better than a fool. Certainly "Bull" Sumner did

Below: For many years this was thought to be a photograph of actual combat at Antietam. in fact, it was taken during the battle, but more than a mile in the Federal rear, showing reserves waiting.

have more of bravery than brains. Around 9 a.m. when he started his divisions forward, he seemed to have little thought more than to drive straight toward the first enemy he saw. Instead of moving around to back up the remnant of the I and XII Corps, he simply moved straight west. Their route took them across the cornfield, where the advance was retarded by the efforts of the men to avoid stepping on the windrows of dead and wounded. Indeed, to some it looked as if the field – every stalk of corn had been clipped by bullets and charging feet – was somehow alive, a bubbling, crawling thing of waving arms and legs and quivering chests.

Sumner placed Sedgwick on his right, marching his men across the cornfield to run into the remnants of Hill and Hood, while Samuel French's division moved in the center, directly south of the Dunker Church. And on the left came Israel Richardson's division, heading almost southwest directly toward that sunken road that extended perpendicularly eastward from the Hagerstown Turnpike.

Sedgwick came in for a drubbing first, striking the Rebel line just as the first of McLaws' reinforcements were rushed to the scene by Lee. The fighting was terrible, and the Confederates steadily pushed Sedgwick back. Sedgwick himself fell wounded, along with one of his three brigade commanders, and half of his division was put to rout as Lee craftily – and partly by luck – got exactly the right number of men in just the right spots to hit him in front and flank. In the end, the Southerners drove Sedgwick back more than three quarters of a mile, where the battered Yankees set up a defensive line and abandoned the offensive, content merely to hold their position against the Rebels.

Lee was doing an excellent job of bluffing McClellan – not a difficult task in any event. Since throughout the morning thus far the Federal had made not a move to advance anywhere else on the field, Lee took it for granted that he could risk weakening his own center and right in order to support the threatened left. The gamble worked, with the added benefit that when Lee launched his counterattack against Sedgwick, it only convinced McClellan anew that the Rebels must have daunting numbers.

With Sedgwick's repulse, the focus of attention would turn now to the divisions of French and Richardson. They were about to face the balance of D.H. Hill's division placed along that so-called sunken road. In fact, it was nothing more than a wagon track, but years of traffic had so worn it down that the road surface lay three or four feet below the farm land on either side. It offered a natural earthwork fortification, from which the Rebels could fire in perfect concealment at any approaching foe. Hill had two brigades in position along its nearly half-mile length, along with the remnants of several other commands, and more on the way, maybe 2,500 in all, to meet the 5,700 in French's line.[10]

The men in the Sunken Road were told to hold their fire until the enemy was close enough for them to see the buckles and badges on their belts, and then to aim at them. The Rebels did as they were told, and when their first volley poured out, it devastated French's lead brigade. In the 4th New York, 150 men went down at a single volley. Even Colonel John B. Gordon, commanding an Alabama regiment in the road, felt sickened by what he saw. "The entire front line, with few exceptions, went down in the consuming blast," he

Above: The Sunken Road, shortly after the battle, attests to the confused and brutal fury of the fighting here. For two hours and more the desperate struggle raged, giving this road to posterity as Bloody Lane.

recalled. Another Rebel colonel thought he saw the Yankees fall "as grain falls before a reaper."[11]

French's first brigade recoiled in tatters, only to be followed by the second and third brigades in turn, each to meet the same beating. In little more than half an hour, French took 33 percent casualties in his division. The Sunken Road had almost put him out of the battle for good.

Like a magnet, the sudden fury of the fighting along Hill's line attracted to it men from adjacent areas, especially with the battle on the Confederate left now quietening to a stalemate. Richardson was soon rushing toward the field at the same time that Lee was sending Anderson's division up to Hill's support. Rapidly the Sunken Road was shaping up to be the central contest of the battle, and all the while, as Lee gambled everything, McClellan kept tens of thousands inactive, unwilling to gamble at all.

There was no lull in the savagery as reinforcements arrived. Rather, the volleys from the road, the din of artillery brought up by the Yankees, and the savage assaults, continued almost unabated. Very quickly, despite their good position and their reinforcement to perhaps more than 5,500, the Southerners began to take desperate losses. General Richard H. Anderson fell almost as soon as he arrived with his division. One of Hill's brigadiers fell with a mortal wound, his successor being killed instantly immediately afterward, and the gallant Colonel Gordon soon fell with a facial wound that bled profusely into his hat. Unconscious, he might probably have drowned in his own blood had another bullet not fortuitously opened a hole in the bottom of his hat that allowed it to drain.

Still they mauled Richardson's men as they came into the fight, brigade by brigade. The Sunken Road devoured the first outfit sent in as it had everything before. But then Richardson's second brigade moved around to the far right of the road and outflanked Hill's position. Since the road formed a salient extending outward from the main Southern line, this right flank was "in the air", and dangerously exposed. Coincidentally, confusion in the main line along the road

was running high. At the same time, Federals in front of the Rebel line launched yet another limited attack. The combination of such influences panicked the right of the line and it collapsed. At almost the same time, the left began to give way, too. In a few minutes the whole Southern line began abandoning the position that had visited such havoc on the foe for the past two hours.

Fortunately, over on the left Longstreet was mounting a weak yet determined charge that took Richardson's advancing Yankees in their own right flank, and though they pushed it back, still the Rebel movement slowed the Federal advance beyond the Sunken Road. Then Hill and others put together a much stronger counterattack, this time themselves charging against a foe now taking cover in that same road. Hill himself carried a rifle into the fight, as they all desperately tried to hold the Yankees at bay or drive them out while awaiting the arrival of more reinforcements. Finally enough artillery arrived to stabilize the pitifully thin Rebel line. Fortunately for Lee, McClellan had sent almost no cannon to support his own assaults, and after a while

Below: Every fence line, like this one along the famed Cornfield, became a defensive position as the Rebels struggled to hold their ground. The heaps of dead gave evidence of the fury of the fighting.

Richardson had no choice but to pull back, away from the hard-won little road that was soon to be called Bloody Lane. Behind lay 5,600 Americans of Blue and Gray, dead, dying, or wounded. Within only a few minutes, Richardson would join them with a wound that eventually proved mortal. And still it was only about 1 p.m.[12]

A strange relative quiet settled over the field during the next several minutes. The fighting in the area of the cornfield was over for the day, and Sumner and McClellan now tacitly abandoned any idea of pushing through Lee's center at Bloody Lane. Inevitably, then, the focus of attention shifted to Burnside and his IX Corps, who had spent almost the entire morning as spectators. After all the reinforcements rushed to support the rest of his line, Lee had only the 3,000 men of David R. Jones' division left to protect the southern half of his line. Fortunately they had excellent ground on their side of the creek, and McClellan on the other. With all the ground he had to protect, Jones could spare only 400 to guard the Rohrbach bridge. But the bridge itself worked to his advantage. The ground leading down to the creek was steep on either side. This gave the Rebels a wonderful field of fire on any foe marching down the opposite slope to the bridge. Then the span itself would allow no more than four men at best to move abreast, exposing them for its entire 125-foot length to concentrated fire from rifle and well-placed artillery fire. It might have been possible to move through the creek itself, but with water four feet deep and more in places, it would slow the columns even more than the bridge. Nowhere else on the field at Antietam did Lee enjoy such an advantage of position and terrain. A handful could withstand a legion on ground such as this.

That is what they did. Burnside did not get orders from McClellan to start his movement until 10 a.m., hours after the fight at the other end of the line commenced. When Burnside sent forward the first of his brigades, with orders to rush over the bridge under fire and secure it, they got lost and then pinned down by Rebel fire without ever seeing the span. Meanwhile, a whole division under Rodman was sent far downstream to find a ford and get across, then come up and take the foe in flank, but nothing could be heard from them. In fact, McClellan's and Burnside's reconnaissance had been so shamefully inadequate that they hardly knew where the fords were, and despite all the time at their disposal on the day before, no one seems to have thought of finding

out just how deep they were. In the end Rodman had to march two miles downstream before he found a crossing, and that cost valuable time.

Well after 11 a.m. Burnside sent in the next assault, and this time the Yankees found their way to the bridge, but not one of them ever set foot on it. Confederate volleys and artillery threw them back in confusion. But by now Jones' men were desperately low on ammunition, and exhausted from three hours of firing. When the next wave hit the bridge, it did not get across at first, but then slackening Rebel fire gave the bluecoats an opportunity and they started to pour across. In the face of their advance, Jones' tired little band had to withdraw. The bridge – ever after called Burnside Bridge – lay securely in Yankee hands.

During the next hour more Federals poured across, soon to see Rodman and his division marching toward them from the south after making their crossing. Now McClellan ordered Burnside to press the advance without delay. But Burnside lost another hour in shifting his troops about, and further time thanks to not having prepared reserve ammunition earlier. As a result, the IX Corps was not ready to move again until about 3 p.m. McClellan, meanwhile, was doing no better. Desiring that the attack on Sumner's front be pressed again, and now supported by the arrival of Franklin and the VI Corps, he finally went himself to the front from his headquarters east of the creek. But he easily allowed the now-beaten Sumner's pessimism to influence him into abandoning any further thought of an offensive on that part of the field. One young lieutenant on McClellan's staff was so disgusted by seeing the commander's timidity frittering away the battle that he briefly proposed to others that Hooker should oust "Little Mac" and take command of the army himself if he could operate in spite of his foot wound. There were no takers at the suggestion of mutiny, however.

Lee, of course, had enough audacity for a host of generals combined. By now he was hoping to mount a counterattack to take advantage of the

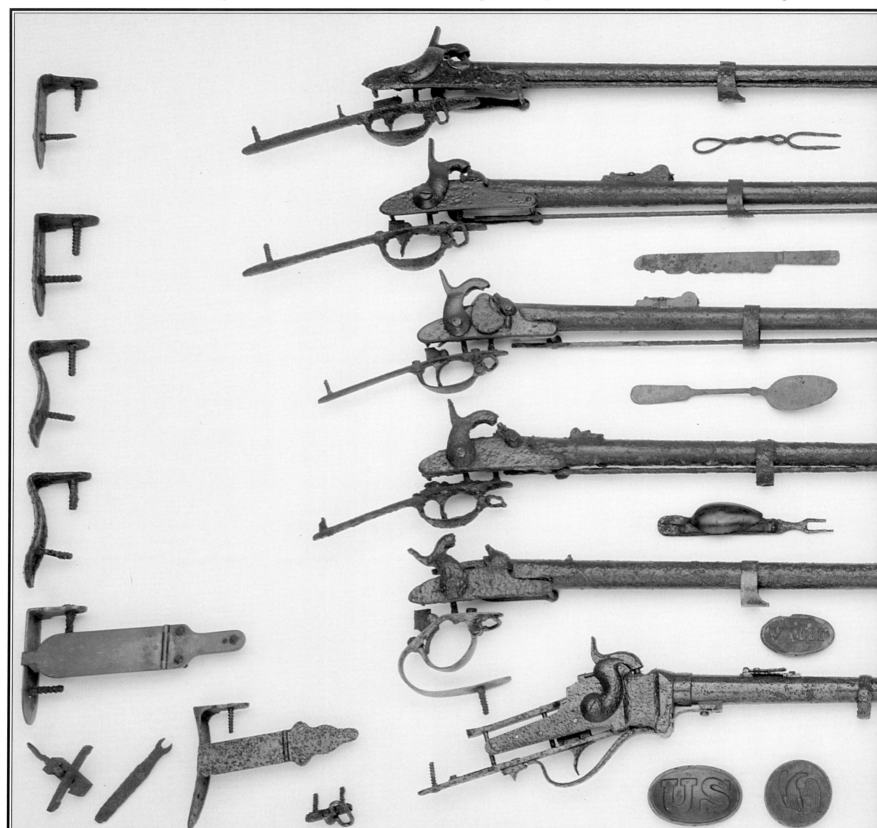

Excavated Union Artifacts

1 Model 1816 musket
2 Model 1858 canteen
3 Model 1833 foot artillery short sword
4 Fork made of wire
5 Model 1816 bayonet
6 Model 1842 rifle, found at Shiloh, Tn
7 Common knife
8 Brass drum stick holder
9 Model 1842 bayonet, found at Spotsylvania Court House, Va
10 Model 1855 rifle, found at Kennesaw, Ga

11 Bayonet scabbard tip
12 Common spoon
13 Pocket knife
14–15 Gun tools
16 Model 1855 socket bayonet, found at Chancellorsville, Va
17 Model 1861 rifle musket, found near Bethesda Church, Va
18 Bayonet scabbard tip
19 Eating utensil set
20 State of New York plate
21 Model 1841 bayonet
22 Model 1841 rifle, found

in the Wilderness, Va
23 Musket tompion
24 Maine volunteer militia belt plate
25 Ohio volunteer militia belt plate
26 Ohio state seal shoulder belt plate
27 Gun tool – spring vice
28 Combination gun tool
29 Sharps model 1859 rifle
30 Belt plate
31 Shoulder belt plate
32 Saber bayonet
33 Tin plate and cup

Artifacts courtesy of: Wendell Lang Collection, Tarrytown, N.Y.

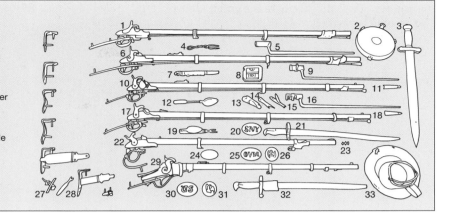

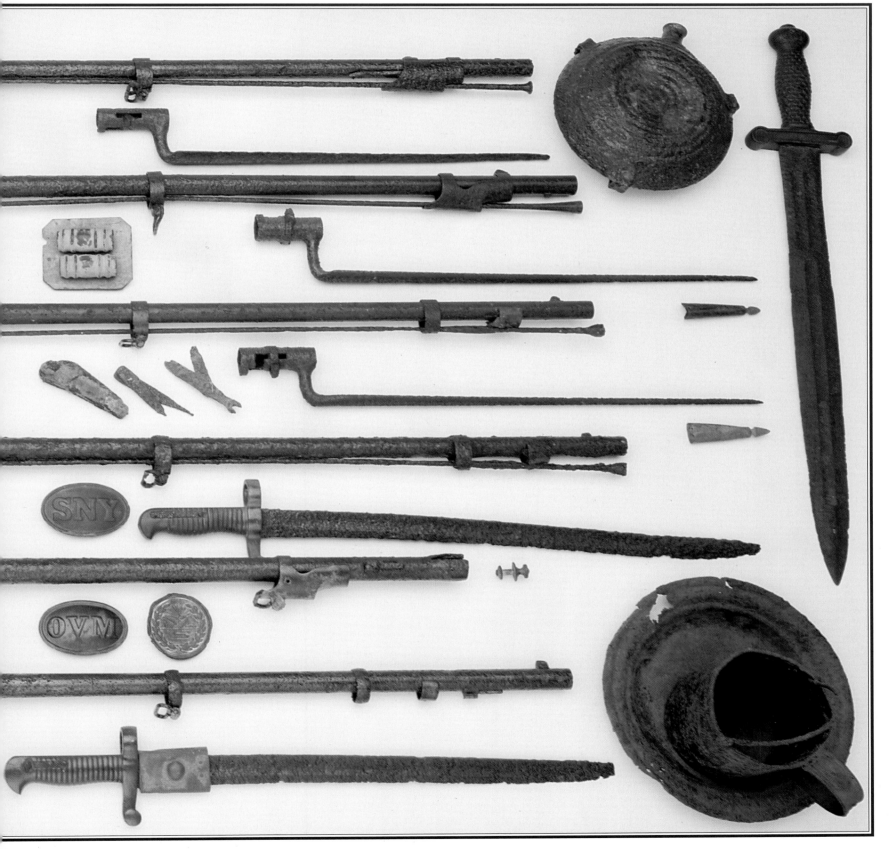

Officer and Enlisted Men, 83rd Pennsylvania Volunteer Infantry U.S.A.

The men who raised and equipped some Union regiments strained their imaginations to create new looks, and borrowed shamelessly from the styles of other regiments, sometimes creating almost bizarre combinations.

Probably no Federal regiment presented a more mixed bag of elements in its private soldiers' costume than the 83rd Pennsylvania Infantry. The enlisted man at left wears a cap much like that seen in Scottish Highland, as well as this, he also sports regiments, epaulettes normally only worn by general officers, a somewhat shortened version of a blouse similar to those worn by some zouave outfits, and exceptionally baggy zouave pantaloons gathered in at the ankles by gaiters. In full dress, when not in the field, he also wore a high shako.

The lieutenant, by contrast, is entirely standard regulation, with nothing distinctive.

stall in Union operations. He allowed Jackson to plan a flank attack on the enemy right, combining Stuart's as-yet unblooded cavalry with whatever remnants of the morning's fight he could muster. At the same time Lee received the cheering news that A.P. Hill's division was even then crossing the Potomac at Boteler's Ford, and was rushing toward the field. They could arrive right where Burnside threatened to press Jones, and Lee was content to count on Hill arriving in time while he worked on his counterstroke against McClellan's right.

Finally at 3 p.m. Burnside started out again, outnumbering Jones by three to one. In spite of gallant resistance, the Federals gained ground almost constantly, driving northwestward toward Sharpsburg itself. Some elements of Rodman's command actually reached the Harpers Ferry road itself where the fighting became hand to hand in places. Lee's right was about to collapse, and some Yankees even penetrated into the outskirts of Sharpsburg itself by about 4 p.m. Lee personally viewed the desperate situation, seeing only one thin brigade now standing between Burnside and the rear of the battered and exhausted Army of Northern Virginia. Should the Yankees break through, the Confederates would probably collapse completely even if the timid McClellan did not press them on the rest of their front. And given Burnside's position, all retreat across the Potomac would be cut off. Lee could only try to escape to the north, away from his supply lines and from safety, with an army exhausted, out of food and nearly out of ammunition, to face a pursuing Federal horde. The other alternative would be surrender.

Then he saw troops marching off in the distance to the south. It was A.P. Hill. In a case of timing unparalleled in warfare, he was bringing his exhausted division up at exactly the right place and time to save Lee, after marching seventeen miles in the past eight hours. Without pausing to rest, Hill slammed into Burnside's exposed left flank, and within half an hour the IX Corps offensive came to an end. Lee sent reinforcements from the quiet spots elsewhere on the line, especially hurrying battery after battery to Hill's assistance. Meanwhile, seeing his advantage, Lee tried to launch his offensive with Stuart over on the far left. It never really got off the ground. Meade, though battered, was still too strong for the Rebels to make any headway. The best Lee could do in the waning hours before nightfall was to push Burnside back almost to the bridge once more, and there the armies settled into the exhausted and dazed sleep of men who have raced through hell.

It had been the bloodiest day of the Civil War – of all American history, in fact. It would be some time before all the bodies were counted, the gaps in the ranks identified and tallied. In the end, McClellan lost 2,108 killed, 9,500 wounded, and several hundred missing, more than 12,400 in sum. Lee was not far behind with 1,546 dead, 7,750 wounded, and more than a thousand missing. On both sides, many of the missing were probably also among the dead, though never identified.[13]

With an audacity that only Lee could display – backed by a streak of undoubted stubbornness – the gray chieftain refused to withdraw from the field that night, though all good sense should have dictated that he leave quickly. He had been lucky to survive September 17, having through circumstances and overconfidence gotten him-

self and his army into the most dangerous spot they would ever experience during the war. Only his own skill, the bravery of his men, and the indescribable folly of McClellan kept him from disaster. But Lee had to get his wounded off the field and on their way south to the Shenandoah. He still had to evacuate much of the captured materiel from Harpers Ferry. And he would not admit defeat easily. In fact, of course, the battle was tactically almost a draw, but considering the overwhelming odds against Lee all through the day, his merely holding out must be considered a victory of sorts, the more so considering the terrible damage he inflicted on his foe. But the battle certainly put an end to his northern invasion far short of his hoped-for goal of reaching the Susquehanna, and short of visiting some decisive defeat on an enemy on his own soil. Strategically, Lee had been beaten. He had no choice but to move back to the safety of Virginia. That would be fine with McClellan, whom Lee contemptuously dared to attack him again on September 18. But McClellan was himself psychologically a beaten man, defeated by his own paranoia and timidity. He would not make a move to follow up his footholds on both the Confederate flanks, nor would he do a thing to impede Lee's subsequent retreat, despite having two fresh corps at his disposal. Instead, he would be content to boast. "Those in whose judgment I rely tell me that I fought the battle splendidly", he would write to

Below: In the end, Antietam came to this, the solitary gave of a Yankee soldier who had seen all the war he would ever see. His comrades, the living, have done all they can for him.

his wife, "and that it was a masterpiece of art." An ego like "Little Mac's" needed flatterers, and he had a host of them in Porter and others. No one else ever accused him of generalship.[14]

On September 19 Lee recrossed back into Virginia, and the campaign was over. With it died Confederate hopes of influencing fall elections in a war-weary North. Perhaps with it, too, died the always slim chance of European recognition and assistance, for both Britain and France, after considering an offer mediation - first step in what would still be a long process toward outright military intervention – decided after Antietam to wait and see. In short, they would return to the position they had taken from the first. They would not risk backing a losing side. If the South was to achieve its independence, it must do it on its own before they would help.

References

1 Dennett, *Lincoln and the Civil War*, p.45.
2 McClellan, *McClellan's Own Story*, pp.535, 566.
3 *O.R.*, I, 19, pt.2, pp.590-603.
4 *Ibid.*, I, 19, pt.2, pp.264-65.
5 McClellan, *McClellan's Own Story*, p.573.
6 *Ibid.*, pp.588-90; *O.R.*, I, 19, pt.1, p.30.
7 Stephen Sears, *Landscape Turned Red* (Boston, 1983), p.174.
8 John Gibbon, *Personal Recollections of the Civil War* (New York, 1923), pp.83-84.
9 *Report of the Joint Committee on the Conduct of the War* (Washington, 1863), I, p.582.
10 Sears, *Landscape*, pp.236-38.
11 John B. Gordon, *Reminiscences of the Civil War* (New York, 1903), p.87.
12 Sears, *Landscape*, pp.253-55.
13 *Ibid.*, pp.295-96.
14 McClellan, *McClellan's Own Story*, p.612.

CHAPTER FIVE

FREDERICKSBURG

DECEMBER 13, 1862

Slim though McClellan's victory was at Antietam, it was still enough to put cheer into the Union war effort, enough of a hook from which Lincoln could suspend his Preliminary Emancipation Proclamation. And it seemed to anchor a change in Union fortunes generally. Out in the West, Grant was once more on the move. Indeed, just two days after Antietam, his forces met and defeated Confederates at Iuka, Mississippi. Then, two weeks later, in the much hotter and more significant Battle of Corinth, Blue and Gray met for two days of fighting before the Rebels retired from the field, thoroughly beaten, this time by portions of Grant's forces commanded by a new Yankee hero, William S. Rosecrans.

Perhaps the tide was turning at last. If only McClellan would move against Lee in Virginia now, Union arms would be on the advance all across the map. Patiently President Lincoln urged his general to press the advantage. Persistently, "Little Mac" resisted. If he did not move, and move soon, the campaigning season for 1862 would be past with a long winter ahead. Lincoln had already known one winter of discontent. He did not want another.

I T WAS SAID of Ambrose Burnside that, when he was born, at first he refused to breathe. The infant could only be coaxed into taking air – and thus life – by tickling his nose with a feather. Whether or not that really happened on May 23, 1824, is a matter of conjecture. That the story exemplified the career of this engaging yet sadly ill-starred man cannot be denied. He attended West Point with McClellan, the two becoming fast friends. Yet while McClellan won glory in Mexico, Burnside arrived just momentarily too late to participate. Thereafter he served a few years on the southwestern frontier before re-signing his commission in 1853. He had invented a new breechloading cavalry carbine, and he began an ill-fated manufacturing enterprise that turned out excellent weapons, but found few takers. When the business folded, his old friend McClellan found a job for him with his own railroad firm.

Indecision seemed to dog Burnside's every move. In civilian life he was left standing literally at the altar by a fiancée who changed her mind at the last moment. Then when the war came, he took the colonelcy of the 1st Rhode Island, and very quickly afterward became a brigade commander, seeing his first action at First Manassas, where his performance, while personally brave enough, showed little of imagination or enterprise. Nevertheless he rose quickly, certainly not harmed by his close ties to the meteoric "Little Mac". And Burnside looked like the era's idea of a bold commander. There was about him somewhat of the air of the corsair. Tall, hearty, with a big smile and inevitably winning, cavalier ways, he sported on his face massive muttonchop whiskers to which a bit of word play with his name gave a lasting sobriquet, "sideburns". Following the Bull Run fight, he soon became a brigadier general, then major general, and had led a successful operation on the North Carolina coast early in 1862 that won Lincoln's notice. Indeed, after McClellan's dreadful performance on the Peninsula, the president had offered command of the army to Burnside, who turned it down with expressions of unfitness for such a high responsibility. A few months later in early September, with Lee threatening the North and no one other than McClellan to turn to, Lincoln again offered the command to Burnside instead, and again he declined. "I was not competent to command such a large army as this", he would protest. Lincoln should have listened to him, for in this war, on those rare occasions when an officer said he was not equal to army leadership, he usually proved to be right.[1]

But in the weeks following Antietam, as McClellan did nothing while the country cried for action, Lincoln increasingly saw Burnside as his only alternative. McClellan had to go, of that Lincoln was sure. He knew it in his heart long before he decided to act. He visited the army early in October in the hope of impelling McClellan to take the offensive once more, yet while outwardly he remained affable and encouraging, inwardly he seethed, especially at the patrician "Little Mac's" obviously condescending manner to his rough-born president. One morning when out looking at the troops in the field with one of his friends, Lincoln bitterly asked if the other

Though this image was made in early 1863, the Confederates standing on the end of the bridge pier and posing for the Yankee camera are some of the same men who held Fredericksburg in December 1862.

recognized what lay before them. "It is the Army of the Potomac," said the other.

"So it is called," the president replied through his teeth, "but that is a mistake; it is only McClellan's bodyguard."[2]

It could not remain so for long. During the next month all McClellan achieved was to take eight full days to get his army across the Potomac – Lee had done it in a single night after Antietam. Lincoln had hoped that McClellan might drive straight south, placing himself between Lee in the Shenandoah and his communications with Richmond, thus cutting the Rebels off from their base of supply and succor. That could force Lee to give battle on McClellan's terms or risk fatal isolation. But by November 4, almost seven weeks after Antietam, the Army of the Potomac was only twenty miles into Virginia, while Lee was speeding around in a great arc to cut off any approach to Richmond. Indeed, that same November 4, Lee had Longstreet and his corps in place at Culpeper Court House, squarely between McClellan and Richmond. The opportunity was lost, and Lincoln decided that this would be the last lost opportunity lost by McClellan. The next day Lincoln issued the order relieving McClellan from command. An emissary from the War Department went first to Burnside and practically forced him to agree to succeed to the command. Then they called on McClellan and handed him the order of removal. He took it well, though he would never see the justice of it, putting his downfall down to the spite of petty men like Lincoln and Halleck who were jealous of his genius and popularity. On November 11 he left the army, never to return to it, or the war, again.

Behind him McClellan left a Burnside almost on the verge of tears, and not just out of sadness for his friend's dismissal. Burnside *knew* that he had no business commanding an army. At Antietam he had showed not a whit of imagination, rigidly following every order given him to the letter and no more, refusing to think or act without explicit instructions. Like a great, cuddlesome, stuffed bear, he looked grand but had little depth. Thankfully he had a few days after receiving the news of his elevation before he actually took the reins from his predecessor, and this gave him time to try to take in the situation before him.

Lee had boldly split his army yet again, keeping Jackson in the Shenandoah to guard that vital back door to central Virginia, while moving Longstreet east. Thus while the latter could provide a roadblock to a Yankee move toward Richmond – which Lee did not expect, given the known sloth of the Yankee commanders – the former posed a seeming threat of yet another invasion sweeping out of the valley and into Maryland once more. Of

Above: The stately Rappahannock River swept broadly past Fredericksburg, seen here in 1863 from Stafford Heights. The town itself was not particularly defensible, but the heights beyond were another matter.

Below: Looking upstream, Fredericksburg stretches westward. The stark bridge piers testify to its depth and the necessity for pontoons for a crossing – a terrifyingly dangerous enterprise.

course, the Confederates just then were entirely incapable of such a movement, but the Yankees did not know that, and Lee was happy to use the time their fears gave him to rest and rebuild his own army.

McClellan had had a plan of sorts for his snail's-pace campaign, but upon considering it in his new position as army commander, Burnside decided to abandon it for one of his own. He intended to concentrate his corps in the Warrenton area, just west of the old Manassas battleground, then make a diversionary move toward Culpeper Court House, about forty miles southwest. Immediately thereafter, he would in fact move the entire army due south to Fredericksburg on the Rappahannock River. That river was one of the few great natural barriers in his way, with Richmond just fifty miles beyond. He believed that he could move quickly enough to accomplish this before Lee could bring up Longstreet to get in his way, and certainly before Jackson could arrive in case a battle should develop.

Of course, Burnside was revealing the worn-out thinking that somehow taking the Confederate capital would put an end to the Confederacy, which was utter nonsense. When McClellan threatened to take the city back in May and June, Davis and his government made considerable preparations for abandoning Richmond if they had to, but no one even mentioned the loss of the

cause as a contingency or a possible result. Lincoln knew better, and tried to persuade Burnside that Lee, and not Richmond, was the strategic objective to be desired. But the president was anxious not to demoralize a commander so freshly come to the job, and in the end gave in to Burnside's proposal, only pointing out that if it were to succeed at all, "Burn" must move and move quickly.

Like his old friend before him, Burnside first turned his attention to organization, though not with McClellan's flair. Nor did he learn any of the few lessons that "Little Mac" seemed to have gained from seeing certain men in action. Burnside returned to the "wing" concept, though he called them Grand Divisions, created three, and gave them to the unlikeliest of men. The Right Grand Division he gave to Sumner, who almost all agreed was a poor commander at best. McClellan's opinion was on record. He would have his own old II Corps, now led by General Darius Couch, Burnside's IX, under General Orlando Willcox, and most of Pleasonton's cavalry. The Center Grand Division commander posed an even more odd choice, for it was Hooker, and Burnside hated him, believing him responsible for the breakup of his own wing prior to Antietam. But Hooker was a senior officer now, recovered from his wound, and an acknowledged fighter. Burnside had almost no choice but to give

him such a command, to include the III Corps led by General George Stoneman, the V Corps now led by General Daniel Butterfield after Washington relieved Porter from command just days after McClellan's fall, and some attached cavalry. Finally, the Left Grand Division went to Franklin, who had showed positive sloth and bewilderment during the Antietam Campaign. He took Hooker's old I Corps, now under the brilliant General John F. Reynolds, the VI Corps led by William F. Smith, and a brigade of cavalry. Additionally Burnside had the XI Corps commanded by General Franz Sigel back at Centreville in reserve, and Mansfield's old XII Corps under Slocum way out at Harpers Ferry.

Significantly, not a single one of all these generals would be exercising the same command as at Antietam. Every single corps commander was new on the job, and so were two of the three Grand Division commanders, and even Sumner had never actually exercised such a level of responsibility in action. And, of course, Burnside himself had never led an army of 100,000 or more men before. In fact, all told, with a host of new recruits and units swelling the ranks, Burnside could count more than 120,000 of all arms, more than even McClellan had managed in actual field operations.[3]

Perhaps because of his consuming self-doubts, Burnside felt he needed to divorce himself from

McClellan by a new organization and a different plan of campaign. Certainly he set a different tone by his willingness to move quickly. Just the day after Washington approved his plan, he had Sumner on the move. Early on November 15 old "Bull" set out and showed that without McClellan to slow them, these veterans could march. In just sixty hours they covered the forty miles to Falmouth, on the Rappahannock, just opposite Fredericksburg, with Lee nowhere in sight and clearly unprepared for such a decisive swiftness. Indeed, Sumner actually reached the Rappahannock nearly a full day before Lee knew with certainty which direction old "Bull" was headed. By then it was too late. Burnside showed just as

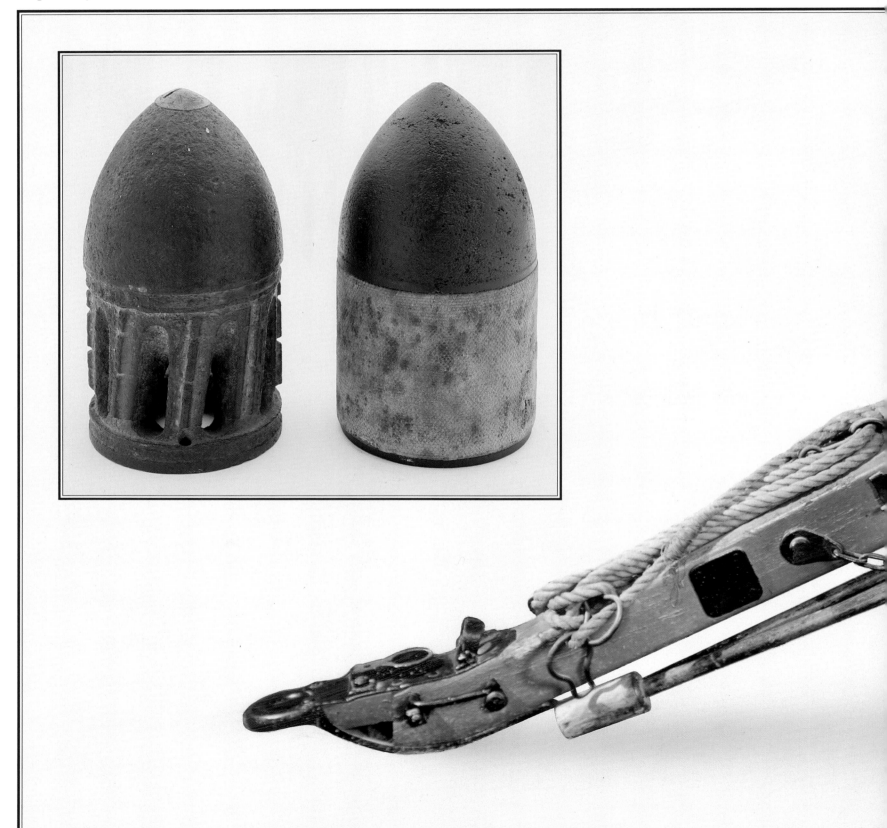

U.S. Model 1841 6-Pounder Smoothbore, James Rifled

Although the projectiles shown here are not of the same caliber as this specific gun, they illustrate a very successful pattern developed by General Charles T. James of Rhode Island, and patented in 1856. The James projectiles were also made in much larger calibers than the one shown here. It was these larger types which were instrumental in breaching the walls of the Confederate-held Fort Pulaski, Georgia, on April 10, 1862. This action was one of the first in which rifled artillery was used at long range against a fortification. The result was a resounding success for the rifle and its explosive shell

1 James shell of 3.8-inch caliber of the pattern used in the accompanying field piece. Note the ridges and slits on the base to take the rifling.

2 James solid metal shot, 3.8-inch caliber, of the pattern used in the accompanying field piece. Note canvas cover over base, around lead-filled rifling slits as seen on 1

3 Model 1841 6-pounder bronze-cast smoothbore, James rifled to 3.67-inch caliber

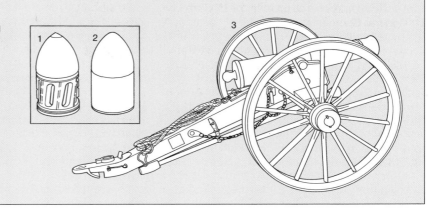

Artifacts courtesy of: Milwaukee Public Museum, Milwaukee, WI: 1; U.S. Army Ordnance Museum, Aberdeen Proving Ground, Md: 2; West Point Museum, West Point, N.Y.: 3

much alacrity in getting the rest of his army on the march, and by November 20, the entire Army of the Potomac was on Stafford Heights, staring across the river at the spires of Fredericksburg.

And there he stopped. While Lincoln and the War Department began to breathe a long-needed sigh of contentment at finding a general who would move swiftly and decisively, Burnside figuratively shot himself in the foot by being inflexible, unable to think and adapt on the spot in reaction to circumstances. There were no bridges at Fredericksburg any more, though there were fords upstream some distance. But even a good ford can slow down the crossing of large bodies of troops. Burnside, of all people, knew after his hours and hundreds of men lost trying to get across a bridge just how vulnerable men were in midstream, especially a wide river like the Rappahannock. Thus, from the first he never contemplated using the fords. Rather, he ordered from the War Department a massive train of pontoon boats. Lashed together and anchored in the stream, overlaid with planks and earth, they would form floating bridges for his army. With good engineers to construct them, such bridges would take only a few hours to erect, and since Burnside expected to beat Lee to the river, he anticipated little problem for the bridge-builders from hostile fire.

The trouble is, when Sumner reached Falmouth, there were no pontoons to be seen, even though Burnside had Halleck's assurance that the pontoon train would arrive at the same time as the Right Grand Division. "Bull" Sumner may not have been an imaginative man, but he could think in a straight line, and his only line of thinking at the moment was to get across the river and take the heights above Fredericksburg before Lee could beat him there. Finding no pontoons, he sent word to Burnside that he could cross at one of the upstream fords and still take the town without a fight.

Burnside refused. Storm clouds threatened, and he feared that a heavy rainfall could isolate Sumner on the south side of the stream before the balance of the army arrived to cross. Besides, the pontoons had been promised, they had the jump on Lee, and they could afford to wait. The bridging material would arrive any minute.

In fact, another week passed after Sumner's arrival before the pontoons finally appeared, a fatal week lost that lays a heavy burden of responsibility on Burnside for what was to follow. For in that week Lee rushed Longstreet to Fredericksburg, arriving late on November 21, more than three days after Sumner first stared across the river from Falmouth, and almost 48 hours after Burnside and the rest of the Federals began to arrive. If only he had sustained his initiative a little longer, Burnside could have been a full day on the road toward Richmond ahead of Longstreet, forcing Lee to rush to battle with Jackson still a hundred miles away. Not for another twenty months would a Yankee commander manage to steal such a march on the masterful Lee.

Caught by surprise, Robert E. Lee would not waste a moment in attempting to recover the ground lost. He had not been happy about McClellan's replacement. Why should he, when he had so easily and consistently outwitted the Young Napoleon. "I fear they may continue to make these changes", he told Longstreet, "until they find someone whom I don't understand."[4] Certainly he underestimated Burnside at first, but as soon as he received definite word of Sumner's movement, he sped Longstreet on his way and called to Jackson to leave the Shenandoah. As soon as Longstreet actually reached Fredericksburg on November 21, he immediately began digging in on the long ridge behind the city, preparing to meet the Federals if and when they tried to cross.

But for Burnside, crossing meant pontoons. Then when they started to arrive on November 24, he proceeded to wait another seventeen days before putting his plan in motion, even though he had 120,000 to not more than 40,000 with Longstreet. Meanwhile, Lee wisely advised the citizens of the city to evacuate. The Federals had not shelled the town as yet, but should a battle

Below: Though this image was made in 1863, it shows pontoon bridges still spanning the Rappahannock near Fredericksburg, and provides much the same scene as in December 1862. Well organized, armies could send thousands across.

start, the townspeople would be directly between the fires of the two armies. For the first time in the Civil War, the conflict had come to the doorsteps of a significant civil population.

Why Burnside waited is a matter of some conjecture, but as much as anything it was mental paralysis. He had only one plan, and the delay of the pontoons and Longstreet's arrival upset that. He seemed unable to think. Sensing the stalemate, Lincoln summoned Burnside to two meetings late in the month to discuss what could be done next. They produced little satisfaction, and in subsequent days, as Burnside discussed plans with his generals, their confidence in him seemed to wane – if, indeed, they had ever felt any. Early in December Burnside hit upon the idea of sending one of his Grand Divisions about twelve miles downriver to Skinker's Neck, to cross there and then move up against Lee's flank at Fredericksburg. No one felt great enthusiasm, and Burnside himself soon abandoned the plan, deciding instead that he could cross the river immediately below Fredericksburg.

On the face of it, the idea seemed utter folly. Any such crossing would be done under the eye of Lee himself, who was hardly likely to sit back and quietly allow the building of pontoon bridges. Furthermore, his position behind Fredericksburg was positively formidable. Fredericksburg sat on a descending plain that stretched less than a mile from a long, steep ridge down to the river. The ridge, in fact, was a series of steep hills – Stansbury's, Cemetery Hill, Telegraph, and Prospect. Directly behind the town proper sat Marye's Heights. Along its face ran another of those slightly sunken roads, this one fronted by a stone wall. Below the town the countryside gradually opened up away from the river, offering more room for maneuver, but still favoring any defender.

Lee at first scattered his divisions widely to cover all contingencies, expecting that Burnside would naturally try to cross either upstream or downstream in order to take him in flank, and in order not to have to cross in the face of Rebel artillery fire from those heights behind the town. He placed Jackson, when he arrived, on his right, with D.H. Hill's division far downstream, Ewell's old division now commanded by General Jubal

Early at Skinker's Neck, A.P. Hill about three miles from Longstreet's flank and fully six miles south of Fredericksburg, and William Taliaferro's division several miles beyond Hill, all designed to provide maximum protection against what Lee most anticipated, a downstream Yankee crossing. Longstreet's five divisions stretched out over four miles, from Anderson immediately behind Fredericksburg, through Robert Ransom, Lafayette McLaws, George Pickett, to Hood. Even though reinforcements from other theaters and hurried recruitment had swelled Lee's numbers to almost 90,000 – the largest army Lee would ever field – still this abnormally long defensive line meant that his legions were spread thin.

Burnside would boast to Franklin that he believed he knew where all of Lee's divisions were placed, and felt confident that he could get his flanking movement across and on to high ground below Fredericksburg before Lee could act. Having gotten the jump on the Confederate master once, Burnside showed uncharacteristic self-confidence in believing that he could do it again. But then, Burnside had little choice. His sloth and inflexibility had squandered all the benefits of his lightning advance to the Rappahannock. Now, unless he was simply to turn back and abandon the campaign, to meet the wrath of Lincoln and the Union, he must give battle. A crossing upstream offered – to his mind – no opportunities, while a downstream crossing, if successful, would give him a chance to push Lee's right flank back and away from a line of communication and retreat to Richmond. If it all seemed reminiscent of his own role at Antietam, Burnside seems not to have been bothered by the comparison. He had to act.

Finally late on December 9, Burnside wired the War Department of his intention to attack two days later. The fluid state of conditions on the other side of the river had changed his plans once again. Burnside now believed that Lee was too well prepared for a crossing below the city. As a result, Burnside now proposed to cross directly in the enemy front right at Fredericksburg. There was some minimal logic to it. From his own position on Stafford Heights overlooking the Rappahannock, Burnside's artillery could completely command the city itself, making it almost

Above: Now a major general, Burnside devised one of the war's best campaign plans, and conducted one of its worst battles, showing again that he simply was not smart enough for high command.

impossible for Lee to mount any substantial infantry maneuver to hinder the bridge-builders. The engineers would only have to brave distant Rebel cannon fire from Marye's Heights, and the attentions of snipers and sharpshooters placed in some of the buildings of the town itself. Burnside believed that such an attempt would take Lee by surprise, and several of his generals agreed – or he said they did.

Below: This battery on Stafford Heights is posing for the camera, and not really engaged in action, but the appearance would be much the same. These guns could command Fredericksburg and the ground beyond, but they could not dislodge Lee's veterans.

He proposed to run three sets of bridges at once. The uppermost would cross directly at the northern edge of Fredericksburg, the next a mile downstream at the city's southern edge, and the third a mile beyond at a slight bend in the river, just below the mouth of a stream called Deep Run that flowed in from the Confederate side. Sumner and his Grand Division was to lay the uppermost bridge and move directly against Fredericksburg. Burnside gave Franklin the task of crossing on the lower sets of bridges, to move across the more open ground to strike Lee where Burnside thought him most vulnerable. Hooker and his entire Grand Division were to remain on the east bank in reserve, a move that could be interpreted as Burnside's revenge for Hooker's part in the breakup of his old wing before Antietam.[5]

There was more than a little outspoken criticism of the plan. Most of Sumner's officers frankly told Burnside that to cross the river and then move straight up against Lee's entrenched infantry and artillery on Marye's and the other heights would be nothing less than suicide. Nevertheless, Burnside was not to be deterred by dissent. He spent much of December 10 drafting the final attack orders to his commanders. Sumner was to move straight across his bridges and up against the Rebels emplaced on the heights beyond. Hooker was to cross after him, but without any specific part in the battle other than to be ready to support either Sumner or Franklin. Franklin was to bridge the river, get across, and then move directly west to get to the roads connecting Lee with Richmond. Once they were in his possession, he would turn northward to strike Lee's flank. Yet these were all the orders his commanders got. Burnside gave them no specific instructions on how or where to fight, and offered no coordinating influence between them. Having told them where to go, he blithely left everything else up to his generals.

Late on December 10, the engineers moved their pontoons and other equipment up close to the river in the darkness. Some even tried a diversion by ostentatiously cutting trees and making a great deal of noise down near Skinker's Neck, hoping to deceive Lee into thinking that the bridging attempt would be made there the next day. It did not work, however. In Fredericksburg

itself, McLaws could see and hear enough to conclude that the enemy would try a crossing right in his front, and at around 4:30 a.m. on December 11 he fired two signal guns that alerted the rest of the army that the fight was about to begin. With more than 200,000 men at arms in the vicinity, it was to be, in terms of the size of the armies, the biggest battle ever fought in the hemisphere.

The bridge-builders needed all the protection they could get, and an early pre-dawn fog helped them at first. Before long, however, McLaw's sharpshooters posted in lofts and upper storys through the town began to take aim as the brave engineers started their work of placing and anchoring their pontoons in full view and range

of the foe. Casualties began to fall immediately, and for the next several hours almost no progress was made on the upstream bridges as Rebel fire drove the engineers back time after time. Even artillery bombardment ordered by Sumner failed to drive out the pesky sharpshooters, though 9,000 shells were thrown into the town, setting ablaze several of its buildings.

Finally one general suggested that infantrymen use the pontoons as boats and row themselves across to the other side to stop the sniping long enough for the bridges to be completed. Regiments volunteered for the work, and almost without a casualty the first wave rowed across, outflanking McLaws' advanced sharpshooters

and tying down their fire. That allowed more men to row over, and finally the engineers themselves were able to complete their work. The same plan worked both for the upper bridges for Sumner, and the middle bridge at the south edge of town. Nevertheless, the attempt to cross had consumed most of the day, and what remained of daylight after the bridgeheads were secured had to be devoted to driving the Confederate snipers out of the town, house by house. Only after nightfall did the remaining Rebels pull back to their main lines on the heights, leaving Fredericksburg to Burnside. A mere 1,600 Southerners from William Barksdale's brigade had held up Burnside for an entire day.

Distinctive Unit Flags of the Confederacy

1 Flag of the 4th Missouri Infantry, carried at the Battle of Pea Ridge (Elkhorn Tavern), Arkansas, March 7-8, 1862. This pattern flag is typical of those flown out West by units serving in the Trans-Mississippi Department of the Confederacy, under the command of General Earl Van Dorn, and is commonly known as the Van Dorn Pattern flag

2 Flag of the Van Dorn Guards of Texas. Two Confederate units are known to have carried this name: Company A, 8th Regiment Texas Volunteer Infantry, and the 4th Battalion, Texas Artillery. Of special note on this flag is the large central star, a device quite common to flags of Texas units, which copies that state's well-known emblem

3 Flag of the 8th Regiment Virginia Volunteer Infantry. This flag was presented to the unit by General P. G. T. Beauregard, in recognition of their valor in combat against Union forces at the Battle of Balls Bluff (Leesburg), Virginia, October 21, 1861. The flag was actually made by General Beauregard's wife from one of her own silk dresses

Artifacts courtesy of: The Museum of the Confederacy, Richmond, Va

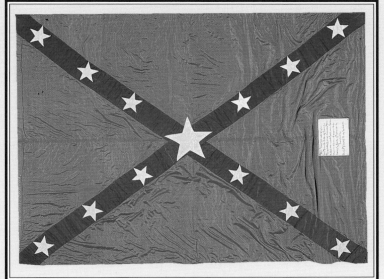

At least Franklin's men did not encounter any difficulty getting their bridges "thrown" across the river near Deep Run. Lee had not expected a movement at that spot, and thus the engineers managed to work almost unimpeded. By 11 a.m., while Sumner's men were still pinned down by Barksdale, Franklin had completed two bridges, and later that afternoon he sent his first brigades across. Unfortunately, virtually unopposed as he was, he did not push getting more of his command over with speed, for he might have turned northward and severely worried Lee, perhaps even relieving the pressure on Sumner to allow a crossing on that front earlier. But Franklin had no specific orders from Burnside – none of them did – and so once across he simply protected his bridgehead and settled down to wait until the next day, when with Sumner finally across, they could renew the advance in concert.

Of course it was only yet another gift of time to Lee, who needed it. He had posted himself on the summit of Telegraph Hill, from which he could see almost all of his line except the far right under Jackson, and here he watched with satisfaction the check given to Sumner throughout much of the day. Still he did not act decisively to keep Franklin from getting across, and while it is possible that he wanted to have the foe between his high ground and the river, there was little point in such a maneuver since, even if he should defeat Burnside, the Federals could simply retreat back across their bridges and defy him from the heights on the other side. More likely, Lee felt secure with only Franklin coming across between Longstreet and Jackson, hoping to pinch him between the two, and also he still seems to have feared that the main Yankee flanking move would come at Skinker's Neck, where he still had the bulk of Jackson's power poised for defense.

Only after nightfall, once satisfied of Burnside's real intentions, did Lee start to shift his troops for the battle that was shaping for the morrow. He brought A.P. Hill and Taliaferro up from their extended positions to extend Longstreet's right, but still left Early and the other Hill far to the right, just in case Skinker's Neck should still pose a problem.

He need not have worried. Even with all of his sloth of recent weeks, Burnside did still have a bare advantage. After all, he heavily outnumbered Lee, and by nightfall of December 11 he had the means to get most of his army across the Rappahannock. Even Halleck, from Washington, sent an urgent wire begging that Burnside get as many people over as possible during the dark hours, in case of a counterattack in the morning. Burnside, however, would not be rushed. Nor did he give his Grand Division commanders much more in the way of detailed orders for the next morning. Afterward he would simply say that it was his intent that Sumner should hit the heights as a diversion, while Franklin rolled up the Rebel flank. Yet instead of being ready to do so at dawn, Burnside occupied much of the morning daylight hours by giving the Rebels a grand spectacle as they watched division after division march in full view down to the bridges and across to the other side. The II Corps crossed at Fredericksburg first, followed by the IX Corps, Couch's command taking position as the far right, with Willcox on his left extending almost down to Deep Run. On Franklin's front, the VI Corps crossed first, followed by the I Corps, but it took until well into the afternoon before they were all over and in line, Smith's veterans resting with their right at Deep Run, and Reynolds on his left.

Lee gave them all the time they wanted. Secure in his positions atop the high ground, he could take advantage of Burnside's tardiness to move Early and D.H. Hill the ten miles or more to the impending battle now that Franklin's might on the Deep Run line relieved anxieties that the enemy might be committing more troops farther downstream. Well before nighfall on December 12, he knew that this was where the battle would be fought.

By dawn on December 13 Lee had almost all of his Army of Northern Virginia united once more. The arrival of Hill and Early extended Jackson's line to nearly two miles in length. Stonewall took position on a wooded crest running from one to two miles back from the river. Stuart's cavalry protected his right, while his left bent backward to take advantage of better ground. Hill held the left, while D.H. Hill, soon to be joined by Early, held the right, with Taliaferro in reserve. Lee needed depth in his line here, for the ground favored the defensive less than elsewhere, and he expected Franklin to make the major push. Longstreet, meanwhile, extended his line from A.P. Hill's left, for almost five miles, through the successive divisions of Hood, Pickett, McLaws, Ranson, and Anderson.

By nightfall of December 12, Sumner had about 27,000 men, backed by 30,000 reserves under Hooker, facing Longstreet's 41,000 on the heights behind the town. Jackson had barely 39,000 in his deep lines, looking across at almost 51,000 under Franklin. And up along Stafford Heights to support the Federal advance would be 147 cannon lined up in neat rows, their bulging caissons and ammunition chests ready for a day's work, while another 190 guns actually went across the river with the attacking columns.

With this enormous mass of power at his disposal, Burnside proceeded to try to win the battle without using it, and without risk. He kept his

Above: An unusual panoramic image made from two photographs taken at the same time by the same camera, showing the ground above Fredericksburg as it slopes upward towards Marye's Heights in the distance.

Below: This 1864 photograph shows the damage still visible in Fredericksburg as a result of the Union shelling in 1862, and the Confederate return fire when the Yankees occupied the city.

Meade moved forward with dispatch, his skirmishers encountering enemy fire almost at once. Before long he came in range of Jackson's artillery pieces on the heights beyond. Taking casualties, slowed by sunken roads and fences that had to be borne down, Meade and his division still continued in their halting progress, followed now by Gibbon, whom Reynolds had placed on Meade's right rear as a support. Seeing Stuart's cavalry off in the distance to his left, Reynolds had to position Doubleday's division to face the Rebel horse and protect the left flank of his infantry. Already, though still only Meade was principally engaged, the whole I Corps was in motion.

There followed an artillery duel of an hour or more as Rebel batteries played heavily on Meade and Gibbon. Still they got within less than half a mile of the crest of the ridge in their front before the enemy on top opened on them in earnest, stopping them in their tracks. There they stayed for the next two or more hours, unable to move until renewed Yankee artillery fire softened the foe enough for Meade and Gibbon to press on. Finally, well after 1 p.m. they reached the foot of the slope and started pushing their way up into the face of A.P. Hill's infantry.

The initial impact of Meade's advance took Jackson by surprise. He slammed into General Maxcy Gregg's brigade and dispersed it, killing Gregg in the process. Portions of brigades on both of Gregg's flanks were broken up as well before a Rebel countercharge struck a gap between Meade and Gibbon and drove them back. The Yankees pulled back as fast as they had advanced. Gibbon fell with a painful wound, and Meade lost a brigade commander killed on the spot. Fortunately a reserve division from the III Corps rushed up and stopped the Southern counterattack, buying time for Meade and Gibbon to retire nearly to the river

commanders ignorant of their attack orders until just after dawn on December 13, giving them no time to prepare, and actually contradicting the plans he had discussed with them the previous day. He instructed Sumner to use but a single division, and that to march straight into the mouths of Longstreet's guns to seize Marye's Heights. Franklin, with more than 50,000 at his disposal, was also to use but a single division, this one to seize Prospect Hill, an eminence on Jackson's far right. He reasoned to taking these two high points, backed by artillery placed thereon, would force Lee to abandon all of the line in between, which meant virtually his entire position. The plan was ludicrous, conceived in a vacuum

that apparently did not realize that Lee might take advantage of the limited assaults to concentrate his own numbers against them. As for the rest of the army, its sixteen divisions were essentially to act as a "reserve" for two.[6]

Despite his disappointment at the orders he received around 7 a.m. that morning, Franklin — who favored a massed attack straight at Jackson's center — gave Reynolds the task of sending out the division for the job, and Reynolds chose his friend and fellow Pennsylvanian Meade. Even here choices were bad, for Meade's was the smallest division in the corps, but he took the assignment and prepared with speed. By 8:30 he was ready to advance.

after their severe mauling. By 2 p.m. the fighting quietened, and though Franklin's men maintained a somewhat advanced position from where they started, the fighting on his front was essentially done for the day. He had tried to execute Burnside's inane order, had failed, and would not move again without positive orders that never came. Meade lost 1,853 and Gibbon 1,267. Combined with other units involved, their casualties totaled 4,861, while A.P. Hill and Early, who did the bulk of the fighting, lost 3,400.[7]

All of that blood had been essentially wasted, but it was nothing when compared to the bloodletting taking place at the other end of the battlefield. Longstreet had dug his divisions into

a fearfully strong position. He put his artillery in the best positions, clearing for them wide fields of fire. His infantry took advantage of every terrain feature to provide cover, especially that stone wall along the forward slope of Marye's Heights. McLaws and Ranson held most of this ground. Immediately behind the stone wall, the strongest position of all, sat the brigade of General Thomas R.R. Cobb of Georgia, a troublesome politician who first served in the Confederate Congress but got so fed up with President Davis that he sought a field commission instead.

Against this line, Sumner ordered Couch and the II Corps to make the advance. Couch, in turn, entrusted the mission to French's division,

already emplaced in the streets of Fredericksburg itself. They were not to advance until word came of success by Franklin, but by late morning, when Burnside still heard nothing favorable from his left, he told Sumner to go ahead.

It was just past noon when the first lines of blue marched out of the outskirts of town and started up the soft incline towards Marye's Heights some 800 yards distant. Their first obstacle, not quite halfway to the stone wall, was a drainage ditch that, unfortunately, could be crossed only on bridges. The moment the Yankees left the cover of the town's buildings, they came under the fire of Longstreet's artillery, suffering substantial losses almost from the first.

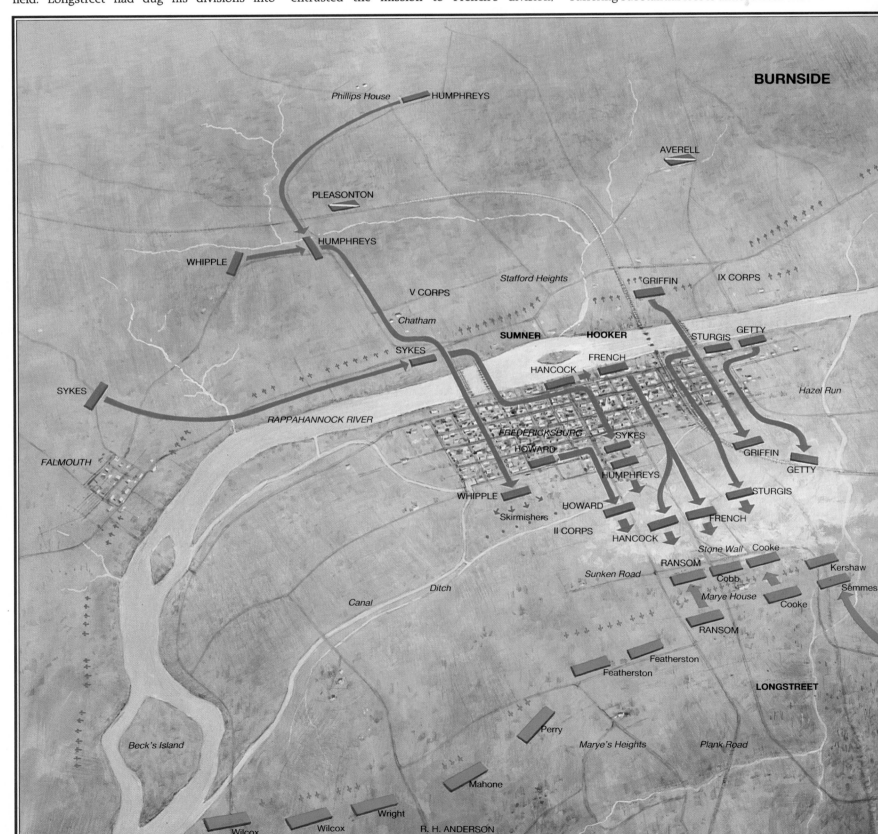

Battle of Fredericksburg, December 13, 1862

In the spring of 1862, Lee visited Fredericksburg with President Davis and pronounced the position indefensible. But by December observation and circumstances had changed his mind.

When the Federal attack came, it could not have played into Lee's hands more if he had planned it himself. Burnside put his hapless bridge-builders to work early on the first morning trying to "throw" their pontoon bridges

across the Rappahannock right in front of Fredericksburg. Only daring and ingenuity got the job done, but it took all day. Downstream, however, the bridges went across without great difficulty and Franklin began crossing his brigades. Alas for the Federals, no one pressed anything that day, and Lee managed to position Longstreet on his left behind the city, and Jackson on his right, in position for the attack expected for the morrow.

On December 13, Franklin sent Meade and his division forward. The fighting became furious and prolonged, but by mid-afternoon had accomplished nothing, and Franklin did not press it further. The battle on the Confederate right was all but over.

In front of Fredericksburg itself, the Federals had massed all night in the streets of the city. Finally at noon portions of the II Corps started to move out of the outskirts

and up the long gradual slope towards Marye's Heights. They were met by a wall of flame, yet on they went. When Rebel fire stopped them, more Yankees took their place, all of them striving toward the stone wall along the sunken road. Division after division went up the slope, only to be chewed by lead and spat back down again. Not one bluecoat ever reached the wall, yet the attacks went on almost until nightfall mercifully put the battle to an end. His

foolishness had cost Burnside 12,653 casualties out of his magnificent army, more than 120,000-strong. Lee, by contrast, took just 5,377 of the nearly 90,000 in his command. The armies remained in place, and Burnside had another division pinned down on the ground in front of the wall, without attacking. They had been sent there under cover of darkness to launch an attack at dawn, but found themselves as exposed as the dead and dying comrades

who surrounded them on every side. The next day Burnside abandoned the field.

The Army of the Potomac would not abandon the town of Fredericksburg. Burnside at least wanted to hold on to some ground on the Rebel side of the river. For two days he remained in the town. Then at last he asked the Confederates on the heights for a truce to collect his casualties. This done he retired back across the Rappahannock.

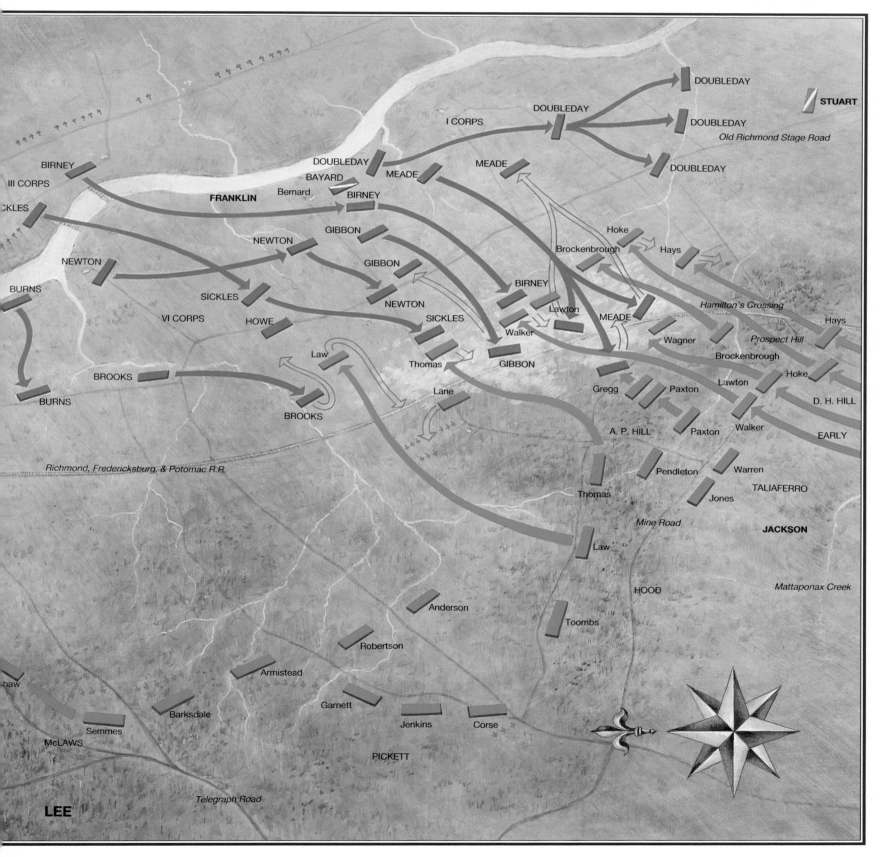

Soon thereafter those who got past the initial barrage came into range of the Confederate rifles. They, too, started to exact their toll. Cobb's Georgians, joined behind the wall by a North Carolina regiment, poured forth sheets of flame from behind their stone rampart.

Incredibly, French's men continued to advance, despite the great gaps opening in their lines. They actually got within 200 feet of the wall before they were propelled backward at last. As they retired, they passed through French's second brigade, which went into the advance only to meet the same treatment, and so did a third that followed. In barely an hour, French's division was put out of the battle for good. The

next II Corps division, this one led by General Winfield Scott Hancock, took their place. He took even worse casualties than French, though he pushed the advance to within 150 feet of the wall. Behind that stone barrier the Rebels were standing four deep now, firing by ranks and almost without let-up. In front of them lay literally thousands of dead, dying, and wounded, along with scores of others who were not touched, but who did not dare rise to run to the rear for fear of being hit. Those who could get away left behind them some 3,200 casualties, with more yet to come.

Couch's third division, led by First Manassas veteran Oliver O. Howard, met nearly the same fate when it followed Hancock forward. When

supporting units from other corps tried to move in on Couch's left to take some of the heat off him, they came under a withering Rebel artillery fire that left them, too, badly blooded. Nevertheless, Burnside, badly out of touch with a battle he never had under control from the moment of conception, continued to expect his right to press on until it took the heights. To support another push, Burnside sent word to Franklin to make an all-out assault over on the left, while telling Hooker to send two F Corps divisions against the stone wall. After seeing the front for himself — which Burnside never did — Hooker went back to his commander and begged that the order be withdrawn. Burnside refused. Meanwhile,

Excavated Union Struck Plates, Identification Badges and Stencils

1 Eagle shoulder plate, struck by bullet

2 Identification pin, private purchase

3 Cartridge box plate, struck by shell fragment

4 Identification pin of Sergeant H.W. Lewis, Co. C, 15th N.Y. Cav. private purchase

5 Identification disc of Henry N. Styer, Co. H, 119th Pa. Inf.

6 Identification pin of Sergeant William Rice

Co. C, 14th Conn. Vols.

7 Struck eagle plate

8 Identification pin of an infantryman named Andrews, 14th Maine

9 Struck U.S. oval plate

10 Identification pin of M.S. Nickerson, C. A, 10th Mass. Vols.

11–12 Private purchase gilt identity disc

13 Identification/corps badge of a soldier in Co. B, 5th Vermont Vets.

14 Stencil of a member of

Co. G, 2nd Maine

15 Identification/corps badge of J.J. Whittier, Co. A, 6th Vermont Vets.

16 Identification tag, private purchase, of a trooper in the 5th Michigan Cavalry

17 Private purchase identification pin

18 Stencil of M. McCormick, 58th Mass. Inf.

19 Identification disc, obverse

20–25 Various stencils

Artifacts courtesy of: Wendell Lang Collection, Tarrytown, N.Y.

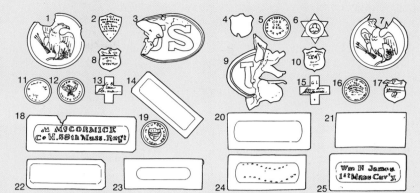

Officer, 39th New York Volunteer Infantry, "Gavibaldi Guard" U.S.A.

His own most distinctive uniform feature is his plumed hat with chin-strap. Unlike most Yankee officers, he also carries no shoulder straps as insignia of rank, and has an unusual piping of gold on his shoulder seams. There is also just a hint of decorative "frogging" on his cuffs, but otherwise his is a subdued, yet colorful, variant on the Union norm. Incidentally, field glasses were not issued by regulation to Yankee officers, and most provided their own. The Italian patriot Garibaldi being a hero of the time, these men tried to copy him, and in doing so created a uniform, which like that of so many regiments in this war, revealed a strong European influence. Officers inflicted very few wounds on their foes in this war, and despite the menacing Remington revolver in his hand, this field officer of the 39th New York "Garibaldi Guard" is likely no exception.

His was one of the more colorful Union units.

Franklin never made his ordered assault, perhaps receiving it too late to do any good as the early December twilight began to settle. But by then two more divisions were on their way to doom against the stone wall, and in their wake after repulse the grisly total carnage before that rude wall came to more than 6,300. After six successive assaults, each one in division strength, the 6,000 men behind that wall had repulsed almost 40,000 Federals. Watching it all from his command post atop a nearby height, Lee felt torn between the elation at what his men were achieving against such odds, and the awful price in suffering being paid out on that field. "It is well that war is so terrible", he said to those around him. Otherwise "we should grow too fond of it."

Hooker echoed Lee's sentiments, though in the bitterly brutal words of a man who knew he was getting orders from an incompetent. "Finding that I had lost as many men as my orders required me to lose", he reported, "I suspended the attack." Nevertheless, only full nightfall made it safe for his men trapped by enemy fire before the wall to creep back to safety.

Sporadic firing continued along the line even after dark, but it came without direction or purpose other than to discomfit the Yankees trying to reform their disorganized regiments and brigades. Burnside sent one unblooded division of the V Corps forward under cover of darkness to bivouac close to the stone wall preparatory to an anticipated dawn assault, but they found they could hardly find a place to lie down for the carpet of bodies of dead and wounded beneath their feet. In the sharp cold of that night, wounded and dying men suffered intensely as they lay bleeding into the near-frozen ground. The living took the frozen bodies of the dead and rolled them around into human breastworks against the fight to resume in the morning. Meanwhile, elsewhere along the line, ill-clad Confederates crept out to scavenge coats and blankets from the dead.

Dawn brought terror to the Yankee division when the lifting fog revealed how close they were to the wall. The Rebels fired immediately, pinning them down, and there they stayed for the rest of the day, not returning fire, but only trying their best to disappear into the hard earth. The night before Burnside and his generals discussed the situation. At first "Burn" intended to renew the attack, reportedly even claiming that he would lead his old IX Corps into the assault in person. Sumner argued strenuously against more fighting, however, joined heartily by the disgusted Hooker. Franklin's position almost certainly echoed the other Grand Division commanders', and in the end Burnside gave in, his resolution weak as always. But having taken Fredericksburg itself, the commanding general decided not to yield it by pulling back to Stafford Heights. Should he want to launch a new offensive after reorganizing his army, he would just have to cross and take the town again. So there they sat in the relative safety of the city for two days while their general pondered his next move.

Only on December 15 did Burnside ask for a truce to retrieve the dead and care for wounded who had now spent almost 48 hours suffering out on the field. Later that night he finally decided that it served no purpose to remain in Fredericksburg any longer, and during the dark hours recrossed his battered legions and took up his bridges behind him. There would be no more contesting the enemy on this front. Lee could have the victory.

The cost to the Army of the Potomac had been truly awful: 1,284 killed, 9,600 wounded, and 1,769 missing and either dead or prisoners. Against this 12,653 total, Lee suffered just 5,377, including 603 killed and 4,116 wounded. They were the most lop-sided casualty figures of any major battle of the war, reflecting just how strong was Lee's position, and how unimaginative Burnside's handling of the fight.[8]

There was cold cheer in the Union camps a few days later when Christmas came. A month later Burnside would try to move again, this time sending his divisions upstream with the hope of effecting an unopposed crossing and then sweeping down on Lee's left flank but heavy rains left the army literally stuck in the mud, a fitting end to Burnside's soggy leadership. As 1863 dawned, the Army of the Potomac was still waiting for the man who could lead it to victory over Lee.

Above: The sunken road and the stone wall at the base of Marye's Heights, taken a year after the battle. Simple things like this rude wall could deicde the outcome of battles.

Below: A target that thousands of Union soldiers saw but none managed to achieve, the Marye house still shows its part in the battle with the bullet marks in its columns and its battered bricks.

References

1 *Report of the Joint Committee*, I, p.650.
2 Sears, *Landscape*, p.325.
3 Edward J. Stackpole, *Drama on the Rappahannock* (Harrisburg, Pa., 1957), pp.77-78, 276.
4 James Longstreet, *From Manassas to Appomattox* (Philadelphia, 1896), p.201.
5 Stackpole, *Drama*, pp.126-28.
6 *O.R.*, I, 21, pp.76-80.
7 Stackpole, *Drama*, p.194.
8 *Ibid.*, pp.196-7

CHAPTER SIX

STONES RIVER (MURFREESBORO)

DECEMBER 31, 1862 – JANUARY 2, 1863

In the end, Fredericksburg had become just another in the long string of Federal failures in the East. The shock to the Union war effort, not just of another defeat, but even more of the dreadful casualties, was almost incalculable. Lincoln might well wish he had never heard of the name Virginia.

Yet for all the misery north of the Potomac, on the south side there was rejoicing as befitted a glorious victory. As 1862 came to an end, the Confederacy had survived almost two years of constant hammering from the enemy, and turned him back almost everywhere except in the Mississippi Valley. Virginia was safe, the Confederate heartland south of the Cumberland River still offered its agricultural riches to the South alone, and Grant seemed stalled in his long-contemplated advance toward Vicksburg. The outlook, though never greatly optimistic for the Confederacy, looked far better than any people of such limited resources and manpower could reasonably have expected.

But both presidents needed the next victory. It would be out in the West, out in Tennessee, where two armies stood opposed in the center of the thousand-mile-long line between Blue and Gray. A decisive victory there could buy the South a full year of life. A victory for Lincoln could buy him the opportunity to keep his nation behind him through the darkest hour of the war.

IN THE "OLD ARMY," as Civil War officers called the service, as it was before 1861, a hilarious story was told and retold at the mess table about one of their more distinctive old comrades, Captain Braxton Bragg. Everyone knew of his irascible nature. Men often and easily came to hate him. During the Mexican War one unknown enemy actually placed a lit artillery shell under his cot in an attempt to assassinate him. The try failed, but everyone knew about it. The story they all told, however, was of a different sort. At one of his Old Army posts, it maintained, Bragg had been serving as post quartermaster when the post commandant was temporarily called away. As senior officer, Bragg became interim commandant as well. Representing the interests of his actual field command, Bragg submitted a request for certain supplies. In his capacity as quartermaster, Bragg approved the request and submitted it to . . . himself as post commander. And in that last capacity, he denied the requisition! The episode led one of his superiors to exclaim to him that having argued with everyone else in the army, "you are now arguing with yourself."[1]

It was certain that Bragg did not get along with people, especially subordinates. His execrable

health, which included undoubted mental instability, affected his attitude and demeanor, making him unable to brook dissent or well-meant criticism. Instead, he was an easy mark for sycophants and flatterers. The same personality traits also made him a virtual martinet, wedded to regulations and discipline, and absolutely ruthless in the punishment of offenders. A host of apocryphal stories gained credence when he became a Confederate general, some maintaining that he had a soldier shot for an infraction as minor as killing a chicken.

Yet behind every myth may lay a seed of truth. Bragg was not inherently a cruel man, only a sick one, and sicker still in his mind after the failure of his October 1862 campaign into Kentucky. After taking command of the troubled Army of Tennessee as it was styled in the months following Shiloh, and Beauregard's relief from command in June 1862, Bragg ruthlessly set about reorganizing the brigades and divisions and instilling his form of iron discipline. That done, he led them north in the attempt to wrest Kentucky from the enemy, an abortive attempt that failed through his own errors and those of subordinates. But when Bragg retreated back into Tennessee after

There are few images surviving from the Stones River Campaign, but this scene of Union soldiers crossing Big Black River Bridge in Mississippi might well have been repeated as the armies marched toward Murfreesboro.

ward, he could not accept the responsibility for his defeat as Lee did after Antietam, or even Burnside following Fredericksburg. Rather, Bragg sought others to take his blame, and he turned to his subordinates Polk and Hardee, thus commencing a war within his own high command that never abated.

He also blamed a man who was not in the campaign, Breckinridge, who through administrative errors and inter-commander rivalry – none of it of his doing – was not able to join Bragg for the invasion. It had been thought that the presence of this most prominent of all Kentuckians would rally the men of the Bluegrass to the Southern banners. When Kentuckians, in fact, proved to be little better than indifferent to Bragg's "liberating" army and gave it neither material support nor manpower, he quickly formed a lasting resentment toward them, including even Breckinridge and the Kentucky brigade in his division. There could not have been a worse time for a man in Breckinridge's 6th Kentucky Infantry to leave ranks to go home and help his widowed mother get in a crop. He was caught, returned, and quickly tried for desertion. On December 20 Bragg approved the court's finding of guilty and ordered the man shot on the day after Christmas. No amount of pleading from Breckinridge and other officers could dissuade the general, who supposedly remarked that Kentuckians were too independent, and he would shoot them all if he must to maintain discipline. Breckinridge accused him of murder, and men in some Bluegrass regiments came near to mutiny. In the end, on December 26, the unfortunate private went to his death. He might have died only a few days later anyhow as the Kentucky brigade fought for its existence, and Bragg battled viciously to hold on to middle Tennessee. But thanks to incidents like this, when Bragg went into the fight, he was already at war with his own army.[2]

Bragg was a man who, by temperament, saw enemies everywhere, and given his personality and character, he might have been right. But as 1862 waned, and as he looked over his shoulder for foes in his own camp, he had to keep a watchful eye on a more formidable opponent in his front. While the failure of his Kentucky offensive had not produced any decisive victory for either

Above: General Braxton Bragg, the irascible, unstable, general who sowed the seeds of dissension and mistrust in his own army, and right on the eve of a major battle that could settle Tennessee's fate.

side, even after abandoning the Bluegrass state, Bragg still occupied must of northern Alabama and a great deal of the central portion of Tennessee that, previously, had been in Yankee hands. Worse, many in the North believed that Bragg and his whole army could have been taken out of the war by a more aggressive commander than Buell. In the end, Buell became the highest ranking casualty of the campaign, as public outcry and his own army's disaffection led to his

Below: General William S. Rosecrans the Army of the Potomac's next commander sits fourth from left, surrounded by his staff and officers, including General Garfield, seated in the middle, and Philip Sheridan at far right.

replacement by Lincoln on October 24. It did not help that Buell was also very tight with the out-of-favor McClellan. Six days later a new commander took over, with his mission very clear – follow Bragg and retake what had been lost.

William S. Rosecrans brought a varied background to high command, yet like almost all army commanders in the war, there had been little to prepare him for this. After doing commendably at the Military Academy, he served eleven years in the Old Army before poor health forced him to resign. He devoted the next several years first to the coal industry, then to making coal oil. Like so many other former officers, he came out of retirement when the war commenced, and rose quickly to become one of McClellan's principal subordinates in the western Virginia campaign that made "Little Mac's" reputation. Indeed, the minor victory by which western Virginia was taken from the Rebels in 1861 was actually more Rosecrans' work than McClellan's. Nevertheless, "Old Rosey" as his men began calling him, also saw his fortunes rise. His successful defense of Corinth, Mississippi, in October 1862 won further laurels, and thanks to this, when LIncoln needed a successor to Buell, Rosecrans got the nod.

Yet he had his peculiarities. Rosecrans was an incessant worker, often pushing his nervously active mind well through the night, and his exhausted staff along with it. A devout Catholic, he often pre-empted younger officers' free time with lengthy discourses on religion. Sometimes his pronounced stutter made him difficult to comprehend, especially if he became excited or flustered. And, like Bragg, he could fly into a rage and severely berate a subordinate, even in front of others. Yet none denied that he possessed perhaps the most brilliant mind of any army commander they ever saw.[3]

He took command of an army in some disarray, with enlistments expiring in some units, and others whose men had not been paid in months. Its old designation as Army of the Ohio was discontinued when he was assigned, and instead it was to be known temporarily as the XVI Corps. Ever afterward, however, it would be called by another designation to come early in 1863: the Army of the Cumberland.

It was a splendid army, though barely half the size of the giant assemblage that Burnside took over in the East. Rosecrans divided it into three wings. The Right Wing went to Major General Alexander M. McCook, one of Ohio's amazing family of "fighting McCooks" that provided three brothers who became generals, and another cousin who wore the stars. Following McCook would be three divisions, each with their particular distinction. The first belonged to a hot-tempered officer with the ironic name of Jefferson C. Davis who, in an altercation earlier that year in Louisville, shot and killed a fellow general. The Second Division under General Richard W. Johnson was made up entirely of western men from Illinois, Ohio, and the like, except for one lone regiment of Pennsylvanians. The Third Division went to General Philip H. Sheridan, a short, combative, ruthless little fellow destined to make a great name for himself later in the war.

The Center Wing belonged to Major General George H. Thomas, a Virginian who placed loyalty to his Old Army uniform and flag above his state sympathies, though it cost him the undying enmity of his entire family for the rest of his life. Lovell Rousseau commanded Thomas' First Division, James Negley the Second, Speed Fry the Third, Robert B. Mitchell the Fourth, and Joseph J. Reynolds the Fifth. Thomas' wing was thus the largest in the army. The Left Wing, like the Right, had but three divisions, commanded by Thomas J. Wood, John M. Palmer, and Horatio Van Cleve, all under the command of Major General Thomas Crittenden. While each division in the army had its own artillery, and some a smattering of cavalry, the bulk of the mounted units made up a division of cavalry led by General Davis S. Stanley. Rosecrans had to struggle mightily to get stragglers in, units reorganized and armed, and the army in marching trim once more, and he had little time in which to do it. Still, by late December he had about 64,000 of all arms at hand. He would leave one of Thomas' divisions to guard his base at Nashville, but that still gave him just short of 60,000 for his coming campaign.

He must start that campaign soon, too. Rosecrans felt pressure from Washington to take an offensive against Bragg within barely two weeks of assuming command. "The government demands action", said Halleck, and when Rosecrans had not moved by early December, the general-in-chief threatened him with removal.[4]

Well aware that Lincoln wanted his armies to advance all across the map before winter set in, Rosecrans was finally forced to start forming offensive plans of his own. Clearly his objective was to be Bragg and his army. Following the retreat from Kentucky, Bragg concentrated his

---

Private, "Guthrie Grays", U.S.A.

It was commonplace among pre-Civil War local militia to adopt "cadet gray" as a uniform color. Naturally, this led to confusion on the battlefield when units like the Guthrie Grays come into play. Some of them later served in the 6th Ohio, while others were absorbed into Neff's Independent Detachment of Infantry.

Formed in 1854, this Cincinnati outfit looked magnificent on parade. The gray uniform was trimmed with black on the collar, and frogging all across the breast, as well as on the sleeves. The short shako was particularly striking, with its white pom pom, gold braid cord, and black band with the letters "G G" in brass. The coattee was typical of many pre-war militia units, as were the brass shoulder scales and the white webbing crossbelt. While different weapons were issued to different companies, this soldier has the predominant three-banded Enfield. The unit still exists today.

Army of Tennessee in and around Murfreesboro, Tennessee, some thirty miles southeast of Nashville. An important road intersection, the little town also sat astride the Nashville & Chattanooga Railroad, a chief supply line for the Confederates. From this position, Bragg offered a constant threat to Nashville, but could also easily move east or west, either to outflank the capital, or else to move against Grant to the west, or far to the east to reinforce Lee, though such a move was never in his thoughts.

Indeed, Bragg himself was busily putting his own army to rights, despite being forced to work with subordinates who despised him and for whom he felt little affection. He had but two corps in the Army of Tennessee. The I Corps was more commonly known by the name of its commander, Leonidas Polk, the wire-pulling old friend of President Davis who had yet to demonstrate any justification for wearing a uniform at all, much less commanding half an army. Happily, at least Polk had competent division commanders beneath him. Jones M. Withers led four brigades of Alabamians and Mississippians, and brought excellent Mexican War experience to his position. So did General Benjamin F. Cheatham commanding four brigades made up almost exclusively of fellow Tennesseeans. Cheatham was a proven fighter, though too fond of the bottle. The II Corps belonged to Hardee, the seasoned old professional, and the only man present who probably deserved to command the army itself. Breckinridge commanded his first division, a mixed one of five brigades composed of men from seven different states. The second division answered to General Patrick R. Cleburne, Irish by birth and Arkansan by adoption, and the most brilliant division commander in this army. A third division under John P. McCown was temporarily attached to Hardee, adding three more brigades to Breckinridge's five and Cleburne's four, and making Hardee's the larger half of the army. Finally, Brigadier General Joseph Wheeler, a favorite of Bragg's whose abilities were overstated, led the three brigades of cavalry with the command.

Confederate First and Second National Flag Variants

1 Confederate National Flag, First Pattern Variant. Probably the flag of the 25th Virginia Volunteer Infantry. This flag was captured by Federal forces in an engagement at Phillipi, western Virginia, in 1861. The engagement, though a small one, was an early popular victory for the North and an ignominious rout for the Confederates

2 First Pattern Flag Variant. The flag of the Flat Rock Riflemen, Company C, 20th Virginia Volunteer Infantry

3 First Pattern Flag Variant. Probably the flag of Company E, 1st Regiment, Kentucky Volunteer Infantry. This regiment mustered in at the beginning of the war, went east, and served for only a short time in Virginia. It had mustered out by 1862

4 Confederate National Flag, Second Pattern Variant. The flag of the 9th Regiment, Arkansas Volunteer Infantry. This flag was carried at engagements against Federal forces throughout the conflict in the western theater: namely Corinth, Missouri; Franklin, Tennessee; Atlanta, Georgia; and Bentonville, North Carolina

Artifacts courtsey of: The Museum of the Confederacy, Richmond, Va

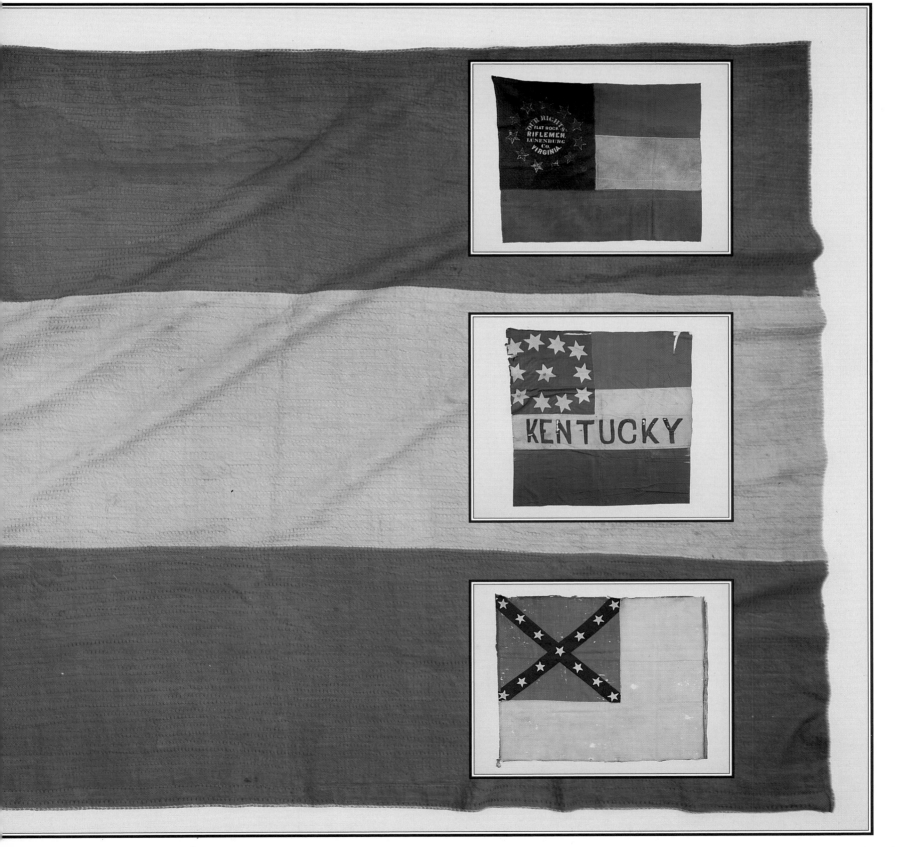

Bragg would continue to tamper with and adjust his organization well through the middle of December, but then he had little else to occupy him, for his only strategy at the moment was simply to stay put at Murfreesboro, and wait for Rosecrans to come to him. Besides, he had almost perfect confidence in the natural defensibility of his position, and of the ability of his more than 50,000 men to hold on to it.

Murfreesboro itself may not have been as good a place as Bragg thought. Certainly strategically it left much to be desired, for it was easily out-flanked on either side, as good roads led from Nashville to points well below Bragg's position, points from which his communications could be threatened and his army approached from flank or rear. Bragg did not overlook this, and by late December had Cleburne and most of his cavalry out guarding the routes that offered Rosecrans an opportunity, while keeping Polk and the balance of Hardee with him at Murfreesboro. Thus in case of a threat he could shift his forces from one side to the other to meet an enemy advance.

Rosecrans offered Bragg just such a choice when finally he put his men on the march out of Nashville on December 26 after spending most of his Christmas planning and talking with his generals. What they settled upon was a plan designed to confuse Bragg as to their intentions. The three wings of the army would move in three different directions. Crittenden would take his wing and march straight toward Bragg on the Murfreesboro Pike. Meanwhile, McCook would move almost parallel and several miles to the west toward Triune, fifteen miles west of Bragg. Thomas was to move straight south, on McCook's right, then turn east to strike what Rosecrans believed would be the Confederate flank. "Old Rosey" felt confident. "We move tomorrow, gentlemen", he told his top generals that Christmas night. "Make them fight or run! Strike hard and fast! Give them no rest! Fight them! Fight them! Fight, I say!"[5]

McCook moved out first at 6 a.m. the next morning. It was not to be quite as pleasant a

110

Union Button Dies, and State and Service Buttons

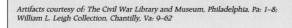

1 Steel die for artillery officers' buttons
2–3 Steel die for engineer officers'
4–5 As 1
6 Artillery officers'
7–8 Steel die for staff officers' buttons
9–10 Staff officer's
11 Artillery officer's
12 Cavalry officer's
13 Dragoon officer's
14 Infantry officer's
15 Rifle officer's

16 Voltigeur officer's
17 Enlisted man's button
18 Staff officer's
19–20 Engineer officer's
21 Topographical engineer's button
22–23 Ordnance buttons
24 U.S. Military Academy
25 Goodyear's patent rubber infantry button
26–27 variations of 25
28 Plain button
29 U.S. Revenue Marine officer's button

30–31, 33, 35, 36 Naval officer's buttons
32, 34 Goodyear naval
37–41 Diplomatic service
42 California state seal
43–44 Connecticut seal
45 Maine state seal
46 Maryland state seal
47 Mass. militia button
48 Michigan state seal
49 Missouri state seal
50 New Hampshire militia
51–62 Various state seal buttons: see Appendix

Artifacts courtesy of: The Civil War Library and Museum, Philadelphia, Pa: 1–8; William L. Leigh Collection, Chantilly, Va: 9–62

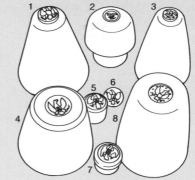
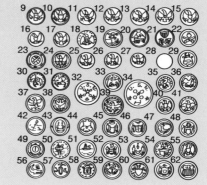

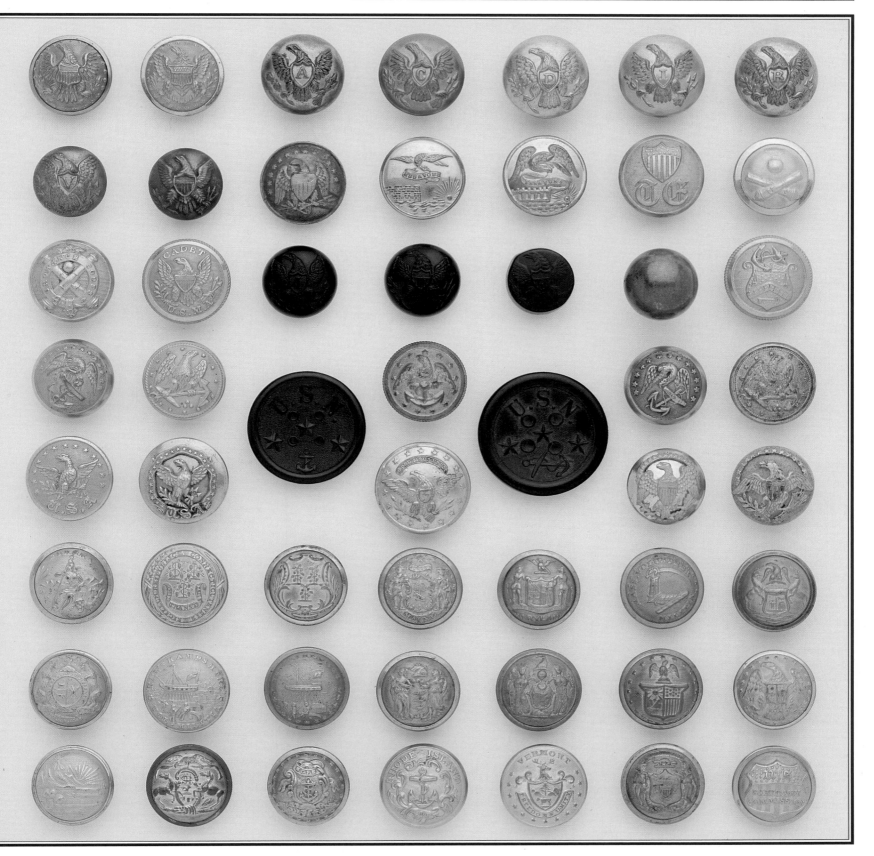

march as the men in the ranks expected, for the unusually mild December weather they had been enjoying suddenly came to an end, replaced by cold winds and rain showers. Nevertheless, the Yankees made good progress, and by nightfall McCook's advance elements were almost within sight of Triune, while Crittenden approached Lavergne, halfway to Murfreesboro. The movement left Bragg confused and uncertain as to the foe's real goal, and as a result he was hesitant to act, though he continued to believe that Rosecrans would make his main thrust at Murfreesboro. Consequently he planned his few movements accordingly. The next morning a dense fog prevented the Federals from getting an early start, and it was past noon before McCook resumed his advance, only to have it stopped minutes later by a driving storm of sleet. Nevertheless, by nightfall they had taken Triune without a fight and extended their lines somewhat beyond, with Thomas coming up on their rear, and Crittenden pressing easily through Lavergne and on to Stewart's Creek, less than ten miles from Murfreesboro.

It was all going Rosecrans' way so far, despite the weather, but by now Bragg no longer felt any doubt that Murfreesboro would be his objective, and that night the irascible Confederate set about arranging his defensive lines. Murfreesboro sat on the east side of Stones River, little more than a mile from the banks of the twisting stream that ran generally north and south. The Nashville-Murfreesboro Pike came in directly from the northwest, crossing the river a mile and half above the town, while three other roads approached from a westerly direction, and the Salem Turnpike came up from the southwest. If Bragg was to hold onto the town and the rail line that served it – and him – he determined that he had better meet Rosecrans on the other side of the stream where possible, and there he posted the bulk of his army. He put Cleburne's division of Hardee's Corps on the far left, covering the Franklin Road, with their left resting on a westward bend in the river, and a brigade of cavalry extending to the south for cover.

On Cleburne's right, posted through a series of woods, Polk's Corps extended for a mile and a half across the open side of a wide eastward bend of the river, their right resting on the stream. Revealing his fear of vulnerability on his right – the river could be forded almost anywhere – Bragg posted Breckinridge's division of Hardee's Corps on the east side, its left meeting with Polk's right across the river, and its line extending almost at right angles eastward across the northern approach to Murfreesboro. Besides providing protection on that flank, this also put most of Breckinridge's brigades in a good position to cross the river at any of the bridges to reinforce Polk in the center. It was a poor position, not least because several hundred yards in front of Breckinridge sat Wayne's Hill, a commanding eminence that even rudimentary reconnaissance should have told Bragg he must have.

There Bragg sat for the next three days without molestation. Rosecrans would not move or fight on a Sunday, and wanted to rest his army besides, so they remained comparatively idle on December 28. The next day the Federals continued their march, and by that afternoon came in sight of the Confederates drawn up in their positions. Before nightfall skirmishing began when Federals approached Breckinridge's outposts near Wayne's Hill after wading across Stones River.

Above: Governor Andrew Johnson, though a Democrat, supported the Lincoln Administration, lending political support to the Yankee armies that struggled to keep the Volunteer State in the Union.

Below: An unnamed young Confederate soldier turns his face to the camera. The full kit on his back suggests that this western boy was probably serving with Bragg's Army of Tennessee.

Quickly the sporadic firing took on intensity, and then an advance by a line of blueclad infantry threatened to take the hill itself, together with the Kentucky battery that Breckinridge had posted there after seeing Bragg's error in not occupying the eminence earlier. Only a timely countercharge by regiments from Breckinridge's old First Kentucky "Orphan" Brigade turned back the Yankee assault. Nightfall put an end to any further contact, and the Federals withdrew to the west bank of the river and their main van once more.

If Bragg had only known that at the moment he faced only a third of the Army of the Cumberland! Crittenden's wing had arrived, but Thomas and McCook had not, and the latter lay more than a mile away. A swift Confederate attack might have done incalculable harm to Rosecrans, but Bragg could not know it just then. Rosecrans worked repeatedly during the night to get McCook up, but he delayed and did not arrive until well into December 30, with the result that another day went by with nothing more than isolated skirmishing. But when night fell on that day, the two armies were virtually complete, and in place not more than a few hundred yards apart in the darkness. There was no doubt that the morrow would bring a great battle.

Rosecrans' army occupied a line almost four miles long, stretching from a position opposite McCown's division, now in advance of Cleburne on the Confederate left, thence along the edge of a dense wood, through fields and more woods, to the Nashville Turnpike and the bank of Stones River nearly a mile downstream of the spot where Breckinridge's line hinged on Polk's. McCook held the Federal right, opposite Cleburne and McCown. Thomas occupied the center, and Crittenden the left. There was much that Bragg could now tell about the enemy's line as he peered across at the moving brigades, but he believed that Rosecrans might be weak on his right flank, just where Bragg had himself shifted McCown and Cleburne from Hardee some time before. That settled his mind on taking the offensive early the next day. Hardee's two divisions would strike the enemy right, while Wheeler's cavalry rode around and got in his rear. Polk would advance as well, and simultaneously, the purpose being to keep Polk's right anchored on the river, and advance the rest of the army, wheeling around to the right and pushing Rosecrans before them until he had his own back to Stones River. Wheeler would cut off his retreat via the Nashville Turnpike, and then they could destroy him in detail. It was, of course, an utterly impossible plan, calling for perfect coordination, no breaks in his own lines, and an advance of some five miles by Bragg's own left wing. On the best of ground, with perfectly open terrain, it would have been extremely difficult with the best army on the continent; with this army, and over ground repeatedly broken by woods, cornfields, and fences, it was completely unrealistic. Nevertheless, none of his commanders objected, and the orders were so given.[6]

As frequently happened in this war, Rosecrans was planning almost the same battle as his foe. He intended to launch his assault early in the morning, and to move against Bragg's right flank. He would send elements of Crittenden's wing across the river once more, smash into Breckinridge, and push him back into the town. Then Thomas would attack in his front to keep Bragg's center occupied, and in the end they would roll

up his right completely while forcing him back across the river. His plans set, "Old Rosey" went to bed along with his army, and all of them, North and South, on either side of Stones River, did their best to take warmth and shelter from the winter cold. The initiative in the battle to come on the dawn, given the similar nature of the attack plans, would go to the general who moved first.

They were all up at about the same time, well before dawn, but Bragg struck first. Shortly after 6 a.m., McCown got his 4,400 Texans, Georgians, Tennesseeans, Arkansans, and a single North Carolina unit, into line and moving forward through the morning fog. Cleburne put his division, just as strong, 500 yards behind the first line as they set off for Rosecrans' right flank. It was exactly 6:22 a.m. when the Federals of Brigadier General August Willich's brigade, Johnson's Division, of McCook's wing, saw them coming. The Rebels got within 200 yards of the Yankees' line before firing broke out, and by then it was almost too late for the Federals to resist them. Within five minutes the brigade of General Edward Kirk, on Willich's left, virtually collapsed, leaving the wounded Kirk himself on the field. The Confederates swarmed over his artillery and then pressed on to hit Willich, now badly exposed by Kirk's retreat. The retreating and demoralized men of the other brigade badly disrupted Willich's men as they raced through his camps. Willich himself was away and his brigade lay in the hands of a temporary commander. With all this going against them, his Indiana, Illinois, and Ohio regiments gave way quickly one after the other. Willich himself, riding hard to rejoin his command, fell into Rebel hands, along with fully 1,000 other prisoners taken from the two brigades, as well as eight cannon. Rosecrans' right

Above: The men in the western armies, Blue and Gray, were cut from much the same cloth. These hardy young Federals could have been the brothers of the men they battled at Murfreesboro in 1862-1863.

almost ceased to exist and it was not yet 7 a.m.[7]

But even now the impracticability of Bragg's wheeling plan manifested itself. When an enemy retreats precipitately, an attacker follows to press the advantage. Only Willich and Kirk's brigades did not cooperate by pulling back in the direction of the wheel. Instead, they withdrew off to their rear and right. McCown, in following them, was pulled off course, actually lengthening the Confederate line and as he moved forward while Cleburne tried to wheel behind him, it naturally brought the latter into the front line. Worse, all along the line men could not keep their pace properly, for to make the wheel men on the right were to move at a walk, while those on the left nearly had to run. The result was soon a ragged line broken by a number of gaps, with men and officers fumbling in the mist to reform their lines. For some time Cleburne did not even know that he was no longer behind McCown, but up on the front, and did not discover it until he smacked into Jefferson Davis' first brigade commanded by Colonel Sidney Post.

The disintegration of the original right flank brigades gave Post time to prepare his brigade to meet an attack. Thus he was ready when the Rebels hit, and managed to offer at least a passing resistance before half his brigade gave way, soon to be followed by the rest. The Confederates, though blooded now and starting to lose their momentum, pressed on. By 7:30 the fleeing Federals had started to reform behind a few fresh regiments and held for a few minutes before a

renewed Rebel onslaught literally ran over them, putting all five of the Yankee brigades on Rosecrans' far right in precipitate retreat. Within just ninety minutes of opening the battle, McCown and Cleburne had pushed the enemy flank back more than a mile.

Now John Wharton's brigade of Wheeler's cavalry rode around the end of the shaky new Yankee position and struck the already demoralized enemy. Whole regiments simply gave up in the face of it, while others tried their best to melt still further into the woods. Wharton even came close to capturing McCook's entire ammunition train until a fortuitously timed Yankee cavalry countercharge stopped the Rebel horsemen.

By now the din of firing grew greater off to the Confederate right, as Polk's attack began to encounter the next division in the Federal line, Sheridan's. Things were not going quite as well here for the aggressors. Cheatham's division was supposed to move out at the same time as Cleburne advanced, but its commander, whom many witnesses that day believed to be drunk, did not act until prodded by Bragg, and then nearly an hour late. That gave Sheridan in his front time to prepare after seeing the debacle over on his own right. Worse for the attackers, Cheatham failed to make his advance in strength, as ordered. Instead, he sent his brigades forward one at a time, with the result that Sheridan could concentrate his fire on each in turn. The result was a severe check that sent Cheatham's first brigade reeling back, though not without cost. The immensely popular General Joshua Sill, commanding one of Sheridan's brigades, fell with a bullet through his brain.

Minutes later Cheatham sent in his next brigade, only to receive like treatment, though in a follow-up charge the Rebels gained some ground on Sheridan's right. Then Cheatham sent in his third brigade at about 8 a.m., to strike the remnant of Sill's brigade. The coming attack was presaged by a virtual stampede of rabbits flushed by the advancing Rebels, and running back and through the Federal line. This time the Yankees were ready. They stopped the Rebel brigade cold with a withering volley not fifty yards from their line. Half an hour later the Rebels struck again, but Sheridan had feverishly strengthened and repositioned his lines to meet the attack. Once more the Southerners came to a halt in the face of heavy fire, and then Sheridan sent forward a fresh brigade to counterattack. The enemy in his front rapidly withdrew.

With the Federals on his own right continuing to fall back in the face of McCown and Cleburne, Sheridan had to pull back to a new position while the Confederates regrouped for another assault. Then he had to do so yet again, and in the end selected a line along a cedar forest so thick that when the Confederates advanced once more and drove their way into the woods, no one could control the battle. Exercising command became impossible. No one could see even an entire regiment in the dense growth, and it became one of those confused, intensely private battles between men in small groups or alone. But Sheridan held on, and by 9 a.m. or thereafter, the momentum of Cheatham's ill-managed attack was spent. The

cost to Sheridan had been terrible, but he finally provided an anchor for Rosecrans' ravaged right, and brought to a halt the seemingly irresistible Confederate juggernaut.

Rosecrans had been at his erratic best thus far in the battle. At first he hardly paid any attention to the sound of firing from McCook's front. His mind was set on his own attack on Bragg's right. But when he heard the intensity of the firing grow too rapidly, and spread too quickly toward the center of his line, he realized something was wrong. As the first intelligence of the collapse of Davis' division reached him, he immediately sent a supporting division from Thomas off to reinforce the threatened area. Then his excitement

Battle of Stones River (Murfreesboro), December 31, 1862

The Battle of Stones River, like several major engagements of the Civil War, displayed the irony of both commanding generals conceiving the same battle plan prior to the fight's commencement. Both Bragg and Rosecrans planned a major movement against their opponent's right flank.

Bragg struck first. Early on the morning of December 31 he launched McCown's Division against McCook's Federal flank, in a savage assault that swept the defenders back steadily. It was to be a "rolling" attack, each Rebel division in turn going in on the right of its predecessor. Cleburne went next, then Cheatham. As the morning progressed, almost Hardee's entire corps become engaged.

The Federals under McCook tried their best, but the foe caught them unawares, in unprepared positions, and never gave up the advantage. The whole Yankee right flank gave way in time, being pushed back nearly two miles, almost to the Nashville Pike. Had Bragg managed to sieze that road, the enemy line of retreat would have been cut off, and Rosecrans severely endangered.

In the end, however, a determined stand by Sheridan anchored the failing Federal right, providing a pivot which at least retained more or less cohesive lines.

Meanwhile, on the other end of the field, elements of Crittenden's corps had crossed Stones River, posing a seeming threat to Bragg's own right. Thus, Breckinridge's division remained there to cover that flank through all of the morning. But Crittenden made no move against him, and by afternoon Bragg began calling for the Kentuckian's brigades on the west side of the river.

In the center of the lines, Bragg had launched a series of spirited, but uncoordinated attacks against an area called the Round Forest. Combined with the failure of the Federal right, the Round Forest attacks gained good ground, but chewing up several Rebel brigades in the process thanks to Polk's inept handling. Once Breckinridge's leading brigades arrived, they, too, were sent into the wasteful assaults, and though they finally achieved some ground, still Rosecrans' center, though battered, held on to their positions.

By the end of the day, the Union army had been severely mauled and Bragg wired to Richmond that he had a victory. But there was still fight in these bluecoats, and Bragg himself had suffered heavy casualties.

That first day's fight ended at dusk, with exhausted Confederate soldiers falling back from the Federal line, unable to press home any more attacks. Both sides had taken a beating, neither were willing to quit the field while the other remained.

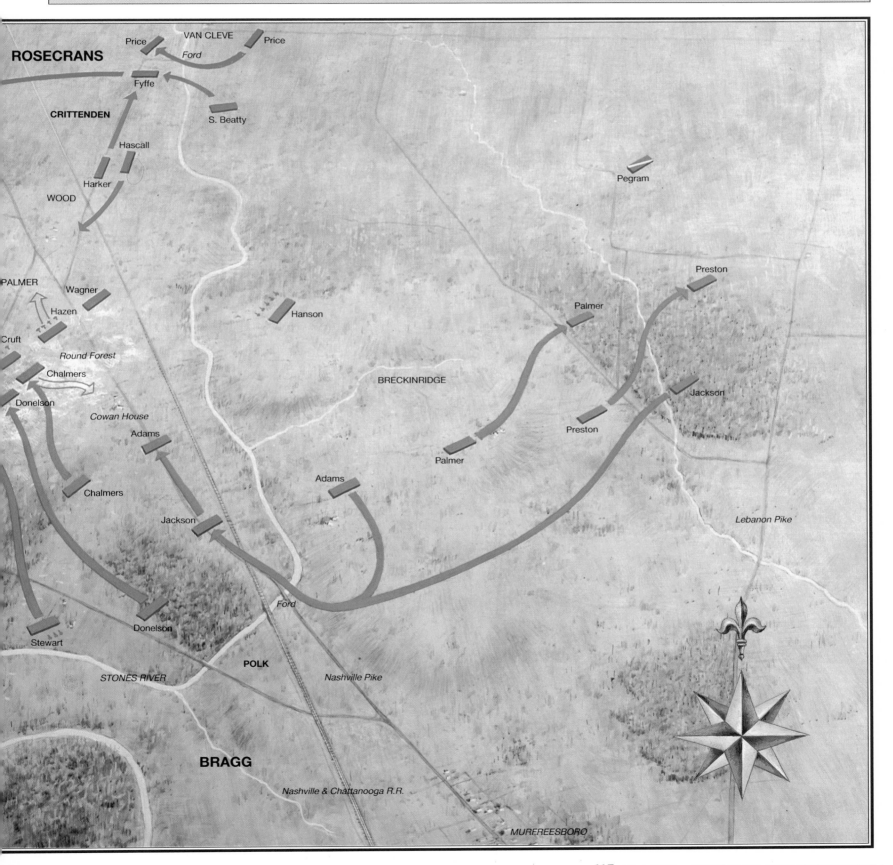

overtook him and he began riding everywhere, giving a flurry of orders to anyone at hand, sometimes to cross-purposes, and certainly confusing his subordinates. As his own right withdrew further and further, he finally committed almost all of his reserves in building a line a mile or so below the Nashville Turnpike, which was obviously the enemy's goal.

Inevitably the Rebels came on again, this time moving more in concert, and with the addition of a fresh brigade to the three that Cheatham had already sent into the fight. At about 10 a.m. they struck the hinge formed by Sheridan's right, and the newly arrived reinforcements sent by Rosecrans, while McCown's and Cleburne's veterans continued their swing around to press the right half of the hinge further back. The attack pressed on with renewed success, aided now by ammunition running low along the Yankee line and McCook's supply trains nowhere within reach. Before long Sheridan found himself fighting the foe on three sides, and dangerously close to being

surrounded. There was no alternative but to get out, and quickly. That left the division of Lovell Rousseau, just arrived to hold the far right, alone and isolated. He was no coward, but speedily realized that he would be overwhelmed from all sides if he, too, did not pull back. Only a briefly determined stand by one of his brigades gave a momentary check to McCown's advance. But by then the attackers themselves quickly ran out of steam. They had been engaged and on the move almost constantly for more than three hours. They were tired, hungry, running low on ammunition themselves, and badly broken up and confused by losses and the rough terrain. One brigade commander had been killed, and the toll in lesser officers elsewhere had been high, with one of the brigades now reduced to no more than 500 men, barely half a full-sized regiment. Only the exhaustion of their opponents allowed them to continue to press forward their advance successfully – that and the fresh brigades Bragg tardily started to commit to the fight.

Bragg had done little in the way of managing his battle so far that morning. When initial reports indicated that McCown, Cleburne, and Cheatham were progressing successfully, he turned his own attention to the center of the Yankee line directly in his front. There the enemy had a firm foothold in a wood locally known as the Round Forest. As the right half of "Rosey's" line fell back further and further, this point became increasingly important, for if a breakthrough could be made there, virtually astride the Nashville road, then the faltering right would be cut off from the rest of the army and caught between two fires, while the other half of the Yankee line would be trapped between Bragg and Stones River. If he could take the Round Forest, Bragg could win the battle and destroy an enemy army utterly.

The first real activity in the vicinity did not come until 8 a.m., well after the battle on the Federal right was going badly. General Palmer, of Thomas' wing, realized that he was in a

vulnerable position and moved his division forward a few hundred yards toward better ground. Negley was supposed to be advancing on his flank, but once Palmer was on the move he found the other officer's division moving away instead, having just seen Sheridan's line off to the right about to give way. Palmer, too, tried to move back, but in so doing created a gap in the line just at the Round Forest. Into this he sent the brigade of Colonel William B. Hazen of Ohio. Hazen moved quickly, and soon had his men erecting breastworks of fallen brush and branches to resist the oncoming Rebels.

Within a few minutes the first wave of Confederates from Polk's Corps ran into the fortified

Below: Dressed in their ranks, their officers before them, the men of this regiment in Tennessee give an idea of the size of an average unit. Multiplied by scores, they made up armies, and rarely did they fight better – under worse leadership – than here.

wood. The defenders had an excellent, unrestricted field of fire from their wooded shelter, and they tore apart the first wave, General James Chalmers' Mississippi brigade. For half an hour the brave Confederates stood in the open, trading volleys with Hazen, before losses forced them to retire. Chalmers himself fell wounded, and his dead and wounded so littered the ground that later that patch of field became known as the "Mississippi Half-Acre."[8]

Though some of the Mississippians held on, by 9:30 Chalmers' brigade was finished. There followed a half-hour lull during which Hazen received reinforcements and he used the time well to enhance his impromptu breastworks. Then the Tennessee brigade of General Daniel S. Donelson, of Cheatham's division, came on the field, the only one of Cheatham's brigades not already embroiled in the fight with Sheridan. Moreover, this was the last fresh brigade west of Stones River. It achieved marked success at first, driving away some of Hazen's supports, but

others soon appeared in their place, and though the situation looked critical for a time, Hazen held out. Suddenly the Round Forest took on the same magnetic aspect of the Sunken Road at Antietam, or the Hornets' Nest at Shiloh. Units from all sides seemed attracted to it almost by telepathy as the Federals dug in to hold on, knowing that if they were routed here, the game was lost and they would all be, in the slang of the day, "gone up".

General Milo Hascall of New York arrived with a regiment and, being senior officer in the forest, took command. "The position must be held, even if it cost the last man we had", he wrote a few days later. They all fought on desperately, losing track of the number of times Polk's battered brigades came at them until, by about noon, the Southern assault finally exhausted itself. Once again, the failure to concentrate their numbers on large-scale attacks had wasted Rebel strength.[9]

The responsibility ultimately was Bragg's. All through the morning he exercised little or no

control over the fight, leaving it to his subordinates to carry out their attack orders of the night before. Yet all the while he had Breckinridge's five brigades on the east side of Stones River doing nothing. At first this seemed justified, because portions of Rosecrans' left wing had recrossed the river just as they had the night before, and reports of their presence came to Breckinridge around 7 a.m. This seemed to threaten a possible attack. Soon after, things began to crumble on the Yankee right; however, these Federals recrossed to their own side of the river, but Breckinridge's scouts failed to report this to him. Thus a lack of good information kept him tied on that portion of the field until late morning, when intelligence

finally revealed that the ground between the Kentuckian's division and the river was clear of all but skirmishers.[10]

Shortly before noon Bragg finally called on Breckinridge for one of his brigades to come assist Polk's attack on the Round Forest. Two were immediately dispatched instead, but then Bragg changed his mind. Suddenly fearful once more of an attack on Breckinridge, Bragg cancelled the order, and instead started to send two brigades to support the Kentuckian. By 1 p.m. Breckinridge had definitely determined that he faced no threat. Fortunately, he had not recalled his first two brigades, and now Bragg ordered him to bring two of the three remaining brigades

across to join with Polk. He obeyed instantly, but when Breckinridge brought the last two of his brigades across, he found that Polk had not waited for his entire division to arrive, but had sent the first two brigades in as soon as they appeared. Once more a piecemeal approach allowed Rosecrans' concentrated fire to take telling effect, and Beckinridge's first two brigades were reeling back in disorder even as he came on the scene.

Polk now sent Breckinridge in with his two fresh units, and they finally did little better than those before, but as the daylight was waning rapidly, they came out of the advance with substantially fewer losses. Had they but known what

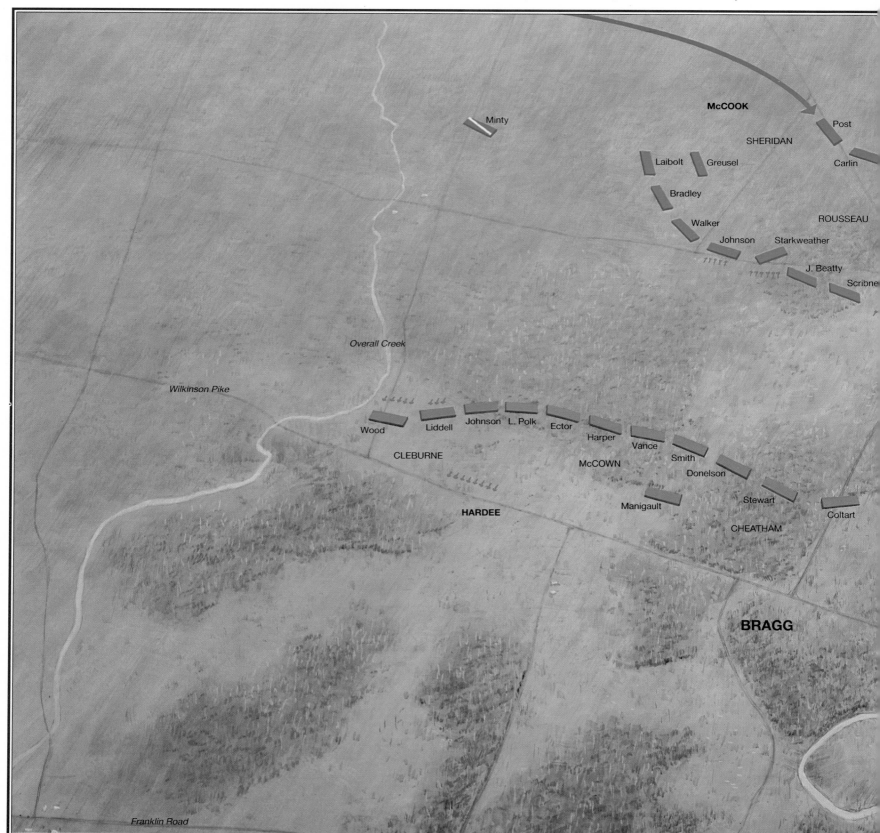

Battle of Stones River (Murfreesboro), January 1–2, 1863

On January 1 the armies stayed relatively quiet, each side waiting for the other to make the next move. Bragg reorganized his battered center and left, and established a solid line to hold the ground taken the day before. Rosecrans evened his newly contracted lines, reformed his disorganized divisions on his right, and prepared to hold his ground. More than this, he anticipated that the enemy, having crushed his right, might next attempt to do the same on his left. As a result, he had Crittenden heavily reinforce his hold on the west bank of Stones River, bolstering it with a massive emplacement of more than forty artillery pieces.

Bragg wasted the morning of January 2, 1863, unfortunately failing even to perform adequate reconnaissance. He believed, however, that Rosecrans' right was now too well placed and reinforced to allow a successful renewal of the attack toward the Nashville Pike. Instead, he planned to send Breckinridge and his entire division in an attack against Crittenden – Breckinridge's childhood playmate. If successful, it could push the Yankees back and take the Nashville Pike from the north, leaving Rosecrans virtually surrounded.

But when Breckinridge got his orders, he argued vehemently that it was suicidal, that those guns of Crittenden's would destroy his command. But Bragg remained adamant, so much so that one of Breckinridge's brigadiers wanted to shoot the commanding general!

Breckinridge sent his division forward at about 4 p.m., in a driving downpour of sleet and freezing rain. Initially the Confederates were successful, driving some advance units of Crittenden's back across Stones River. But then what Breckinridge feared happened. His men found themselves in an exposed position, and made it the worse when they pursued the fleeing Federals down to the river and themselves started to wade across. The massed artillery on the heights on the other side opened up on them with a dreadful barrage that mortally wounded one general and sent the attackers reeling back with 1,500 casualties. Crittenden did not follow up the repulse, and there the battle ended. Rosecrans lost 13,000 out of his 60,000-man army. Bragg lost 10,000, from his 50,000 Confederates. And for all of that, neither side truly gained a decisive advantage. Rosecrans held the field, and Bragg had retreated his army south to the Duck River, there to set up another line of defense. Strategically the Federals now held the initiative, and the senior officers of the Army of Tennessee could spend the winter warring with their own commander.

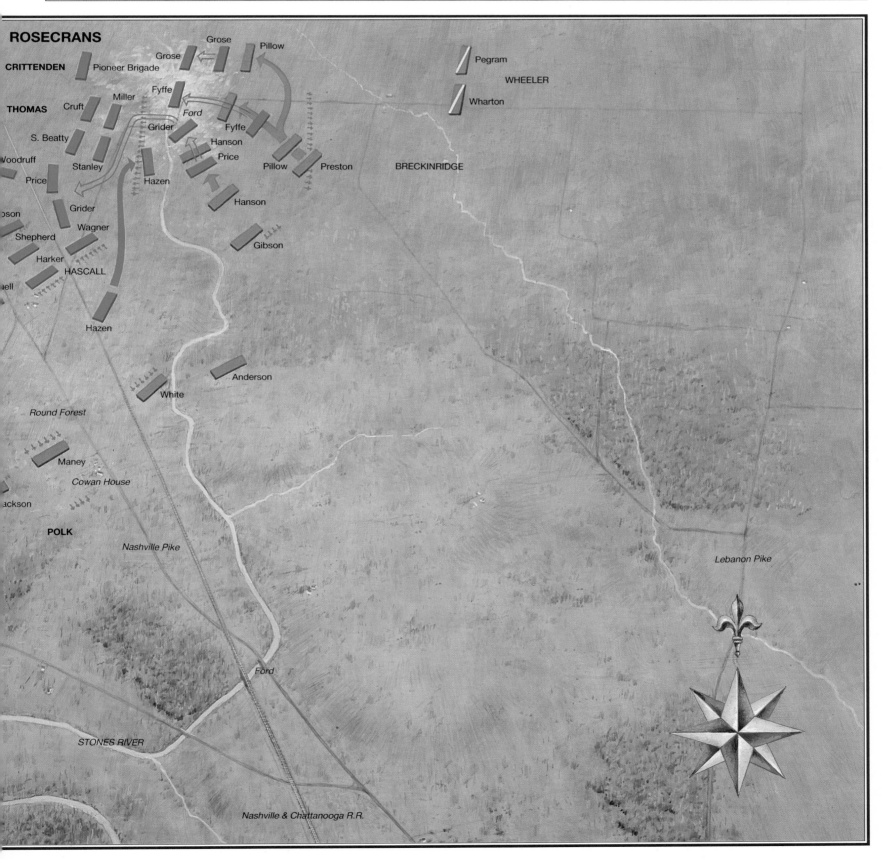

First Lieutenant, Co. A, 5th Georgia Infantry, "Clinch Rifles", C.S.A.

This particularly handsome Rebel officer serves with the 5th Georgia's Company A. The unit, also known as the "Clinch Rifles", was one of the South's most colorful, thanks to most of its companies having existed before the war, each bringing its own distinctive uniform in 1861. Company A wore green broadcloth with gilt buttons, and gold lace and trim on the collar and sleeve. The trousers matched, though of a slightly darker green, and so did the kepi, on which the letters "C.R." appeared inside a gold wreath. On full dress uniforms a pom pom topped the cap.

A distinctive feature is the use of U.S. Army insignia for officer rank on shoulder straps instead of on the collar. This man's two bars indicate a captain in the U.S. service, a first lieutenant in the Confederate. In either case, his mein of calm, cool competence, reveals that he is destined to rise to higher command in this war.

was happening on their own left flank, they might have pressed harder, for almost at this same time, Cleburne and McCown, and portions of other units, had made one more massive push that finally forced Rousseau back right to the Nashville Turnpike. One last rush, especially if supported by a firm push at the Round Forest, and the road was theirs for the taking, with all the fruits of victory that might attend. But those battle-weary Rebels did not have it in them. With darkness falling, they had literally fought from dawn to dusk. They simply could not go on, and instead almost spontaneously they began to fall back, leaving behind them the Federals and a much relieved Rosecrans.[11]

The battle was done for the day, and Rosecrans was lucky to have an army at all. Bragg had battered him almost to pieces along fully two-thirds of his line, and the only consolation – though he did not then know it – was that he had battered Bragg just as badly. "Old Rosey" called his generals to him and conferred on what they should do. The only choices were either to stand where they were and try again, or else retreat. Reports differed on who advocated what, but at the end of the conference the commanding general decided to remain. Well after midnight he ordered Crittenden to occupy a high ridge on the west bank of Stones River, roughly opposite the original position of his childhood playmate Breckinridge. Elsewhere he had his men dig in where they were after straightening their lines. Rosecrans seems not to have had much of a plan for renewing the battle; he only knew that he refused to leave the field unless Bragg drove him away.

Bragg, meanwhile, felt confident of victory, and at first expected January 1, 1863, to dawn with the fields and woods in front of him cleared of the foe. But when he found Rosecrans still there, he did almost nothing, allowing his army to rest after the exertions of the day before. That was fine, but all the while his men could see the enemy digging in, which meant that when the attack was renewed, the Yankees would be even harder to drive away.

By early morning on January 2, Bragg's reconnaissance indicated to him that Rosecrans might be intending yet again to advance on Breckinridge's position, the only direction from which a threat to the Confederate army could now come. Not that Rosecrans was in a position to mount an offensive, but Bragg did not know this with certainty. Further, still hoping to drive the enemy from the field, Bragg concluded not illogically that having failed to do so everywhere else, this area off to his right was the last logical place to try. Unfortunately, Rosecrans anticipated this when he ordered Crittenden to occupy those heights opposite Breckinridge. The Kentuckian's own reconnaissance this morning confirmed the strength of Crittenden's position, and especially the presence of a considerable amount of artillery on that ridge.

Nevertheless, Bragg summoned Breckinridge shortly after noon and ordered him to take his division and drive the Federals out of his front and back across the river. It was madness. Both Hardee and Polk opposed the plan, arguing that the Federals in the Kentuckian's front offered no real threat, while the Federal strength building on the west bank of the river's high ground certainly did. If Breckinridge drove the enemy back to the river, he would have to stand his ground under the fire of those Yankee guns. "My

Above: Stones River was an equivocal victory at best, yet after Fredericksburg it came at a vital time for the Union. President Abraham Lincoln was grateful to Rosecrans for the remainder of his days for the one bright spot in a winter of gloom.

information is different", said Bragg, and insisted on the attack. Some later believed that the assault was a punishment to Breckinridge for failing to join in the Kentucky campaign, or even for standing up to Bragg in the matter of the executed soldier. A few believed that Bragg secretly hoped to see the general killed.

When Breckinridge went back to his brigade commanders with the orders, they were close to mutiny. General Roger Hanson, commanding the Kentucky "Orphan Brigade," fumed that he would personally kill Bragg, while others urged Breckinridge to challenge Bragg to a duel. Even worse, Bragg now, at the last minute, forced the Kentuckian to accept General Gideon Pillow, a hopelessly incompetent old politician and schemer, to supersede one of his veteran brigade commanders. Breckinridge did as he was told, readied his brigades, and at 4 p.m. led them forward at Bragg's signal.

Just over 5,000 men commenced the advance; fewer than 3,500 came back. At the very outset Pillow turned coward. Breckinridge found him cowering behind a tree. Then two-thirds of the way across the ground to be covered, Hanson went down with a mortal wound, Breckinridge himself standing beside him trying to hold the severed artery in his leg to prevent him bleeding to death. Almost from the first the advancing division came under heavy fire, and it got more intense the closer they came to the river. Crittenden had massed 45 pieces of artillery on the heights across the river, and as soon as Breckinridge pushed the Yankees in his front off a hill some distance in advance of Wayne's Hill, his men came in for the full force of the shelling.

Worse, by now it was 4:30 or thereabouts, and already the light was failing. Sleet started coming down, hitting the men in the face and blurring their vision. Then when the Federal brigade in their front pulled back entirely, while others retreated across Stones River, the Confederates spontaneously became caught up in the pursuit. Despite Breckinridge's efforts to keep them in hand, they raced down to the river after the fleeing foe, only to bring themselves under the murderous fire of that artillery high on the opposite bank. It cut them to pieces. Within a few minutes, the attackers started to stream back to their starting point, and by 5 p.m. it was all over, Breckinridge himself riding along his reforming lines with tears in his eyes as he saw the great gaps. He never forgave Bragg.[12]

There the battle ended. Late that night Bragg met with all of his ranking generals, and they were not yet decisive about what to do next. But on January 3, when Bragg received reports that Rosecrans' cavalry was moving around his flank, he finally decided to abandon his position and retreat. The movement took all night, and soon the Army of Tennessee was on its way several miles south to the Duck River, and a new line of defense that it would occupy for much of the winter. For his part, Rosecrans was more than happy to hold his position, having recovered some of middle Tennessee and seen the enemy leave his front.

It was a curious battle. Neither commander fought it well. Tactically Bragg held the upper hand all through December 31, and finished January 2 with the same ground he started it with. Nevertheless, he felt forced to abandon the field, which gave Rosecrans, almost by default, the strategic upper hand. Certainly Lincoln was anxious for the news of anything resembling a victory, following the failure at Fredericksburg and a stalemate out west on the Mississippi. He gave Rosecrans his heartfelt thanks, and to the day he died remained grateful. Earning him that gratitude were the 13,000 killed, wounded, and missing, from the Army of the Cumberland, losses substantially greater than the 10,000 suffered by Bragg. But the incalculable damage done within Bragg's army by Bragg himself more than made up for the disparity. In the aftermath of the battle, Hardee, Polk, Breckinridge, and a host of other ranking generals literally went to war with Bragg over his conduct of the fight, and his subsequent attempt to lay blame for the failure to destroy Rosecrans at their feet. Those battles in headquarters tents and in the newspapers and halls of the War Department in Richmond would have repercussions far outlasting the military impact of the bloody winter days along Stones River.

References

1 Glenn Tucker, *Chickamauga* (Indianapolis, Ind., 1961), p.75.
2 William C. Davis, *Breckinridge: Statesman, Soldier, Symbol* (Baton Rouge, La., 1974), pp.331-32.
3 Peter Cozzens, *No Better Place to Die* (Urbana, Ill., 1990), pp.17-19.
4 *O.R.,* I, 20, pt.2, pp.60, 64, 123-24.
5 Cozzens, *No Better Place*, p.46.
6 *O.R.,* I, 20, pt.1, pp.773, 966.
7 *O.R.,* I, 20, pt.1, pp.316, 325-26.
8 Cozzens, *No Better Place*, p.153.
9 *O.R.,* I, 20, pt.1, pp.560-61.
10 Davis, *Breckinridge*, p.336.
11 *O.R.,* I, 20, pt. pp.678-81, 722, 784, 884-85, 891-94.
12 Davis, *Breckinridge*, pp.343-44.

CHAPTER SEVEN

CHANCELLORSVILLE

MAY 1-4, 1863

Neither side got the victory it really needed at Stones River, but at least Lincoln could look on it as a triumph for Northern arms so far as public opinion was concerned. He needed more, of course. He needed to beat Robert E. Lee on his own ground in northern Virginia. He needed to show that the South's premier army was not invincible.

As the winter of 1862-63 passed on into spring, Union arms had much to encourage them out on the Mississippi, where Grant, though stalled, was still making general progress and expected soon to start his final overland march against Vicksburg. Lincoln now needed pressure on Lee in Virginia. If he could take Vicksburg and Richmond in the same summer, the back of the rebellion would be broken. Elsewhere, in South Carolina, in Tennessee, and in Louisiana, everything seemed static, stalled. Lincoln needed movement in Virginia, and quickly, decisively. He might not be able to rely on the discredited Burnside to lead the Army of the Potomac, however, but there were usually plenty of others ready to try a hand at army command.

The problem was to pick the right one.

"MR. PRESIDENT, I am not 'Captain' Hooker, but was once Lieutenant-Colonel Hooker of the regular army," said Lincoln's visitor in the summer of 1861. "I was lately a farmer in California, but since the Rebellion broke out I have been here trying to get into the service, and I find that I am not wanted. I am about to return home, but before going I was anxious to pay my respects to you, and to express my wishes for your personal welfare and success in quelling this Rebellion. And I want to say one word more. I was at Bull Run the other day, Mr. President, and it is no vanity to me to say that I am a d . . . sight better general than any you had on that field." Almost on the spot Lincoln made him a brigadier.[1]

At least, this is how "Fighting Joe" Hooker remembered the episode more than eighteen years later, but like so much of his bombast during his military career, the story is probably more weighted toward conceit than accuracy. For one thing, he had never been a lieutenant-colonel in the Old Army. Indeed, he almost did not make it into the service at all. His intemperate mouth almost got him evicted from the Military Academy before his eventual graduation in 1837, then later during the Mexican War, though his field performance was good and won him *brevet* promotions to higher rank, his open criticism of General Scott made even more trouble for him in the post-war army. That may have accounted for his banishment to the tranquil but almost soporific service on the Pacific coast, the same posting that nearly made an alcoholic of Grant and bore Sherman into leaving uniform. In 1853 he

finally resigned his commission – as a *captain* – and bought a farm near Sonoma, California. Despite the lushness of the soil, Hooker was not molded for farming, and dabbled in a number of small enterprises and minor public positions. In 1859, as California was being torn along with the rest of the nation, he took the colonelcy of a militia regiment – a commission that he misrepresented to Lincoln as being in the old Regular service – and two years later, when war broke out, he borrowed money to make the trip to Washington to scramble for a place in the new army. Unfortunately, when he went to the War Department, he found that the general-in-chief at that time was his old antagonist Scott, who was quite happy to put Hooker's ambitions right at the bottom of his already long list of priorities.

If his later account of the interview with Lincoln is at all accurate, Hooker also misrepresented something else – his lack of vanity. In an era of towering egos, still his conceit loomed monumental. It was what gave him the self-confidence and – he felt – the right to go over the heads of his superiors as he chose, just as he did in going directly to Lincoln. Hooker's ends always justified his means, in his mind, and if jumping channels, disobeying orders, intriguing, wire-pulling, or outright lying would accomplish his goals, he did not shrink from the task. Then there were other rather compatible elements of his character that did not endear him to many. His was a personal as well as professional vanity. He was undeniably good looking by the standards of the time, six feet tall, robust, fair, with a rakish air that easily attracted the ladies. Indeed, his

Captain A.J. Russell made this image of the stone wall and sunken road in front of Marye's Heights only hours after the Confederates withdrew in May 1863. The dead and debris testify to the fury of the defense.

intemperance in that line soon became the talk of the army. Contrary to an old myth, the term "hooker" as applied to a prostitute does not derive from Joseph Hooker's fondness of the ladies. It predates him by many years and comes from a section of New York locally known as "the Hook," where they were once concentrated. But the fact that the myth took hold so quickly and tenaciously is ample evidence that such an association surprised no one.

As a result, Hooker was not very well liked by his fellow officers, other than those on his personal staff. Men like Burnside and Meade saw him as dangerous, too ambitious, too ruthless, too imprudent. He flaunted the system by which they lived, too flagrantly, as when he ignored Scott and went to Lincoln hiding his ambition for a commission behind that transparent veil of wishing the president good luck. Unfortunately, such behavior was only encouraged when Lincoln actually made Hooker a brigadier. Thereafter, despite his able performance in divisional command on the Peninsula, at the head of the I Corps at Antietam, and as Center Grand Division commander at Fredericksburg, he still did not enjoy widespread trust from most of his fellow generals.

They may also have resented his instinct for newspapermen and headlines, especially after the North fell in love with his sobriquet "Fighting Joe," which originated, in fact, in a typographical error. A compositor at the New York *Courier and Enquirer* received a report of an engagement on the Peninsula with the heading "Fighting – Joe Hooker," which was meant only to indicate that the text following was to be integrated into an existing article as up-to-date information on the general's part in the battle. Instead, the compositor left out the hyphen, and set up the article as a free-standing piece with a headline reading "FIGHTING JOE HOOKER." Though the general disliked the sobriquet, the public loved it, and after a time his objection to the title was probably nothing more than an exercise of his occasional false modesty.[2]

It was his ruthless scheming that most offended Hooker's associates. Chaffing at being a part of Burnside's wing during the campaign leading up to Antietam, Hooker was widely believed to have politicked with McClellan to get himself and his corps an independent status just before the battle. Then when Burnside later assumed command of the Army of the Potomac, Hooker almost flagrantly set out to undermine him. He wrote to influential men in the War Department, complained about Burnside to important Congressmen on key committees, circulated stories – probably exaggerated – about the inefficiency within the army, and especially after Fredericksburg lost no opportunity to disparage his commander. At the same time, he sowed discontent within the Army of the Potomac itself, especially in his own command. By January 24, 1863, Burnside's outrage was so great that even in the act of resigning his own command of his army, he sought to strike out at his greatest internal foe. Calling Hooker "guilty of unjust and unnecessary criticism of the actions of his superior officers, . . . and having . . . endeavored to create distrust . . . , made reports and statements which were calculated to create incorrect impressions, and for habitually speaking in disparaging terms of other officers", Burnside issued an order dismissing Hooker from the service "as a man unfit to hold an important commission."

Below: "Fighting Joe" Hooker seemed to give every promise of a great improvement: braver than McClellan, smarter than Burnside. His confidence infused the army with confidence in victory.

Above: Winter quarters before the launch of a spring campaign always presented a scene of dilapidation, as here, where Union boys have abandoned their cabins and firesides to take the war across the Rapidan.

Nevertheless, the next day, when Lincoln accepted Burnside's resignation, he offered the army command to Hooker in his place and quashed the dismissal. After almost two years of trouble from Democratic high commanders like McClellan, the Republican Lincoln and his cabinet liked the apolitical Hooker. And after the battles of 1862, there was no doubt that he would fight. Lincoln had tried a man like Burnside who doubted his own abilities, and who went on to justify those doubts. Maybe a supremely confident fellow like Hooker could actually achieve something.

Just the same, the president felt little comfort in his choice, and some doubts of his own that he did not hesitate to share with the Army of the

Potomac's new commander. Just the day after receiving the command, Hooker received a summons to meet privately with Lincoln. Very probably Lincoln told him in person what he later that day put in a confidential letter to the general. "There are some things in regard to which I am not quite satisfied with you," said the president. Hooker was ambitious which was not of itself a bad thing, "but you have taken counsel of your ambition, and thwarted [Burnside] as much as you could, in which you did a great wrong to the country." Moreover, Lincoln had heard more than once Hooker's boastful claims that the country needed a dictator to see it through its crisis. In spite of such sentiments, Lincoln gave him the command, reminding him that only "those generals who gain successes can set up dictators." He expected such success from Hooker, "and I will risk the dictatorship." Yet Lincoln made it clear that he feared Hooker far less than he feared *for* him. "The spirit which you have aided to infuse into the army, of criticizing your commander and withholding confidence from him, will now turn upon you", he warned. "Neither you nor Napoleon, if he were alive again, could get any good out of an army while such a spirit prevails in it." Lincoln would help as much as he could, but Hooker must be careful not only to his front, but also his back. "And now beware of rashness", the president concluded in an almost paternal tone. "Beware of rashness, but with energy and sleepless vigilance, go forward and give us victories."[3]

Despite its stern words, Lincoln's letter won Hooker's heart, at least for the moment, but then he was already riding high. He has just passed his forty-eighth birthday, and now he commanded the largest mobilized field army in the world, the Army of the Potomac, almost 150,000 men and officers as of the end of January 1863. As did every commander before him Hooker set about first to do a little reorganizing, especially since he made it clear that it would be foolhardy to attempt any sort of winter campaign, especially after the disaster at Fredericksburg and the humiliating "Mud March" that followed. He would wait for good weather before commencing campaigning.

Hooker maintained the Grand Division organization inherited from Burnside, but made some changes in commanders. Couch now succeeded to command of Sumner's old Right Grand Division, with Oliver O. Howard now leading its II Corps and John Sedgwick the IX Corps. Hooker's own Center Grand Division went to Meade, with Daniel Sickles now in command of the III Corps and George Sykes taking over the V Corps from Butterfield, whom Hooker had made his chief of staff. The Left Grand Division went to William F. "Baldy" Smith, with Reynolds still leading the I Corps and Newton in charge of Smith's old VI. Sigel's XI Corps and Slocum's XII Corps became a Reserve Grand Division, led by Sigel. Cavalry and artillery Hooker parceled out to each of the Grand Divisions, for a time breaking up the former unified mounted command. Less than two weeks after taking over, however, Hooker did away with the Grand Divisions altogether. He recombined all of the cavalry into a single corps under Stoneman, while the abolition of the divisions meant that Couch, Meade, Smith, and Sigel all reverted to their old corps commands, respectively the II, V, IX and XI. All of this displaced other officers, and through seniority Sedgwick now had to be given the VI Corps, and when Sigel objected to being reduced in command, he went on extended leave and Howard took his XI Corps. At the same time, Hooker dispatched Smith and the IX Corps to Fort Monroe, Virginia, as a diversion, hoping that Confederate fears of another move up the Peninsula would lead Lee to send men from his own army to protect Richmond. It worked, as Hood and Pickett and the 13,000 men in their divisions were soon on their way to the capital. Yet this and some threats to Confederate North Carolina constituted practically all of Hooker's strategic maneuver in the final weeks of the winter.

Hooker inherited a virtually static military situation when he took command. The Army of the Potomac still perched on Stafford Heights, overlooking the Rappahannock, glaring across at Lee. The hard winter made each suffer alike, and the inactivity led to a relaxation of drill and discipline on both sides of the stream. Indeed, when the river was not frozen over, the enlisted men carried on a brisk trade back and forth, Rebels trading tobacco for Yankee coffee, all of the commerce passing in little sailboats blown across on the winds. Still Hooker worked tirelessly to rebuild the equipment and morale of the men in the ranks, as well as their numerical strength, for absenteeism and desertions by the disheartened

Colonel and 1st Lieutenant, 16th Virginia Volunteer Infantry C.S.A.

These officers of the 16th Virginia Infantry well illustrate the divergent uniforms to be found within many Confederate units.

The standing colonel, commanding the regiment, wears essentially the regulation field officer's blouse in Confederate gray, with sky blue trousers. His saber, however, is just as likely to be an Ames of Chicopee, Mass., as one of Southern manufacture, especially if he was a militia officer before the war and brought his favorite sidearms with him into Confederate service.

The 1st lieutenant seated before him has in common with his commanding officer, the regulation kepi and the trousers. His blouse is very much the old Federal blue, however, and he appears to be wearing an old Virginia state beltplate, or perhaps one from the Virginia Military Institute. No wonder there was confusion on the battlefield.

rose to alarming proportions, approaching ten percent of the army at one stage.

Meanwhile Hooker slowly, methodically, and very secretively made his plans for the spring campaign. Lincoln had given him a relatively free hand. The only requirements were that he take the offensive as soon as possible, and that he not move so as to leave Washington unshielded. Almost from the first he seemed to fix on the idea of repeating Burnside's success early in the Fredericksburg Campaign, by stealing a march on Lee and getting between him and Richmond. As the plan evolved, however, Hooker simply reversed it. Instead of flanking the enemy by moving down the Rappahannock, Hooker fixed on the idea of

moving upstream instead. He could make a major demonstration against Fredericksburg itself, enough to keep most or all of the Army of Northern Virginia in its earthworks, while sending a large movement secretly upstream to cross over, then use the generally good roads to march first west, then south to get around Lee. In March Hooker started putting logistical plans for such a campaign in motion.

Lincoln came to visit the army in person as April dawned, and there Hooker paraded his magnificent legions for the president. Everyone noted in Hooker's conversation his repeated use of expressions such as "when I get to Richmond", or "after we have taken Richmond." To some the

general boasted that he almost felt sorry for the enemy. "I have the finest army the sun ever shone on", he boasted. "My plans are perfect, and when I start to carry them out, my God have mercy on General Lee, for I will have none." The president found it disturbing. "That is the most depressing thing about Hooker", he told a friend. "It seems to me that he is overconfident."[4]

Again, what Hooker and Lincoln said in private is not known, but within days of Lincoln's visit the general presented in written form his final operational plan. He would send his cavalry upriver, to cross some twenty miles northwest of the armies, at or near Kelley's Ford, then to drive south, cutting off Lee from Richmond, after

Federal Cavalry and Confederate Infantry Uniforms

1 Hooded talma or cape, with tassel. Amongst Civil War regiments, the talma was an item of uniform unique to the 3rd New Jersey Cavalry

2 Forage cap of a member of Company C, 1st Massachusetts Cavalry

3 Unique shell jacket of an enlisted man in the 3rd New Jersey Cavalry. Note the extremely elaborate yellow piping. This regiment became

known as "the Butterflies"

4 Distinctive shell jacket of a sergeant, 11th New York Cavalry, a unit also known as "Scott's 900"

5 Confederate issue shirt, taken to England as a souvenir, by an English participant in the Civil War

6 Louisiana nine-button frock coat complete with state seal buttons

7 Pair of white cotton

gloves found in the tail pocket of Louisiana frock coat *6*

8 Framed image of the owner of Louisiana frock coat *6*

9 Louisiana shell jacket. Like item *5*, this was taken to England as a souvenir by an English Civil War participant from England

10 Storage bag for framed image *8*, of owner of Louisiana frock coat *6*

Artifacts courtesy of: Don Troiani Collection, Southbury, Conn

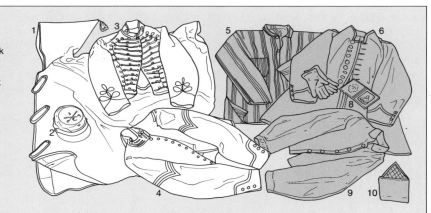

which Hooker would cross at Fredericksburg with the main army. Threatened front and rear, Lee would have to retire to the west, his only avenue of retreat. But no sooner did Hooker put his horsemen on the road than torrential rains fell, raising the river level, and making the whole strategy impracticable for the moment. The general called off the campaign, and for the next two weeks men both blue and gray slogged about in almost non-stop downpours. Hooker decided simply to make occasional threats along the river at its several crossing places, all the while keeping an eye out for a weak spot where he might cross quickly without loss, and for weather to allow him to move. Nevertheless, all along Lee confidantly predicted that if and when the Yankees tried to cross the Rappahannock, they would do so upstream of Fredericksburg.

While Lee anticipated at least that much of Hooker's design, he did not at all foresee the radical change in the Yankee commander's overall strategy that developed during those dreary wet weeks in April. By about the 24th or 25th at the latest, he had completely reformed his plans. No more did he think of pushing Lee aside and striking for Richmond. Now he wanted Lee's whole army itself. He would send infantry marching up to Kelley's Ford, where they would cross, then march down the opposite bank, driving the Rebels away from United States Ford and Banks' Ford, respectively just seven and four miles upstream. Then he would send more infantry pouring over those now undefended crossings and the whole group would hit Lee's left flank on the heights at Fredericksburg, while other legions crossed the river below the city and moved up on Lee's other flank as well as cutting him off from Richmond. Hooker hoped to use his superiority in numbers, along with rapid movement, to trap Lee with his back to the river.[5]

It was an excellent plan, *if* the Yankees moved swiftly and silently, and *if* Lee was slow to respond and remained content to stay at Fredericksburg facing the Federals left to occupy his attention. Certainly Hooker wasted no time in setting things in motion. On April 26 he directed

that the XI, V, and XII Corps leave at dawn the next day for Kelley's Ford, expecting them to arrive by afternoon April 28. Couch was to take two of his three divisions up to the vicinity of United States and Banks' Fords, while leaving one division very ostentatiously in their camps opposite Fredericksburg to make Lee believe no one had left. Hooker hoped that Lee would sense the threat to the two lower fords and detach men to protect them, without spotting the main crossing to be made at Kelley's. This would also weaken the remainder of the Rebel army to be dealt with at Fredericksburg itself. That portion was to be the responsibility of Sedgwick, whom he temporarily placed in command over the I, III, and VI Corps. Reynolds was to cross the river at Fitzhugh's Crossing four miles downstream from the city, before dawn on April 29. The VI Corps would move at the same time over Franklin's Crossing, two miles closer to Fredericksburg, while the III Corps would follow one or the other of the first two, depending upon which crossing had been more easily and completely affected. They were all to drive inland toward the Telegraph Road, Lee's chief link with Richmond. Meanwhile, if informed that Lee had detached much of his army to meet the threats from the upper fords, then Sedgwick was to attack the remainder at once.

Just where the actual fighting would commence depended upon how quickly Lee ascertained what Hooker was about, and how quickly he responded. The men crossing at Kelley's Ford would have to move straight south, once across, and after six or seven miles would come to the Rapidan, which flowed into the Rappahannock from the west near United States Ford. They must get to the Rapidan and across it at Germanna or Ely's Fords before Lee moved to stop them. If they did, then they would march almost immediately into a dense, tangled forest known

Below: In an era of colorful regiments, none were more eye-catching than the several varieties of zouaves, including this squad of Yanks with tasseled fezzes on their heads. Rarely did such headgear last the war.

locally as the Wilderness, and crossable only on the few turnpikes and "plank" roads – literally roadbeds of wooden planks laid over log rails to evade the mud beneath. Just a few miles south of Germanna and Ely's Fords they would come to the Wilderness Tavern on the Orange Turnpike. Then they would turn due east. If Lee and his army did not confront them by that time, they could still expect to meet with him anywhere along the turnpike, perhaps a few miles further at Wilderness Church, perhaps two miles beyond, at Chancellorsville.[6]

With luck, Lee might not meet him at all, for the Rebel chieftain, for all his gifts for outthinking enemy commanders, seems not to have anticipated Hooker's strategy. On the same day that Hooker was issuing marching orders to his corps commanders, Lee still held to his notion that the enemy would send his cavalry on a wide upriver sweep, while advancing the rest of the Army of the Potomac directly across against him at Fredericksburg – essentially the plan Hooker had abandoned several weeks before. There may have been something of wishful thinking in this, too, for Lee was hardly in a position to counter the more ambitious movement that Hooker had set in motion. The detachment of Hood and Pickett to meet the threat of the IX Corps on the Peninsula reduced Lee's tally by 13,000, and he had sent Longstreet to command them as well. This left Lee personally directing the remaining I Corps divisions of Anderson and McLaws, with an effective strength of less than 18,000. Jackson's II Corps was intact, with the divisions of A.P. Hill, Robert Rodes commanding in D.H. Hill's absence, Early, and Raleigh Colston standing in for the absent Isaac Trimble. This corps totaled just over 38,000, and with the artillery and Stuart's cavalry division, gave Lee a grand total of just under 61,000. Even though Lee discounted reports that put Hooker's strength at 150,000 or higher, he still would not have been surprised by morning reports from mid-April that showed just under 134,000 Yankees present for duty. Thus Lee knew that Hooker could meet him two for one. As it was, he had Jackson spread out over about two

miles of the works immediately below Fredericksburg, while McLaws' division held the heights behind the city itself. Anderson's men were spread out as far to McLaws' left as seemed prudent, with only elements of Stuart's cavalry thrown several miles west trying to watch the upper fords. In the face of Hooker's might, Lee was in no position to do more than await the enemy's first move. Fortunately he still enjoyed that impregnable position at Fredericksburg itself, another reason he hoped that the enemy could come straight at him as Burnside had done.[7]

The movement to Kelley's Ford, despite heavy rain, went brilliantly. The XI and XII Corps arrived by the late evening of April 28, to find that there would be no repeat of the administrative bungling that ruined Burnside's campaign. Their pontoon bridge materials were already there waiting for them, and so was Hooker himself. By 10 p.m. the first regiments were marching across the bridges, and the few Rebels on the other side were quickly captured without getting news back to Lee. Orders for the next day went out quickly, with Meade to take the V Corps straight for Ely's Ford on the Rapidan, while Howard and Slocum marched for Germanna. Movement on April 29 was slower, hampered by the dense country and the clogging effect of tens of thousands of soldiers on the limited roads. Even though the Yankee advance was now being observed by Stuart's scouts, they still misinterpreted it as a move to make a wide sweep around Lee rather than a direct march on his army's left flank. Thus by nightfall, Meade had crossed the Rapidan unhindered while Howard and Slocum straddled Germanna Ford.

Hooker had achieved his first goal successfully. He was south of the Rapidan without opposition, with virtually all of Lee's army still in its works at Fredericksburg. One more good day like the past two, and it might be too late for the Confederates to do anything, especially with Sedgwick starting to put elements of his half of the army over the river at Franklin's and Fitzhugh's Crossings that same day. The best Lee could do was warn Richmond of the movements on both his flanks and ask for any available reinforcements. He still did not divine Hooker's true intentions, or the real strength of the column advancing toward his left. Only after dark did he learn of the crossing of the Rapidan, and only then did he realize the immediate danger. At once he knew that the enemy would be heading toward Chancellorsville, for only there could Hooker pick up the Orange Plank Road, or stay on the Orange Turnpike, the two routes that led to the rear of his position at Fredericksburg. That same evening he sent urgent orders to Anderson to take two brigades and part of another and race to Chancellorsville to hold those roads.

Anderson himself reached Chancellorsville at about midnight, and around 8 a.m. the next morning, April 30, army engineers arrived to help his start erecting a line of defenses. Since the turnpike and the plank road ran roughly parallel to one another, and actually came together again about four miles east of Chancellorsville near Tabernacle and Zoar Churches, he pulled his weak line back to this point. It still protected both roads, and also put him that much closer to assistance from Lee.

Hooker's right wing corps marched on throughout the day, Slocum in the lead, followed by Howard, while Meade moved south from Ely's

Above: The great gray commander, General Robert E. Lee, had a penchant for daring moves involving splitting his army in the face of a larger foe. Never did it serve him better than at Chancellorsville.

Ford to take a position in their front, occupying Chancellorsville by nightfall. Meanwhile, Couch finally crossed the II Corps at United States Ford and came in on Meade's left. By dark they had a well established line nearly four miles long along the turnpike between Chancellorsville and Wilderness Church. In fact, Meade reached the

Below: Officers of the 16th Virginia, one of the regiments with Lee in the battle, strike a deceptively peaceful pose for the camera. They, like their outfit, will suffer in the battle.

objective at around 2 p.m. that afternoon, and showed uncharacteristic elation when Slocum came up. "This is splendid", he said. "We are on Lee's flank, and he does not know it. You take the Plank Road toward Fredericksburg, and I'll take the Pike, or *vice versa*, as your prefer, and we will get out of this Wilderness." Meade smelled victory. But Slocum, the senior officer, stunned him with news that, on orders from Hooker, he was now assuming command here, and they were to halt their advance with hours of daylight ahead of them.[8]

Hooker was starting to lose his nerve. The previous evening he had Butterfield send Sedgwick a message that Hooker hoped to make Lee fight him "on his, Hooker's, own ground." Such terminology naturally implied going on the defensive and waiting for Lee to come to him, and now the orders to Slocum to stop the advance sounded like more of the same. Commanders and soldiers alike were dumbfounded by the order, and not a little chagrined. That same afternoon Hooker gave even more evidence that he would not attack. He issued a congratulatory address to his army proclaiming that given the position they now occupied, Lee must either "fly" or leave his entrenchments and attack them. Furthermore, in the orders he issued for May 1, the only references to attacking related to Sedgwick's portion of the army. Hooker seemed to intend standing idle and letting Sedgwick do the work of driving Lee back to him.

He need not have bothered, for that was exactly what Lee decided to do on his own. By late on the afternoon of April 30, when no strong movement had been made against his Fredericksburg lines, Lee decided that the main threat must be from Hooker's corps advancing toward Chancellorsville. He determined to leave one brigade of McLaws' and one division of Jackson's, under Early, to face the masses under Sedgwick. To the remainder he gave orders to move at daylight to join Anderson at Tabernacle Church. By midnight, the advance elements were nearly there, with the rest marching through the night. Lee would take the battle to Hooker.[9]

The III Corps started to cross at United States Ford in the morning, and shortly before noon Meade started slowly moving forward once more, though Hooker had no intention of attacking. Rather, he was simply planning to take up a more forward defensive position.

Meanwhile, when Stonewall Jackson arrived on the scene at Anderson's line, he almost immediately began to think of taking the offensive. No sooner did he relieve Anderson at Tabernacle Church than he began to push his first units forward, along both the turnpike and the plank road. By 11:30 he had covered about the half the distance to Chancellorsville on both roads when finally he came in sight of the enemy at last.

Before him on the turnpike was Sykes' division of the V Corps, with the balance of Meade's corps more than a mile north on a back road, and separated by dense growth. Facing Sykes Jackson had the brigade of General William Mahone. A mile southwest, on the plank road, the brigade of General Carnot Posey faced practically the entire XII Corps. It was not an enviable position for the Confederates, no more than 12,000 men – including others rushing to the scene – facing 40,000 under Meade and Slocum. Nevertheless, it suited Jackson. He kept right on advancing, and around 11:40 a.m. the first firing finally broke out. By early afternoon, however, more and more of Jackson's divisions arrived, until there were

40,000 on the spot or near at hand, with still more coming up behind. All told, Lee took the bold gamble of committing 50,000 men, five-sixths of his army, to the move to meet Hooker's threat. And now, with the battle just in its first minutes, they were even on the deception score. Hooker had fooled Lee and massed more than half his army on the Rebel left flank. But today Lee had left a sixth of his to hold down Sedgwick, while Hooker still believed that the bulk of the Army of Northern Virginia remained at Fredericksburg. It was to be a battle of surprises.

"I trust that God will grant us a great victory," Jackson said to 'Jeb' Stuart just after noon. It was indicative of Jackson's mood. Despite being

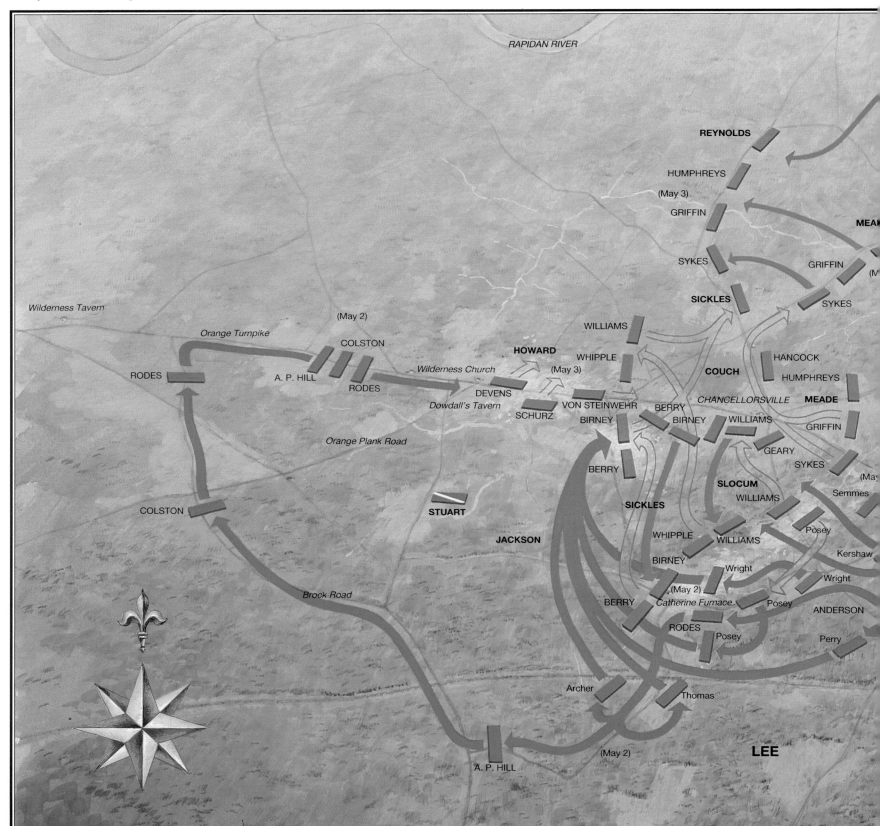

Battle of Chancellorsville, May 1-4, 1863

Lee had a remarkable facility for taking campaigns that started out badly against him, and turning them into victories. He did it at Fredericksburg, and here on nearby ground five months later, he did it yet again, and more spectacularly.

Hooker's crossing of the Rapidan took Lee somewhat unawares. After passing Ely's and United States Fords, Hooker's legions advanced into the Wilderness. he had intended for Sedgwick, facing Fredericksburg, to make a demonstration to hold the bulk of Lee's forces there, while Hooker moved against Lee's rear via the Orange Turnpike.

Unfortunately, Sedgwick failed to act with his usual alacrity, allowing Lee to detach substantial portions of his army to advance to meet Hooker's threat.

On May 1, little fighting occurred as Hooker took Chancellorsville and tentatively pushed beyond. But Hooker was already losing his resolve, and now Lee seized a chance to take the battle away from him. He sent Jackson with several divisions on a wide sweeping flank march via the Brock Road, to reach the turnpike in Howard's flank and rear. Shortly after 5 p.m., May 2, Jackson was in place and drove like lightning into the Federals. He sent Howard reeling back in shock and demoralization. Only darkness and disorganization stopped Jackson's advance.

Hooker found himself in deep trouble. All the next day he struggled to hold his position at Chancellorsville, and even gained back a little ground. But just as Meade was ready to launch a very promising attack on Stuart's own exposed left flank, Hooker called off further offensive operations. Now he was concerned only to get his army out of this place intact.

The scene of action shifted to Fredericksburg on May 3, as Sedgwick finally began to launch a more spirited attempt. He sent in attacks in the city itself against Early's Rebels posted on Marye's Heights, and then the next day drove very deliberately toward Lee's rear. But by now Lee was able to take men from the Chancellorsville line and send them to stop Sedgwick. In and around Salem Church on May 3, Lee stopped the Yankee advance, and the next day himself attacked. Though desultory skirmishing continued over the next two days, Hooker had had enough, and pulled his army back north of the river. The campaign had been a failure, and had gained the North nothing but casualties. The South had once again triumphed, but at terrible cost to herself. For among the thousands she could ill-afford to lose was one of her finest generals: Thomas J. Jackson. The debacle had cost Hooker 17,000 out of his 134,000, while Lee's 61,000-men army suffered 12,800 casualties.

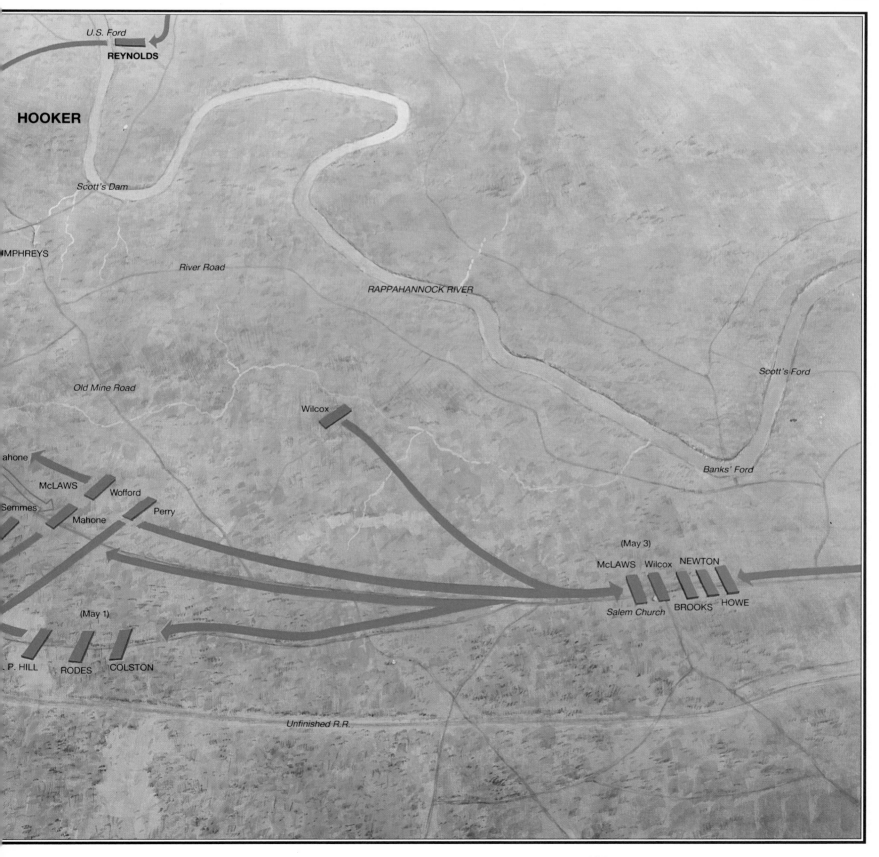

heavily outnumbered, he thought only of winning the battle. During the next hour or so, the units on both sides rapidly came up as the battle-lines began taking shape. On the Plank Road Jackson kept his left well advanced, while his right on the turnpike fell back to less than a mile in advance of the Zoar Church, and soon extended into the Wilderness in an arc covering the church and the road to Fredericksburg behind. But the Federals were not for giving him much fight. Instead, Hooker ordered Sykes to pull back towards Chancellorsville, and gave similar orders to Slocum on the Plank Road. Unfortunately, Sedgwick had not yet launched the demonstration called for at Fredericksburg, and in the absence of that, Hooker feared – quite rightly – that Lee would have too many divisions free to meet him here in the Wilderness. Thus, again, the commanding general shifted from a tentative offensive to the defensive. By 2 p.m. Hooker gave up any idea of offering battle. "Have suspended attack," he notified Butterfield. "The enemy may attack me."

Soon thereafter critics would accuse Hooker of being drunk that afternoon, and that this accounted for his erratic actions and loss of nerve. But several of his generals steadfastly denied that he had consumed anything alcoholic all day. Rather, he simply lost his nerve. He had been counting on Lee to passively fall back before his advance. Indeed, his whole campaign was predicated on the Confederates behaving exactly as he expected, and up until the previous day, they had. But Jackson coming straight for him on the two roads upset Hooker's predictions. Never seriously considering that an outnumbered foe might attack him, he now thought of nothing but getting his corps back to their fortified positions of the day before in and around Chancellorsville.

Late that afternoon Hooker gave orders to re-establish the lines of the previous night, with some small modifications. Trees were felled to provide fields of fire for the artillery, obstructions were placed in advance of the infantry lines to slow an attacker, and the men dug themselves in to the hardscrabble soil. By midnight they were stretched out on a line five miles long. Commencing on the far left it was anchored by Meade extending from Scott's Ford on a bend of the Rappahannock, almost due south. The V Corps line ended just short of the Turnpike, about three-quarters of a mile east of Chancellorsville. There

Below: Oliver O. Howard was a hard-luck general. His command started the rout at Bull Run. At Gettysburg, his XI Corps would he demoralized on the first day's fight. Here at Chancellorsville, it simply dissolved.

Meade linked with portions of Couch's II Corps that completed the line to the Turnpike. But at that point Couch's men formed a salient, as the line turned abruptly west along the pike itself to Chancellorsville. Slocum's XII Corps then took up the line, extending it in a wide bulge south of Chancellorsville, swinging back up to the road something less than a mile east of Wilderness Church. There Howard's XI Corps took it up, Howard's front running westward along the road, past the church and nearby Dowdall's Tavern, and halting nearly a mile west of them. The whole line was based upon the assumption that all of Lee's strength in any attack would be concentrated against Chancellorsville from the Plank Road and the Turnpike. As a result, Howard and the far right flank, seemingly unreachable, were left literally "in the air" and unprotected. This did not seem to bother Hooker. "It is all right, Couch," he told the II Corps commander that evening. "I

Above: A view looking toward Chancellorsville, with one of the Chancellor houses in the distance. Such wide fields were the exception in that country, limiting movement to narrow country lanes.

have got Lee just where I want him; he must fight me on my own ground." Couch was not comforted. "I retired from his presence with the belief that my commanding general was a whipped man."[10]

The fighting had been little more than skirmishing during most of the day, with casualties few, but Lee anticipated that on the morrow, May 2, he would have to make a real push now that he had most of his forces on the field. Indeed, several of his generals, including Jackson, believed that Hooker would actually pull back across the Rappahannock during the night, but Lee held no such fears. Indeed, he seems to have believed that Hooker might launch a renewed offensive on the morrow, probably coordinated with a massive push by Sedgwick. Only a Confederate stroke beforehand could pre-empt such a movement, and that evening Lee and Jackson conferred as to how and where to move. Jackson knew the ground a little better, and had already made some reconnaissances earlier in the day, and even an attempt to get around the Yankee right that failed. Now reports came in to the two generals as their scouts moved along the enemy front toward the west, feeling the strength of answering fire from pickets and outposts. Finally the cavalry sent in a report that Howard's right appeared to be on the Orange road, at its intersection with the Brock Road. Then a man in Jackson's command who grew up in the vicinity came to them with word of the Brock Road. It snaked up from the south, well below the two armies, and wound northwest almost to the Wilderness Tavern, intersecting the Orange road more than a mile west of Howard's flank. It offered a perfect opportunity to march unseen across the whole front of Hooker's army, and then suddenly appear in his right rear. The result could be devastating. Lee left the entire operation in Jackson's hands, and the mighty Stonewall, smiling, announced that his men would be ready and on the move at 4 a.m.

Hooker seems to have felt a little insecurity about Howard's flank, for that same night he ordered Reynolds and the I Corps to leave their

Above: In this 1863 photograph, the ruins of the Chancellor house pay ample testimony to the damage inflicted on civilians unlucky enough to have their homes caught in the way of the armies.

position and come join this half of the army, but that movement would require many hours. If Jackson moved quickly, he could easily beat Reynolds to Howard's right. And Jackson would strike with massive force. Lee assigned him fully three-quarters of the men at hand, all of Jackson's own II Corps plus enough elements from other commands to total almost 32,000 and 112 cannon. The boldness of the plan was breathtaking. Lee was dividing his army in the face of an enemy that outnumbered him by about three-to-two. With no more than 12,000, he would stay and demonstrate against Chancellorsville, buying time and diversion to allow Jackson to make a march of nearly nine miles from their existing right flank, all of it in full daylight. If Hooker should strike while Jackson was somewhere off on the Brock Road, the result could be disaster. But Lee knew how to smell fear and irresolution in a Yankee commander. Not only was such a movement indicative of the gray chieftain's boldness; it also announced his contempt for his opponent.

Hooker began to get a glimmer of what Lee might be about when his pickets reported numbers of Rebel troops seen marching off that morning. He even warned Howard to look out for his right flank. Unfortunately, Hooker did not know, as he should have, that there was a two-mile gap between Howard and their Rapidan River crossing at Ely's Ford. Should the enemy pass through that gap, the army would be in great danger. Unfortunately, too, Howard never had been, and never would be, much of a general.

Neither was Daniel Sickles. As the morning wore on, he, too, got reports of sightings of Rebel soldiers marching off toward the right. The trouble was, as he viewed it, this meant they were probably retreating. As a result, with Hooker's permission, he advanced his III Corps from Slocum's right on the Turnpike, more than a mile south to try to intercept Jackson's columns. Unfortunately, by that time virtually all of Stonewall's divisions had passed well to the south, on their way to the Brock Road, and Sickles' movement only succeeded in stretching the Yankee

Below: Nothing excited more pride in a regiment than its regimental flag and its national colors. This stand of colors of the 19th Massachusetts shows the ravages of battle on the banners.

line into a new, deeper, salient, and leaving the joint between Slocum and Howard thin and very vulnerable.

Throughout the morning and afternoon, Lee remained content to make his presence known by sporadic artillery firing and threatening movements in Hooker's front, without actually offering battle. Later in the afternoon he did send in McLaws to harrass the Yankees on the Turnpike, but again only to hold Hooker's attention to this front. Meanwhile, Jackson's tired and hungry veterans pressed on. He seemed everywhere, eyes flashing, leaning forward over his horse as if bending into the wind. "Press forward, press forward," he called out incessantly. He even placed guards with bayonets at the rear of every regiment to prod forward the stragglers.[11] Incredibly, despite his hopes for secrecy, reports of Jackson's march came into Hooker and Howard all through the morning and the afternoon, but were either

misinterpreted or ignored. Howard himself, around 3 p.m., laughed derisively at reports that a substantial body of the foe was seen moving in his front. They could not possibly be there, he replied, and did nothing more. As for Hooker, by now he had convinced himself that all the information of movement by the Rebels indicated a retreat, and he sent Sedgwick a peremptory order to attack and take Fredericksburg. "We know that the enemy is fleeing," he said in triumph.

By 3 p.m. Jackson's first division neared its destination, with the other two closing up quickly. As the legions came on to the Turnpike from the Brock Road, they turned right, to the east, and he put them into a formation stretching two miles in length and three lines deep. There were to be no bugle calls, no shouted orders, no cheering Jackson as he rode along the lines. By about 5 p.m. they were ready, with still a couple of hours of daylight ahead. Howard's unprotected right flank lay half a mile ahead of them, with his corps stretched out along the Turnpike beyond. Jackson could hit him front, flank and rear, all at the same time. "Are you ready," Jackson asked Rodes, commanding the front line. The simple reply "Yes" launched the attack.

Never before or after in this war was there anything like it. Within seconds they encountered Howard's scanty outposts and swept them over. And now, from every point on the charging Southern line came cheers, bugles, and the high-pitched "Rebel yell". The first thing Howard's men saw was hundreds of deer, rabbits, and other forest creatures, scampering toward them to escape the wall of men and iron sweeping through the wilderness. "It was a terrible gale," Howard later remembered. "The rush, the rattle, the quick lightning from a hundred points at once; the roar redoubled by echoes through the forest; the panic, the dead and dying in sight, and the wounded straggling along; the frantic efforts of the brave and patriotic to stay the angry storm." From all points, though there was isolated and heroic resistance, the men of Howard's corps simply melted away in the face of the Rebel onslaught.[12]

After barely fifteen minutes, batteries were being captured, regiments and whole brigades put to rout, and entire divisions disrupted. Less than an hour after Jackson launched the attack, Howard's men were streaming back toward Wilderness Church, and they did not stop when they reached it, but kept right on going. Jackson pressed on after them, now with renewed vigor because they could hear the sound of artillery several miles to the east, sign that Lee had commenced his demonstration against the other end of Hooker's line. Nothing could stand before the Rebel onslaught. Jackson pushed the foe back to the church, down to Dowdall's Tavern. Incredibly, just two miles east, at Chancellorsville, Hooker was sitting on the porch of his headquarters unaware of the calamity on his right. No one seems to have thought to send him word of it, and atmospheric conditions temporarily kept him from hearing the mounting sounds of the fighting. They only got their first hint of the disaster when a staff member stood in the road and happened to turn his field glasses to the west, toward Dowdall's. "My God", he shouted, "here they come!" Hundreds of fugitives could be seen streaming towards them in the distance.[13]

Elements of the II Corps and Reynolds' I Corps, just arrived at United States Ford, rushed to stem the rout, literally pushing their way through masses of XI Corps fugitives clogging all the roads, but still no semblance of a defensive line was finally established until close to nightfall as Sickles pulled back from his southern salient to form on Slocum's right, while elements from Meade and Couch reversed themselves to face the new threat from their rear. But shortly after 7 p.m. Jackson's advance stopped anyhow as his own units became badly disorganized by their long march, precipitate advance, and the tangled countryside. Still, Hooker was in a desperate situation. His command at Chancellorsville was literally formed into a circle, with Jackson to the west and Lee south and southeast. His only remaining line of communications was a single

road leading northeast to Scott's Ford and United States Ford, and vulnerable to either of the two wings of the Rebel army. About 9 p.m. Jackson gave A.P. Hill orders to do just that, then Stonewall himself rode off to his left, north of the Plank Road, to reconnoiter. In the gathering gloom, edgy men of the 33rd North Carolina accidentally fired on the general and his party. Three of them actually hit Jackson – one in the right hand, another at his left wrist, and the third in his left arm between the shoulder and elbow. The first two wounds were painful, but minor; the third shattered the bone and severed an artery. Jackson's panicked horse started to run toward enemy lines, dashing Jackson's face into a low-

hanging tree branch before he regained control. It seemed to take forever before others got him from his horse, brought up a litter, and took him to the rear, pale and in pain. His only words were to encourage them to keep up the fight and complete the victory.

With Stuart acceding to command of the II Corps, efforts were made to renew the push, but by this time night had fallen so completely that little remained to do. By midnight the lines lay about where they were at Jackson's wounding. There was no sleep on either side of the lines that night, as Confederates raced to bring everything up to the front, and Hooker and his generals frantically sought to consolidate their precarious

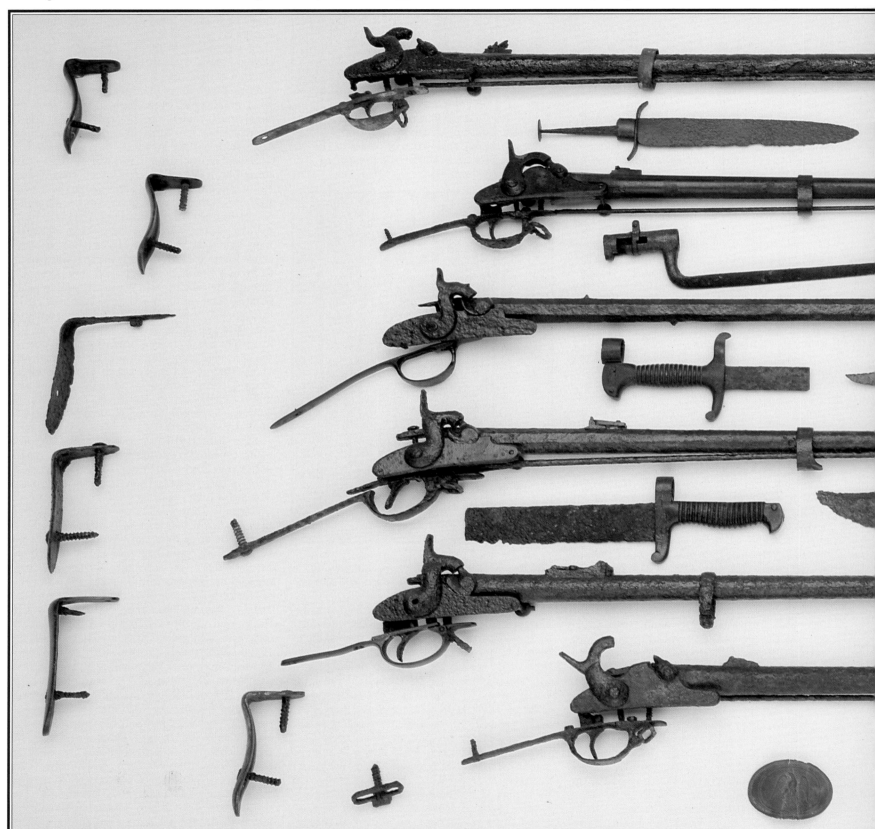

Excavated Confederate Artifacts

1 Early Fayetteville 1862 rifle, found at Cold Harbor, Va
2 Foot artillery sword
3 Side knife
4 Fayetteville 1862 brass-hilted saber bayonet found at Gaines Mill, Va
5 Richmond rifle musket, found in the Wilderness, Va
6 Raleigh bayonet, found at Resaca, Ga
7 Boyle, Gamble and Macfee of Richmond, Va

saber bayonet, found at Seven Pines, Va
8 Country rifle found at Barnesville, Va
9–10 Buckle variants
11 Boyle, Gamble and Macfee saber bayonet, found at Chancellorsville, Va
12 Side knife
13 Iron-hilted saber bayonet, probably from Tylor, Texas, found near Williamsport, Md
14 Model 1854 Austrian

Lorenz rifle found near U.S. Ford, Va
15 English snake buckle
16 Fork tongue buckle
17 Royle, Gamble and Macfee bayonet, found near Fox's Gap, Md
18–19 Bowie knife variants
20 English pattern 1853 rifle
21 Richmond carbine
22 Two piece belt plate
23 Mississippi belt plate
24 Side knife
25 Tin drum canteen found at Totopotomy, Va

Artifacts courtesy of: Wendell Lang Collection, Tarrytown, N.Y.

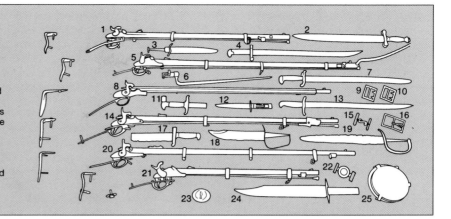

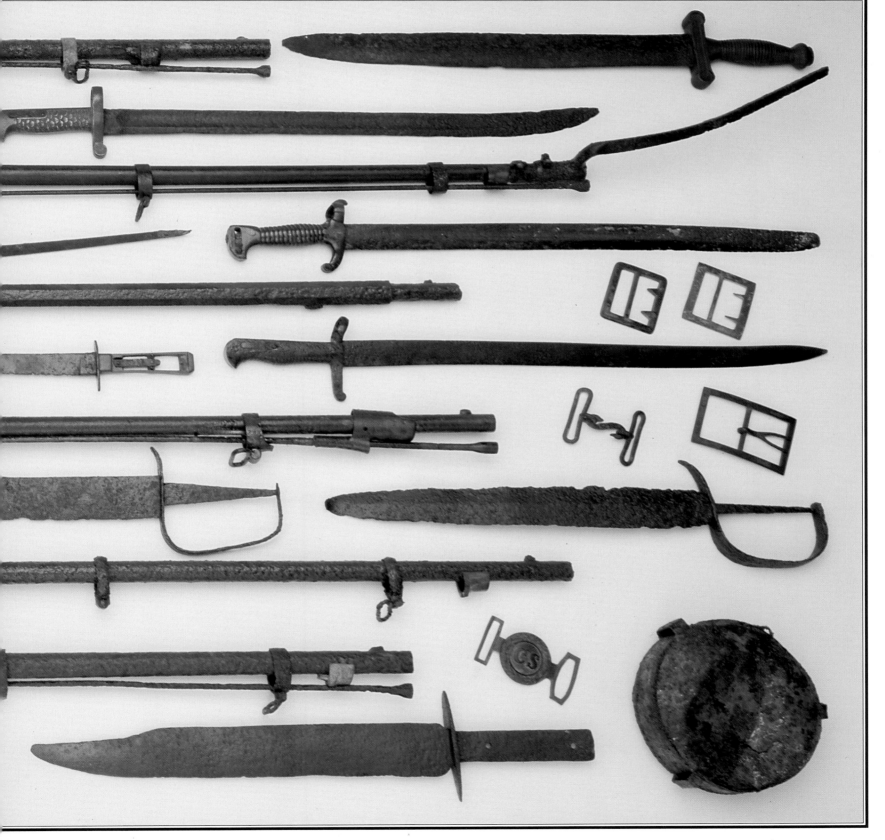

position and get more reinforcements on the field. Lee urged Stuart to press on around south of the Yankees and push them out of Chancellorsville, allowing the two wings of the Army of Northern Virginia to reunite. When dawn came, however, it was apparent that Sickles still stood in the way of such a junction, and though Sickles eventually pulled out of his own accord, Stuart would not link with Lee for some time.

Instead, for the balance of the day Stuart slugged it out along the line of the night before, and some Federal units even managed to push him back with spirited counterattacks as they struggled to hold on to Chancellorsville. Shortly after 9 a.m., Stuart drove a dangerous wedge into the center of the line facing, penetrating to within a quarter mile of Chancellorsville. Hooker himself was standing on the porch of the Chancellor house in the village just now, when a Rebel artillery projectile struck a post against which he had been leaning, splitting the post and sending half of it bouncing off his head. At first his men thought him dead, but he soon revived, though only to suffer intense pain thereafter. Before long he put Couch in command of the army, and never afterward exercised any positive control over it.

By late morning, the Federals were starting to recover, and Meade actually sat poised upon what was now an exposed Confederate left flank. When he met with Couch to discuss what might have been a devastating attack on Stuart, however, Couch informed him that in turning over the command, Hooker's last order had been to leave the field. With Sedgwick making no progress in front of Fredericksburg, Hooker was not willing to remain here taking a beating any longer. To the combative Meade, the news came as a shock, but there was no choice but to comply, and from that moment onward, the Federals directed all their efforts at pulling out of

Private, 140th Pennsylvania Volunteer Infantry, U.S.A.

The 140th Pennsylvania Infantry carried with it the unusual nickname, the "Walking Artillery". They got the name because of the old Belgian-copied French .69 caliber "vincennes" muskets that they were issued. The huge bore of the guns led to the joke about these infantrymen being "artillery", and fellow soldiers taunted them with questions such as "do you shoot solid shot or shell out of those pieces?"

Worse than this, the Pennsylvanians also carried massive sword-bayonets that others teasingly called "cavalry sabers." No wonder the "Walking Artillery" was so delighted when finally issued new weapons in January 1863.

Their uniform was a standard pattern dark blue tunic, mid-thigh length, over sky blue trousers. The brass shoulder scales were discarded as the war progressed, but otherwise the uniform remained little changed.

Chancellorsville and back toward the Rappahannock. Behind them, Lee and Stuart once more joined forces, and around noon were preparing to continue the pursuit and press the enemy into the river when Lee got word of heavy fighting far to his right rear. At last Sedgwick was attacking Fredericksburg.

Sedgwick's had been a difficult and frustrating role through most of the past few days. His orders from Hooker were to move against the enemy in his front when the proper opportunity presented itself, which it never seemed to do. Worse, Hooker's instructions had been either too discretionary, or else not discretionary enough. All through May 1 Sedgwick awaited any action in his own front until assured that Hooker was achieving success, yet at the same time Hooker said he withheld pushing his advance any further until he learned of Sedgwick taking Fredericksburg. Thus May 1 was entirely wasted. Not until the next afternoon did Hooker issue a direct order for Sedgwick to cross the river, and then it was to follow up what Hooker then thought was a retreating foe. It was too late to do anything effective that night, but Sedgwick did push in the midnight hours, moving forward his units already across at Franklin's Crossing until they hit the Confederates placed on the high ground. Slowly they pushed the thin line of Rebels back into the defenses of the town itself, and then Sedgwick started preparing for a general assault, and at 10:30 finally sent his divisions forward.

Once more Federals charged up toward Marye's Heights and the stone wall, and after a bloody and hotly contested half an hour, the wall was gained. He drove on toward Chancellorsville, having penetrated the center of the thin Confederate line, but only got as far as Salem Church, about three miles west of Marye's Heights, before he ran into Lee with most of the I Corps. Though he had nearly 20,000 fresh troops that morning in moving against the Fredericksburg line, Sedgwick broke through with only about 5,000, and then ill-advisedly raced ahead of the rest of his command, so that when he hit Salem Church with his 5,000, he was no match for Lee, whose 10,000 would otherwise have been outnumbered had Sedgwick taken more time. But being kept out of the battle for the past two days by Hooker's wavering, Sedgwick was anxious to get into the fight now and not to wait about.

Lee stopped him cold at Salem Church, and meanwhile the rest of the Confederate line he had torn through now lay in his left rear, threatening his hold on Fredericksburg. When night fell and more and more Rebels rushed to his front, Sedgwick sensed that he might be in

Above: Chancellorsville covered a great distance, being actually two battles. Here at Fredericksburg, only the hesitation of Union commanders prevented what could have been a decisive counterstroke.

Below: The inevitable toll of war. Burying the dead in and around Fredericksburg occupied both armies for days, and friends and family for months in trying to recover them to bring home.

trouble, and began preparations to abandon Fredericksburg, bring all of his troops up to his current vicinity, and then fall back northward toward the Rappahannock, where he could hold a secure grip on Scott's Ford in case he had to recross. Lee, meanwhile, did not have everything in position to mount a major assault on Sedgwick until about 5:30 on the afternoon of May 4, but when he did, despite spirited resistance, Sedgwick felt forced to withdraw, and by nightfall stood with his back to the river and the ford. Hooker, still occasionally giving orders despite being in a delirium much of the time, could have come to his aid, for the 75,000 men in his wing sat idle all day long, content to catch their breath. But he did nothing, too stunned by the defeat of May 2-3 to move or act. Thus Sedgwick was on his own.

Late that night Hooker decided to order a withdrawal north of the Rappahannock. He had no more fight left in him. Indeed, in later days his own summation of his conduct of the campaign was that for once in his life, Hooker had lost faith in Hooker. He had let Lee beat him by beating himself, and after what started as a brilliant campaign, he had nothing at all to show for the 17,000 casualties suffered, as against 12,800 for Lee. Lincoln was almost beside himself when he got the news. "My God, my God, what will the country say!" he wailed. "What will the country say!" Yet again the Army of the Potomac had been thrown back by that insurpassable barrier, the Army of Northern Virginia. After fully two years of war,

the Federals stood no closer to Richmond that they had in 1861, and all they had to show for their efforts were the thousands of dead.

But the Confederates suffered, too. On May 10, 1863, at Guiney's Station, ten miles south of Fredericksburg, Stonewall Jackson died of complications from his wound. Ironically, bullets from his own men had inflicted perhaps the greatest injury yet suffered by the Confederacy. Lee could only shake away his tears and remark with anguish that he had lost his right arm. It was a dear price to pay for a battle won.

References

1 San Francisco, *Chronicle*, November 1, 1879.
2 John Bigelow, Jr., *The Campaign of Chancellorsville* (New Haven, Conn., 1910), pp.4-6.
3 John G. Nicolay and John Hay, *Abraham Lincoln, A History* (New York, 1890), VII, p.88.
4 Noah Brooks, *Washington in Lincoln's Time* (Washington, 1879), pp.50-51; Bigelow, *Chancellorsville*, p.108.
5 *Report of the Joint Committee*, IV, p.145.
6 Bigelow, *Chancellorsville*, pp.173-79.
7 *Ibid.*, pp.132-33, 136.
8 R.M. Bache, *Life of General G.G. Meade* (Philadelphia, 1884), p.260.
9 Bigelow, *Chancellorsville*, pp.232-33.
10 *Ibid.*, p.259.
11 *Ibid.*, p.275.
12 Oliver O. Howard, *Autobiography* (New York, 1907), I, pp.368-70.
13 A.C. Hamlin, *The Battle of Chancellorsville* (Boston, 1884), p.148.

CHAPTER EIGHT

VICKSBURG

MAY 18 – JULY 4, 1863

Yet again Lincoln faced the bleak fact of defeat, of a Rebel army that seemed invincible. Could he never expect victories in the East? Even out in the West there seemed to be little or nothing happening. He had been eternally grateful to Rosecrans for the, admittedly qualified, victory at Stones River. But as the summer of 1863 approached, Rosecrans failed to follow up on the promise of that earlier success, and now a whole campaigning season in middle Tennessee seemed about to go to waste.
Only on the Mississippi was there promise, and even there Grant faced great odds, not in men and guns, but in the land. Vicksburg was there waiting for him. But could he get to it, and once there, could he get through the tons of dirt and masonry erected to defend the city on the hill?

O NE OF THE greatest ironies of the Civil War is that while most of the world's attention then and later was focused on the scant 100-mile corridor between Washington and Richmond, a region in which the status quo barely changed for the first three years of the conflict, whole states and regions were being won and lost out in the so-called west. The fall of Forts Henry and Donelson claimed Kentucky and half of Tennessee for the Union. The Yankee victory at Shiloh solidified that gain and added to it part of northern Mississippi, while the Stones River draw at the dawn of 1863 gained more of middle Tennessee for the Union. Meanwhile, the seizure of New Orleans closed the mouth of the Mississippi to the Confederates, and thereafter the Yankees moved steadily to take control of more and more of the great river. Baton Rouge fell to them in the summer of 1862, though they later abandoned it, and to keep them downriver, the Rebels erected a formidable river bastion on the bluffs at Port Hudson. Upriver, meanwhile, the Yankees did better, moving steadily south from that point. By the summer of 1862 they controlled everything as far down as Memphis.

Indeed, for one brief moment that summer, they almost took the entire river. A combined army and navy movement from both above and below came perilously close to meeting in the middle, threatening to capture Vicksburg, Mississippi, the greatest Rebel stronghold on the river. Had the city fallen, the entire western Confederacy – Texas, Arkansas, and western Louisiana – would have been cut off, and the resources of men and material that poured from the Trans-Mississippi would have been denied to a hungry Southern war effort. Happily for the South, the attempt failed, and by late fall the Federals had pulled back, below Baton Rouge on the south, and to Memphis on the north. That left just 300 miles of the river in Rebel hands. In

infantryman's terms, that was a long way. For any invader that could move by water, however, it was a matter of barely two days' travel.

This was hardly lost on the unassuming little man who took as his task the capture of Vicksburg. U.S. Grant was an authentic national hero following Shiloh, even despite the criticism of having been taken by surprise. When Halleck was transferred east to become general-in-chief, his departure left Grant with virtually a free hand in determining his course, though there was never a doubt that he must make Vicksburg his ultimate objective. "To dispossess them of this," he said, "became a matter of the first importance."[1]

Unfortunately, before leaving, Halleck divided his army of some 120,000 by more than half, including the sending of Buell into eastern Tennessee and, eventually, galloping after Bragg in Kentucky. Left with only 50,000 placed throughout western Tennessee and northern Mississippi, Grant actually found himself briefly on the defensive as Rebel attempts to strike the divided portions of his command led to an engagement at Corinth in early October, where Rosecrans repulsed General Earl Van Dorn and won for himself promotion to the command of the Army of the Cumberland.

The Southern failure at Corinth led Richmond to replace Van Dorn with a new man, Lieutenant General John C. Pemberton. It is one of the perversities of this war that a remarkable number of the highest ranking men on each side were natives of the other. Samuel Cooper, the senior general in the entire Confederate Army, came from New Jersey. Winfield Scott, first general-in-chief for the Union, was a Virginian. Pemberton fitted the mold, being a Pennsylvanian by birth. He finished at the Military Academy in 1837, and remained in the Old Army thereafter, fighting in Mexico and out on the Plains where he probably first attracted the notice of then Secretary of War

Never in American history did an American city so fully experience the rigors of war as during the 47 days of siege at Vicksburg. Ringed by batteries such as this, the fortress city on the Mississippi stood bravely.

Jefferson Davis. When the war came, his political conservatism and his marriage into a Virginia family persuaded him to go against his section and join the Confederacy, after which he served both in Virginia and South Carolina. In command at Charleston, he made himself particularly unpopular with the citizenry, yet continued to enjoy the growing confidence of now President Davis. As a result, when he had to relieve Pemberton in the face of the outcry against him, Davis did what was for him a typical move by promoting him to lieutenant general and giving him an even more important command. No one then or later ever adequately explained the grasp that this general had on his president's affection and confidence, but to the very end of the war, in the face of disaster and worse, Davis never abandoned the Pennsylvanian.

Pemberton found things in a dismaying state when he reached his new command. Making his headquarters at Jackson, fifty miles east of Vicksburg, he soon discovered that very little had been done in the way of protecting and fortifying the river city. He had only about 24,000 men, many of them of very questionable reliability, and far too few heavy cannon to emplace in the batteries he would build on the bluffs overlooking the Mississippi. In addition, he also had to concern himself with Port Hudson as well. Within a few months he put the defense of the latter in the hands of Major General Franklin Gardner (a native of New York!), which freed him to concentrate on his primary responsibility, Vicksburg.

From the first, Pemberton had two internal foes – Joseph E. Johnston and Jefferson Davis – and the two of them combined made an obstacle almost as imposing as Grant's army. Despite his dislike for Johnston by late 1862, Davis had to find some post for such a senior general after Johnston recovered from his wounds gained at Seven Pines. The month after sending Pemberton west, Davis created a "super department" comprising everything from the Mississippi to the Alleghenies, and assigned it to Johnston. This meant that he would now direct the armies of both Bragg in Tennessee and Pemberton in Mississippi. But Johnston, showing the same lack of moral courage that he displayed through the entire war, declined to direct either. Instead, he would spend his entire tenure showing an obtuseness over his actual powers that can only have been intentional in the face of Davis' repeated assurances that his military authority

was almost absolute. Johnston would spend so much time in petty haggling with Richmond that he proved to be worth nothing to Pemberton during the ensuing campaign. As for Davis, despite all his good intentions, despite his lifelong personal attachment to Mississippi, he would never give the war on the great river the attention that it deserved. John C. Pemberton would learn to his sorrow that, when it came to holding Vicksburg, he would be on his own.

Pemberton worked almost night and day during his first month in command, and he put everyone else to work, too. No home or front yard was immune from being commandeered for his purposes, whether for an earthwork or a new battery emplacement; no idle soldier was safe from

being given a spade and made to work alongside the slaves at the digging. Van Dorn, unpopular thanks to his vanity and insobriety – not to mention the filandering that would see him assassinated in a few months by a cuckolded husband – was not missed by anyone, but the incessant toil that his successor visited upon Vicksburg was not welcomed either. Still, Pemberton raised morale in the city's civilian population as well as in its small army. Even his opponent Grant, who had known him in the Old Army, expressed respect for his new adversary. "He was scrupulously particular in matters of honor and integrity", said Grant, remembering an episode in Mexico when despite great discomfort, Pemberton refused to disobey an order prohibiting junior officers from riding, even though almost all rode anyhow. "This I thought of all the time he was in Vicksburg and I outside of it; and I knew he would hold on to the last."[2]

Grant did not let knowledge of Pemberton's determination hinder his own resolve to take the river fortress. Within only weeks of the change of Confederate commanders, Grant was planning his movement south. His first move was to shift his headquarters from Jackson, Tennessee, south to Grand Junction, almost on the Mississippi border. Then he started a massive buildup of supplies across the border at Holly Springs, which he would use as a base for his invasion of Mississippi. With 30,000 he marched there, and soon ordered his trusted subordinate Sherman to join him with two more divisions. Pemberton was facing them some fifty or sixty miles to the south, having advanced from Vicksburg to meet the threat, but Grant had to postpone his own further advance until his base at Holly Springs was sufficiently built up that he could depend entirely upon it rather than a long supply line that currently ran clear back to Kentucky.

Furthermore, as he looked long and hard at the situation before him, Grant decided that a lone overland advance offered too many dangers. Sherman had been proposing using the Mississippi instead, making a direct campaign against Vicksburg. Grant did not favor it at first, but by early December he modified his own plans to include Sherman's. He would make a combined campaign of it. He sent Sherman back to Memphis to organize a separate wing that would soon total some 32,000. With this small army, Sherman would move downriver on transports and, with the assistance of gunboats commanded by

Private (and mascot) of the 8th Wisconsin Volunteer Infantry U.S.A.

Quite a number of regiments went to war with special mascots, either little drummer boys, vivandiers – women in military costume – and most of all pets. Dogs, cats, raccoons, even small bears, went to war, but no mascot was as distinctive as "Old Abe", the "war eagle" that went into battle with the 8th Wisconsin Infantry of the old Iron Brigade. All across the battlefields of Tennessee and Georgia, Old Abe soared into the air when the bullets started to fly, hovering over the fighting until it was done. In camp and on the march, he stayed on a special perch made in the shape of the Union shield, with the stars and stripes painted on the surface, and tending to the eagle was a special duty in the regiment.

His tenders wore very much the regulation uniform and equipment, excepting their headgear, some wearing slouch hats like this private.

Captain David D. Porter, make a landing at or near Chickasaw Bluffs just north of Vicksburg. Grant, meanwhile, would move overland, engaging Pemberton's army and thus weakening any resistance to Sherman. It was an able plan, each wing advancing in support of the efforts of the other.

And it probably would have worked but for the disliked Earl Van Dorn. He had made his own headquarters at Holly Springs back in October, and knew the ground well. Now, though he no longer commanded the department, but led only Pemberton's 3,500-man cavalry division, he could still influence the campaign. On December 20, 1862, with virtually no warning, he swept down on Holly Springs out of the winter dawn. In a lightning raid, he swiftly overwhelmed the 15,000-strong garrison Grant had guarding his supply base, and then put the torch to all of the Federals' carefully assembled provisions and munitions. Perhaps more than $1,500,000 worth of *materiel* went up in smoke. At almost the same time, more Rebel cavalry, led by the peerless Nathan Bedford Forrest, raided deep into northwestern Tennessee and tore up some sixty miles of track on the single railroad line that connected Holly Springs with Columbus, Kentucky, the embarkation point for all of Grant's supply and communications. Thus in a single stroke, the Southern horsemen put an abrupt end to Grant's proposed winter overland campaign. His supply base was gone, and until the track was repaired he had no way of building it up once more.

Unfortunately, by this time Sherman was already on his way down the Mississippi to carry out his half of the program, and Grant could not get word of the disaster to him. Thus, expecting that Grant would have Pemberton fully occupied in northern Mississippi, Sherman was not prepared for what he met on December 26 when he marched his men ashore after steaming into the Yazoo River a few miles north of Vicksburg. Rain hampered his march toward the city, and after two days he came up against Chickasaw Bluffs. The next day, after a morning-long shelling, he

Union and Confederate Coehorn Mortars with Projectiles

The Coehorn mortar was a light siege weapon, used mostly in trench warfare, and designed to be carried in battle by four men. Its name derives from the seventeenth century Dutch soldier and siege engineer Baron van Coehoorn who first developed the weapon in 1674

1 Coehorn, Model 1841 bronze 24lb mortar, used by Federal forces. This mortar has a bore of just under 6 inches; its tube weighs 164lb and is approximately 16 inches in length

2 Twenty-four pounder shell for 1. Note hole in side for time or percussion fuse. The projectile weighs 17lb, and holds a bursting charge of 1lb of powder. It took ½lb of powder to send this shell to its maximum range of 1,200 yards

3 Confederate 24lb iron Coehorn mortar. This particular piece has a reproduction wooden 'bed' or support. Note the metal carrying handles on the bed. A feature also to be seen on the bronze Federal mortar, artifact 1

4 Confederate 24lb mortar shell. Note indentations either side of the fuse hole to allow loading by lifting tongs

Artifacts courtesy of: Gettysburg Museum of the Civil War, Gettysburg, Pa: 1, 2; Wendell Lang Collection: 3, 4

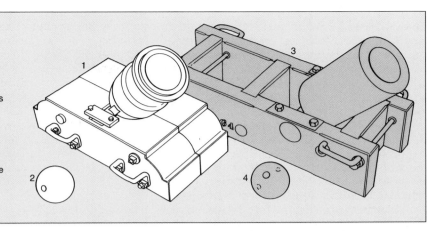

sent his men forward in a pointless assault that saw 1,700 of them fall as casualties, against a mere 187 Confederates. All of those defenders were supposed to be occupied elsewhere by Grant, but now they stopped Sherman cold, and the next day, as he heard the sounds of trains bringing more and more Rebels into Vicksburg, he decided to abandon the campaign.[3]

There was more misfortune immediately to follow. Lincoln had unwisely yielded to political pressure by making an important Democratic politician, John McClernand, a major general senior to Sherman, and assigned him to command on the Mississippi, though subordinate to Grant. McClernand arrived just as Sherman was steaming back upriver from his unsuccessful expedition. Intensely ambitious and scheming, McClernand immediately took command and planned and put into operation his own operation against Fort Hindman in Arkansas, without consulting Grant or informing him until it was too late. Sherman and Porter fought the entire affair in January, took the fort, and thus gained a good foothold on the west side of the river near the mouth of the Arkansas River, but McClernand immediately took credit for the entire affair. An indignant Sherman and Porter begged Grant to come and take command personally. Otherwise they would find it unbearable to work with McClernand, and there was no telling what the inept fellow would do next in what was obviously a personal campaign not to win the war, but to win sufficient glory to win the White House one day in the future.

Finding that so many were "distrustful of McClernand's fitness to command", Grant saw the wisdom of their wishes, and on January 30, 1863, he came to Young's Point, just a few miles north of Vicksburg, and on the opposite bank of the river from the city. Here he superseded an irate McClernand, and now, he later recalled, "the real work of the campaign and siege of Vicksburg commenced."

Once he had his forces from Tennessee downriver with him, Grant's overall command totaled just over 60,000. Leaving his XVI Corps back in Tennessee for the time being, he reorganized the forces at Young's Point into three new corps – the XV commanded by Sherman, the XIII by McClernand, and the XVII to be led by James B. McPherson. That done, he proceeded to take a fresh look at how he might take the prize. Any idea of a simple assault across the river was out of the question. A crossing would have to be made under the fire of those massive batteries now staring over at him from the bluffs on the east side. The only practical approach was to strike at the city from the Mississippi interior, and preferably from its rear, so that the garrison would have its back to the river with no line of retreat. Unfortunately, the ground north of the city was so favorable to the defenders – as Sherman had discovered – that a campaign from that direction stood little chance of success, or at best would lead to a long siege, which Grant did not want. The country south of the city was more open and level. He could move and maneuver there. The trouble was that Porter's fleet of gunboats and transports had to be there to ferry his army across, and Porter could not get his flotilla past the several miles of batteries on the bluffs.

With the onset of winter and the rainy season, Grant had time to work out the problem before the ground would be dry enough for active operations, and this allowed him to try a succession of

Above: A portion of the mighty river fleet commanded by Porter in the bombardment of Vicksburg. One of the mighty *City* class ironclads sits second from left, with others in the far distance.

Below: Porter made his own headquarters on his flagship U.S.S. *Blackhawk*, a converted river steamer heavily armed and clad with sheathing. It became a familiar sight on the western rivers throughout the war.

plans for getting Porter downriver. Just below Young's Point, the Mississippi made a deep bend eastward, then abruptly doubled back on itself, forming a peninsula of Arkansas land that jutted some seven miles into Mississippi. Vicksburg sat on the east bank, just opposite and slightly below the tip of that peninsula. Grant could easily march his army across the base of the peninsula, bypassing Vicksburg entirely. But his fleet could not go overland. An earlier commander (one General Thomas Williams) back in the summer of 1862 had faced the same dilemma, and he tried to solve it by cutting a canal across the base of the point of land that would allow ships to get below Vicksburg without steaming around that wide bend. The canal was never finished, but now Grant set men to work once more on the digging. He never actually expected much from the enterprise, but it kept the men busy. Even with the Mississippi running high due to rains, it refused to send sufficient water through to float Porter's boats. Ironically, years later the Mississippi shifted its course on its own, carving out a new

path directly through Grant's unsuccessful canal, and leaving Vicksburg practically high and dry. But that was much too late to help the Federals.

Even while his "ditch" continued to defy completion, Grant looked to other possibilities, always with a view to getting his army and Porter's fleet below the city. A series of streams and bayous, often little more than sluggish swamps, meandered through the Louisiana interior west of the Mississippi. Some thirty miles upriver from Young's Point, Lake Providence sat a scant half mile west of the great stream. Bayous flowed into it from the south, connecting in turn with the Tensas River, and then eventually led to the Red River, which merged into the Mississippi near Port Hudson, more than eighty miles below Vicksburg. The bayous were shallow and filled with trees growing out of the bottom, while thousands of others, dead or dying, further blocked any passage. Nevertheless, Grant set McPherson to work clearing a path through all of this, and when a passage was finally cut to connect the lake with the Mississippi, the river flowed in with such force that it flooded wide areas and helped to clear its own path. Nevertheless, Grant would never use the route. It still required Porter's vessels to chart a tortuous route more than 400 miles long as it wound back and forth, and should the river fall, so would the level of the bayous, leaving the gunboats stranded in the mud. It could not be risked.

But another risk, formerly unthinkable, suddenly looked much better, and it came about almost without anyone noticing. Late in January, a Confederate supply steamer arrived in Vicksburg, having come upriver from the Red and run past a few Yankee guns on the west bank below the city. Though hit once or twice, she was hardly injured. The move taught Grant and Porter two lessons: they must somehow stop supply traffic from reaching Vicksburg; and it was possible for a vessel, moving swiftly, to run past batteries with minimal damage. Porter decided to send one of his gunboats, the *Queen of the West*, on a daring run past Vicksburg's batteries to get below and interdict enemy supply. On February 2 she ran down just after dawn, suffered a mere three hits

from all the batteries above the city, set fire to a Rebel steamer, and then passed the batteries below the city with only twelve more hits, none of them serious. Though the ship was lost a few days later when run aground, still she had shown what could be done. Porter sent another past the batteries, this time at night, without difficulty. This one, too, was lost soon thereafter, but neither casualty negated the salient fact that Porter could expect to get his ships past Vicksburg after all, probably under cover of dark, and with an acceptable risk of damage. A few days later he even sent a barge disguised as a gunboat floating down the river past those same batteries, and though they did their best, they could not inflict more than minor damage. Now Grant knew he could get his army across the river once he marched it south.

The question was when. Meanwhile, other opportunities presented themselves. Grant found more than 150 miles upriver from Vicksburg a place called Yazoo Pass, once a route inland on the east side to the Coldwater River. If flowed south into the Tallahatchie, and that in turn kept on

south until it flowed into the Yazoo, which ran right down to Vicksburg. If the route could be cleared, transports could carry Grant right to Vicksburg's back door, and all that had to be done was break through a 100-foot natural levee that separated the Mississippi from the pass. The whole journey for his men would be almost 700 miles, when they started from Young's Point and went upriver, only to turn around again and come down the new, winding route. On February 3 a charge of explosives blew a hole in the levee, and the Mississippi spent the next several days pouring a flood of millions of tons of water through, widening the opening and inundating the Coldwater and Tallahatchie. Grant then sent 4,500 men aboard transports on their way south. Unfortunately, almost from the first they encountered natural obstacles and mile after mile of trees felled into the streams by Confederates who anticipated the move. In nearly a month of cutting their way south, the Federals got no further than the mouth of the Tallahatchie, where a hastily erected Confederate Fort Pemberton stopped them cold. By late March Grant ordered them to return.

Now it was Porter's turn for a bayou expedition. He found that, in the flooded countryside north of Vicksburg, he could get up the Yazoo a few miles to Steele's Bayou, steam northward through to Black Bayou, then north on Deer Creek, to the Sunflower River, which then flowed back south into the Yazoo. This roundabout route bypassed the portion of the Yazoo above Vicksburg that the Rebels had fortified. It also cut into the Yazoo below Fort Pemberton, allowing a way to get troops ashore north of the city but without passing any batteries. It would take Porter 200 miles on two long sides of a triangle, to finish at a point just thirty-five miles east of where he started, but it appeared to be a good gamble. In the end, it turned out to be a fiasco, as Porter got well stuck in the thick undergrowth of Deer Creek.

Four separate schemes for getting on the inland side of Vicksburg had come to nothing, but Grant did not waver. It was the essence of his peculiar genius. His mind once set on a goal, he did not turn back to give up. He just kept looking for another way. Some of his officers felt infused by the same spirit. One of them, when captured on Porter's last expedition, heard a Confederate

Above: As Grant's grip on Vicksburg tightened, camps like this one spread around the city throughout May and June, pinching every artery of supply or succor until Pemberton had no choice but to yield.

Below: Many rugged western men like these had friends, even brothers, across the lines in the Confederate trenches, and informal local "brothers" truces were sometimes called that they might meet between the lines.

exclaim that Grant had "tried this ditching and flanking" so many times he should have given up. "Yes," replied the Yankee, "but he has thirty-seven more plans in his pocket."[4]

The one that finally worked was a variant of ideas already tried. Those bayous running through the interior of Louisiana, even if not deep enough to accommodate his gunboats, could still handle the much more shallow draft transports and barges. Part of the army could move aboard them while others carried ammunition and supplies and the balance of the infantry marched across the marshy land. Once below Vicksburg, they could be ferried across by the steamers, and all with the assistance and cover of Porter's gunboat fleet. Porter, Grant decided, could safely risk running his vessels past Vicksburg during the night. Grant would cross his army to the Mississippi side somewhere near Grand Gulf, sixty miles below Vicksburg, march inland, then turn north and attack Vicksburg from the rear.

Grant set the plan in motion on March 31, when the first of McClernand's corps began their march into the bayous. But then another disaster struck. The level of the Mississippi, so high from the winter rains, suddenly fell. The bayous now would not even handle the most shallow draft vessels. As always, Grant refused to be deterred. Instead, Porter would simply have to take everything with him past the batteries. Well after dark on April 16, the flotilla started downstream. They were not yet abreast of the river when the Confederates spotted them and opened fire. What followed Porter could only liken to a trip through hell. One by one, during almost two hours, the ships ran the gauntlet of fire. But when Porter counted bows afterwards, only one transport had been lost. Six nights later the six remaining transports made the run, and five made it. So far so good.

Grant had intended using the gunboats to force Rebel fortifications at Grand Gulf to yield so he could cross his army. Porter wisely persuaded him not to risk the transports in a river crossing under fire, and Grant decided instead to cross at Bruinsburg, ten miles further south. It was a good idea, and by April 30 Grant was ready to start sending McClernand and McPherson and their corps on the most difficult one mile of Rebel

Above: Colonel Benjamin Grierson and his 1,700 bold troopers made one of the most daring cavalry raids of the war through Mississippi to act as a diversion for Grant's move to invest Vicksburg.

territory a Yankee army ever had to cross – the Mississippi. And after all the time and effort it had taken to embark those troops, the crossing was almost anti-climax, for Grant made it virtually unopposed. By evening on May 1, both corps were on the east bank, while Sherman remained above the city, making a demonstration against the bluffs along Chickasaw Bayou to keep Pemberton from sending any men south to resist the landing. It had all worked brilliantly. At the same time, Grant had launched another diversion, sending Colonel Benjamin Grierson and 1,700 Federal cavalry on a raid from La Grange, Tennessee, all the way down to Baton Rouge. The troopers were to travel 600 miles in sixteen days, tear up nearly sixty miles of railroad, disrupt telegraph lines, and destroy tons of enemy supplies

and weapons. Moreover, the lightning raid kept all of the Confederate cavalry in Mississippi busy looking for Grierson, and away from Grant as he made his crossing.[5]

The entire operation had been one of the most brilliant in the history of warfare, but Grant did not pause to congratulate himself. He moved, and quickly. He marched immediately to Port Gibson, pushed aside the small force of defenders, and that, in turn, forced the Grand Gulf garrison to evacuate to avoid being cut off from any line of retreat. Then he moved on inland and northeasterly, heading for the state capital at Jackson. If he was successful and took it, he would break Vicksburg's only remaining rail line of communications. Then he could turn west, knowing that Pemberton would be cut off from not only supply, but also any reinforcements, and march straight west against Vicksburg.

Pemberton knew that he stood increasingly in a bad way now. He wired to Johnston for reinforcements, or at least to move with whatever forces were at hand to threaten Grant's rear and keep Vicksburg from being cut off. Johnston, true to form, did next to nothing, though Bragg's army was then sitting around Tullahoma spending its most idle summer of the war. All Pemberton could hope to do was keep his army in the field, avoid being bottled up in Vicksburg, and pray that Johnston would develop some courage, or that President Davis would intervene. Both proved to be idle dreams.

Grant waited for Sherman to make the long trip south through the bayous and across the river. When he arrived on May 8, the reunited army totaled more than 40,000, and with it the Yankees continued their drive. "The road to Vicksburg is open", he told Sherman, a glory road on which he believed nothing could stop them. Indeed, it could not. By May 12 they were at Raymond, just fifteen miles southwest of Jackson. There they met and repulsed a ragtag force trying to stand in their way. The next day Grant sent McPherson straight north to cut the Southern Railroad line at Clinton, Vicksburg's link with Jackson. Then McPherson would turn east as Sherman himself marched on Jackson. Again it all worked beautifully, and two days later the state capital flew United States flags. Leaving the city's factories and warehouses in smoldering ruins, Grant launched McPherson and McClernand on the road to Vicksburg the next day.

All Pemberton could hope to accomplish now was to stop Grant somewhere along the road before he could reach Vicksburg. On May 16 he met the foe on and around Champion's Hill, twenty miles west of Jackson. Finally he had heard from Johnston that help was on the way, and he must delay Grant to give time for them to arrive. To do so he took a grave risk, and brought 23,000 of his troops – virtually the whole army – out to meet the foe. Combined forces of McClernand and McPherson were there to meet him, and in a battle lasting from mid-morning until well into the afternoon, Pemberton actually achieved some early gains, but could not hold on to them. A Federal counterattack put one whole division to rout, and it could not make its way back toward Vicksburg. Instead, it moved below

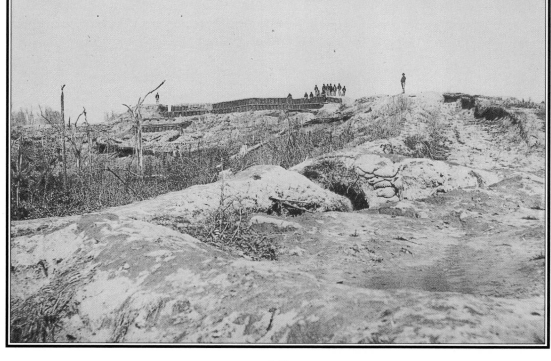

Left: A portion of the siege lines surrounding Vicksburg, photographed shortly after the surrender, attests to the depth of the fortifications ringing the besieged city before its fall.

Modified Officer's Grimsley Saddle, Harness and Tack Box (of then) Major General Ulysses S. Grant

1 Wooden, iron-mounted tack box, with hinged lid (shown in the open position) and lock. The box is stencilled at both ends with the name and service of the owner "U.S. Grant U.S.A."

2 Brass-bound Grimsley saddle with hand-tooled, padded leather seat. The stirrups are wooden with tooled leather hoods. The stirrup leathers also appear to have been lightly tooled. This saddle is in the western style, but it was common for officers of both sides to also use the flat English type. In contrast to this fine saddle, the regulation saddle for enlisted men in the U.S. Army was the McClellan pattern saddle, that had been developed by that officer in the 1850s.

3 Leather-mounted heavy fabric girth for Grant's saddle

4 Brass plated steel bit, double bridle and reins belonging to General Grant. U.S. Army regulations called for a curved bit (as here), but with single reins. Officers', and particularly generals' riding outfits often differed from this and were non-regulation.

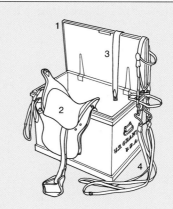

Artifacts by courtesy of: John G. Griffiths Collection: 1; U.S. Army Quartermaster Museum Collection, Fort Lee, Va: 2, 3, 4

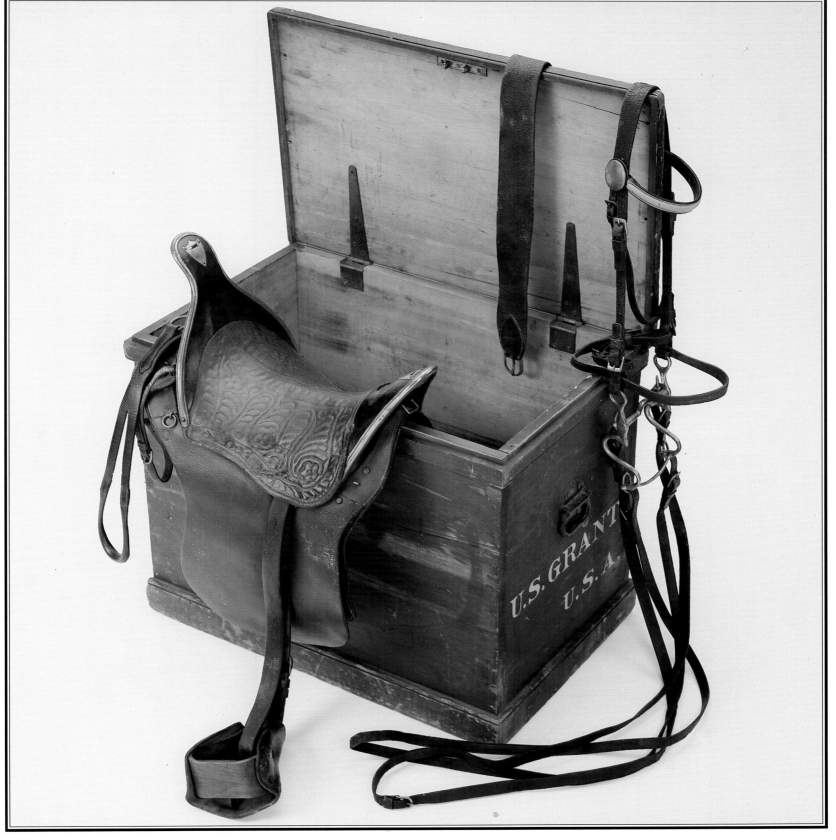

and east of Grant in hopes of meeting the relief column.

Pemberton suffered heavily. He took 3,800 casualties in the fight, and lost that whole division. With the balance, he could do nothing but retire. The next day he mounted a brief rearguard action at Big Black River Bridge, but had to retire once more after little more than an hour. That afternoon Pemberton and his dispirited soldiers filed back into their earthworks surrounding Vicksburg and dug in even deeper, realizing that now there was nothing left to them but to withstand a siege until Johnston should arrive.

Grant was to get there first. On the evening of May 18, McClernand and McPherson began to see the spires of the city, topped by the cupola on the Warren County courthouse. By the next morning, the Federal line had spread itself along the entire eastern face of the defensive lines held by Pemberton's 31,000 remaining soldiers, a line more than five miles long. Sensing the demoralized state of the defenders after their reverse at Champion's Hill, Grant did not want to waste time. A resolute assault could, and would, carry the works and take the city in a single blow. He gave orders that at 2 p.m. his entire army would charge all along the line.

The works Grant faced at Vicksburg were among the most formidable to be seen anywhere in the war up to this time. They had been planned eight months before, on three quarters of an elliptical arc stretching from Fort Hill, overlooking the Mississippi directly north of the city, northeast to the Stockade Redan, then south through a series of other redoubts and lunettes, to South Fort, two miles below the city. Every road into Vicksburg had a fort guarding it, and in between all of these works ran a deep line of trenches with firing ramparts, and often obstructions like sharpened stakes or felled trees, even wire, to slow or halt an enemy charge. Pemberton had placed batteries in commanding positions, and thanks to the hills surrounding the city, there were many places in which the works sat on such an elevation that any attempt to attack would be suicidal.

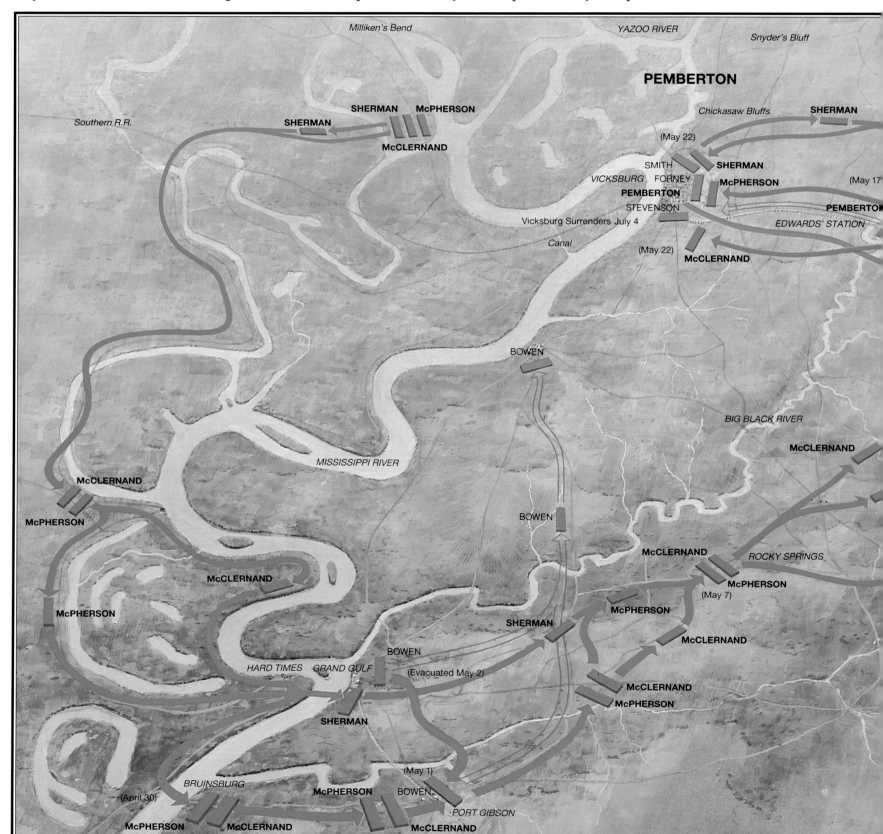

The Vicksburg Campaign, March 31-July 4, 1863

The sweep of Grant's Vicksburg Campaign is readily evident in the area needed to show his movements. Its beginnings in the intricate series of bayous parallelling the Mississippi in the Louisiana interior appear at upper left. On March 31, 1863, McClernand's corps, followed by McPherson, began their muddy voyage through these bayous, heading south, sometimes overland, until they emerged on the west bank opposite Bruinsburg, at bottom left. Porter's gunboats and transports, meanwhile, had come down the river past Vicksburg's batteries at upper left center. Now these vessels ferried the Yankees across. A week later Sherman's corps made the same journey, and Grant was ready.

He began the overland campaign with a brief sidestep to Grand Gulf to clear it of Rebels, then drove for Raymond, at right center. Defending Confederates stalled him for only a day before he drove on. From here he sent McPherson north to strike the railroad at Clinton, while Sherman and the remainder of McPherson's corps attacked and took Jackson. The state capital fell after two days.

Pemberton, knowing the desperate situation that faced him, moved out of Vicksburg to try to halt Grant's progress. On May 16 he took a position on and around Champion's Hill on the road between Jackson and Vicksburg, at upper right center. Grant hit him hard and drove him from the field. The next day Pemberton tried again at Big Black River Bridge, upper center, but to no avail. There was now no choice but to pull back into the defenses of Vicksburg itself and hold out, waiting for aid from Johnston that never came.

That same evening Grant moved his first elements up to the defenses, and the next day, May 19, he launched his first assault. It was a spirited and in places desperate affair, but the Confederates held out, as they would through a succession of assaults on subsequent days until Grant finally decided that he would have to lay siege. For the next six weeks the Yankees laid a constant barrage on the beleaguered defenders, meanwhile spreading their lines to close off every avenue of supply or escape. By July 1 Pemberton could hold out no longer. His force was on the brink of starvation and its numbers were decreasing all the time through disease and desertion. Though he hoped until the very end that Johnson would arrive, his position was such that a surrender to Grant's army was inevitable. A meeting between the two commanders on July 3 settled the details of the surrender of the city, the next day. Of Pemberton's army of 28,000, few were casualties but all surrendered. Grant's forces, by contrast, had grown to 77,000 or more.

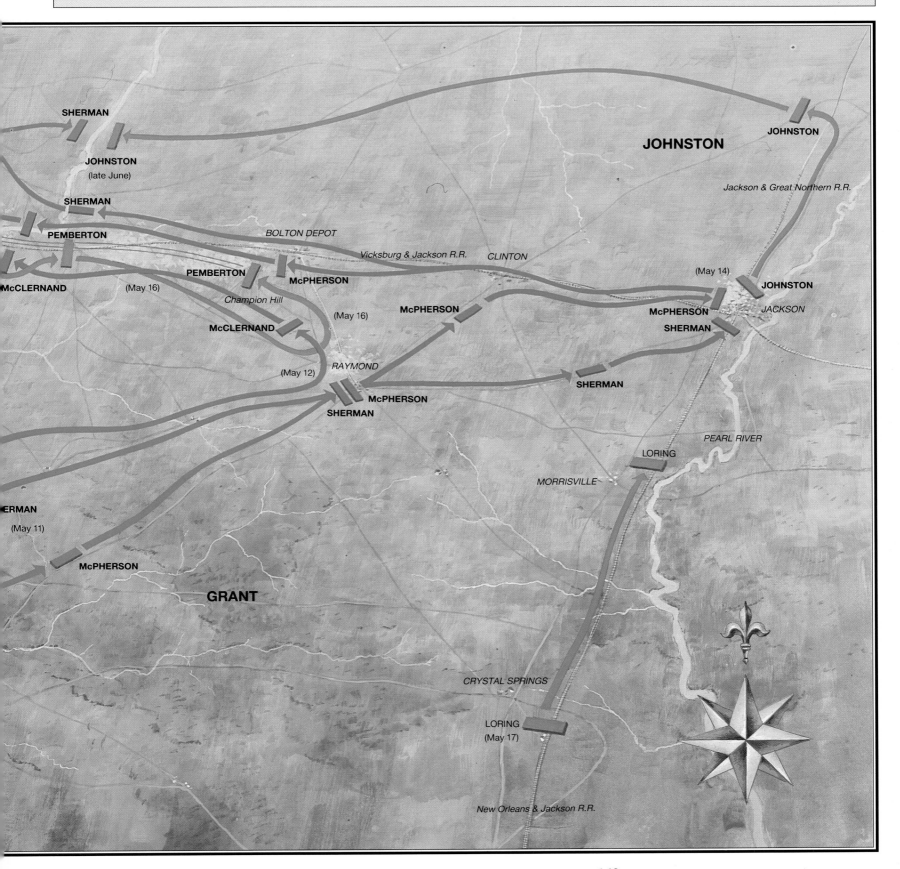

It was another reason why Grant wanted to take it now before the Rebels could dig in even further. At 2 p.m., May 19, his signal guns launched the attack. The fighting became at once furious, and rare for Civil War engagements, hand to hand in places. All along the line isolated Federal units actually reached the outer edge of the enemy parapets, only to be unable to go further or hold on. Sherman saw men falling "as chaff thrown from the hand on a windy day." The repulse did not deter Grant, who ordered another, better coordinated attack for May 22. He let his artillery bombard the defenses continually beforehand, both to soften them up, and to keep the Confederates from sleeping. Porter's gun-

boats were out on the river adding their long-range shells to the din of the cannon.

At 10 a.m. the signal sounded once more, and in went all three corps: Sherman from the northeast, McPherson on *his* left, and McClernand on *his* left. Some men actually carried ladders, so steep were the outer walls of the enemy parapets, while others brought planks to throw across ditches and trenches. Sherman's men met the worst of it, since they moved against the most difficult terrain. Enemy fire soon halted them, though they struggled valiantly time after time to get into Stockade Redan. When they were forced to halt at the bottom of the slope leading to the works, Confederates lit artillery shells and threw

or rolled them over the parapets as grenades.

McPherson met a galling fire, too, and got no farther. Only McClernand enjoyed a measure of success when he took a redoubt guarding the railroad. Yet resistance was so great that he had to ask that Grant order Sherman and McPherson to renew their stalled assaults to take the pressure from him. Grant did as asked, but it availed nothing. McClernand held on to the Railroad Redoubt through most of the day, but by evening a Confederate counterattack drove him out. Once more night fell with no advantage gained, but Grant had more than 3,200 casualties in the bad bargain "without giving us any benefit whatever", he later lamented.[6]

Federal and Confederate Handgrenades

1 Federal Ketcham percussion 5lb grenade. Wooden tail and paper wings are to assist the bomb in flight
2 Confederate Rains handgrenade. Of the same basic design as the Ketcham, but with a modified plunger head
3 Ketcham 1lb handgrenade
4 Confederate spherical handgrenade
5 Confederate Rains handgrenade variant
6 Printed instructions for the charging and use of the Ketcham grenade
7 Ketcham 3lb handgrenade
8 Confederate spherical handgrenade
9 Adams handgrenade
10 Haynes Excelsior patent (1862) handgrenade. The weapon came in two halves. An outer casing surrounded an inner sphere containing the charge. It was detonated by percussion caps on the sphere hitting the outer casing
11 Confederate spherical handgrenade
12 Internal bursting charge container of Haynes grenade
13 As 12. Note nipples
14 Exterior casing of Haynes grenade, open, exposing internal bursting charge sphere with nipples removed

Artifacts courtesy of: Gettysburg Museum of the Civil War, Gettysburg, Pa

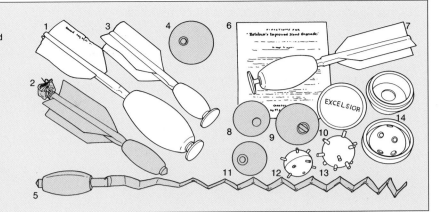

Private, 4th Kentucky Infantry, C.S.A.

The most colorful aspects of many a Johnny Reb's garb were the bits that he brought himself from home. In the case of soldiers of the famed Orphan Brigade, like this private in the 4th Kentucky, there was only one chance to bring something from home, for once Kentucky Confederates left the bluegrass, they never returned until war's end. Thus they were called "orphans." They proved to be among the best soldiers in the Army of Tennessee.

Their 1853 Enfields were seen all across the battlefields of the South, as were their distinctive short jackets, the whole uniform made of cotton. This soldier's color comes from his blanket, probably a quilt made by his mother. Things changed dramatically for the 4th kentucky and its sister regiments late in 1864 when the War Department turned the depleted units into mounted troops. They finished the war as cavalry, something they had wanted almost from the very first.

It would be the last wasted attack. That night Grant concluded that "the nature of the ground about Vicksburg is such that it can only be taken by a siege." He hoped it would take no more than a week.[7]

In fact, it took another six weeks. From May 23 onward, Grant maintained an almost constant bombardment of the works and the city itself. The soldiers burrowed even deeper into their trenches, while the townspeople soon sought shelter in impromptu caves dug into hillsides. The streets became a virtual no man's land, dangerous for everyone alike, including animals and even pets, for as the supplies of food dwindled, soldiers and civilians were forced to turn to anything they could find. When the cattle were gone, they turned to horses, then mules, and finally even dogs and cats. By the beginning of July, even the city's mice and rats found their way into pots. As Grant steadily extended and strengthened his line around the city, the plight of those inside his steel coil became ever greater. Pemberton was not strong enough to attempt to break out, and besides he had orders from Davis to hold Vicksburg at all costs. All he could do was hope that Johnston would arrive in time, and by late June that tardy general was believed to be approaching Jackson. Perhaps he would arrive in time. But then Grant sent Sherman with more than 30,000 back east to meet and stop the relief column, and in fact Johnston would never get any closer than Jackson. "The fall of Vicksburg and the capture of most of the garrison can only be a question of time," Grant told Washington late in June.

The siege wore hard on blue and gray alike. Grant's army grew through reinforcement to 77,000, greater than at any time during the campaign. Pemberton, by contrast, saw his own numbers dwindle through exhaustion, death, and desertion, to a mere 28,000 or more. Even the mules and rats ran in short supply, and the Vicksburg *Citizen* had to start printing on wallpaper. Pemberton set some of his men at work building flatboats and skiffs with a possible view toward ferrying his army across to Louisiana rather than see it forced to surrender. But always he continued to hold on in hope of Johnston. A tantaliz-

Above: The Sherman battery, one of the ring of gun emplacements encircling Vicksburg, in a photograph taken shortly after the siege concluded. It shows the power of the armament arrayed against Pemberton.

ing report came into the Rebel lines on June 20 that Johnston would be attacking Grant's rear in four or five more days, and it caused considerable excitement in Pemberton's weary lines, but of course it proved to have no substance, and in the following days more realistic word came from Johnston, word with no hope attached to it for saving Vicksburg, and little hope for the garrison unless Pemberton could break out on his own, a movement now manifestly impossible.

By the end of June, Grant's engineers were digging tunnels under the Rebel works and explod-

Below: With the siege done and the garrison surrendered, much of the captured Confederate artillery sits parked in a city lot, now silent and never again to hurl iron at attacking Yankees. The Confederates now had little control over the Mississippi.

ing massive powder changes that sent men and earth flying into the sky. It became increasingly obvious to Pemberton that time was running out. On July 1 he asked his generals what they advised, and the next day they replied with near unanimity that there was no alternative. Even if a way to evacuate could be found, the men were so reduced by hunger that they had not the stamina for a fight. That night Pemberton asked Grant for an armistice to discuss terms of surrender.

The two met the next day in a tense interview that almost saw a proud ar d hurting Pemberton call it all off. Only tact by his subordinates, and the generous and humane terms offered by Grant, saved the proceedings. Rather than make prisoner of the Vicksburg garrison, Grant would release them on their paroles not to take up arms again until exchanged officially for released Federal prisoners of war. As a final gesture of respect, Grant also agreed to allow the garrison to march out of the defenses before the Federals marched in, thus honoring a vow of Pemberton's that no Yankee soldier would set foot in Vicksburg while his army was there. What Pemberton could not get was an agreement to postpone the ceremony for a day. Grant wanted it to happen on July 4, a symbolic day for the Union now to be made the more so.

It all happened the next day. Vicksburg changed hands. A Confederate army simply ceased to exist, and all that remained in the way of the Yankee takeover of the Mississippi was Port Hudson, which fell within the week. In Washington a jubilant Lincoln exulted that "the Father of Waters again goes unvexed to the sea." The Confederacy was cut in half, and in Richmond a War Department official lamented that "the Confederacy totters to its destruction."[8]

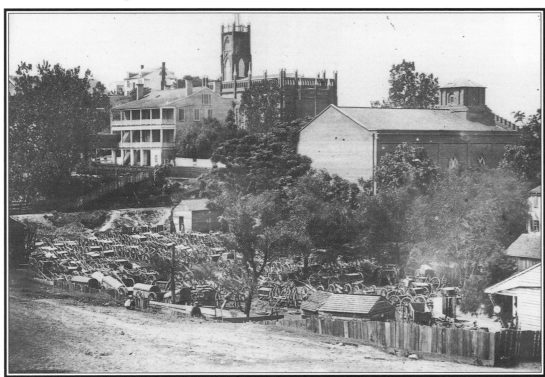

References

1 Grant, *Memoirs*, I, pp.392-94.
2 John C. Pemberton, *Pemberton: Defender of Vicksburg* (Chapel Hill, N.C., 1942), p.14.
3 William T. Sherman, *Memoirs* (New York, 1875), I, p.292.
4 Samuel C. Carter, *The Final Fortress: The Campaign for Vicksburg 1862-1863* (New York, 1980), p.149.
5 *O.R.*, I, 24, pt.1, pp.33-34
6 Grant, *Memoirs*, I, p.531.
7 Carter, *Vicksburg*, p.233
8 Frank Vandiver, ed., *The Civil War Diary of General Josiah Gorgas* (Tuscaloosa, Ala., 1947), p.146.

CHAPTER NINE

GETTYSBURG

JULY 1-3, 1863

Grant's triumph at Vicksburg meant more than people at the time could realize. Obviously, with the Mississippi now in Yankee hands – once Port Hudson surrendered later in July – Confederate control of the great river was at an end. Men could still steal across in the darkness, and operate on its tributaries in Louisiana and Arkansas, but no more could the South use the river to shift great numbers of troops or *meterial* from one side to the other. The Confederacy was effectively cut in two.

But the emotional damage to the Southern morale was perhaps equally as great, though less appreciated. Vicksburg symbolized something – the South's idea of its impregnability, its confidence in victory. When that city fell, the shock was enormous.

Now Confederates placed their remaining faith in Robert E. Lee and the Army of Northern Virginia. Never beaten on equal terms, it still stood for something invincible in the Southern spirit. As long as Lee remained unbeaten, there was still hope.

THE ACCIDENTS of history give nobility to the commonest of men and turn rude country crossroads into names that echo through the ages. Men like Alexander of Macedon and places such as Rome may have been ordained from inception for greatness, but for most people and locations such recognition comes as nothing more than the fickle choice of chance. But for the events that made them, Napoleon would be forgotten, U.S. Grant would never have risen above being a face in the crowd, and even Lincoln would have remained an obscure prairie lawyer. But for chance, little-known places like Bull Run and Shiloh would still be nothing more than wide spots on forgotten country roads. Even when men tried to plan in advance those locations that the events they controlled were about to immortalize, chance robbed them of the choice.

If one of the army commanders involved had had his way, the greatest engagement of the Civil War – and perhaps of all American history – would have come down to us as the Battle of Pipe Creek. But men could not plan for greatness, and "Pipe Creek" would never be emblazoned on the battleflags of Blue or Gray or stir the memories of old veterans and serve as a rallying cry for the patriots of later generations. Armies sometimes move on a momentum all their own in their lurching, swaying paths to meet, and the tens of thousands who might have let their blood to make Pipe Creek immortal missed each other there and collided some twenty miles north instead, giving their nomination for fame to a dull Pennsylvania crossroads at Gettysburg.

Despite the fact that no one could predict where the armies were to meet next after Chancellorsville, a certain logic virtually dictated that they should clash again on Northern soil, and that the victorious Lee should be the aggressor. In part, the reason was personal and emotional. Lee, after a year in command of the Army of Northern Virginia, had become accustomed to victory. McClellan, Pope, Burnside, Hooker – he had beaten them all. Only once did he taste defeat, and that was on Yankee ground at Antietam. Lee was a very human man for all his growing stature as a Confederate demigod. To himself and the South he had proven almost everything, except that he could meet and beat the foe on their own ground. Lee, the man, needed a successful invasion of the North.

Lee the general needed it, too. He had beaten Hooker, but he had not removed in any way the continuing threat to Virginia. The Army of the Potomac was still massive. With the return of Longstreet with Pickett and Hood, Lee's own army still mustered great numbers. But they had to be fed and clothed by an increasingly overburdened Confederate supply system. Moreover, when Hooker pulled back across the Rappahannock in early May 1863, there still remained at least six months of good weather for active campaigning. Given Northern resources in men and material, Lee could only expect that if he waited for the enemy to make the next move, it would only be a renewed, restrengthed, and redoubled effort to march on Richmond. The only way to prevent that was to preempt the offensive by seizing it himself. If he marched into the North again, Hooker's army would have to follow him to protect Washington, Baltimore, Philadelphia, perhaps even New York. That would relieve Virginia from invasion. And if Lee could bring Hooker

The tangled, rocky front slopes of Little Round Top give ample evidence of the rugged ground that blue and gray had to contest in the struggle to possess the decisive terrain in the Battle of Gettysburg.

to battle and decisively defeat him on Union soil, the humiliation and dismay might discourage the Northern population from continuing what was already an increasingly controversial war. Off-year elections that fall might see Lincoln's majority in Congress reduced. Also, such a threat might force Grant to divert some of his forces then approaching Vicksburg and relieve the pressure on the Mississippi. And then there was the still-tantalizing – though rapidly fading – hope of convincing European powers to intervene on behalf of the Confederacy.

In fact, Lee had an old plan worked out a year before with Jackson, and perhaps even originally inspired by the lamented "Stonewall". A variation of it led to Antietam, but now Lee worked on the original notion of moving the entire army into the Shenandoah Valley, then closing off all gaps to keep the enemy in the dark as to his movements. While Hooker still sat in northern Virginia, Lee could slip quietly and quickly northward through the Valley, cross the Potomac, and suddenly appear in Maryland. Before Hooker could react, Lee would have his divisions marching into Pennsylvania, heading for Harrisburg and the Susquehanna River. It meant leaving Richmond exposed, of course, but Hooker would not dare march on the Confederate capital with a Rebel army threatening his own. He would have to abandon Virginia to go after Lee.

It was a tremendous gamble, the kind of chance that Lee could make with conviction, but

it took him much of May to convince President Davis to go along. Only after a long meeting on May 26 did Lee finally have approval, and still Davis wavered in the days that followed. Lee did not, however, and arduously made his preparations, mindful that time could be blood. He had already obtained maps of the routes through Maryland and into Pennsylvania the previous winter. He even had some thoughts as to where he might have to do battle with Hooker, probably in south-central Pennsylvania, and most likely at Chambersburg, York, or Gettysburg, since the available roads would direct the armies through those towns. Lee never seriously considered actually attacking and taking Philadelphia or Baltimore. His objective was the Army of the Potomac. If he could do to it in the North what he had done to it at Chancellorsville, he could almost walk into any city he chose.[1]

Feverishly Lee gleaned reinforcements from throughout Virginia and the Eastern seaboard. He reunited the old I Corps with the return of Longstreet, who now once more commanded three divisions: McLaws' Georgians, South Carolinians, and Mississippians; Pickett's all Virginia division; and Hood's Alabamians, Georgians, Texans, and the one and only regiment from the Trans Mississippi in Lee's army, the 3rd Arkansas. The death of the irreplaceable Jackson eventually led Lee to replace him at the head of the II Corps with the solid though unimaginative Richard S. Ewell, just barely recovered from the

Right: Major General John F. Reynolds, shown here as a brigadier, was fighting the Confederates almost on his home soil when he commanded the Union resistance during the early hours of the great battle's first day.

Below: Major General George G. Meade had commanded his army for only a few days when fate propelled him into the biggest battle of the war. He stands, thumbs in belt, at center, surrounded by his staff.

loss of a leg the previous summer. He, too, had three divisions. The cantankerous yet combative Jubal Early led a mixed command of Louisianians, Virginians, Georgians, and North Carolinians. Edward Johnson commanded the division that had been Ewell's, and Jackson's before him, almost all Virginians, with one Louisiana brigade and some North Carolinians and one Maryland unit. And Robert Rodes, who brilliantly led Jackson's great flank attack at Chancellorsville, led a division of three North Carolina brigades, plus one from Georgia, and one from Alabama. Furthermore, Lee created a new III Corps made up of units from the other two corps, along with new additions, to be commanded by the superlative A.P. Hill. Richard H. Anderson led five brigades hailing from as many different states – Alabama, Georgia, Virginia, Florida, and Mississippi. Major General Henry Heth – said to be the only general that Lee addressed by his given name, though no one knew why – led North Carolinians, Virginians, Mississippians, Alabamians, and the only Tennessee regiments with the army. And William Dorsey Pender, at a mere twenty-nine one of the most loved and respected generals in the army, commanded two North Carolina brigades, and one each from South Carolina and Georgia. Every division in the army had its own artillery, and each corps an artillery reserve. Meanwhile, "Jeb" Stuart continued to command the army's cavalry, seven patchwork brigades of varying size, all but two from Virginia, and six batteries of light and mobile horse artillery.

All told, and including small units that he would pick up during the early stages of the coming campaign, Lee could count 80,000 or more of all arms, though he still knew himself to be outnumbered by those Yankees glaring at him from the other side of the Rappahannock.

Despite his humiliating defeat at Chancellorsville, Hooker still commanded those Yankees. The Army of the Potomac, though discouraged by its defeat, remained strong. Except for the XI Corps, the men had fought better than ever before, and most acknowledged that what Lee defeated in the recent battle was their commander, and not themselves. The chagrin was greatest at the several corps headquarters, where Hooker lost the confidence of almost all of his commanders. Speculation that he would be relieved ran wild almost as soon as the retreat from Chancellorsville was finished, and Couch, Slocum, and Sedgwick started what amounted to an incipient cabal when they approached Meade with assurances that they would gladly serve under him instead. Hancock, Sickles, Sedgwick, Couch, Reynolds, and others, were all spoken of as possible successors. yet Lincoln stood by

Lieutenant Colonel, 44th Georgia, C.S.A.

Georgia was almost a small nation in itself, making most of the things needed to arm and equip its men. This lieutenant colonel of the 44th Georgia Infantry is "Georgia" from head to toe. His uniform is a standard gray pattern piped in black. He wears a two-piece Georgia state belt plate, and it supports a sword from the works of W.J. McElroy. Even the stars on his collar, cut from brass sheets, are of local manufacture. Most distinctive of all, however, is the Griswold and Gunnison revolver in his hand. Made at Griswoldville, Georgia, it is a blatant copy of a Colt model, the only real differences being in its brass frame. Almost the only non-Georgia bit of equipment on this officer is the canteen he carried, definitely Yankee issue, and undoubtedly taken from some unfortunate foeman who fell by Confederate forces in battle or suffered capture. Though a later war regiment, the 44th, as evidenced by the equipment on this man, was still well outfitted.

Hooker through the balance of May, not so much out of conviction that he could bring a victory, as from an unwillingness to have yet another change of commanders so soon. By waiting to relieve him, he hoped psychologically to diminish the emotional impact of the recent defeat in the North. Still, on June 2 Lincoln asked Reynolds confidentially if he would take the command. The Pennsylvanian declined, chiefly because he feared too much political interference in his handling of the army. Receiving this refusal, Lincoln apparently decided to back away briefly. Hooker was handling the army well again. Perhaps he could retrieve his own and his army's fortunes if given another chance.[2]

Indeed, in June, just as Lee was starting to put his plans in motion, Hooker launched a small offensive with his cavalry, sending Pleasonton with 11,000 cavalry and 3,000 infantry on a raid across the Rappahannock. Hooker had word that elements of the Army of Northern Virginia had been pulling out of their camps on June 2 and the days following. Fearing a raid – not a major invasion – Hooker sent Pleasoonton toward Culpeper Courthouse, where he believed Lee was concentrating supplies for the raid. The result was an engagement on June 9 between Pleasonton and Stuart's 10,000 cavalry near Brandy Station. The inconclusive fight failed to reveal to Hooker the true extent of the mounting threat, and it proved

to be his only attempt to take control of affairs. For Lee, it only served to make him more determined to move and move quickly, lest Hooker or some successor might strike again with more determination. The very next day he rushed Ewell and his corps toward the Shenandoah, with the rest soon to follow.

The move came with lightning swiftness. In three days Ewell's advance approached Winchester; in a week Lee's entire army was in the Shenandoah. Hooker, reacting to the enemy movement, shifted his army west toward the Blue Ridge, but lack of information prevented him from taking any more decisive steps, while his own tardiness and hesitation made the situation

The Whittier Group of Uniforms and Personal Effects

1. Rhode Island state issue gray wool trousers, worn by Private E.N. Whittier, Co. C., 1st Rhode Island Volunteer Militia
2. Whittier's non-regulation knitted wool bonnet
3. Forage cap worn in Battle of First Manassas (First Bull Run) by Private Whittier
4. Rhode Island state issue 'pullover' blue wool field blouse; worn by Private Whittier
5. Civilian blanket used by Private Whittier
6. Civilian coverlet used by Private Whittier
7. Wood foot locker used by Private Whittier
8. Blue wool short jacket worn by Lieutenant Whittier after his commission and transfer to the 5th Maine Battery
9. White linen trousers worn by Lt Whittier at Gettysburg
10. Framed carte-de-visite

taken in 1863 of Lt Whittier
11. Artillery officer's trousers, worn at Chancellorsville by Lt Whittier
12. Artillery officer's forage cap, worn by Lt Whittier at Chancellorsville and Gettysburg
13. Leather bag with 6th Corps badge, property of Lt Whittier
14. Private purchase, off-duty smoking cap, used by Lt Whittier

Artifacts courtesy of: Don Troiani Collection, Southbury, Conn

even worse. The Confederates captured almost the entire Federal garrison at Winchester, dispersed the Yankee defenders at Martinsburg, and clearly were on the move northward in strength. Yet Hooker faced with this overwhelming body of evidence, could not decide what to do. "His role now is that of Micawber", wrote Hooker's provost marshal, "'waiting for something to turn up'." Even after 4,500 men were lost at Winchester, Lincoln continued to stand by Hooker, not out of confidence so much as unwillingness to replace a commander now that the enemy was on the move. Changing horses in midstream would be dangerous, said the old adage; changing generals could be disastrous.[3]

Instead, Washington tried its best to reinforce Hooker from other commands in Virginia and Maryland, until by late June Hooker had at his disposal over 100,000 men. But by that time, Lee had crossed the Potomac and was marching through Maryland once again. No one could question that he now intended a major invasion of the North, yet still Hooker remained indecisive, Lincoln and Halleck having to direct most of his moves. Lee meanwhile, even as Longstreet and Hill crossed the Potomac on June 23-25, and Ewell surged far ahead, moving toward York, Pennsylvania, decided on yet another bold maneuver. He sent Stuart and most of the cavalry on a sweeping ride around the eastern side of

Hooker's army, between him and Washington, to sever communications, capture and destroy supplies, and put even more panic into an increasingly nervous Union government.

By now, because of Hooker's reluctance to move decisively, Lee stood considerably north of him, with his army spread out on the roads between Frederick, Maryland, and Chambersburg and York, Pennsylvania, while the Army of the Potomac still remained in northern Virginia. Lee was at least two days ahead of him, maybe more. On June 25 Hooker finally put his own army in motion northward, and sent Reynolds rushing ahead in command of the I, III, and XI Corps. The next day the balance of the Yankee army started

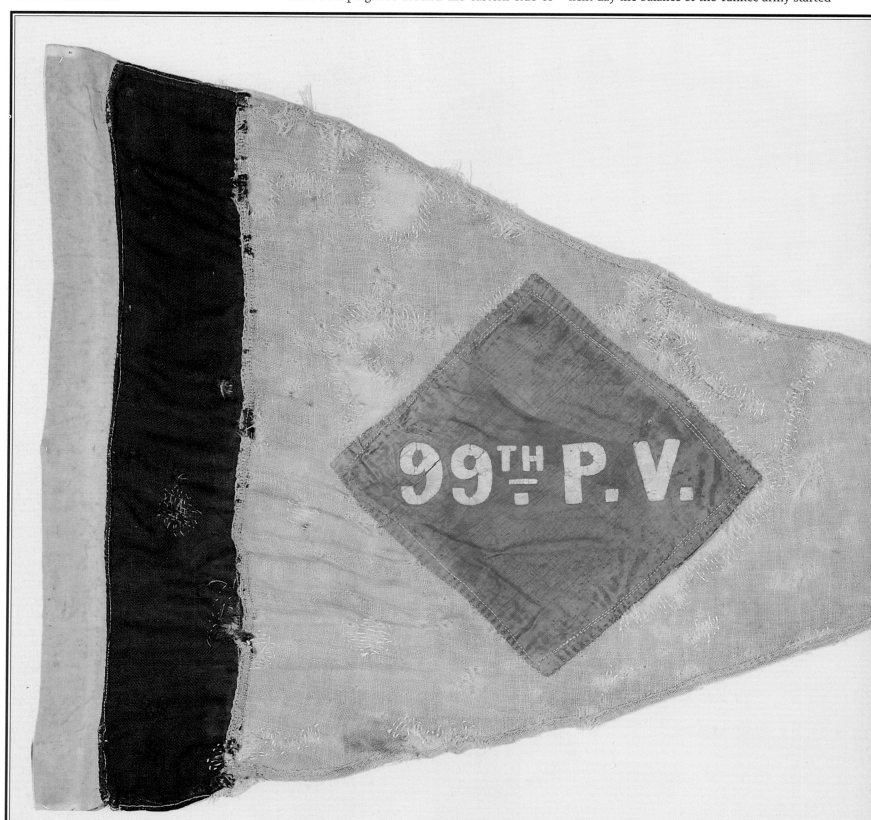

Union Camp Colors and Field Markers

1 Small color of the 99th Pennsylvania Volunteer Infantry. The red diamond in the center indicates that this regiment was assigned to the 1st Division, 3rd Corps. The blue stripe of the hoist signified that the regiment was part of the 2nd Brigade of that Division. This evidence suggests that the color may originate from the time of Gettysburg. The regiment

served in Sickle's 3rd Corps, 1st Division, 2nd Brigade (under Brigadier General J. H. Hobart Ward) and fought in and around Devil's Den during the Battle of Gettysburg

2 Camp color of the 56th Pennsylvania Volunteer Infantry, bearing battle honors and trimmed with a gold fringe. The battle honors, are an unusual application to this type of

regimental flag which normally only measures about 18 × 18 inches

3 Swallow-tail camp color of the 91st Pennsylvania Volunteer Infantry. The regimental number and abbreviation is seen painted inside the red square of cloth in the center. The flag itself is non-regulation in shape, but features a white field in accordance to the regulations of 1836

Artifacts courtesy of: The Civil War Library and Museum, Philadelphia, Pa

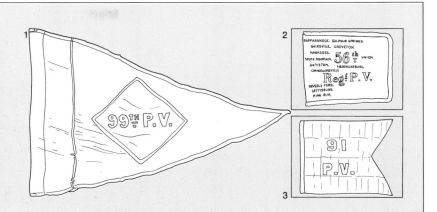

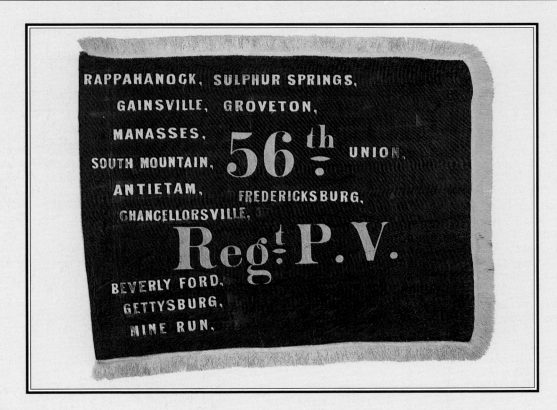

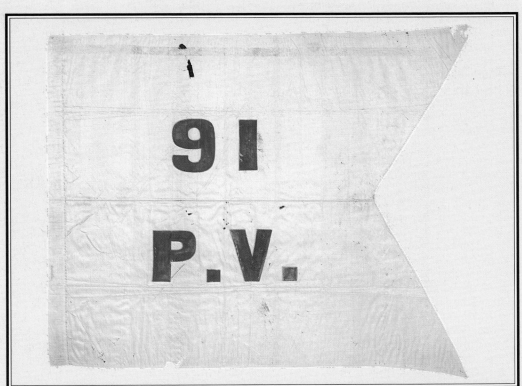

crossing the Potomac on the race to catch up with Lee. In two days they covered up to forty-five miles, reaching the vicinity of Frederick by evening on June 27.

But that same evening, other events came to a head. From a variety of motives – fear of fighting Lee again, feuding with Halleck, bluffing that went wrong – Hooker sent in his resignation. Lincoln had no choice but to accept it, and probably felt little or no reluctance, except for the act coming in the middle of a crucial campaign with the armies on the move. Wisely, he did not consult his cabinet or political advisors about a replacement. There was not time, and even asking such a question would be to invite political turmoil. Instead, Lincoln took counsel only of himself, and that night issued an order placing Meade in command. With little enthusiasm, but with an unquestioning soldier's obedience to orders, Meade accepted the post – one he would hold for the next two years, making him the Army of the Potomac's sixth and last commander.

Certainly Meade presented nothing at all imposing or inspiring to those who saw him. Tall, slender, with thin graying hair and a generally schoolmasterly look, he utterly lacked the sense of style that made men like McClellan and Hooker the favorites of the men at times. Moreover he possessed a wickedly acid tongue, with a temper that gained him a reputation as a "snapping turtle" with many of his associates. Neither timid nor impetuous, he was a cool, methodical old professional who enjoyed almost universal respect among his peers.

At once Meade met with Hooker to take the command and learn the latest of the situation. Though he took over at a most difficult time, at least Meade had the advantage of assuming command of a magnificent army. It was little changed from the days at Chancellorsville just a few weeks before. Reynolds led the I Corps, with three divisions. Hancock had the II Corps with three more. The inept Sickles still held the III Corps and its two divisions. George Sykes now assumed command of the three divisions making up Meade's old V Corps, while Sedgwick remained at the head of the VI Corps' three. Howard, who never on merit deserved command of the XI Corps, still led it, though after Chancellorsville its three divisions were the smallest in the army. Slocum still

Below: Some of Meade's men referred to him as an old snapping turtle, in part due to his less than handsome visage, as well as for his much feared temper, which could be turned on anyone.

had the XII Corps' two divisions, and Pleasonton commanded three divisions of cavalry. Every corps had a brigade of artillery attached, while there was also an army artillery reserve with an additional five brigades of batteries. In all, what with detachments left along the march to date, its numbers now ran to about 95,000.

The very date of his appointment, Meade told Washington that he would march directly for the Susquehanna, keeping always between Lee and Washington. He did not intend to move directly for Lee, but rather wished to be in a position to force Lee to turn and fight him on ground of his own choosing. By dawn on June 29 Meade had the whole movement planned and under way, intending to cross into Pennsylvania the next day if necessary. Then he learned that Longstreet and

Above: Gettysburg's Lutheran seminary stands at far left in this 1863 view showing some of the ground that the armies contested on the first day, as viewed from Gettysburg along the Chambersburg Pike.

Hill were at Chambersburg, moving toward Gettysburg, while Ewell's corps occupied both Carlisle and York. Thus Lee was spread out on a triangular line more than fifty miles long at its base between Chambersburg and York, and extending twenty-five miles northward to its apex at Carlisle. Elements of Ewell's corps were within striking distance of Harrisburg, the state capital, barely twenty miles above Carlisle. If they took it and destroyed the Baltimore & Ohio bridge they would disrupt the union's major east-west transportation route. Should Lee get his whole army north of the Susquehanna, he could prevent Meade from crossing in his rear and wreak havoc on the North. Thus Meade's rapid march toward Lee's rear, to make of his own army such a threat that Lee must turn from his invasion to deal with Meade. And now, seeing the Army of Northern Virginia spread out at those three points of the triangle, Meade could also see that if Lee should try to concentrate his army for a battle, the shortest route for all three arms would be for them to move to their center – almost exactly at the Gettysburg crossroads.

Rapidly the momentum of events began to take over. Just as Meade put himself in a position to threaten Lee's rear, Lee himself had decided to abandon attempts to get across the Susquehanna, though he came close enough for elements of Ewell's command actually to cool their feet in its waters. Lee now decided that he must concentrate his scattered corps. Meade, meanwhile, deciding that his foe was gathering his forces to strike a blow at the Army of the Potomac, chose several possibilities of battlefield, preferring a line along Pipe Creek in northern Maryland that offered excellent defensive terrain. Nevertheless, in the hope of striking elements of Lee's army and defeating them in detail before they were reunited, he issued orders directing a concentration of his own corps at Gettysburg. The result is that by dawn of July 1, both armies were rapidly converging on the same little town, Lee swooping down from the north by four major

Above: An 1863 view of Gettysburg, made not long after the battle, and showing the town much as the Confederates saw it when they advanced into the streets on that first day – July 1.

Below: General Reynolds fell in or near the woods on the right, while McPherson's Ridge stands in the background. The man in the straw hat has long been believed to be photographer Mathew Brady.

Parallel to it, and commencing immediately below the town, ran Cemetery Ridge, its northern end punctuated by Cemetery Hill, and its southern end, two miles distant, by a rocky, wooded eminence called locally Little Round Top, and just below it by the even bigger Big Round Top. The roads over which Lee's corps were converging would bring them in from the north, northeast, and northwest, directly toward the town and the upper reaches of Seminary Ridge. Meade's columns, marching from the south, southeast, and southwest, would almost all be arriving on either side of the Round Tops, with the high ground of Cemetery Ridge before them immediately for the taking. Only Reynolds with his I Corps was marching in from almost due west, on a route bound to make him collide with Heth as he returned on the Chambersburg Pike from the northwest.

The first guns sounded shortly after dawn, as Heth moved toward the outskirts of town only to encounter pickets from General John Buford's brigade of cavalry posted on the Chambersburg Pike. Obviously heavily outnumbered, Buford sought only to fight a delaying action, to slow Heth's advance while Reynolds came up. Reynolds himself arrived several hours later, around 10 a.m., and begged Buford to hold on until his first division should arrive. Reynolds at once apprehended how important Gettysburg would be for Meade, and started issuing orders designed to erect a defensive line to hold out until the rest of the army could come up. Reynolds was practically fighting on home ground here, being a native of Lancaster, a town just forty miles to the east.

When the first regiments began arriving, the general himself led them into place one after another. All along it was evident that he could do no more than fight a holding action, and then a chance bullet from a Rebel gun struck the valiant Pennsylvanian in the head, killing him in the saddle. His subordinate Abner Doubleday immediately assumed command. Each side gained some success, turning back the right wing of the other, and Heth found himself in a quandary. There were more Yankees here than he expected, and quickly the fighting escalated as such engagements tended to do. Hill sent forward reinforcements, and before long, what had been

routes, and Meade moving up from the south on five separate roads. All told, there were eighty-eight brigades of infantry and over 600 cannon on their way to Gettysburg – 93,500 Federal soldiers and about 75,000 Confederates. Because of their advanced positions, Reynolds and A.P. Hill would likely be the first to reach the little town that had never expected the war to come to its doorsteps.

In fact, men both blue and gray had skirted Gettysburg the day before, when Heth and his division approached in the hope of obtaining supplies and shoes, and some of Pleasonton's cavalry approached from the opposite end of town. Only Heth's withdrawal prevented the small detachments from coming to blows, and when the Rebels backed away, the Federal horsemen held the town. But then that evening Heth asked his

commander Hill if he could return to Gettysburg the next morning, still intent on there being material there worth the taking. Hill agreed.

Gettysburg sat some thirty miles southwest of Harrisburg. It was not an old town by Pennsylvania standards, nor a large one, and enjoyed few advantages to its benefit other than its being the chance converging place of a number of important roads leading more or less directly to most of the major cities in the region. Indeed, it was nearly impossible to move east to west or north and south through Pennsylvania without passing through the sleepy town, and all traffic south from Harrisburg had to tread its dirty streets. Ridges and hills dominated its landscape. Less than a mile to the west ran Seminary Ridge, running north to south for more than two miles.

intended as merely a foraging raid on a scantily defended town had turned into a substantial firefight, magnetically attracting more and more men to the front. By noon, as Reynolds spent the last minutes of his life expanding his defensive line, Hill and Heth, wisely or unwisely, were already deciding for Lee where the great battle of the campaign would be fought.

Not long after Doubleday assumed command, and while the Federals enjoyed better than expected success in holding back the growing gray lines, Howard arrived on the field. Being senior, he immediately assumed command, with hardly any idea of what to do. He had seen Cemetery Hill on his way to the front, however, and well

recognized the strategic importance of its commanding height. Making it his headquarters, he hurried his divisions on their march to the fight, no doubt mindful that after their disgrace at Chancellorsville, the veterans of the XI Corps had a score to settle with Lee, and something to prove to themselves and the army. By early afternoon the entire I Corps was on the field, and two of the three XI Corps divisions, with the third soon to arrive. But by now Heth's full division was on the line, with Rodes' rushing into the fight as well, and Early's division of Ewell's corps coming in from the northeast. The result was that, by 3 p.m., the Federals were faced with enemies on a wide front in a semicircle extending from due

west, all across to the northeast, and in numbers that for the moment outnumbered their own.

Rapidly the fighting became intense as the Rebels pressed to envelop both Yankee flanks. Rodes led the attack, sending all five of his brigades – 8,000 strong – forward, with Heth on his right and Early on the left. In the first action, Howard's line stopped the attack and severely mauled Rodes' center. But the Confederates pressed again, and in time Early got around the right rear of the XI Corps, striking them at their most vulnerable spot, just as Jackson had done less than nine weeks before. The attack crushed the right of the Federal line, leaving the defenders no choice but to pull back from their position

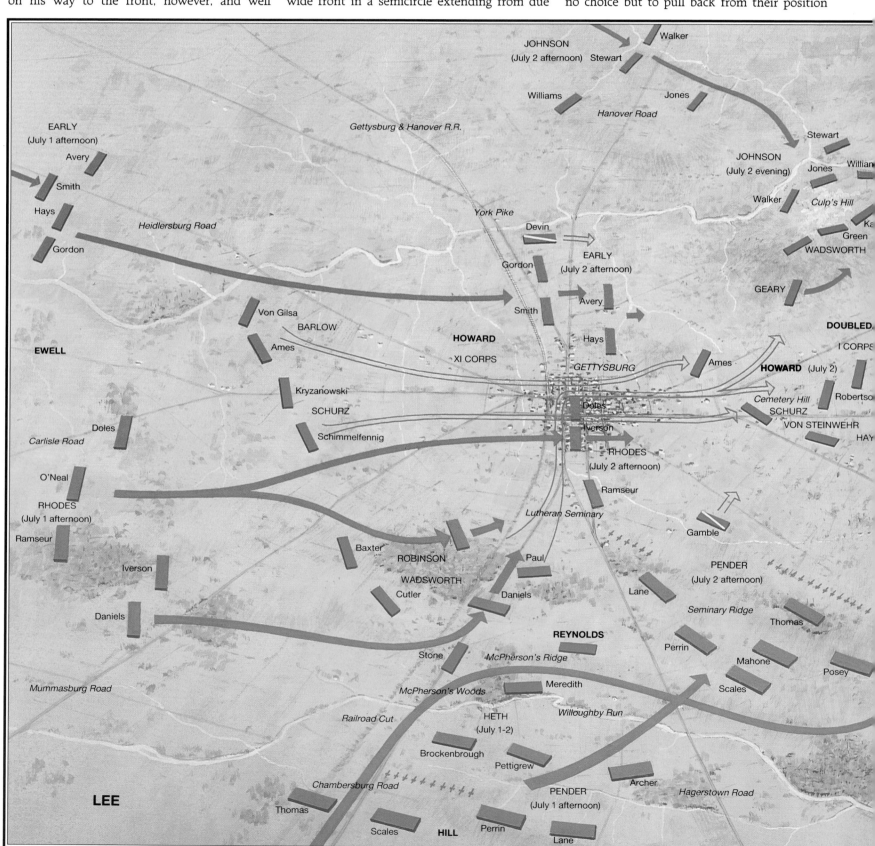

Battle of Gettysburg: the first two days. July 1-2, 1863

The first two days at Gettysburg were a study in the gradual development of a major battle, as Lee sought to develop a successful attack and Meade sought to counter each move and prepare for the next.

Early on July 1 advance elements of the Army of the Potomac first entered the town from the south, and soon discovered A.P. Hill's Confederates approaching from the north and northwest. The Rebels were merely on a

foraging raid, but as soon as they first encountered the Federals north of twon, at far lower left, they determined to fight. Here in and around McPherson's woods and along the Chambersburg pike, the battle rapidly developed as more and more troops gravitated toward the sound of the guns. Early's Division approached from the northwest, upper left, and Rodes from the north, left center. meanwhile the Federal I Corps came in on the

Fairfield Road, and the XI Corps rushed up along the Emmitsburg Road, lower right center.

The fighting north of the town became furious, Reynolds fell in action, Doubleday took command, and despite a brutal punishment inflicted on the Rebels in the Railroad Cut, left lower center, the Federals were finally pushed back through the town in disorder to Cemetery Hill, center. Here, at last, Hancock took over and

began setting up a desperate line of defenses on the northern and eastern slopes of the hill, with his right extending back along the west face to Cemetery Ridge.

Reinforcements arrived throughout the night, and by July 2 the armies were substantially complete. Neither Lee nor Meade had planned to give battle over this particular piece of Pennsylvania, but the troops on the ground – and perhaps the hand of fate – had chosen

otherwise. Their positions were such that neither could maneuver to another position of their own choosing.

Having both reached the field late on in the day, the commanders of North and South alike spent that first night in reconnaissance and organization of their forces. Both knew that the fight on the following day would be crucial. Fighting did not start until late in the afternoon of the second day, when Longstreet launched his

attack on Meade's left and Little Round Top. Sickles' injudicious placement of his III Corps in advance of the Federal line made him a perfect Confederate target, and in the Peach Orchard fighting, lower right, the III Corps was almost destroyed. Rebels then moved on to Little Round Top, right center, but spirited resistance held it for the Union. Late that evening another attack on Cemetery Hill and Culp's Hill, upper left center, also failed.

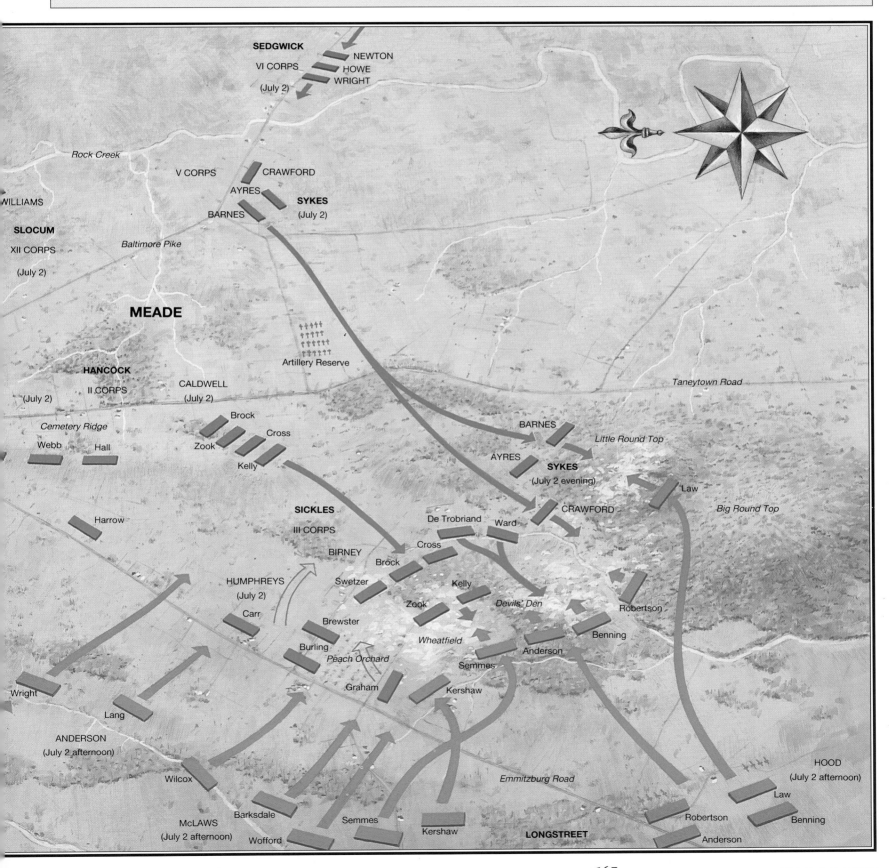

north of Gettysburg. Otherwise the Rebels could get between them and the town and cut them off from the rest of Meade's army advancing from the south. But then the simultaneous advance of the rest of the Confederate line turned any attempt at an orderly withdrawal into a shambles. It was not the panicked rout of Chancellorsville, especially for the I Corps, but neither was it a calculated affair. Instead, the men simply broke up in regiments and companies and raced back to the streets of Gettysburg, sometimes stopping to deliver fire on the way. Howard sent instructions to reform on the slopes of Cemetery Hill, but he did not need to. It was the only obvious choice in the direction in which the Rebels were driving them.

The rush through the town proved to be even more disorganizing, as the men scrambled to reach the high ground below. The narrow streets broke up their units. Some left ranks to plunder, others to hide. Wagons and animals jammed crossroads. Men simply found themselves lost and unable to find their way out. And all the while the pursuing Rebels were right behind them. Hundreds were captured, and others had to take refuge to escape being made prisoners. Brigadier General Alexander Schimmelfennig, commanding Howard's XI Corps third division, was cut off from his command during the retreat and forced to hide in a small pig barn off Baltimore Street, where he remained for the rest of the battle, being brought food by its owners when they came to feed the swine.

Happily, when the ragged remnants clambered up Cemetery Hill, they found a fresh brigade on the summit, with the cool composed Winfield Scott Hancock there in person setting up a new line of defense. Meade had sent him ahead with

Above: The very unlucky General Alexander Schimmelfennig of the equally unlucky XI Corps got cut off in Gettysburg when his command evacuated, leaving him trapped in a pig barn for the rest of the battle.

Below: Culp's Hill, the eastern anchor of the fishhook-shaped Federal line, as viewed in 1863 from Cemetery Hill. Though seriously threatened, Culp's Hill held on, saved by nightfall and timidity.

orders to take command on the scene, since Meade wisely had little use of Howard's skills. To his credit, Howard uncomplainingly yielded and cooperated admirably with Hancock thereafter. The II Corps commander's presence put spine back into the disheartened men of the I and XI Corps as he dug in determined to hold his ground. He sent a division off to the east of Cemetery Hill, to Culp's Hill, upon which he decided to anchor the right of the Federal line, and then, assisted by Brigadier General Gouverneur K. Warren, Meade's chief engineer, Hancock prepared to fight to the end if need be.

By now it was perhaps 5:30 in the afternoon, and more reinforcements were approaching, the XII Corps coming in towards Culp's Hill, and Sickles and his III Corps at last marching up the Emmitsburg Road from the south towards Hancock's much more vulnerable left, now trying to stretch down to Cemetery Ridge. Nightfall would be upon them in a couple of hours or more, and Hancock felt some assurance that he could hold his position until dark, by which time Meade would arrive to take command and decide whether to fight or retire. It had been a near thing for the Yankees that afternoon. Only the valiant yet costly fighting of the I Corps had held the line long enough to have Cemetery Hill ready to occupy. It cost them dearly, with nearly 65 percent casualties out of the 8,500 Yankees who started the battle. Yet they managed to inflict almost 40 percent losses on the 16,000 men of Hill's corps who opposed them. The I Corps almost ceased to exist, but won imperishable glory by its heroic stand.[4]

Lee himself reached the battlefield around 2 p.m., but did not exercise much direct command of the following assaults became he had not yet

seen the ground, whereas Heth and Rodes knew it better. By late afternoon, with Hill now present, too, Lee started to exert more control, deciding that he would have to fight Meade here. Yet he did not yet have a battleplan – there had been no time to formulate one – and not all of his army was yet on the scene. He needed time to make a plan and for Longstreet and Ewell to arrive. Nevertheless, in looking at the Yankees as they withdrew up Cemetery Hill, he realized immediately that he must drive them off that important elevation before Meade got sufficient reinforcements on the scene to make the hill impregnable. But when he sent the just-arrived Ewell a request to attack the hill, it was just that – a request and not an order – and Ewell decided that he could not achieve the goal and did not try. Considering that his own men were disorganized after passing through Gettysburg, and that he was receiving constant reports of Federal reinforcements advancing toward Culp's Hill on his left, he was not unjustified in deciding not to risk a hastily organized assault, late in the day, against a formidable position.

Thus the first day of the gathering fight came to an end, and rapidly balances started to shift. Though barely 12,000 Federals remained of the 18,000 who started the day, more rapidly arrived. The XII Corps came on to the east side of the field. Sickles started coming into view on the Emmitsburg Road, and shortly after twilight settled in, there were perhaps 27,000 in the blue line, now commanded by Slocum. Meade himself rode through much of the night, only arriving on Cemetery Hill before the next dawn. The contrast between him and some of his predecessors in army command became immediately apparent. While McClellan or Hooker would have struck a

Above: The gatehouse of Evergreen Cemetery on Cemetery Hill. Fighting on July 2 and 3 raged just over the brow of the hill in the rear. Four months later Lincoln would deliver his immortal address here.

Below: This small house behind the east slope of Cemetery Hill served as Meade's headquarters after he arrived onto the scene of battle. Here, in council of war with his staff, he decided to stay and fight on July 3.

heroic pose and uttered some bombastic boast to show his courage and inspire the army, Meade simply asked his senior generals on the field about their positions. Hearing that they felt themselves strongly placed, he said simply that this was good, for they were all too committed to pull out now. He would fight Lee here and that was all there was to it.[5]

The previous day Meade had directed his efforts to getting all of his corps on the several roads towards Gettysburg, leaving it to his capable subordinate Hancock to manage the fight. Once on the scene, the commanding general immediately made a personal reconnaissance of his position and the available ground for the coming battle. He rode almost to Little Round Top, then turned north along Cemetery Ridge, to Cemetery Hill, and then east to Culp's Hill. He did not fail to see that along this route he could occupy an almost unbroken ridge of high ground, roughly in the shape of a fishhook: his left near the Round Tops at the "eye" and his line running north up the shank until it curved to the right to the barb at Culp's Hill. It meant a line three miles long, but it would have the advantage of "interior lines" – Meade could shift troops from one end of his line to the other by cutting across the axis of the arc, while Lee, whose line would naturally have to be longer, and with fewer men, would have to march men the long way around the arc to reposition his forces.

When the sun began to peer over the back slopes of Cemetery Ridge, Meade had his I and XI Corps remnants on Cemetery Hill, and two divisions of Slocum's XII Corps on their right at Culp's Hill. On Howard's left Hancock put his three II Corps divisions in line along Cemetery Ridge, and on his left sat Sickles and the III Corps. Only Sickles was not sitting where Meade wanted him. Through a combination of imprecise orders from Meade, and Sickles' own palpable lack of competence, the III Corps was not placed on the ridge south of Hancock where it would enjoy the best defensive position. Instead, for reasons that made sense to few but himself, Sickles advanced his two divisions nearly three-quarters of a mile beyond Hancock's line, extending from Hancock's left out into the lowland

to a peach orchard barely half a mile from the Confederate positions on Seminary Ridge, and then arcing back in front of a wheatfield on a tangled rock outcropping known locally as Devil's Den near the foot of Little Round Top. Meade did not learn of it precisely until nearly 4 p.m., when he rode to inspect his left and saw for himself what Sickles had done. He was furious, for Sickles had put himself in a position where he was terribly vulnerable to the enemy, and in which he could only be reinforced with difficulty. He jeopardized the entire left of Meade's line, virtually making pointless the plans the commanding general had been making for an offensive. Sickles expressed regret and offered to move his

corps back to the ridge. "I wish to God you could", said Meade, just as Confederate cannon announced the coming of an attack from Lee that belatedly started the day's fighting, "but the enemy won't let you."[6]

The Confederates had waited far too long to commence the battle. Lee still enjoyed a numerical advantage, as well as one in morale, when the sun set the night before, and even though Meade's reinforcements in the dark strengthened his position, every hour that Lee delayed only gave the foe more time to choose the best ground. The trouble was, unlike Antietam or Fredericksburg or Chancellorsville, where Lee had enjoyed plenty of time to study the terrain, Gettysburg

was entirely new to him. He needed time to study it. Longstreet did not want to fight there at all, not liking the look of Meade's position. But Lee felt that they had to give battle now, while Meade's army was incomplete. It might be the best opportunity they would ever have to defeat him in detail. All through the night he talked with his generals and considered plans, finally deciding that the place to hit Meade was on the Yankee left, exactly where Sickles would foolishly expose his corps. This task would fall to Longstreet, whose corps was slowly coming up and was to extend itself south all along the length of Seminary Ridge. Hill's corps, battered from the previous day, would hold the center on the ridge,

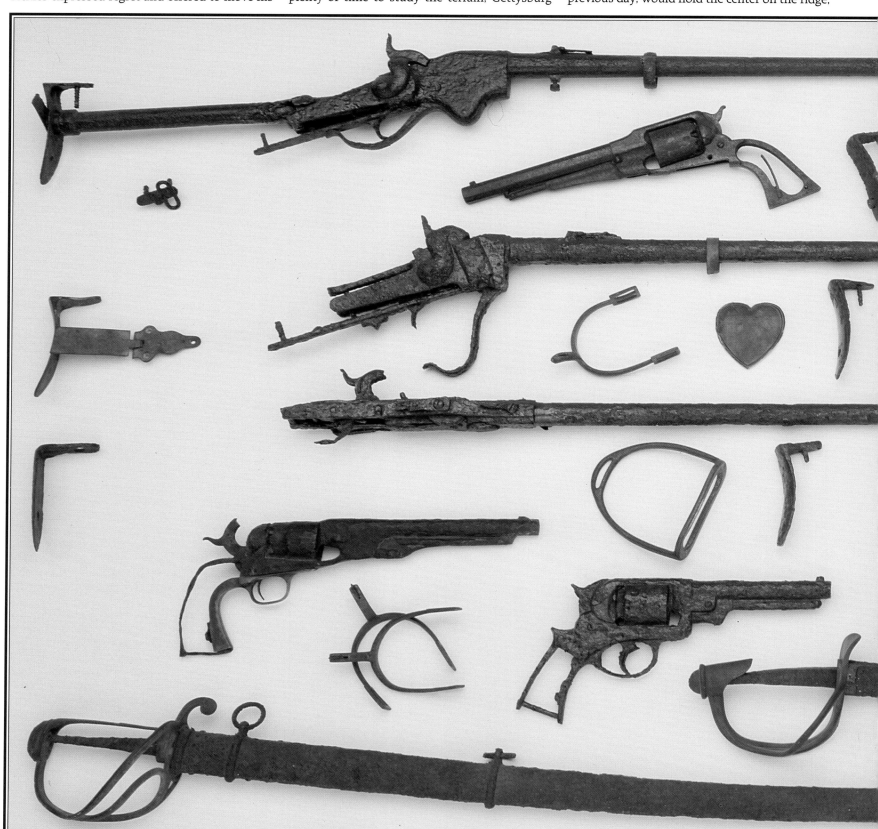

Excavated Union Cavalry Relics

1. 1861 Spencer carbine with full magazine, found at Dinwiddie Court House, Va
2. Smith single shot carbine, found at Cold Harbor, Va
3. Remington New Model Army revolver
4. Horse bit
5. Sharps Model 1863 carbine, found in the Wilderness, Va
6. Sharps Model 1863 carbine, found at Cedar Creek, Virginia
7. Brass spur
8. Brass martingale plate
9. Brass buckle and tip and iron hook for a carbine sling
10. Hall-North Model 1843 carbine, found at Union Mills, Va
11. Wesson carbine, found near Guinea Station, Va
12. Iron stirrup
13. Colt Model 1860 Army revolver
14. Pair of brass spurs
15. Starr Army revolver
16. Curry comb
17. Brass stirrup
18. Horse shoe
19. Complete horse bit
20. Model 1840 Heavy Cavalry saber, found at Bermuda Hundred, Va
21. Model 1833 dragoon saber with scabbard, found at Cold Harbor, Va
22. Sword belt plate
23. Coffin-handled Bowie knife
24. Picket pin

Artifacts courtesy of: Wendell Lang Collection, Tarrytown, N.Y.

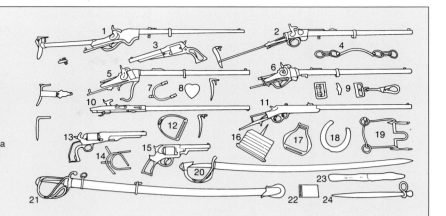

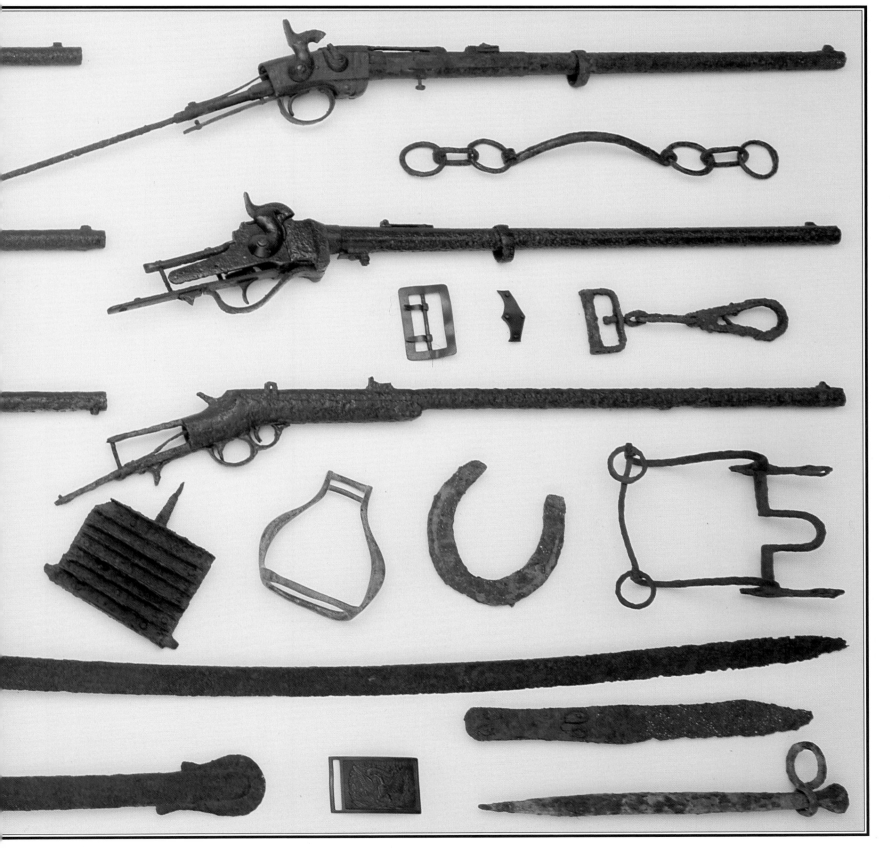

reaching almost to the town, where Ewell took up the line, swinging around in an arc matching Meade's. Ewell would make a demonstration against Cemetery Hill and Culp's Hill when he heard Longstreet's attack, both to occupy and keep Yankees in his front from going to Sickles' aid, as well as to press on in a concerted attack should an advantage be gained.

But so much depended upon time, and Longstreet's corps proved agonizingly slow in arriving. Still, both Hood and McLaws and their divisions were on hand by 9 a.m. Lee gave Longstreet his instructions for the manner of the attack to come, but then "Old Pete" balked. Opposed to the attack since the night before, he argued. Some later claimed that he intentionally delayed in hopes of getting Lee to call off the attack, though with McLaws and Hood still four miles from Seminary Ridge, even such a short movement required four or five hours, including final alignment, placement of artillery, and the other details involved in an assault. Confusion of troops on the road made the whole process take even longer, and in the end Longstreet was not in position and ready to attack until 4 p.m.

When the screaming Confederates rushed forward, their attack was devastating. Hood, on the far right, swept up from the southwest, crossed Plum Run, and smashed into Sickles' left between Devil's Den and the wheatfield, while McLaws hit the jutting angle of Sickles' line in the peach field from two sides at once. Longstreet battered the Federals with a barrage from fifty-four of his cannon. Immediately Sickles' men put up a stiff resistance that in the ensuing three hours would take on a grim determination. Almost from the first, their position turned critical, as one of Hood's regiments spotted the importance of Little Round Top. Big Round Top was too wooded to allow artillery to reach its summit. The smaller hill was different, however, and it appeared to be unoccupied. Guns placed there could fire down on the entire Yankee line as far north as Cemetery Hill. But as the Rebels raced up its slopes to secure it, the timely arrival of a brigade from the V

Above: Sometimes called the Saviour of Gettysburg, Major General Gouverneur K. Warren, then an engineer, was among the first to spot the weakness on Little Round Top and rush men there to hold it in time.

Below: Little Round Top, photographed in 1863, as seen from the valley between it and Seminary Ridge. Cannon on its heights could command the entire Union position, making it vital to Meade.

Corps, just now rushing to the field after a long march, allowed Meade to hold the position until more reinforcements could reach the scene. In a furious little fight from behind boulders and fallen trees, blue and gray battled bitterly for the hill before the Rebels were forced eventually to retire from the scene.

Even while fighting for their lives, the Yankees on Little Round Top could see the III Corps being cut to pieces out on the lower ground to the northwest. Six full brigades and portions of a seventh hammered at Sickles. Almost immediately the casualties became dreadful. An artillery shell carried away one of Sickles' legs. Hood took a desperate wound that left one of his arms useless for the rest of his life. The men fought on almost until dark, the Federals slowly giving way, then collapsing entirely when men from McLaws' division broke through and into the rear of the right center of the shaky Yankee line. In a near panic, the survivors rushed back to the safety of Little Round Top and more freshly-arrived Union reinforcements. The III Corps had almost ceased to exist. Shortly after the battle it would be disbanded.

Despite his success in destroying Sickles, Longstreet saw that it was too late for him to follow up the advantage and mount a new attack on Little Round Top itself. His men were exhausted after four hours of solid fighting, and Union reinforcements appeared to be too numerous. Besides, the fighting had spread up along Cemetery Ridge for some distance, and he had no fresh troops to lend strength to his own battered divisions. Anderson's division of Hill's corps went forward on McLaws' left, and spent most of the afternoon in bitter fighting with elements of the I, II, and III Union corps, but especially with the intrepid Hancock, who managed the defense of the ridge brilliantly and even managed a spirited counterattack late in the day that reversed some of the gains made against Sickles. When darkness fell, the Union line, though battered, was intact from Cemetery Hill south to Little Round Top. A simultaneous threat had come on the right of the

Headquarters Designation Flag, Army of the Potomac, July, 1863

This is the headquarters designation flag of the 2nd Brigade, 2nd Division, 2nd Army Corps, Army of the Potomac, commanded by Brigadier General Alexander Stewart Webb. The brigade of approximately 1,220 men was made up of four Pennsylvania infantry regiments: the 69th, 71st, 72nd, and 106th. On the third day of the Battle of Gettysburg, July 3rd, 1863, the brigade held a position on Cemetery Ridge near "the copse of trees", a position which was to become the focal point of the Confederate assault later known to history as Pickett's Charge. The brigade was organized in two lines, the 69th and 71st Regiments were posted behind a stone wall and breastwork to the front, while the rest of the brigade held ground some 40 yards behind on the ridge overlooking them. The Confederates of Pickett's Division (1st Corps A.N.V.) led by General Armistead marched up the ridge and reached the stone wall but got no further. Withering fire from the ridge and into the Rebel right flank (from regiments of Colonel Norman Hall's 3rd Brigade) broke the attack. This flag was carried at that action: the high watermark of the Confederacy

Artifact courtesy of: The Union League of Philadelphia

line when Ewell moved against the east slope of Cemetery Hill and Culp's Hill. He sent his whole corps forward around 6:30 p.m., not in a demonstration as ordered, but in a general attack, haphazardly supported by Rodes' division of Hill's corps. The whole affair was ill-managed and uncoordinated, and at the end of the day's fighting, Culp's was still securely in Yankee hands, though threatened, while only a small piece of the down slope of Cemetery Hill fell to Early and his Virginians. The firing of skirmishers and artillery continued long after nightfall, not dying away until around midnight when finally the thousands of battered and exhausted survivors could fitfully try to sleep.

There was to be little sleep for the two commanding generals, however. Well after dark Meade gathered his senior commanders at his headquarters behind Cemetery Hill. For almost three hours he listened to their thoughts and experiences of the day. Finally they responded to three basic question: should they hold their position or retire; if they remained, should they attack on July 3 or remain on the defensive; and if they chose the defensive, how long should they stay in place awaiting Lee's offensive? The generals were almost unanimous in recommending that they stand their ground, stay on the defensive, and give Lee no more than another day to attack them before they either took the offensive or else

moved away from Gettysburg. The decision taken – and Meade made it himself – they spent the rest of the night readying themselves to meet what came on the morrow. Since Lee had tried to turn both his right and left flank without success, Meade suspected that the enemy would strike his center next.

He guessed well. Lee was frustrated – frustrated by Longstreet's and Ewell's reluctance on the first two days' fighting, frustrated by struggling for the first time on ground he did not know well, frustrated by his own poor health, and most of all frustrated by the steadfast resistance of a Yankee army that he had so many times before put to rout. As a result, he was fighting arguably

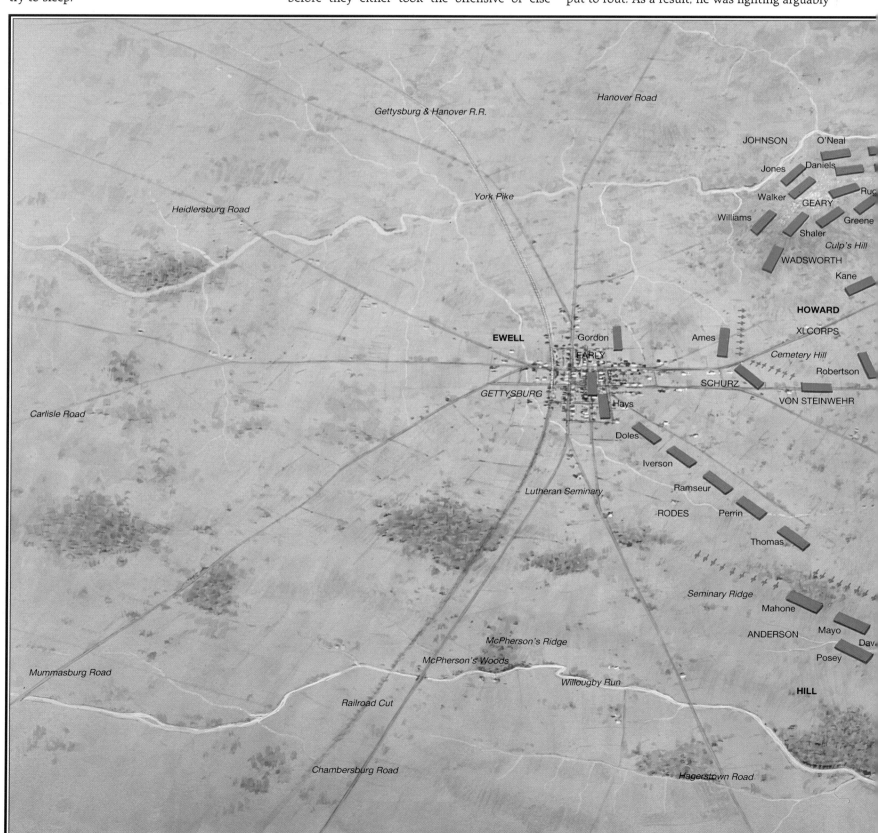

Battle of Gettysburg: the final day, July 3, 1863

All night on July 2, Lee and Meade prepared their armies for what was to come. Meade anticipated that his foe would launch an attack on his center, and heavily strengthened his line and husbanded his reserves. Lee, at the same time, concentrated the mass of his artillery along Seminary Ridge, lower center, intending with it to batter Meade's center before his assault.

When the artillery barrage opened, it aimed at a copse of trees on Cemetery Hill, near right center. But the artillery proved ineffective, largely overshooting the target. Thus, when Longstreet sent forward the divisions of Pickett and Pettigrew, some 15,000-strong including supports from other commands, the Federal artillery was still in place.

Heroically the Confederates marched across a mile of open ground to rush up the slopes of Cemetery Hill and Cemetery Ridge. As soon as they came in range, they began to take a fearful shelling from Yankee cannon, and then vicious volleys of small arms fire from the bluecoat infantry. Still they pushed on until it came time for the final push. Rushing up the rocky slope of the enemy position, the Rebels flung themselves against men posted behind stone walls and hasty earth and rock mounds. Across a battle front half a mile in width, the struggle became intense. In places, for the first time in the war – and for many their last – men actually engaged in hand-to-hand combat. Yet the Federals, tough veterans of the II Corps and elements of the old I Corps, held their ground. Only portions of Armistead's and Garnett's brigades actually penetrated the Federal line, one of those getting through being General Armistead himself, who got far enough to lay his hand on a Yankee cannon before he was hit.

Inevitably, however, the weight of Union fire and their excellent defensive position told, and the Confederate tide receded toward Seminary Ridge, the grandest frontal assault of the war a failure, and with it Lee's battle.

Other Confederate moves – by Ewell's 2nd Corps, and Stuart's cavalry – were also stopped by Federal troops. On July 4, both armies were too exhausted to do little else but collect their wounded and begin burying their dead. The weather deteriorated towards the end of that fourth day, and under cover of darkness and heavy rain, Lee set the Army of Northern Virginia marching back towards the Potomac and Southern soil. The Army of Northern Virginia's last invasion of the North was over. Given the situation, many in the North criticised Meade for not pursuing Lee, but the commander knew that his army needed time to recover. Gettysburg cost Lee over 20,000 casualties, a fourth of his army. Meade lost 23,000, a fourth of his own.

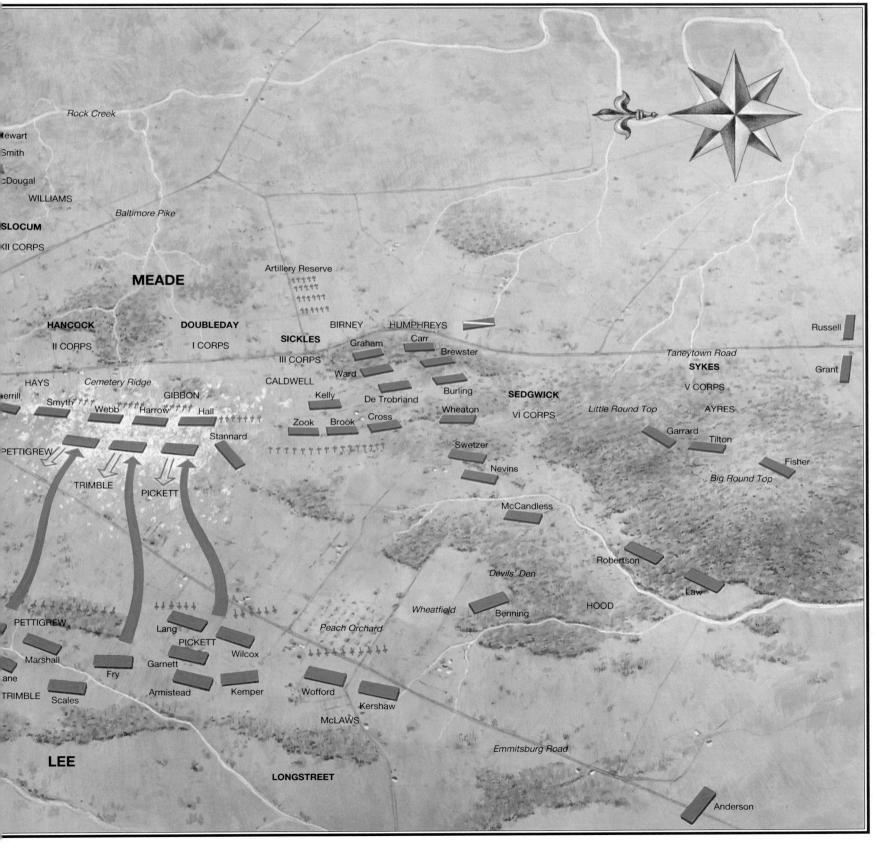

his worst battle of the war. He issued orders late on July 2 for a virtual repeat of that day's efforts, coordinated attacks on Meade's right and left, to launch just after dawn the next morning. But it all went wrong before it was fairly started. Longstreet, still intimidated by the Yankee position, used latitude in his orders to try a wide sweeping flank march instead, and Lee only discovered and countermanded the move at about 6 a.m., July 3. By then it was too late, for Ewell had already started his renewed assaults on Culp's Hill. Instead, Lee decided to have Longstreet make a hammering frontal assault on Meade's center on Cemetery Ridge. His men would have to march more than a mile, largely out in the open, across broad fields of milo, then up the slopes of the ridge and into the face of Yankee guns. Fixing a clump or "copse" of trees on the otherwise bare crest as the aiming point, he told Longstreet to take 15,000 men and drive themselves as a wedge into the Union center. If successful, they would have Meade's right half trapped between themselves and Ewell and Hill. Again Longstreet objected, this time vehemently, but Lee was adamant that the plan go ahead.

Sergeant and Private, 22nd Regiment, New York National Guard, U.S.A.

The 22nd New York National Guard presented the same contrast and confusion of colors in their dress that the 16th Virginia and other Confederate regiments presented. This was largely due to its being a pre-established unit before the war broke out, one that carried its old uniform and equipment into the new conflict.

On close examination, of course, the first uniform of the 22nd bears many similarities to the Confederate standard issue, excepting the red cuffs and collar band. Naturally this caused confusion at First Manassas and elsewhere, and after Antietam the regiment switched to the regulation blue chasseur blouse on the sergeant. They kept their distinctive 2-banded Enfield rifles and sword bayonets, however, and carried their regimental number on their kepi and their company letter on their beltplate. After Gettysburg, the 22nd was not seen again in the field.

Given no choice, Longstreet assembled his freshly arrived division of General George E. Pickett, along with several units on loan from Hill's III Corps, especially Heth's division now led by General James J. Pettigrew following Heth's suffering a disabling wound. Brigades from Pender's and Anderson's divisions would also join in, but Longstreet would command the whole. In the end, it came to 13,500 men. Lee wanted them to advance only after a massive barrage from 159 cannon spread all along Seminary Ridge had silenced or driven away defending artillery on Cemetery Ridge. At 1 p.m. the barrage began.

Almost two hours later the Confederate guns fell silent. Surmising correctly that the shelling presaged a massive attack, Hancock had Federal artillery conserve their fire to meet the assault. But Southern commanders interpreted this to mean that they had silenced the Yankee batteries. Against his better judgment, Longstreet, when pressed by Pickett, finally nodded his head to send the assault forward. All who witnessed it, North and South alike, admitted that it was one of the grandest sights of the war. Across a front more than a mile wide, a sea of red banners fluttered above the glistening bayonets of the flower of Lee's army. It was the greatest and grandest infantry assault of the Civil War.

It was also a terrible mistake. For all his brilliance, Lee suffered one deadly flaw. When desperate, when he had tried everything else, he would resort to the frontal assault, despite his own and others' repeated experience of bloody failure of such a tactic. The deadly accuracy and firepower of the infantryman's weapon in 1863 made such an attack virtually obsolete when directed against well-placed defenders on good ground. But Lee was out of ideas.

Thanks to the momentum of such a mass of men, and to the valor of them all, Pickett's and Pettigrew's men swept in brilliant ranks over the ground before them. When they came in range of Yankee guns, fire hit them from all sides. First came the shelling, then the musketry. Throats shouting the Rebel yell, the Virginians and North Carolinians and others hurled themselves toward the wall of blue flame atop the ridge. The closer they got, the fewer their number. General Lewis Armistead was almost at the forefront of the first few hundred who actually made it to a stone wall in front of the clump of trees. He put his hat atop his sword, held high over his head, and called for his men to follow him onward. Perhaps 150 got over the wall; few returned from it. Behind them Armistead lay with a mortal wound. It was as close as they got. Gradually, then with increased momentum, the defenders pushed back the Rebels, who then had to run a deadly gauntlet of fire for another three-quarters of a mile to get back to Seminary Ridge. Lee watched them stream into his lines, tears in his eyes as his told them that it was all his fault. Behind them lay one general killed, one mortally wounded, and another wounded and captured, and nearly half of the attacking force either killed, wounded, or taken prisoner.

Meanwhile other efforts had failed as well. Ewell's attack on Culp's Hill came to nothing well before noon, thus leaving Meade free to shift troops to support Hancock's center against the great assault. Lee had also sent the tardily arrived Stuart on a ride around Meade's army with a view to striking his center from the rear just as Longstreet's assault hit, but Federal cavalry stopped Stuart several miles away.

Above: Major General Winfield Scott Hancock, one of the true heroes of Gettysburg, selected the Cemetery Hill position and held it on July 1, then bore the brunt of the assault of July 3, himself being wounded in the action. There was no finer Union general.

Below: The aftermath of the battle would remain to be seen for months, even years. Thousands of horses had to be buried, and tons of ruined equipment remained to be cleared from the field. Worst of all, 6,000 men lay dead – the war's worst carnage.

And that was an end to it. The bloodiest battle in American history, and the one to spawn the most enduring controversies, had cost Lee a staggering 20,451 in casualties, with losses among his officers so high that his command system would never entirely recover. For Meade, too, the damage was great – Reynolds dead, Hancock seriously wounded, and 23,049 casualties of his own. Between the two armies, the dead alone numbered more than 6,000.

As a result, the armies simply sat and glared at each other on July 4, Lee too proud to leave the field, and Meade too battered to follow up a clear advantage. Heavy rain set in that afternoon, making further fighting impracticable anyhow, and during the night Lee silently withdrew and marched for the Potomac, his invasion and all its hopes at an end. He would not set foot on Northern soil again. Meade, though criticized for failing to pursue Lee vigorously, had still achieved what no one else had done so far in this war, not even McClellan at Antietam. He had given the entire Army of Northern Virginia a crushing open field defeat. It was to be Lee's last offensive operation of the Civil War.

References

1 Armistead L. Long, *Memoirs of Robert E. Lee* (New York, 1886), pp.267-69.
2 Edward J. Nichols, *Toward Gettysburg: A Biography of General John F. Reynolds* (University Park, Pa., 1958), pp.220-23.
3 Marsena R. Patrick Journal, June 17, 19, 1863, Library of Congress, Washington, D.C.
4 Edwin B. Coddington, *The Gettysburg Campaign* (New York, 1968), pp.306-307.
5 *O.R.*, I, 27, pt.1, p.705.
6 Coddington, *Gettysburg*, p.346.

CHAPTER TEN

CHICKAMAUGA & CHATTANOOGA

‾ SEPTEMBER 19 – NOVEMBER 25, 1863 ‾

The triumph at Gettysburg seemingly exhausted the two great armies in Virginia, and well it should have, with each suffering nearly a third of its numbers in casualties. For months the two battered and bleeding forces simply studied each other, as Meade followed Lee slowly back into Virginia. Meade launched into motion a brilliantly conceived campaign around Manassas Gap that might have yielded yet another great battle, but it misfired. Then that fall he tried again, with yet another ably planned campaign in the Mine Run area. Only the poor performance of subordinates prevented what could have been a crushing blow from falling on Lee, and at a time when Longstreet and his corps were not with the army.

Meanwhile, in the aftermath of Vicksburg, there had been little activity in Mississippi. The half-hearted Joseph E. Johnston, arriving too late to help save Pemberton, fought a small battle against the Federals at Jackson a few days after the surrender, and then withdrew. As usual, he would leave the fighting to others, and as the summer wore on he tardily sent some of his men back to Bragg. Also on the way to Bragg was Longstreet and his I Corps. The Confederates were about to try a daring concentration that could regain for them much of what had been lost in the central Confederacy.

S ETTING aside the often overblown claims of their partisans and supporters, it is more than a little interesting to consider just how few battles most Confederate army commanders' reputations are actually based upon. Lee is the exception to this, for during the conflict he led his army in no fewer than a dozen major engagements, and far more if the Seven Days' fights and those of 1864 are considered individually. No other commander of the war except Grant came even close to this. Of his fellow Confederates, Albert Sidney Johnston had just one battle, Beauregard only portions of two, and Joseph E. Johnston no more than half a dozen, most of less than major stature. Ironically, next to Lee the one Confederate commander to lead a full army in major battle more than any other was the despised Braxton Bragg. Four times he would commit his Army of Tennessee to full-scale engagement: the first two at Perryville and Stones River. With even greater irony, in his last two battles, fought within less than ten weeks of one another, he would inflict the most complete defeat ever suffered by a Union army, and then follow it by taking the most humiliating beating in the brief but glorious history of Confederate arms. First to last, the man presented the most intriguing puzzle of any general of the war.

It all revolved around a place known only by an old Cherokee name that meant something like "crow's nest". In fact, the name Chattanooga was actually aimed at the nearby towering eminence of Lookout Mountain, but when a small city of 5,000 grew up on a bend in the Tennessee River near its base, the name quickly shifted to the town. Of little consequence before the outbreak of war, its vital rail link between the Confederacy east and west of the Alleghenies gave it a strategic importance second only to Atlanta or Richmond. Control of Chattanooga offered the key to a back door into Virginia and the front door to Atlanta. The Confederacy had to hold it at all costs.

Typically, Braxton Bragg lost it in early September 1863 without firing a shot. Following his marginal defeat at Stones River the previous January, Bragg withdrew toward the Duck River. Before long, however, he relocated to Tullahoma, on the Nashville & Chattanooga Railroad, about forty-five miles west of Chattanooga, thus protecting his supply line. Here he sat out the balance of the spring and the summer, the longest period of inactivity during the prime campaigning months of any major army in the war, a period almost matched by Rosecrans and his Army of the Cumberland. Both armies remained content to refresh and refit themselves. But then toward the end of

---

Lookout Mountain, overlooking the Chattanooga valley and the Tennessee River, afforded the most spectacular scenery of any battlefield, and seemingly promised one of the most impregnable positions.

summer, Rosecrans marched tentatively south once more, intent on taking the city more by craft than force. First he conducted a brilliant campaign of maneuver that threatened to cut Bragg off from his base, forcing him to abandon Tullahoma and pull back to the defenses on the mountains around Chattanooga. Then, sending a diversionary movement to make Bragg fearful for his right, Rosecrans moved the bulk of his army to the Tennessee River west of Chattanooga. There he crossed in late August and started to move east along its bank toward Chattanooga. Bragg, his left turned, and fearing that the enemy might now strike for his rear and his communications with Atlanta, abandoned the city. On September 9 Rosecrans marched in without a fight.

But then Rosecrans misjudged Bragg. Thinking that the Rebel army was demoralized and in full retreat, he was not careful to keep his own three corps under Thomas, McCook, and Crittenden, together, or at least within easy supporting distance. Instead, he left Crittenden at Chattanooga and sent the other two on widely diverging routes in pursuit of the enemy. The long high ridges of Raccoon Mountain, immediately below the Tennessee, and the even longer Lookout Mountain further south, shielded Bragg's movements from the Yankees. Moving without sure information of the foe's whereabouts, Rosecrans became reckless. He sent Thomas through Stevens Gap in Lookout Mountain, twenty miles southwest of Crittenden at Chattanooga, and McCook nearly twenty miles farther along at Winston Gap. As a result, his right and left were separated by almost forty miles of tough country. And Bragg, in fact, had all of his forces tightly in control and almost directly in Thomas' path. All that separated them was a gap in Missionary Ridge, a valley known as McLemore's Cove, and the waters of Chickamauga Creek.

It was another Indian word, this one meaning, they said, "River of Death." If so, Bragg was ready to give meaning to the name. This general, though he only ever commanded the Army of Tennessee, never fought with the same army twice. Earlier in the year he had detached the disliked Breckinridge and his division to go on Johnston's abortive Vicksburg relief expedition, and other units had been loaned out elsewhere during the summer of inactivity. But now many, including Breckinridge, were being sent back to him. Furthermore, given the static situation in Virginia with almost no activity between Lee and Meade at all, Richmond had prevailed upon Lee to send Longstreet and most of his I Corps out to reinforce Bragg temporarily. And of course, even among those units that had never left him immediate command, Bragg's continuing war with his own commanders had led to substantial changes. Hardee, who by now loathed Bragg, had been sent to Alabama by Richmond, and partially at his own request. His corps had been broken up, Breckinridge sent west and Cleburne put under the irascible Daniel H. Hill. Hill had so fallen out with Lee and everyone else in Virginia that Davis sent him out to replace Hardee, thus placing two of the worst tempers in the army together. When Breckinridge returned, he would be assigned to Hill as well. Meanwhile, another old Bragg enemy, Leonidas Polk, remained in command of his corps, with one division under the capable but erratic Cheatham, and another led by General Thomas C. Hindman of Arkansas, now in command of what had been Jones Withers' division. Bragg also created a Reserve Corps commanded

Below: Major General Thomas L. Crittenden led the XXI Corps, and though a much better officer than McCook, he, too, would be eclipsed by the disaster that befell him and his command.

Above: Major General Alexander McCook, one of the famous "fighting McCooks" of Ohio, commanded Rosecrans' XX Corps, but gave a poor account of himself at Chickamauga that virtually ended his military career.

by W.H.T. Walker, with two divisions led by General States Rights Gist and St. John R. Liddell.

Once Longstreet arrived, Bragg would assign him command of a separate wing composed of two hastily organized corps. The first would go to Major General Simon Buckner, now returned from a Northern prison camp. Bragg would give him Hindman's division from Polk's corps – a clear blow at Polk – and small divisions led by Generals William Preston and Alexander P. Stewart. Longstreet's other corps would be commanded by Hood, recovered as much as he would ever be from his arm wound at Gettysburg, and now to lead McLaws, his own old division in the hands of Evander M. Law, and another under the Ohioian Bushrod Johnson. Even the cavalry fell

into two separate corps, one of two divisions under the erratic Wheeler, and two others led by the incontestably brilliant Nathan B. Forrest.

Many of these units were not yet with Bragg, and there was some fear that a few might not arrive in time if Rosecrans pressed with vigor. But should they all come together, they would total about 60,000, giving them a wonderful advantage over Rosecrans, then believed to have no more than 50,000 in his three corps. For once, as at Stones River, the Confederates might go into battle with numbers on their side.

The question was where the battle should be fought. Rosecrans believed that Bragg was retreating toward Atlanta, though his basis was little more than wishful thinking, for his cavalry was bringing him almost no intelligence of Confederate whereabouts. Thus he kept his corps going ahead, Crittenden now moving out of Chattanooga and turning due south toward what was, in fact, Bragg's position several miles to his front. On September 9 Thomas and McCook were moving through their gaps, and then the next day General James Negley, commanding the second division of Thomas' XIV Corps, stumbled into the frightening sight of most of Bragg's army awaiting him as he was the first to march into McLemore's Cove. Within a few minutes Negley found himself almost surrounded. Only nightfall and more of Thomas' troops arriving saved him, as well as command confusion between Hill, Cleburne, and Bragg, that resulted in the wonderful opportunity to destroy a full division being squandered. Worse, the next day the Southerners allowed Negley and his supports to withdraw from the cove, letting yet another chance slip away. Bragg would later arrest Hindman for disobedience of orders in the affair.

Even now Rosecrans was not fully alerted to the danger facing him. He thought Negley had only encountered a rearguard. He was unaware of another similar danger narrowly averted when Bragg ordered Polk to attack the isolated Crittenden, only to have the utterly incompetent corps commander fail to carry out the order or even

Sergeant, 33rd New Jersey Volunteers/2nd Zouaves, U.S.A.

There was nothing quite like a colorful zouave to capture the imagination of the public and of young boys anxious to enlist. The 33rd New Jersey, also known as the 2nd Zouaves, was one such unit, and a relatively rare zouave unit in that it fought its war in the West with Sherman, whereas most zouave units marched with the Army of the Potomac.

Their trousers were dark blue pantaloons in the baggy zouave style, matching their short jackets, and both sporting colorful but discrete red trim. The leather leggings contained their trouser cuffs, while the regulation kepi capped their heads instead of the fez often used by zouave outfits.

They carried both the .577 Enfield and the .58 Springfield, though by war's end they had abandoned their zouave garb entirely, adopted regulation Federal uniforms, and equipped themselves entirely with Springfields.

attempt to do so. And certainly Rosecrans had no idea at all that on September 9, the same day Crittenden occupied Chattanooga, trains in Virginia were loading the first of Longstreet's corps for the long ride to eastern Tennessee. At the same time, Breckinridge's and Bushrod Johnson's divisions were rushing to join Bragg. Walker's Reserve Corps was coming from Johnston out in Mississippi, and Buckner had brought his command from Knoxville. Meanwhile, "Old Rosey" marched blissfully onward, unaware of the combinations forming against him.

Only on the evening of September 12 did the commander finally realize his jeopardy and order his army to concentrate, calling Crittenden and McCook to rush to Thomas in the center. Unfortunately, Bragg gave him four days of unhindered time in which to do so, simply keeping his own army in check and awaiting Longstreet's arrival. But then he decided to go ahead. Walker had arrived and Breckinridge and Johnson were coming up. Rosecrans was in a beautifully exposed position. All Bragg had to do was move north along the near bank of the Chickamauga, cross Johnson, Walker, Buckner, and Forrest at any of the available fords, and he could get between the Yankees and Chattanooga. Then he could retake the virtually unoccupied city and Rosecrans would either have to attack him in his defenses, or else withdraw back towards Nashville in order not to risk having Bragg cut his own lines of supply.

Bragg intended to strike on September 18, but delays held it off for a full day, during which Rosecrans moved quickly in response to his own danger. Chickamauga Creek ran roughly north to southwest, with numerous bends and cutbacks. Rosecrans had Crittenden on the west side just behind Lee and Gordon's Mill about two miles above Crawfish Springs, Rosecrans' headquarters, on September 18. In order to speed the concentration toward his threatened left, he had halted Crittenden here while ordering Thomas to move to his support, and that night the XIV Corps marched behind and past Crittenden almost two

Variant Pattern Confederate Unit Flags

1 Flag of the 1st Kentucky Brigade, Army of Tennessee – known to history as the "Orphan" Brigade. The unit consisted for most of the war of the 2nd, 4th, 5th, 6th and 9th Kentucky Infantry Regiments, together with a unit of artillery. The brigade saw action in nearly every major engagement of the western theater, and in that time saw their numbers fall from more than 4,000 to a little over 600

2 Hardee Pattern flag of an unknown regiment, Army of Tennessee. Captured at Lookout Mountain – Battle of Chattanooga – November 24, 1863, by a soldier of the 149th New York Infantry

3 A Confederate hospital site designation flag captured at Waynesborough, Virginia, March, 1865, by soldiers of the 8th New York Cavalry

4 Polk Pattern unit flag possibly of the 16th Tennessee Infantry. The distinctive battle flag of General Leonidas Polk featured the cross of Saint George in its design. This was also the emblem of the Episcopal Church of which Polk was the bishop of Louisiana

Artifacts courtesy of: The Museum of the Confederacy, Richmond, Va

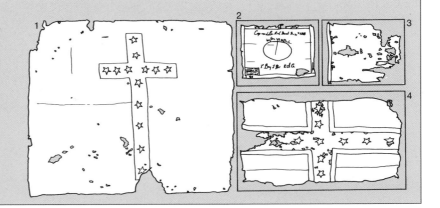

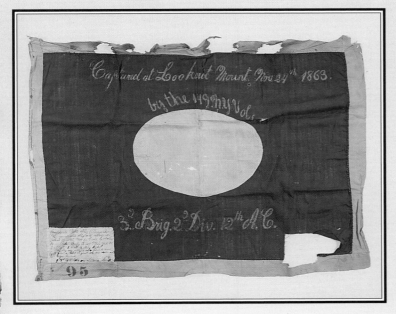

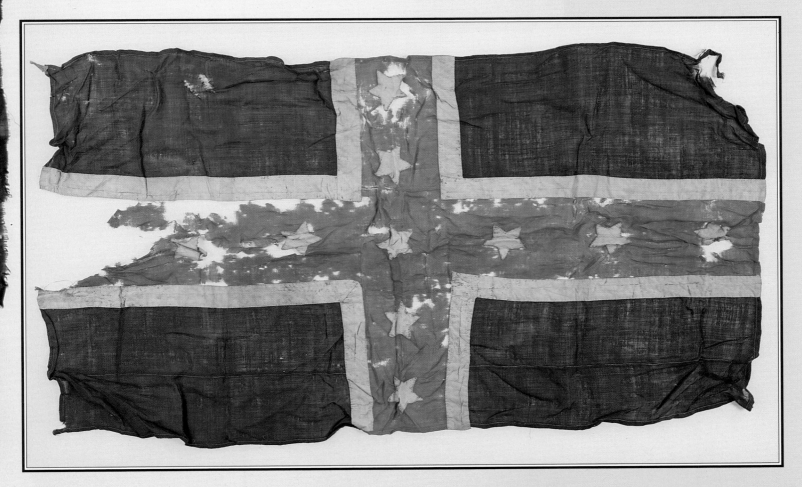

miles in order to meet the anticipated threat from Bragg. As it turned out, no battle erupted on the appointed day either, as spirited actions with Yankee cavalrymen prevented Bragg from crossing where and when he wanted. Instead, Rosecrans used September 18 to continue extending his line northward to protect the road to Chattanooga. On into the morning of the 19th he continued rushing his scattered divisions in their concentration.

It was tangled, sometimes obscure countryside between the two armies. Admitting that he could not keep Bragg from crossing the creek in his front, Rosecrans was arraying his forces to protect his lines of retreat and to take advantage of a few natural advantages, most notably some dense woods and thickets, and a moderately commanding eminence called Snodgrass Hill. On such ground it would be difficult to control the armies for there was no point from which a commander could see all or even a significant part of his command. They would have to fight by "feel". Rosecrans, being outnumbered and caught in a bad position, knew that he would have to stay on the defensive. Bragg, with a slight numerical advantage of perhaps 7,000 and 30 pieces of artillery, would unquestionably be the aggressor.[1]

The fighting really began the next day, almost accidentally, and again heavily influenced by the Federals' lack of knowledge. Word came to Thomas that a single brigade from Bushrod Johnson's division was isolated on the west side of the Chickamauga, midway between Alexander's and Reed's Bridges. Sensing an opportunity to take out an entire enemy brigade at a quick stroke, Thomas immediately sent two brigades off to make the easy kill. He did not know that the rest of Johnson's division was lurking nearby, and when the would-be Federal attackers actually encountered that brigade, they ran into the entire Reserve Corps under Walker, supported by Forrest's cavalry. The Yankees advanced and quickly

discovered their mistake. Johnson's men handily repulsed the first blue advance, but commanders on either side of the line were somewhat surprised at the ferocity of the fighting. As rapidly as possible, more reinforcements were summoned, and soon successive divisions of each army began coming on the scene at about the same time. Neither Rosecrans nor Bragg was in control just now. Rather, the battle was taking on a life of its own as the momentum of the fighting automatically drew more and more units toward the action.

Neither did North and South operate with any conventional battlelines such as Lee and Meade used at Gettysburg. The ground was too confused, and the spread of the fight too haphazard. Instead, units simply came to the sound of the firing and went into action against the first foe they encountered. As a result, the fighting shifted southward from the Reed's Bridge area throughout the day.

Before long Walker committed both of his divisions to repulsing the first attack, and did so

almost decisively, nearly capturing an entire brigade themselves. But then a fresh division from McCook came up on the Federal right and struck Walker's flank with vigor, putting renewed energy into the other battered Federals, and together they drove Walker back more than a mile, only to be themselves stopped cold when Cheatham's division of five brigades suddenly appeared, passed through Walker, and slammed into them.

It was past 11 a.m. by now, and the fighting had been going on for about two hours. Cheatham's new division almost immediately went into action with Federal general Richard Johnson's division, just south of Walker's earlier position. Soon more of Thomas' divisions appeared, being matched almost one-for-one by the arrival of more Confederate units. Late in the day, the Federal line stretched for nearly five miles through the woods and fields west of the Chickamauga. Thomas held the left, from just beyond Reed's Bridge, with two of his divisions. But then came Johnson's division of McCook's Corps, then John

Below: These raw-boned Western men are just an example of the kind of rugged fighters that made up the Army of the Cumberland. At Chickamauga their leadership and their luck let them down.

Above: Lee and Gordon's Mill on Chickamauga Creek, might have been the site of the great battle, for Rosecrans placed Crittenden here in anticipation of meeting Bragg, but it did not come to pass.

Right: An example of the difficult ground along Chickamauga Creek where the battle was finally fought. Though taken months after the fight, this photograph still shows what the men fought through.

Palmer's division of Crittenden's XXI Corps. Then came Reynolds' division from Thomas, followed by Horatio Van Cleve's division from Crittenden, then James Negley's division from Thomas again. Off to the right came Jeff C. Davis' division from McCook, then Thomas Wood's division from Crittenden, then Sheridan's division from McCook's once more. Obviously, the entire line was an organizational jumble, with all semblance of neat corps unity set aside in the rush to get units on the field. As a result, an informal command system went in place instead, with Thomas commanding the divisions on the left, Crittenden those in the center, and McCook the right.

Bragg's units, too, came into the line without much regard for organization. Walker's corps remained on the right, though battered, then Cheatham from Polk's corps came in on his left, followed in turn by Stewart from Buckner's, then Johnson and Law from Hood's corps, and Preston from Buckner. Breckinridge and Cleburne, Hill's whole corps, were still on their way, having been posted down river the day before in anticipation of action at Lee and Gordon's Mill. And Hindman from Polk's corps was just on his way to Bragg's left. Thus the Rebel line ran just over three miles, from Reed's Bridge down to a bend in the Chickamauga just opposite Davis' division in Rosecrans' right center. Bragg had intentionally kept shoveling his divisions northward that morning in the hope of still getting between Rosecrans and Chattanooga, but by noon, with the strength of Thomas' attacks on Walker, it became evident that the foe knew what he intended, and that this approach was not going to succeed. Yet so fixed was Bragg on his original intent that he seems not to have credited, or even looked for, information about a wide gap in the center of Rosecrans' line. If he had sent one or two of his divisions forward at the right place and time, he could have split the Yankee army in two. Such an opportunity rarely came to a general even once in a battle, and Bragg

Above: Longstreet, like A.P. Hill, would make a career of timely arrivals on battlefields. His men's eleventh hour coming onto the field at Chickamauga virtually turned the tide of battle, aided by Yankee ineptitude and the accidents of war.

passed this one by. Miraculously, he would get another in time.

The fighting shifted toward the center where, in spite of the failure to capitalize on the gap in the Yankee line, Bragg's divisions struck good blows, and Stewart in particular drove a deep bulge into Van Cleve's line in a brutal afternoon attack that nearly saw the Federal line break before Thomas sent Palmer and Negley to stem the break. It was a near thing, for Stewart's hammering brigades, especially the one led by

Tennesseean William B. Bate, broke through to one of the two roads connecting Rosecrans with Chattanooga, and got within sight of the other before being pushed back. Only the arrival of Joseph J. Reynolds and his division finally pushed them back to their own lines.

By about 4 p.m. the fighting started to die down on Stewart and Cheatham's fronts, only to be taken up by some new arrivals on the field. Longstreet's corps, moving laboriously over the overburdened and badly worn Confederate railroad system from Virginia, actually started to arrive at Ringgold, several miles southeast of Chattanooga, on September 16. From there it was a march straight west about fifteen miles to reach Bragg on the banks of the Chickamauga. Law's division arrived first, while Hood himself with more men only came on the field about 3 p.m., while the battle was under way, and immediately assumed command of Law and Johnson. Hood was not a brilliant man, but he did have a perfect instinct for combat, and no sooner did he arrive than he pitched into Rosecrans' right with his two divisions. The fighting became brutally vicious, with whole regiments almost disappearing in the melee. Some soldiers saw an owl flying overhead with crows diving and pecking at it, and believed the spirit must be contagious. "Moses, what a country", said one Yankee. "The very birds are fighting."[2]

Hood's fight sputtered out as darkness fell, and the buildup of Federal forces on his front prevented any gains. But there was still one more attack to be made, this one in the gathering gloom of night as Cleburne brought his division on the field after a day-long march from the south. Bragg, still clinging to the idea of turning Rosecrans' right, had marched Cleburne and Breckinridge all the way from Lee and Gordon's, clear past the rear of his entire battleline, to get them on his own right. The result, of course, was two exhausted divisions by day's end.

Bragg ordered Cleburne to report to Polk, commanding the right wing of the army, and when Cleburne went forward into the fight, well after 5 p.m., he exploded out of the trees and thickets to slam into Thomas with ferocity. For half an hour the firing was as thick as any of the veterans could ever remember, and only the settling of absolute darkness brought it to a halt.

At the end of the first day's fighting, neither Bragg nor Rosecrans had achieved a thing, and the men and officers in both armies suffered low spirits that night. Rosecrans and his officers discussed the situation, hopeful that Bragg would retreat, and decided to stand their ground. There was more activity across the lines. That evening

Longstreet himself arrived, with McLaws' division, commanded for the moment by Joseph B. Kershaw, marching behind and due to arrive in time for the next day's fight. Bragg decided to divide his command into two wings. Polk would command the right, consisting of Walker's Corps, Hill's Corps, and Cheatham. Longstreet was to command the left wing with Hood's corps of three divisions, Buckner's corps of two divisions, and Hindman on loan from Polk. As for a plan of battle, Bragg hardly had anything more sophisticated in mind than a general attack all along the line, starting with Polk on the far right, and then spreading to the left. Polk would begin his attack at daylight.

Of course, Polk did no such thing. Supposed to attack before 6 a.m. on September 20, he did not get things moving until almost four hours later. Confusion, bad communications, and Polk's customary obtuseness, delayed the attack inexcusably, and Bragg would not forget. Finally he had Breckinridge placed on his far right, and it fell to the Kentuckian to launch the attack.

The ferocity of the Kentuckian's attack would become legendary. His three brigades almost shattered the Federal left flank, and two of them quickly got around the end of Thomas' line, only to encounter very stiff return fire from behind breastworks. It cost him General Ben Hardin Helm, commanding his old Kentucky "Orphan

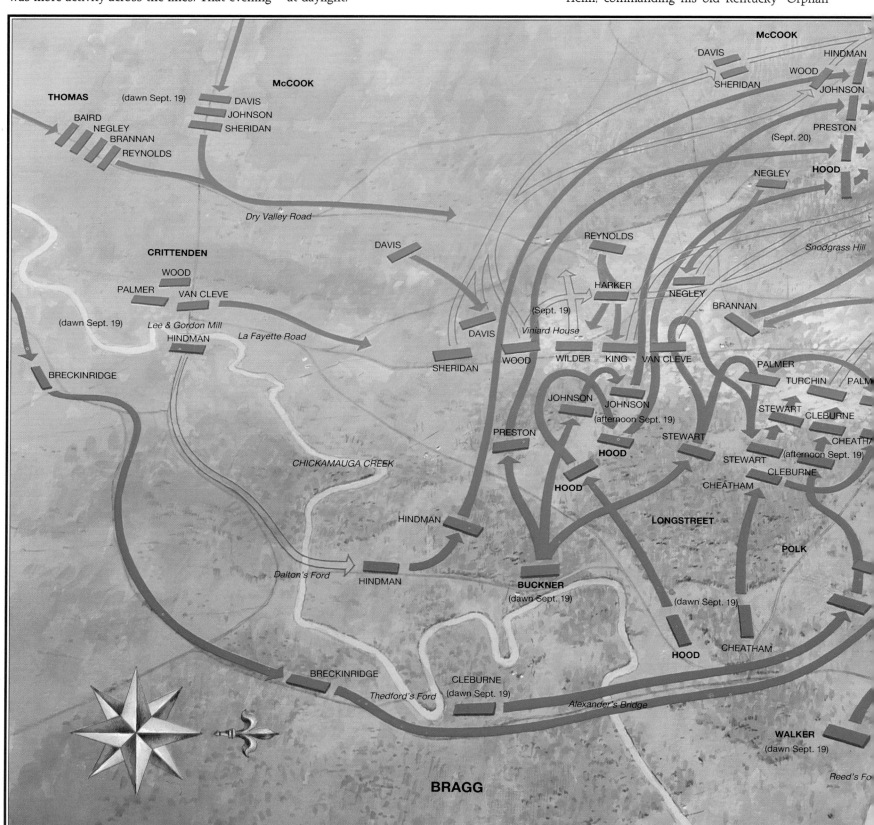

Battle of Chickamauga, September 19-20, 1863

The Battle of Chickamauga began on September 19 when Federals tried to destroy an isolated Confederate brigade on the west side of the creek just below Alexander's Bridge, left lower center. However, the battle quickly developed as both sides rushed more and more divisions to the scene, Bragg moving up the east bank of the Chickamauga, and Rosecrans up the opposite side. Opposing divisions went into line facing each other almost simultaneously,

stretching the battleline northward, until it extended almost five miles.

The fighting rolled gradually from the south end toward the center throughout the day. At one point in the afternoon a hole opened in Rosecrans' line which, had Bragg taken advantage of it, might have put the Yankees to rout, left center. However, Rosecrans filled it and the fight continued. A substantial attack by Stewart nearly penetrated the enemy line,

and later that afternoon a brutal assault by Hood on Rosecrans' right flank inflicted serious damage, but the blue line held. The day then closed with a twilight attack by Cleburne at the opposite end of the field, right center.

Bragg intended a general attack the next day, starting on his right and rolling to the left, the reverse of his Stones River plan. Breckinridge struck first, joined thereafter by Cleburne, and the fury of their assaults led Thomas to call for

more and more reserves, especially after Breckinridge pushed around his left flank, center right. Finally Thomas J. Wood's command pulled out of the center of the Federal line, in obedience to confused orders, opening a massive hole just as Longstreet's men were going in to assault, center left. The effect was electric. The whole Union right collapsed and fled for Chattanooga. The left, commanded by Thomas, pulled back under massive

pressure from three sides and held out on Snodgrass Hill long enough to cover the retreat, upper center. Fortuitously General Gordon Granger came to his aid from the north, right, and Thomas held out until nightfall. It wasn't until then that Thomas left the field to Bragg. His work had been vital in saving the Federals, for by that time most of the Army of the Cumberland including Rosecrans, its commander, were well on their way back

to Chattanooga.

Bragg now had the chance to completely destroy the Union army, but perhaps partly because of the appalling casualties his force had taken he declined the opportunity. Instead he closed in around Chattanooga, sealed off its lines of supply and began a siege of the Federals inside.

The two-day battle produced other, more terrible results. Rosecrans lost 16,000 casualties, and Bragg 21,000.

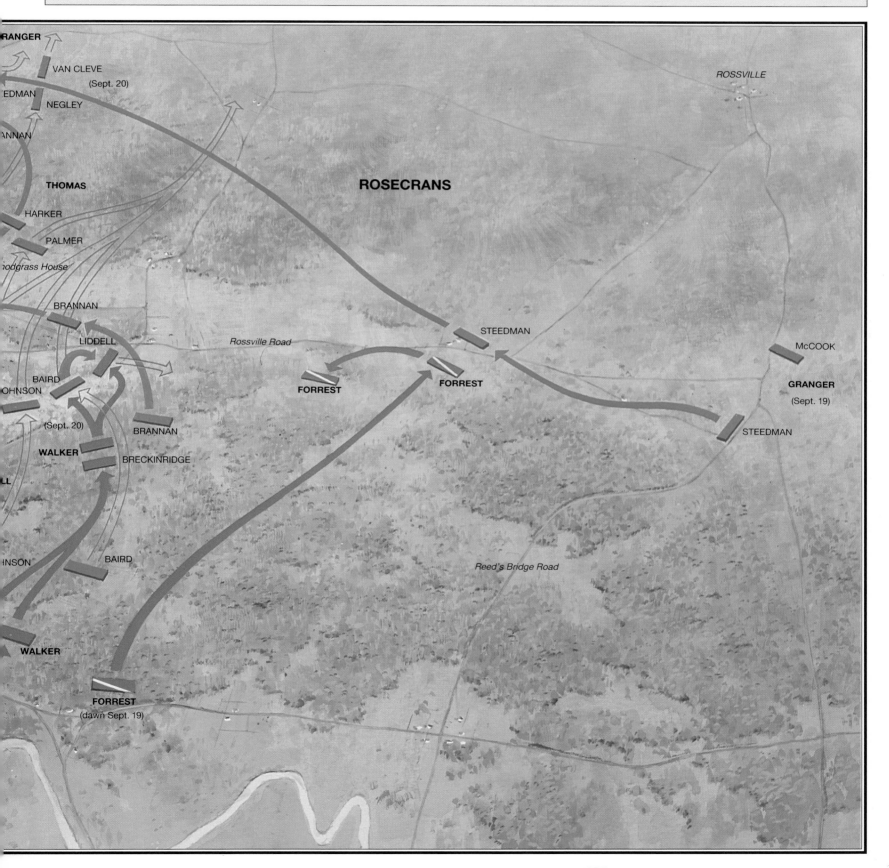

Brigade", who took a mortal wound whose pain would be felt all the way to Washington, the White House, and the home of Helm's brother-in-law, President Lincoln.

Still Breckinridge hammered away, sending the normally cool and reserved Thomas into something as close to overexcitement as he ever got. He began to call on Rosecrans for reinforcements to stem the attack. Then he called again. When Cleburne took up the attack on Breckinridge's left, the pressure only got worse, and Thomas called for yet more. Then Cheatham moved in on Cleburne's left, and Thomas renewed his request for more troops. These three Rebel divisions were battering him badly, and Rosecrans tried to comply. The trouble was, as "Old Rosey" sent more and more units from his center to his threatened left, he did not pay careful attention to the rest of his line or to the precision of his orders – a fault common to Rosecrans. Finally he ordered the division of Thomas J. Wood to shift to its left. Unfortunately, Rosecrans had exploded in temper at Wood earlier for not explicitly following an order, so now Wood did exactly as he was told, even though in so doing he would leave a hole in the Federal line close to half a mile wide.

Here was Bragg's second chance of a lifetime, yet ironically, he would not capitalize upon it intentionally. With the resistance on the Yankee left being greater than he had anticipated, Bragg abandoned his earlier intent of a rolling attack, and now left it up to his line commanders to strike as opportunity offered. That was all the combative Longstreet needed to hear. There was no Cemetery Ridge in front of him this time. He had nearly 11,000 men at his disposal in the divisions of Johnson, Kershaw, and Law, all commanded by Hood, with Hindman on his left, Stewart on his right, and Preston behind him bringing the total to more than 20,000. He intended to make a massive attack.

Through sheer chance, when Johnson went forward as spearhead of the attack, he aimed almost exactly at the spot where ten minutes before Wood's Federal division had pulled out. Incredibly, after battling through some tough resistance on his right front, Johnson saw nothing to his center and left front but open country and a few scattered enemy commands on the march toward Thomas. Followed by the rest of the attacking column, Johnson swept through, pushing more than a mile to the Dry Valley road, one of Rosecrans' links with Chattanooga. Hindman came through behind him and turned left, taking Sheridan in flank and dispersing him. Behind them, more and more Rebels poured through the gap, now widened to more than half a mile. By noon, after an hour of fighting, the Army of the Cumberland was in two pieces, separated by ground it no longer held, and thousands of screaming Confederates.

The breakthrough was coming at a high cost. Breckinridge and Cleburne, who set it up with their relentless pressure on Thomas, suffered heavily. Now Hood fell with a bullet in his right leg, requiring amputation later that day. But in payment, they had the joy of seeing whole brigades of Yankees put to flight. And now as Hindman began to roll up the flank of the right half of Rosecrans' army, even more of a panic set in on the now demoralized Federals. As the remaining Confederates on the left of Bragg's line pressed forward, the divisions of Sheridan and Jeff C. Davis became disorganized and terrified at seeing

Above: Chickamauga would be the last battle for General William S. Rosecrans. The disgrace of his flight from the field and subsequently allowing himself to be besieged in Chattanooga never left him.

Above: General Thomas J. Wood, shown here very early in the war, had been rebuked by Rosecrans for not obeying an order. Thus when he got an order directing him to move his command, he obeyed to the letter.

the enemy coming at them from two sides and their line of retreat to Chattanooga almost cut off. In groups, then by whole regiments, they broke and ran for the rear. In no more than an hour the entire right half of the Yankee army became infused with the panic, which spread even to McCook, Crittenden, and Rosecrans himself. In a scene not repeated since the rout at First Manassas, thousands of demoralized Federals started streaming back toward Chattanooga. Rosecrans, seeing what happened, simply attached himself to the fleeing horde. So did Sheridan and Davis, though the latter managed to rally himself and some of his men and return to the battlefield by nightfall. The rest simply kept on until they reached Chattanooga. Many including Rosecrans, would never hold field command again.

Still the left half of the Army of the Cumberland held out, and now Thomas shone as he stood his ground. Now it was not merely a matter of repulsing the enemy attacks. He had to stand in order to cover the precipitate retreat of the rest of the army. Thomas' almost frantic appeals of that morning for more reinforcements helped lead directly to what was now happening, but his earlier apprehensiveness now disappeared as, in the height of the crisis, he acted with cool courage and a granite immovability. Time after time Breckinridge and Cleburne, joined now by elements of Longstreet's victorious command, hammered away at him. Thomas held his front, where Polk obligingly relaxed his pressure, and turned his main attention to his right, now turned back obliquely to meet Longstreet. On the slopes of Snodgrass Hill, and a little further to the right rear on Horseshoe Ridge, the Federals put up one of the most heroic stands of the war, well earning Thomas his later sobriquet as the "Rock of Chickamauga." He refused to move. Only darkness closed the fighting, by which time General Gordon Granger and his Reserve Corps arrived, acting on no specific orders but largely on Granger's own initiative. Having heard the sounds of the battle from his position many miles north, Granger deduced that it was not going well for his army, and rushed to the sound of the guns. He arrived in time to bolster Thomas sufficiently to hold out until nightfall. Only then did Thomas finally yield the field, now that most of the rest of the army was well on the road to Chattanooga – an army that he may well have saved.

Even Bragg must have been a bit stunned at the result. The losses to both sides were dreadful, 21,000 for the Confederates and more than 16,000 for the Federals, totals not much less than the numbers suffered at Gettysburg by much larger armies in a three-day battle. Bragg's casualties told of the ferocity of Rebel attacks and the stubbornness of Thomas' stand. Still, the field belonged to Braxton Bragg, and with it the glory of having inflicted the most complete defeat ever suffered by an army in blue.[3]

That defeat might have been more complete if Bragg had chosen to make a spirited pursuit of the retreating foe, but he did not. Longstreet was for moving rapidly and attacking the Yankee garrison at Knoxville, cut off Rosecrans' communications between Chattanooga and Nashville, and force "Rosey" to come out in the open again where they could finish the work of destroying him, which had to be done quickly before massive reinforcements could be sent to him from the North. Bragg would have none of it. He decided, instead, to lay siege to Chattanooga and cut off the Federal supply line at Bridgeport,

Union Mountain Howitzer

The barrel of this small gun weighs only about 210lb and has a tube only 33-inches long. The carriage is about 60-inches long and weighs, with wheels, around 180lb. This allowed the carriage and tube to be dismantled and transported over rough country by mules. The ammunition was also carried in this way, as were field forge and other tools. The piece itself had a range of 900 yards. On average a battery would consist of six howitzers plus ammunition and tools, carried by 33 mules. Only a handful of these pieces survive today, but evidence suggests that they were used by both combatants, primarily in the high country of the western theater and in western Virginia

1 Model 1835 12-pounder mountain howitzer, of 4.46-inch caliber smoothbore. Tube and carriage assembled for firing

2 Combined rammer and sponge for mountain howitzer. The sponge for cleaning out the howitzer tube is on the top, the rammer is below

3 Solid shot 4.62 caliber for mountain howitzer, together with sabot and attached powder bag – this artifact is a replica

Artifacts courtesy of: U.S. Army Ordnance Museum, Aberdeen Proving Ground, Md

where it crossed the Tennessee River. Never a daring commander, Bragg now recoiled from taking the kind of risks that Lee would have made, for the chance of greater gains. It was what he had done through the battle just ended, for that matter, exercising a minimum of influence over affairs. "All this nonsense about generalship displayed on either side is sheer nonsense", a Federal in the fight would later say. "There was no generalship in it", and where Bragg was concerned, he was right.[4]

But that was not an end to it. Chickamauga was to have an echo just two months afterward, when these two armies were to meet again in circumstances completely different, yet leading once more to the near-total demoralization and rout of one of the contestants, and this time in a contest of inestimably more far-reaching impact than Bragg's victory along the Chickamauga, the fruits of which he largely discarded by failing to go after Rosecrans.

"Rosey," of course, was delighted to be allowed to withdraw into the defenses of Chattanooga, and once there his men took heart once more, even those from Crittenden's and McCook's corps who had fled so precipitately. Surprisingly, Rosecrans did not immediately lose his command after the debacle. Lincoln always felt rather kindly towards him, chiefly because of the muted victory at Stones River, coming when it did at a desperate time after Fredericksburg. Indeed, Lincoln did not have to do anything directly about Rosecrans. Instead, he and the War Department created a new command, the Military Division of the Mississippi, encompassing Grant's Army of the Tennessee, Rosecrans' Army of the Cumberland, and the small Army of the Ohio now at Knoxville under the command of Burnside. It all went to Grant, and to him fell the decision of what to do about Rosecrans. The decision was a simple one. Though Grant never felt great regard for Thomas, he nevertheless relieved Rosecrans on October 19 and turned the army over to the Rock of Chickamauga.

It was an army that was not in a happy position. Bragg's army appeared on Lookout

Above: Ironically, Major General George H. Thomas emerged as the only great Yankee hero of Chickamauga, even though his earlier near-panic led directly to Wood being ordered from the center, opening the gap.

Mountain and Missionary Ridge in their front soon after the battle, and began a shelling that, while it did little harm, certainly did not encourage calm in the camps. Worse, Bragg soon had them virtually under siege. Chattanooga sat with its back to the east side of the Tennessee River, its only connection to the supply base at Nashville being a single rail line that moved along the same side of the river until it crossed to Bridgeport, twenty-five miles west. The rails went directly beneath the slopes of the northern end of Lookout Mountain, and when the Rebels occupied that height, the railroad was effectively cut. The only other way for supplies to get from the Federal railhead at Bridgeport to Chattanooga was by a rugged wagon road on the northwest

side of the river, through rugged mountains, and always subject to raids by enemy cavalry. It was more than sixty miles long, and so difficult that only the merest fraction of an army's needs could be brought over it. Bragg was very quickly in a position to starve Rosecrans out even before Thomas took command.

Fortunately for the federals, Bragg, as usual, went back to war with his own generals as soon as Chickamauga was over. He arrested Hindman and relieved Polk. In his typical cowardly fashion, Polk blamed his own failings on D.H. Hill, but Bragg was already setting an attack on that general as well. Then Longstreet got involved, hoping to get Bragg's army for himself, and before long Richmond had a petition signed by twelve generals asking for Bragg's replacement. Davis himself came to see the troubled army in October, but only to sustain Bragg. In a series of attempts to quell the unrest, Davis replaced Polk with Hardee, and then allowed Hill to be relieved and replaced by Breckinridge. Then Buckner's corps command was abolished, and he reduced to leading a division. Longstreet was too powerful to attack in such mnaner, however, so in the end Bragg got rid of him by giving in to "Old Pete's" own repeated suggestions that he take his corps and leave the idle lines about Chattanooga to go take Knoxville from Burnside. This was the worst of all the feuds, for by his manner of solving it Bragg sent away nearly a third of his army. Now he was down to just two corps. Hardee commanded the I Corps with four good-sized divisions, and Breckinridge the II Corps, smaller though it, too, had four divisions. All told, Bragg had about 32,000 men in the two corps still with him after he sent Longstreet away on November 4. They had a lot of ground to cover. Missionary Ridge ran almost parallel to the Tennessee, from Tunnel Hill on the north down to its terminus at

Below: Chattanooga in 1863 looked much as this 1864 image depicts it, with the Tennessee sweeping past. Rosecrans let himself be besieged, and it took the coming of Grant to save the army.

Rossville Gap three miles below. Hardee had to cover almost all of it with the three divisions of Cheatham, J. Patton Anderson, and Walker, along with Cleburne, transferred to this corps from Breckinridge's after Buckner was sent off to Knoxville to aid Longstreet. On Hardee's left Breckinridge occupied the so-called Lookout Valley between the end of Missionary Ridge and the tip of Lookout Mountain. He had just his own old division, now led by Bate, along with Stewart and a small division led by Carter Stevenson. They were spread out over a three-mile front, covering the valley, then holding the slopes and crest of Lookout. Almost all believed that they had a virtually impregnable position on those heights, and that Thomas was in serious trouble.

Below: The task of taking Lookout Mountain fell to an out-of-favor former army commander from the east, General Joseph Hooker, standing second from right. This would be his first fight since Chancellorsville.

Above: The view looking down from Confederate positions atop Missionary Ridge, showing the imposing panorama before Bragg's men, and the height that deceived them into thinking their position impregnable.

They reckoned without U.S. Grant. He reached Chattanooga in person on October 23 to see for himself the condition of Thomas and his army, and to do what he did best – turn adversity into opportunity. Of most immediate concern was lifting the blockade of supplies. That same evening he and his generals solved the supply problem. They laid out a line that cut the overland distance from Bridgeport from sixty to thirty miles, successfully planned the elimination of the small Rebel command holding one vital gap on the way, and by the end of the month had the route open for mountains of supplies to start coming in once more. Thanks to all the hardtack that formed a staple of their diet, the soldiers dubbed the new

route the "cracker line." Over that line there also came substantial reinforcements. Sherman brought portions of his Army of the Tennessee, including most of the XV and XVII Corps. From Virginia came old familiar names in the persons of Hooker and Howard. Hooker brought the XI and XII Corps, with himself in overall command. John M. Palmer took over the XIV Corps, now reconstituted from Thomas and McCook's old commands, while other divisions from McCook and Crittenden joined with Granger to make a new IV Corps. Thomas and Sherman maintained their separate army commands, and combined they totaled more than 60,000 by mid-November.

Grant was anxious to attack and force his way out of the siege as soon as possible, and the necessary delays vexed him considerably. By November 21, however, he was determined to wait no more. He had Sherman placed on his own left, Thomas in the center holding the works of Chattanooga itself, and Hooker on the right. Originally Grant wanted to make a diversion against Lookout Mountain, while Sherman launched the main attack on Bragg's right, at Tunnel Hill. Turning the north end of Missionary Ridge, Sherman could then sweep down its length, or pass beyond and get in the enemy rear, perhaps cutting him off from a line of retreat toward Dalton, Georgia, on the railroad to Atlanta, his own base of supply.

It did not work out exactly as Grant planned. Rain on the appointed day kept Sherman from moving out. Then information came in suggesting that the Rebels might, in fact, be pulling out. Before putting more of his plan into operation, Grant ordered Thomas to make a strong reconnaissance on November 23. Thomas was simply to move forward from his positions in Chattanooga's works and advance to the forward enemy outpost line some mile and a quarter west of Missionary Ridge, anchored on an eminence called Orchard Knob. Thomas mustered 25,000 from the Army of the Cumberland and swept them forward that morning in full view of the main Rebel lines atop Missionary Ridge. The sight was magnificently impressive, as even the Confederates

later testified. It proved to be almost a walkover. The thin enemy line could hardly resist this avalanche of moving blue, and within a few minutes the Yankees swarmed over them. Grant moved his headquarters to Orchard Knob almost at once. Bragg, meanwhile, seeing the danger in his front, strengthened Missionary Ridge by drawing Walker's division from Lookout Mountain. It could not have come at a more unfortunate time, for that summit would be the next object of Grant's desiring.[5]

On the morning of November 24, Grant struck at Bragg's right and left more or less simultaneously. Sherman sent three divisions across the Tennessee north of Chattanooga and swept along Chickamauga Creek toward Tunnel Hill, now held by Cleburne and Walker. He ran into less than he bargained for. Expecting substantial resistance, he found very little, in part because he stopped short of hitting Tunnel Hill. Instead, he occupied two nearby eminences that would put him in a good position for renewing the advance the next day. At the other end of the line, however, Hooker made spectacular gains at Lookout Mountain. Thanks to fogs and mists clinging to the slopes of the steep mountain, the fight would ever after be called – erroneously – the "Battle Above the Clouds."

Hooker led 10,000 forward along the far side of the Tennessee, having previously crossed over downriver. This brought him directly against Stevenson's division posted on the summit and forward slopes. They heavily outnumbered the defenders, and even though Stevenson enjoyed high ground, it was so steep and rugged in places that it offered few advantages, and none for artillery on its slopes. Striking around 8 a.m., Hooker pushed them easily back. Finally the brigade posted at the northern base of the mountain, where it met the Tennessee, was forced to retire. With the Yankees now spilling around the base of the mountain, those on the summit and slopes were in danger of being cut off from the rest of their army. Breckinridge rode over to personally direct the defense, but outnumbered more than

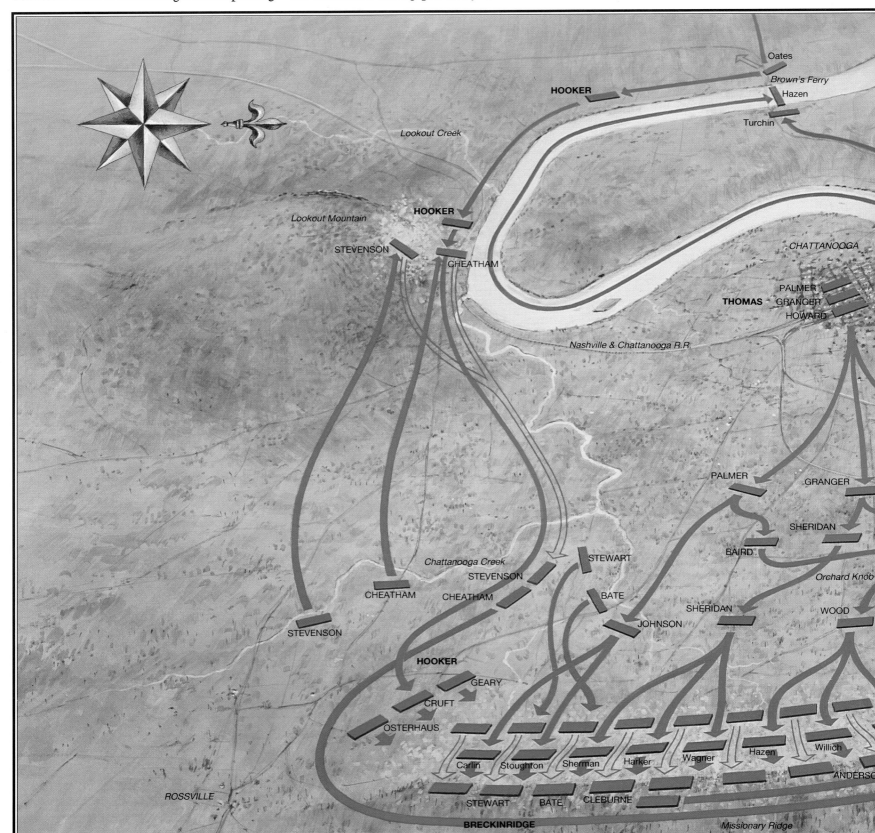

Battle of Chattanooga, November 23–25, 1863

Grant's breakout from Chattanooga proved to be one of the most spectacular affairs of the entire war, though the sheer impressiveness of it all actually worked against the Confederates who supposedly had him bottled up inside the city.

During the weeks following the victory at Chickamauga, Bragg and his army rapidly invested Chattanooga. The city sat with its back to the Tennessee River, at center, and its only usable avenues of retreat or supply were via Brown's Ferry, top center, Tunnel Hill, lower right, or southwestward along the river's south bank, past Lookout Mountain, and then toward Wauhatchie, off the map at top right. Bragg neatly cut off all of these routes by holding the west bank of the river at Brown's Ferry, and by placing his

army in a semicircle stretching from Lookout Mountain, upper left center, east to Rossville, and then along the crest of Missionary Ridge, bottom center, to Tunnel Hill, bottom right center. Additionally, he placed units in the interior north of Chattanooga to cut off any succor.

To break out, Grant first struck Brown's Ferry in late October, sending units both overland and down the river.

With that place secured, a line of supply and reinforcement was open. Next, late in November, when his army was ready, Grant moved forward to Orchard Knob, lower center, on the 23rd to feel Bragg's resistance. The result encouraged him to launch a major assault on Lookout Mountain the next day. Hooker moved from Brown's Ferry and Wauhatchie and in a

spectacular engagement, drove Breckinridge's thin lines from the crest, forcing them back to the main line.

That set the stage for November 25, and the grand assault. Sherman moved across above the city and struck Cleburne and others at Tunnel Hill, where the Rebels stubbornly held on. Then Thomas swept forward to the base of Missionary Ridge, and on up its slopes, while Hooker struck Bragg's

left flank. Breckinridge's corps was soon put to rout, and only Hardee and Cleburne's rearguard action saved the army.

It had been the most crushing defeat ever inflicted upon a Confederate army, a fitting revenge for the Yankee loss at Chickamauga. It would also finally make an end of Braxton Bragg as an army commander, and just further impair the noble, ill-fated old Army of Tennessee.

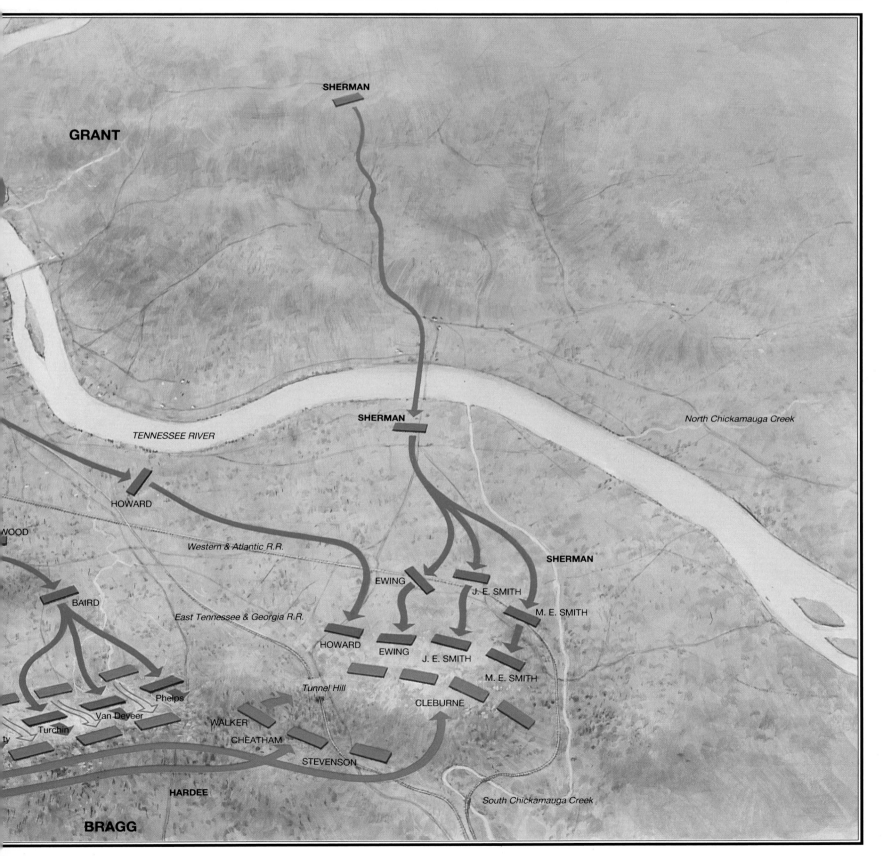

5th Georgia Infantry, "Clinch Rifles" C.S.A.

Company A of the 5th Georgia called itself the Clinch Rifles from the start of the war, and never yielded that sobriquet. Neither did it give up its distinctive uniform. Indeed, most companies of this regiment wore differing outfits, causing General Braxton Bragg to call them the "Pound Cake Regiment." Some wore green, but most opted for the light blue trousers and dark blue frock coat shown on the corporal at right. Distinctive was the dark blue kepi with the "CR" badge, and the Georgia state belt plate. They carry the Model 1841 Harpers Ferry Rifle, and the typical wooden canteens seen on so many Rebels. As the war ground on, the 5th was issued with standard issue uniforms, like that on the soldier at left, though the men might still retain their colorful shirts. However they were dressed, these Georgians were found on many of the major battlefields of the South, and did not yield their arms until the very end.

four to one, there was nothing to be done. The best he could do was hold out until nightfall, and under darkness' cover he withdrew everyone to the south summit of Missionary Ridge.

This left Bragg with but a single position, the three-mile front slope of Missionary Ridge. Breckinridge held somewhat more than half of it from the Rossville Gap northward, with only the divisions of Bate and Stewart. Bragg had sent Stevenson off to the right, where Hardee had Cleburne and Walker near Tunnel Hill and Cheatham and Anderson on the ridge itself. The night before, in council, most of the generals renewed their belief that their position was so strong that they could not be driven off it.

Shortly after dawn on November 25, Sherman advanced once more, and this time he ran squarely into one of the stubbornest fighters in all the Confederacy. Cleburne, called by some the "Stonewall of the West", refused to yield an inch of ground. All day Sherman hammered at him, and all day Cleburne hammered right back. On into the afternoon the Federals kept trying to take that Tunnel Hill, and they never got it. Never, that is, until events elsewhere forced Cleburne to give it up.

Those other events were the absolute collapse of the divisions in the center and southern positions of Missionary Ridge. Grant did not intend Thomas to attack Missionary Ridge with his army. The heights looked too well defended. Besides, since he envisaged turning both Bragg's flanks, he would lose many of the fruits of victory if Sherman and/or Hooker got around a flank, only to find that there was no Confederate rear to strike because Thomas had pushed Bragg off the mountain. Instead, Grant only told Thomas to advance to the base of the mountain and drive in the enemy skirmishers placed there in rifle pits. This should occupy enough of Bragg's attention to facilitate the two flank attacks.

But then something happened. When Thomas paraded his 25,000 out in full view once more, and then sent them forward, the sight of all those

Right: Missionary Ridge, as seen from the lowland over which Thomas' veterans advanced on November 25. They took the battle out of the generals' hands, and carried the crest on their own initiative. The result was a rout that ended Bragg's career.

Above: After the Confederate rout, U.S. Grant, at left, toured the summit of Lookout Mountain to see for himself the scene of the "Battle Above the Clouds." It soon became a popular spot for army photographers.

Yankees had a demoralizing effect on the men in the lines atop the ridge. They knew their own weakness – outnumbered nearly two to one. They knew that Cleburne was taking – and giving – a beating over on their right, and they knew that they had lost the seemingly impregnable Lookout Mountain. Now this wave of blue was coming towards them. When Thomas' men reached the base of the ridge, men in the rifle pits raced back up the slopes behind them with only token resistance, and many did not stop when they got to the summit. Instead, they began disrupting the main lines as they passed through. Then the Army of the Cumberland decided not to stop. Grant did not order it, and neither did Thomas. When the veterans reached the base of the ridge, they found themselves coming under a heavy fire, and with nowhere to take refuge. They could not stay where they were. They could only retire, or go on up the ridge. The soldiers them-

selves and their officers on the scene took the decision out of the hands of the generals. On up they raced, impelled by fear, patriotism, anger, revenge for Chickamauga, and a simple absence of knowing anything else to do.

The result was a virtual stampede. Despite heroic efforts by Breckinridge and Bragg, both of whom narrowly escaped being captured, most of Bate's and Stewart's divisions dissolved, and those on their right began to waver as well. Whole batteries of artillery were captured, then turned against the remaining defenders. Hardee, commanding the right, suddenly found himself facing Sherman on his front and right, and Thomas on the ridge to his left. The best he could do was to hold on for a little longer until the early winter twilight brought operations to a close. Then he pulled out to follow the rest of the shattered army, with the ever-defiant Cleburne covering the retreat.

How ironic that just over two months to the day after giving the Yankees their most humiliating defeat ever, Bragg now suffered a similar fate himself. There was no explaining what had happened, he would later say, and many agreed with him. The men were defeated before the fight began, defeated by the sight of that massive army in full view, and coming toward them, bayonets glistening. Now they could do nothing but retire toward Dalton, Georgia, where they would winter while Grant relished yet another victory in a Chattanooga freed from siege. But at least the discouraged soldiers of the Army of Tennessee could take heart from one particular casualty of their bitter defeat. Three days after the collapse on Missionary Ridge, Braxton Bragg asked to be relieved from command. He would never lead troops in the field again.

References

1 Glenn Tucker, *Chickamauga: Bloody Battle in the West* (Indianapolis, Ind., 1961), p.125.
2 Frank Moore, *The Civil War in Song and Story* (New York, 1889), p.169.
3 Tucker, *Chickamauga*, pp.388-89.
4 St Louis, *Globe-Democrat*, July 15, 1890.
5 James L. McDonough, *Chattanooga – A Death Grip on the Confederacy* (Knoxville, Tenn., 1984), pp.113-14.

CHAPTER ELEVEN

THE WILDERNESS

MAY 5-6, 1864

The disaster at Missionary Ridge set the stage for the spring campaigns of 1864. The most immediate result was Lincoln's appointment of Grant to be general-in-chief of all Union armies, with the rank of lieutenant general. Once Grant had the position firmly in his grasp, he could set in motion his long-held notions of how to win the war. He would press the enemy at every point, making the Confederate armies themselves his objective. There would be no fighting, then resting, as in the past. Rather he would press, and press, and press. If he exhausted his own armies, he knew that it would exhaust the Rebels even more.

He planned four major thrusts. With the Mississippi safely in hand, he would send Nathaniel Banks up the Red River of Louisiana to split the Trans-Mississippi in two. Sherman, now commanding an army group facing Joseph E. Johnston near Dalton, Georgia, would march through that state toward Atlanta, thus splitting the eastern Confederacy. Another, smaller, army was to advance into the Shenandoah and finally wrest that valley from the enemy.

As for Grant, he would move with Meade. At last Grant and Lee must meet in Virginia.

IT is a measure of just how much Gettysburg devastated both armies involved that Lee never again fought an offensive battle, and neither side could bring itself to a major engagement for almost a year afterward. Certainly Meade was willing, but the failures of subordinates spoiled one opportunity after another. As for Lee, ever-pugnacious, he simply could not offer battle on anything like an equal footing, but should the foe attack him, he could still prove to be very dangerous.

By the beginning of 1864 it became obvious that that foe would be under the overall leadership of a new man, one whom Lee had never faced. Meade had performed well, and would stay in command of the Army of the Potomac for the balance of the war. But when Washington made U.S. Grant general-in-chief following the victory at Chattanooga, all thoughts united in expecting him to come to Virginia to direct the campaign there in person. Here it was that for almost three years the Union had strategically gained nothing. Here President Lincoln needed his most victorious of commanders.

When Grant came east, his first proposal for countering Lee was both daring and novel. He wanted to send 60,000 men from the Washington area on a waterborne expedition to the coast of North Carolina and then drive west toward Raleigh through the sparsely defended North Carolina interior. By doing so he would cut most of Lee and Richmond's rail links with states to the south, while the fall of Chattanooga and Sherman's advance toward Atlanta would cut the rest.

Lee would be cut off and isolated from all supply and succor, facing a powerful army under Meade, and having little choice but to fight at a dreadful disadvantage, or else withdraw toward eastern Tennessee, abandoning Virginia to the Federals. The plan also had the beauty of creating a potential new front far to the south and west of Richmond, whereas any advance by Meade over the old routes toward Richmond would mean meeting Lee on familiar ground of his choosing, and which he had well prepared with defenses.

Unfortunately, Lincoln vetoed Grant's plan for a number of reasons, not least being a fear that the daring Lee might take the opportunity to move against a weakened Washington defense. As a result, Grant had no choice but to look once more at that ground between the two warring capitals to search for a road to victory.

He approached the task with seemingly overwhelming power. The Army of the Potomac, augmented by the IX Corps commanded by Burnside (kept independent, chiefly as a consideration to Burnside himself to prevent the onetime army commander from having to serve under Meade, his one-time subordinate), numbered nearly 120,000. There were familiar old names here, along with some new ones. Gone was Reynolds' old I Corps. Hancock, recovered from his Gettysburg wound, was here at the head of his II Corps and its four divisions under Francis Barlow, John Gibbon, David Birney, and Gershom Mott. The III Corps, virtually destroyed at Gettysburg thanks to Sickles, was also gone. Gouverneur Warren, one of the heroes of Little Round Top, now

These seasoned veterans are all that were left of their company after the searing inferno of the Wilderness ravaged them, as so many others during the first meeting of Grant and Lee.

commanded Meade's old V Corps, with four divisions led by Charles Griffin, John C. Robinson, Samuel W. Crawford, and James S. Wadsworth. Sedgwick and his VI Corps were here, Horatio Wright, George Getty, and James Ricketts leading its divisions. And Burnside, reporting directly to Grant and not Meade, had his old veterans in hand under division commanders Thomas G. Stevenson, Robert Potter, Orlando Willcox, and Edward Ferrero. Each corps carried its own artillery brigade. Grant's old associate from the West, Phillip H. Sheridan, assumed command of the three-division Cavalry Corps, his able subordinates being Generals Alfred T.A. Torbert, David M. Gregg, and James H. Wilson. After three years of war and despite having suffered heavy losses over the years, the Army of the Potomac was at its height of experience and efficiency. It was never in better condition.

Denied his plan for a grand strategic movement to remove Lee from Virginia, and literally starve him towards submission without, it was hoped, inflicting or taking heavy casualties, Grant had no alternative but to drive straight for the enemy in his positions below the Rapidan River. Those positions were good ones, though Lee was vulnerable on a number of fronts, and no one knew that better than the gray chieftain himself. Detachments to other more pressed fronts had considerably weakened his army during the winter. Longstreet and two of his divisions had gone to join Bragg in Georgia and Tennessee. Now Lee asked for their return. Pickett's division had gone to North Carolina, and it, too, was returned to Lee. Eventually, by late April, he had built his forces back up to about 63,000, though in many ways it was still rather a shadow of what the Army of Northern Virginia once had been.

"Old Pete" Longstreet still headed the noble old I Corps, though until Pickett returned he had only the divisions of Joseph B. Kershaw and Charles W. Field. Richard S. "Baldy" Ewell still led the II

Above: Grant, the supreme commander, stands third from the left, his look of grim determination fixed on some point in the distance, perhaps upon Richmond, more likely on Lee's Army of Northern Virginia.

Below: Meade, the "snapping turtle", stands fourth from the right, surrounded by officers of his Army of the Potomac, including General John Sedgwick, second from right, who will outlive the Wilderness only by days.

Corps, carrying as it did the spiritual memory of its mighty Stonewall Jackson, now dead nearly a year. Ewell's health was not good since he lost a leg in action in 1862, and some feared that the "fight" that used to animate him had dwindled when he took corps command. But he had fighters under him with division leaders like Jubal Early, Edward Johnson, and Robert E. Rodes. And then there was the magnificent A.P. Hill at the head of his III Corps. He, too, suffered increasingly severe and disabling bouts of illness. Some thought them psychosomatic, induced by tension before battle, when the fact was that complications from venereal disease were slowly killing him. But when he fought, he could be a

tiger in front of the divisions of Richard H. Anderson, Henry Heth, and Cadmus Wilcox. Then there was Lee's cavalry, still led by the boldest cavalier of them all, the brilliant – if erratic – Jeb Stuart. Wade Hampton, Fitzhugh Lee, and William H.F. Lee commanded his mounted divisions, everyone of them a dangerously capable cavalryman.

Despite being outnumbered nearly two-to-one, Lee possessed a few advantages, chiefly his position on the south side of the Rapidan. Grant would have to cross to get to him, and that always gave the defender some edge. Lee knew the ground intimately; Grant and Meade did not. Indeed, he even entertained hopes of being able to

strike an offensive blow at the Yankees, as he had the year before on almost this same ground at Chancellorsville. Should he have to stand on the defensive, his greatest asset would be the tangled growth of woodland and bush known locally as the Wilderness, an almost impenetrable wall of vegetation crossable only on a few roads, some of them barely more than paths. Lee could use this to hold off twice his numbers, he believed. Furthermore, if the armies became mired in this jungle, then Grant's heavy superiority in cavalry and artillery would be neutralized.

Grant, too, foresaw this as he turned his thoughts to the best way to approach Lee. He considered trying to outflank him on the west, but that route suffered the disadvantage of stretching his supply lines from Washington to a dangerous length, even though the more open country would afford him the mobility that he preferred in campaigning. On the other hand, he could move around Lee's eastern flank, which would keep his own back close to the Chesapeake and its tributaries, each affording an excellent opportunity for establishing a supply base close in his rear and a direct water link to the North.

Private, 11th Mississippi Infantry, C.S.A.

The 11th Mississippi Infantry presented one of the handsomer variations on the standard gray uniform of the Confederacy. The regiment was made up chiefly of pre-war volunteer militia companies, many of which wore entirely different garb. Most eventually wore slight variations of the state militia dress, the mid-thigh length gray blouse with red collar and red frogging on the breast, red cuffs and trouser stripes. Their headgear was predominantly the Hardee-style hat worn by this enlisted man, pinned up on right or left side, according to the individual taste of the wearer. Some wore a distinctive state seal beltplate, but this private had only an ordinary buckle.

In action most were armed in time with this Model 1855 U.S. Percussion Rifle with Maynard tape primer. The Mississippians used their rifles well. At Antietam they held the line near the Dunker Church until all their field officers went down, and still repulsed assault after assault.

But to go in this direction, Grant would first have to cross the Rapidan and march through the Wilderness. If he got through before Lee could meet him, his numerical superiority could be used to great effect. But if Lee should meet him in that tangled undergrowth and engage him there, then the outcome not just of the fight, but of the campaign, could risk serious setback.

Grant decided inevitably to risk the march through the Wilderness, even though old hands with the Army of the Potomac tried to warn him that "you don't know Bobby Lee!"[1] He was not certain about either Lee's numbers or his positions, but as April waned and plans for his other armies' advances into Georgia and the Shenandoah approached execution, he could delay no longer. By about May 1 he made the final decision. On May 4 he would send the Army of the Potomac across the Rapidan and into the Wilderness.

The plan depended upon much, including an expectation that the Yankees could get through the Wilderness before the Confederates could stop them. Lee had a long front to cover, and had deployed his forces over a line more than twenty miles in width. Longstreet was in and around Gordonsville, some twenty-five miles southwest of the Wilderness. Lee himself, along with Hill, held the middle of the line near Orange and Rapidan Station, still some seventeen or more miles from the area of Grant's concern. Ewell was closest, positioned right on the Rapidan directly south of Grant's headquarters at Culpeper Court House, but he still lay about ten miles west of the Wilderness. If Grant's leading elements could make the seven miles to Germanna Ford and nearby Ely's Ford, get the pontoon bridges up, and get across before Lee reacted, the Wilderness lay barely more than six miles beyond. One stolen march, half a day's head start, would be all he would need.

The Wilderness itself lay immediately below the Rapidan, and stretched for miles along its south bank, extending several miles southward from the river. Only a few roads offered passage. The Orange Turnpike and the Orange Plank Road

Field Artillery Projectiles and Fuses

1. Federal Eureka 3-inch shell (10 pounder)
2. Federal Hotchkiss pattern 3.76-inch shell (20 pounder)
3. Ordnance Department fuse board showing various types of fuse
4. Federal Schenkel pattern 3.67-inch shell (20 pounder)
5. Frankfort arsenal fuse pack
6. Federal Sawyer pattern 3.76-inch shell
7. Federal James pattern

3.8-inch bolt (14 pounder)
8. English Britten 3-5-inch shell (12 pounder)
9. English Whitworth pattern 2.75-inch shell (12 pounder)
10. English Whitworth 2.15-inch shell
11. English Armstrong pattern 3-inch shell (10 pounder)
12. Confederate pattern 3.5-inch Blakely shell
13. English Britten 3-inch

shell (10 pounder)
14. Confederate Schenkel 2.25-inch shell
15. Frankfort arsenal fuse packs
16. Read pattern 3-inch shell (10 pounder)
17. Archer pattern 3-inch shell (10 pounder)
18. Read pattern 3-inch shell (10 pounder)
19. Tennessee sabot 3-inch shell (10 pounder)
20. Read pattern 3-inch shell (10 pounder)

Artifacts courtesy of: West Point Museum, West Point, N.Y.

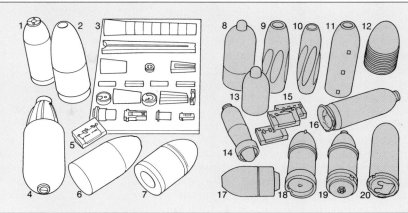

crossed it from west to east, converging near the Wilderness Church and then passing on to Chancellorsville. From the Germanna Ford the Germanna Plank Road extended down to Wilderness Tavern, beyond which it merged with the Brock Road. This, the Ely's Ford Road about two miles to the east, and a lesser track about a mile to the west, offered the only useful routes north to south. Almost everything else was thickets of second-growth oak and pine.

Just after midnight, in the first minutes of May 14, 1864, the Federals started their move. Wilson's cavalry division led the way, taking bridge builders on the road to Germanna Ford where, by dawn, the bridges were in place and Warren's infantry was marching across. Confederate outposts on the south side were quickly dispersed and Wilson took his cavalry slowly forward toward Wilderness Tavern, arriving about noon. Meanwhile, to the east, Gregg's cavalry moved to Ely's Ford with Hancock behind them. By daylight, the II Corps was crossing and Gregg's men rode on to Chancellorsville and beyond, clearing and holding the roads Grant would need for his march. It all appeared to be going very well indeed, and by noon Griffin's division of the V Corps reached Wilderness Tavern. Two hours later Warren had his entire corps in the vicinity and facing west to cover the Orange Turnpike and the Plank Road. Thus positioned, they could cover the passage of the rest of the army behind them as the Federals crossed the Rapidan and moved southeast toward Lee's supply and communications lines with Richmond.

By about 6 p.m. that day, the Army of the Potomac was well on its way. Warren held the important intersection near the Wilderness Tavern. Sedgwick was over the river and in position on his right, while Hancock had his corps in and around Chancellorsville. Burnside was on his way to Germanna Ford, and Lee had apparently been caught napping.

So it seemed. In fact, Lee anticipated that the foe would attempt such a crossing, and was already moving to counter it when Grant's legions left their camps. Still the Yankees had a

Above: The cleared fields and apparent open ground in this 1864 view of some of the Wilderness terrain gives little impression of the tangled hell that most of it became for blue and gray alike.

Below: The Orange Turnpike passing near the Wilderness Church, photographed in 1864 or 1865. Where the roads were good like this, armies could pass, but on most of the ground there was little but paths.

good start on him, but their interpretations of his own movements were faulty, and the reconnaissance conducted that evening, especially by Wilson, was so flawed that they failed to discover that Ewell had moved his II Corps to within four miles of the V Corps on the Orange Turnpike by nightfall. As for Longstreet and Hill, Grant and Meade had no certain information of their whereabouts, though Grant took it for granted that by now Lee knew of his movements and would be speeding to meet him.

It had been a rushed day for the Confederate leader, to be sure. First certain intelligence of the Federal crossings came to him sometime before 9 a.m. He had Ewell in motion by noon. Hill started out even earlier, with much farther to go, and Longstreet moved shortly afterward. Though neither the I nor III Corps could be expected to

reach the Wilderness on May 4, at least Ewell could come close enough to be available once Lee decided upon closer inspection just how he should react. Indeed, even as he rode with Hill toward the scene, Lee was formulating his plans based upon the limited information coming in to him from his cavalry outposts. Grant might be either attempting to reach Fredericksburg and move from there toward Richmond, or he might march through the Wilderness, then turn south and west and attempt to strike Lee's left, turning it and getting between the Confederates and Richmond. Either eventuality would be dire. It was not until well into the early hours of May 5 that Lee finally decided that Grant intended to march into and through the Wilderness, heading perhaps towards Spotsylvania Courthouse, and the decision buoyed his spirits. Grant would bog

himself down just as Hooker had the year before, and when Lee ate breakfast that morning with an old friend, General A.L. Long, Long found him "in the best of spirits" and expecting to be able to repeat the victory at Chancellorsville, perhaps even inflicting greater damage on Grant than he had on Hooker.[2]

At the moment, however, Lee had to face some daunting statistics. The Federals had fully ten divisions in his front; he had just five at hand, and those understrength. Worse, Longstreet could not be expected to be on the scene before nightfall, nor would Anderson from the III Corps. Hill's other two divisions could not get up before noon, and that left just Ewell and the II Corps for him to work with that day. Lee was too outmanned, with his army divided, even in the protective thickets of the Wilderness, to make a serious move against Meade on May 5, yet if he waited until the rest of his forces arrived, the Yankees might keep on marching and emerge in the open country below, ready to turn and strike his own flank that afternoon.

The best he could do was to send Ewell and Hill forward that afternoon in what amounted to a reconnaissance in force. He would give them orders not to bring on a general engagement, though any opportunity to delay the Federal march should not be passed by. Meanwhile he would wait for Longstreet and Anderson to come up on his own right flank, and on the morrow would use them to swing around the Yankee left, he hoped, and smash them if they did not get out of the Wilderness first. Daring as always, Lee sensed that the advantage of the ground might allow him to do what he felt he did best – take the offensive.

Above: This lone cavalryman creates a tranquil picture that no-one would have recognized who fought in the great blazing battle that raged on both sides of this little stream.

Below: Few photographs of the Wilderness terrain were taken during the war, or survive today. Most preferred to forget it in later years, and what remains are largely quiet scenes with no sign of war.

While Lee was making his decision, Grant and Meade arose on May 5 intent upon continuing their basic plan. Indeed, having set this in motion, Grant seems to have left most of the actual planning and detail to Meade, and he in turn left much of it to his staff, quite properly in both cases. Both of the infantry columns were to keep moving on their roads, Warren and Sedgwick moving south from Wilderness Tavern toward Parker's Store on the Plank Road, a distance of about three miles, and Hancock continuing on from Chancellorsville until he could turn southwest on the Catherine Furnace Road. This would bring him out of the Wilderness on a route headed toward Lee's left. If he got out of the woods, that is.

By 5 a.m. the troops were on the march, but within an hour Griffin reported to headquarters that he could see the forward elements of Ewell's corps moving toward him on the Turnpike. Soon Griffin saw the Rebels go into infantry line and heard them start skirmishing. Meade himself arrive on the scene shortly after 7 a.m. and peremptorily ordered Warren to attack at once. Thus the march was halted almost without a shot being fired, something Lee might only have prayed for. Now, instead, the Federals themselves went into line of battle. At the same time, Meade ordered Hancock to halt his own march, in order that he did not inadvertently move out of supporting range. Ironically, even though Meade suspected that the Confederates might only want to delay his march, he stopped it himself now. Nevertheless, Grant accepted Meade's judgment, as he almost always accepted the judgment of a commander on the scene if he trusted him. "If any opportunity presents itself for pitching into a part of Lee's army", he wrote in a message to

Meade, "do so." Grant did not know exactly where all of the Confederate army was just then, but he did believe that Lee could not possibly have it all with him on the Orange Turnpike. This meant that, even with the Wilderness terrain working against him, he would still have an over-whelming numerical advantage and might des-troy a part of the enemy before the rest arrived.[3]

Unfortunately it took Warren some time to get his corps in line and ready for the attack, due in part to the general's own apprehension of acting before he was fully prepared, and also because of the terrain. In front of him was a maze of small streams, marshy bogs, thickets, and dense woods. Nowhere would his men be able to see more than a hundred yards or so to their front, except along the Turnpike. Worse, during the agonizing minutes that it took Warren to get ready, word began to come in from Wilson's cavalry, farther south near Parker's Store, of action in the vicinity of the Chewning farm, action that soon resulted in Wilson being forced back. Before long it became apparent that the advance of Hill's columns was approaching on the Plank Road. Should the III Corps veterans get past the store, they would actually be barely a mile distant from Warren's exposed left flank as he was deployed to advance against Ewell.

This changed the situation at once. While the horsemen fought furiously to hold back Hill's advance, Grant personally arrived on the scene in Warren's rear, not at all pleased with the fact that more than two hours had passed and still Meade had not gotten the V Corps to attack. Now their flank was threatened. Grant knew that this was the time to make a major tactical change. He abandoned the original intent to march through the Wilderness, and instead ordered an im-mediate attack from each of Warren's divisions, supported by other units from Sedgwick's VI Corps. Some of Sedgwick's other units were directed to head for the Plank Road to hit Hill, and he sent orders to Hancock to hurry units of his II Corps along the Furnace Road to its inter-section with the Brock Road, there to join the

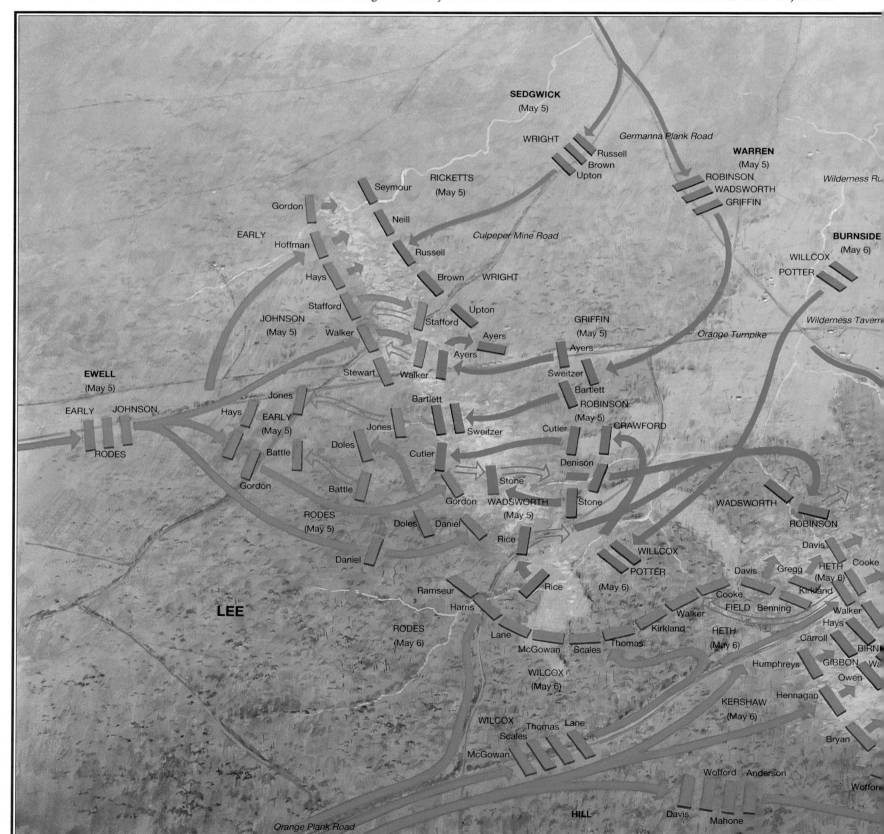

The Battle of the Wilderness, May 5-6, 1864

Running across the center, from left to right, is the Orange Turnpike. The Orange Plank Road, coming up from the southwest, intersects at right center, at Wilderness Church and Dowdall's Tavern. The Germanna Plank Road comes in from the north to meet the Turnpike at the Wilderness Tavern, center, where Wilderness Run crosses it. The Brock Road connects the Orange Plank Road and the Turnpike at center, while the Ely's Ford Road comes in

from the north at right center to Chancellorsville.

Meade had sent Hancock down from Ely's Ford toward Chancellorsville, while Warren's V Corps and Sedgwick's VI Corps moved on the Germanna Plank Road. When it became apparent that Lee would meet them before they could march through the Wilderness, Meade turned Warren west to meet Ewell's rushing Confederates along the Turnpike, left center. Hancock, meanwhile, began

to move west toward Lee's flank, bottom center, and Sedgwick came up on Warren's right.

While Warren was attacking Ewell, Hill's Corps rapidly approached on the Plank Road from the lower left, and Hancock arrived on the scene just in time to help blunt the Confederate thrust at Warren's left flank. During the rest of the day there were essentially two battles, Ewell against Warren and Sedgwick on either side of the Turnpike,

and Hancock against Hill on the Plank Road. Neither proved decisive.

The next day, May 6, Grant planned a major assault, but Lee struck first, hitting Sedgwick's right flank. The attack failed, but soon thereafter Longstreet's providential arrival on the Plank Road slammed into Hancock and soon put Hancock's left in serious jeopardy, bottom center. By noon, Longstreet had pushed around Hancock, seriously

endangering the whole Federal position. Even Burnside's late arrival between Hancock and Warren did not stem the Gray tide, lower center. But Lee's army was spent by this time and could not sustain the momentum any longer. By this time, Grant had also decided that this particular battle had run its course. Darkness was falling, and it was clear that a decisive result was now beyond his reach. Unlike previous commanders of the

Army of the Potomac, however, this was not to be the cue for yet another march back north. Grant had crossed the river to fight and take the war to the Confederacy. Northern Virginia was where the war was, and Grant intended his army to stay there. The stage was being set for that long struggle that was to end at Appomattox. Lee lost 8,700 of his 63,000 men, but he held Grant's 120,000 and inflicted 17,000 casualties.

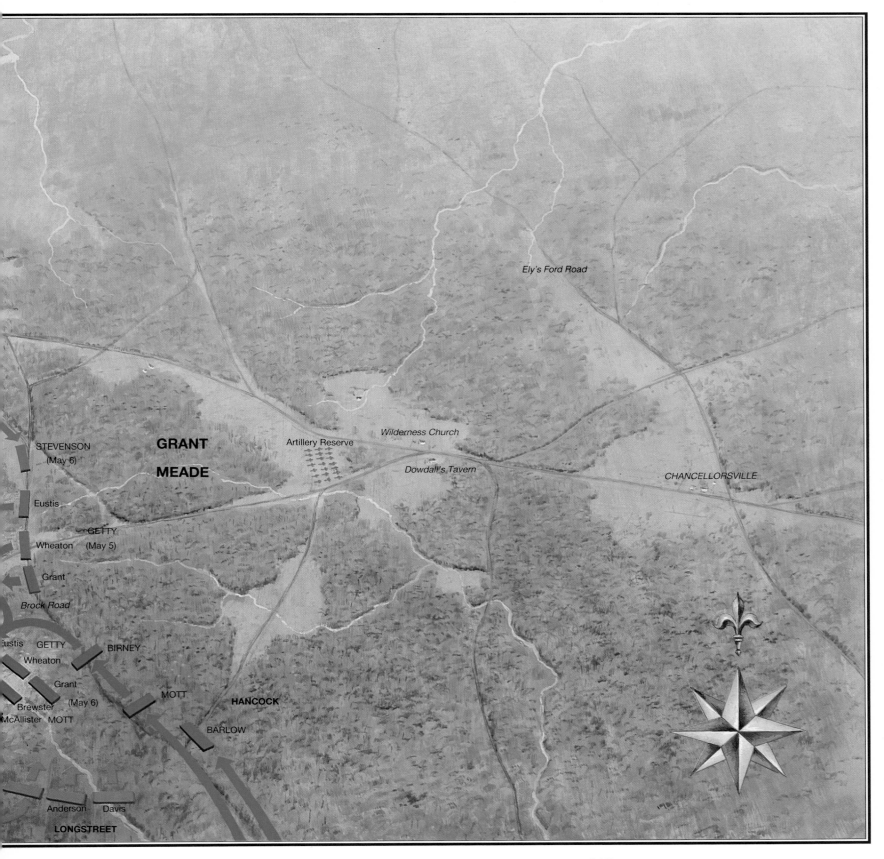

assault on Hill. In a few short minutes Grant's reaction to the threat transformed the campaign.

It still took some time for the Federals to mount their first assault, but when it came the Rebels found it "furious". In front of Ewell's column was Johnson's division. "Old Club Foot", as some of his men called General Edward Johnson, formed his four brigades at right angles to the Turnpike not long before the Yankee wave hit them. When the antagonists joined, his line held except for John M. Jones' brigade of Virginians, who broke and ran for the rear, Jones himself taking a mortal wound as he tried gallantly to rally them. Happily, Early's division came up just as Johnson's line started to waver, around 1 p.m. At once both sides began to understand what fighting in this dense vegetation would be like. One Yankee described getting only "occasional glimpses of gray phantom-like forms." Attacking Federals had to lay on the ground, fire their rifles, then roll over on their backs to reload. Men became so confused in the increasingly dense smoke and the confused terrain that they wandered in the wrong direction. Officers wandered around in a daze looking for their men, and soldiers trying to withdraw from the firing became so disoriented that they "dashed directly into the enemy's fire."[4]

As Early arrived on the scene, Joseph Bartlett's brigade of bluecoats were starting to swarm through the center of Johnson's line. Ewell sent the brigades of John B. Gordon and Junius Daniel to the right of the line, and in a swift counterattack they struck the flanks of the advancing Federal penetrating wave and turned it back. Indeed, Gordon pressed the Yankees so hotly that he pushed them back to their own starting point until he found himself on a line with the Federal front. Behind him, the field over which he had advanced caught fire from gun flashes, and soon the flames began to consume wounded and dead Federals just as Ewell's counterattack consumed Warren's advance. Soon the flames hit the cartridge boxes on the belts of the dead and dying, exploding them like volleys, tearing the bodies in half and leaving in their wake a scene from the *Inferno*. Men rushing to help the wounded could not find them in the choking smoke. Meanwhile, the flames, fanned by a stiff breeze, swept across the field like a wave, so that in a few minutes it had burnt itself out, consuming all the dry grass and sputtering out in leaves and twigs at the forest edge. When the flames subsided, Blue and Gray returned to their work.

Finally Gordon and the others halted their counterattack, seeing Warren's line before them thoroughly demoralized and beaten back. In the breathing time given them, the battered Union veterans of V Corps regrouped, while Ewell turned his attention to readying himself for what he perceived as a threat to his left by Sedgwick's arriving VI Corps. "Baldy" was still heavily outnumbered, with but 13,500 men in line to face Warren's and Sedgwick's combined 39,000 or more.

By about 3 p.m. Grant was trying to get the army in control to send forward a general attack, and was largely directing affairs, having lost some confidence in Meade. Unfortunately, the terrible terrain, the unexpected check to Warren, and the miscarriage of orders doomed his efforts to achieve concert of motion. Instead, the offensive that Grant wanted rapidly slowed. He did not know for certain where Hancock was, and kept sending him orders to attack when, in fact, the II

Corps was not yet on the field. In fact, it was 4 p.m. or afterward before Hancock had his corps in position across the Plank Road, facing Heth's Confederates before him. Even now, Hancock could only get two of his divisions – Birney's and Mott's – on the line, joined by Getty's division from the VI Corps. Heth had only four brigades facing him, but once again the terrain favored the defender, though he was outnumbered some 17,000 to 6,700.

Hancock's attack went in shortly after four o'clock, Getty taking the first fire from Heth's concentrated Confederates, and suffering for it.

Above: This 1864 or 1865 scene taken near the Wilderness Tavern shows some of the ground so hotly contested by Ewell and Warren and Sedgwick – ground over which Grant and Meade never wished to fight.

Below: Brigadier General Charles Griffin commanded the First Division in Warren's V Corps, and bore much of the brunt of the fight. He stands fifth from the right, surrounded by his staff.

Despite their numerical superiority, the Yankees were handled badly by volley after volley from the well-placed Southerners. Getty soon had to call on Birney for more reinforcements, and the whole attack soon degenerated. Mott's units became almost bewildered as they tried to get into line through the tangled maze of growth. General Alexander Hays fell at the head of his brigade as volley met volley along the Plank Road. Hancock himself had to try to rally some of Mott's disorganized men as they raced to the rear, but by 5 p.m. or shortly afterward the attack had failed.

Undaunted, the combative Hancock struggled to renew the attack as his other divisions under Barlow and Gibbon reached the front. Warren, to his right, was extending his line to link with the II Corps, and Grant decided now that the major effort of the day should be Hancock's renewed advance, as he sent an additional division to assist. By about 6:30 Hancock was ready, though fighting had never entirely stopped during the interval since the last advance. By now the sun was setting over the smoke-blackened treetops, but the fury of the firing only grew as the Federals went in again. Hancock arranged his brigades so as to envelope Heth's line on both flanks. Sheer weight of numbers now threatened to overpower Heth and his veterans, badly battered already after their earlier fighting. It was a struggle for them just to hold on until nightfall could close the battle. Lee sent them two more brigades to help hold the line, but they were not enough, and two more were dispatched. It became almost a foot race to see who would arrive first – the relieving brigades, or the Yankees rushing through the brush to get around Heth's flanks. Almost miraculously, the Rebels got there in time. Around 7:30 Barlow's rush to get around Heth's right was met and stopped, and a similar move against Heth's left was stymied soon afterward. The Rebels might not be able to hold their position through the night, but at least they could hold on long enough for darkness to prevent any further exploitation by Hancock. And even as they struggled, the Confederates knew that Anderson and Longstreet were on the way and would be with the army on the morrow.

Over on the Federal right, Sedgwick's attempt to hit Ewell's left flank met with similar fate, though the Confederates halted him at considerable cost, including the mortal wounding of General Leroy Stafford commanding one of Johnson's brigades. They turned back Sedgwick's first assault by Horatio Wright's division, then counterattacked. In the end, the preponderance of Federal might pushed them back again, and there the fighting died down to a steady skirmishing while Grant shifted his attention to Hancock's grand assaults.

By 9 p.m. the fighting had settled to skirmishing all along the line. Tens of thousands of dirty, sweating, exhausted Northerners and Southerners fell where they stood, scrambled for water and rations, and sought sleep amid the terrible scenes of the day's carnage. Neither Lee nor Grant could be very happy with the day's results. The Federal's hopes for a massive general assault never materialized, as the terrain and other factors made it impossible for him to manage his army as he wished. Lee, for his part, could be pleased that he had stopped the Yankees in the Wilderness, but Hill and Ewell were still not linked securely, and Longstreet and Anderson had yet to join the army. Worse, the III Corps had suffered terribly from Hancock, and was saved only by nightfall, his positions so confused and compressed by the foe on his flanks that an aide described his men as being placed "like a worm fence, at every angle."[5] If either side held the upper hand, it was Grant, who still had massive uncommitted numbers, with Burnside on the way, and a good lodgment on the enemy right.

That night Grant, unwilling as ever to abandon a line of attack once commenced, decided that the massive assault that failed to materialize on May 5 should be attempted the next morning, now that almost all of the Army of the Potomac would be in position. He had intelligence that Longstreet still had not joined Lee, and knew that May 6 would be his best chance to strike a telling blow before "Old Pete" brought his I Corps to Lee's aid. He directed that at 5 a.m. Sedgwick would demonstrate to hold Ewell in position, while Hancock hammered once more around

Above: Major General Andrew A. Humphreys served as Meade's chief of staff, and to him Meade gave the task of devising the order of march into the Wilderness – a good plan that went awry.

Below: Seasoned veterans of the Wilderness like these Billy Yanks never forgot the two-day encounter in hell with an army half their size yet determined never to yield. Victory had to await other fields.

Hill's flank and Burnside and other elements drove through the gap now thinly filled between the two Rebel corps. The only serious worry for Grant was that Longstreet might arrive from the southwest and hit Hancock in his left rear while he enveloped Hill. For his part, Lee had little choice but to subordinate his wishes for the battle to the necessity of reacting to Grant's. He toyed with taking the offensive himself, but mature reflection showed that to be impracticable, made the moreso by word received late that night that Longstreet could not be on the scene before daylight, meaning that Hill would have to withstand another pounding on his own.

Nevertheless, as the first hints of dawn came on May 6, Ewell took the initiative by sending forward a modest-sized attack at 4:30 that took the Yankees quite by surprise. He did not have the manpower to make it a decisive strike, but it was enough to stun the enemy. Sedgwick and Warren sent forward a counterattack that saw most of the lost ground regained, but when they came up against "Old Club Foot" Johnson's breastworks, three successive charges came to an end at the muzzles of his guns. By this time, Hancock's attack had jumped off at 5 a.m., and Grant's plan for a massive offensive was well underway.

Sedgwick and Warren were supposed to play an essentially passive role for the rest of the day, though heavily engaged. Yet it did not work out as Grant had planned. Ewell held out sufficiently that he was able to lend a brigade to the fighting elsewhere on the line, which is exactly what Grant hoped to prevent. Worse, Longstreet was approaching the field as the sun rose, and was well within the sound of the guns when the grand attack jumped off. It would be now a race to see if the Yankees could press their advantage before they had to face the fresh I Corps.

Indeed, everything seemed to miscarry. Besides the V and VI Corps roles' coming to naught, the ever inept Burnside failed utterly in his part of the plan. With a wide gap still existing between Ewell and Hill, the IX Corps could have plunged through and cut Lee in half. Instead Burnside ran over ninety minutes late in getting

his men into position, and even then brought only one division. Hancock's men were heavily engaged by that time and achieving dramatic gains against Hill's right flank. If Burnside could have attacked in force and in time, the Army of Northern Virginia might very probably have ceased to exist.

Hancock moved precisely at the appointed time and though encountering heavy resistance, soon saw that they would be able to push over Hill's weakened and outnumbered positions. The Federals struck like a storm on both the Confederate front and left flanks. "Our left flank rolled up as a sheet of paper would be rolled without power of resistance", a North Carolina colonel confessed. Between them Birney and Wadsworth swept over seven of eight defending brigades in their fronts as Hill's line simply collapsed. As Birney pursued, he gradually extended his own left until it overlapped what remained of Hill's right. Hit front, left, and right, at the same time, the III Corps turned in rout by 6 a.m. Lee and Hill were there to see it happen, and Lee at least was reported to have resorted to "rough" language in attempting to rally the men that few ever heard him use. Hancock drove the Confederates for nearly a mile along the Plank Road, and was so elated with his success that he exclaimed to one of Meade's officers that "we are driving them most beautifully . . . be-au-ti-fully."[6]

Unfortunately, the failure of Burnside soon took the euphoria from Hancock's spirits. There was worse to come. As elements of Hill's command streamed back on the Plank Road, they encountered the head of Kershaw's division. Longstreet had arrived. The news put resolution back into Hill's veterans, and before long, though routed, they formed ranks once more. Kershaw at once spread his command out on the south side of the Plank Road while Field came up on his left. Without stopping, Longstreet's tired veterans slammed into Hancock in a vicious assault that stopped the Federal attack in its tracks. Lee himself apparently attempted to lead Hood's old Texas Brigade into the fight, but was deterred by

Excavated Battle-Damaged Union Relics

1. Model 1858 canteen found at Cold Harbor, Va
2. Model 1861 rifle with burst barrel, found at Cold Harbor, Va
3. Model 1858 canteen found at Shiloh, Tn
4. Model 1858/61 ramrod
5. Model 1855 bayonet
6. Model 1861 rifle, struck by bullet, found at Cold Harbor, Va
7. Model 1858 canteen struck by shell fragment
8. Model 1861 rifle, struck by bullet, found at Bethesda Church, Va
9. Model 1861 rifle, struck by cannister ball
10. Breech end of burst rifle barrel
11. Cartridge box tin liner with bayonet holes, deliberately made
12. Brass cartridge box plate
13. Breech end of burst rifle barrel
14. Model 1842 musket struck by 6lb solid shot; found at James Mill, Va
15. Model 1840 heavy cavalry saber scabbard
16. Fragment-damaged tin plate, found in the Wilderness, Va
17. Damaged tin cup
18. Fragment-damaged Model 1858 canteen
19. Bullet-struck Model 1858 canteen
20. Fragment-damaged Model 1858 canteen, found at Shiloh, Tennessee

Artificats courtesy of: Wendell Lang Collection, Tarrytown, N.Y.

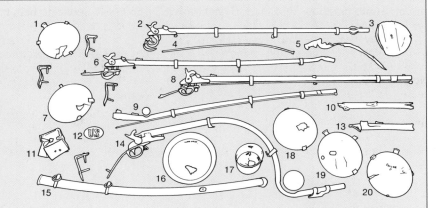

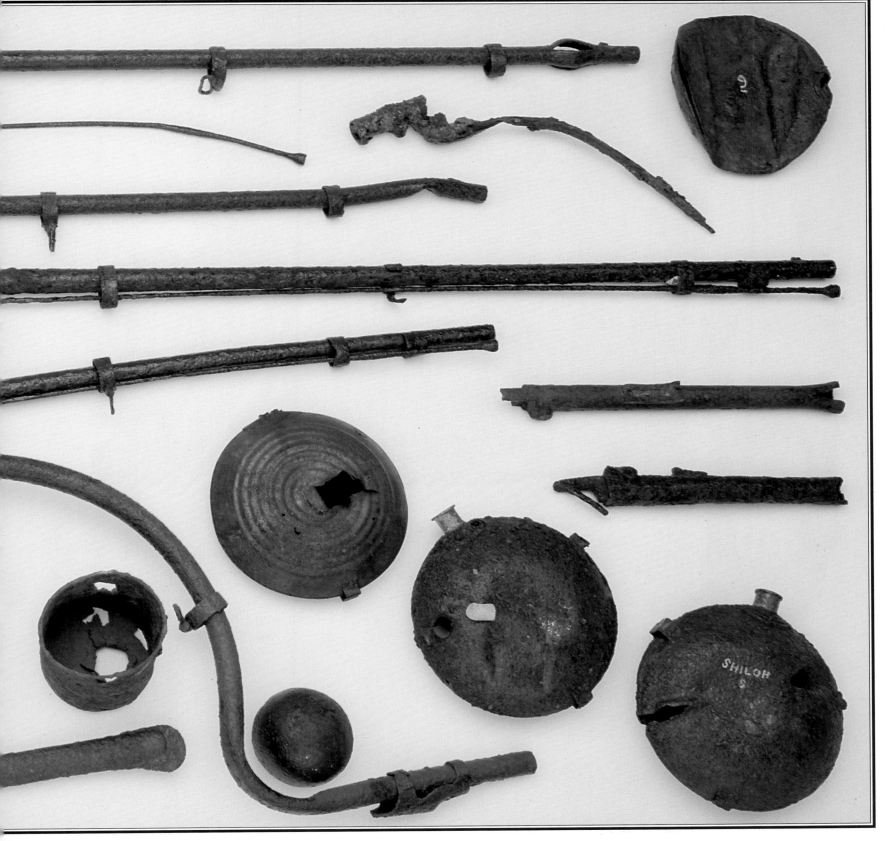

First Sergeant, 3rd New Jersey Cavalry, "the Butterflies", U.S.A.

For real resplendence in the Union army, it would have been hard to beat the 3rd New Jersey Cavalry, so ornamented that their nickname became "the Butterflies", hardly a martial sobriquet. yet they were among the best fighting regiments in Sheridan's Cavalry Corps.

Their armament alone made them formidable, with the repeating Spencer carbines and the Remington .44 revolvers. But then there was their uniform, more typical of European hussars than American horsemen fighting in the 1860s. Indeed, they were also known as the Trenton Hussars thanks to their visorless forage cap, ornately frogged short jacked, and hooded "talma" or cloak. yet none would dare make sport of their firearms could not finish, the heavy Model 1860 saber might. Certainly this sergeant, for all his rather old-fashioned looking finery, is still every inch a soldier.

shouts of "Lee to the rear", and the apparent refusal of the men to go forward until their beloved commander took himself out of harm's way.

By 8 a.m. Hancock's advance had been brought to a halt, and Longstreet was pushing back his own right flank, while Wadsworth's Federals were being shoved back on the left, Wadsworth himself falling with a mortal wound. But Grant and Hancock were not discouraged yet. Grant sent one of Burnside's idle divisions to bolster the II Corps, and Hancock himself speedily worked to reform his battered corps and renew the attack. It was a valiant effort, but hampered by exhaustion, limited visibilty, the timidity of Burnside and Gibbon, and a growing "fog of battle" that hereafter impeded almost every effort at decisive action. The Army of the Potomac simply was not reacting quickly and efficiently to orders, and the morale boost given the Confederates by Longstreet's arrival, with Anderson right behind him, swung the pendulum of initiative on the south end of the line to Lee. Indeed, sometime after 9 a.m. Longstreet started purposefully extending his own right around Hancock's left, and the II Corps had to look out for its own safety.

By 10 a.m. Grant had seen his whole plan thwarted. Consequently he ordered Sedgwick and Warren to halt their holding action against Ewell and start erecting defenses instead. He would send reinforcements to try to get Hancock moving again. But Lee, sensing a momentary advantage, was moving ahead of him now. He quickly sent what he could to strengthen Longstreet, who had held three brigades out of the heaviest fighting in order to use them as an offensive reserve. Now "Old Pete" shifted them to his right under cover of the heavy woods on a route previously reconnoitered. It was almost a repeat of Jackson's Chancellorsville maneuver, though in miniature. Federals detected fragments of the movement, but too late to do anything about it. The Rebels surged forward, yelling at the top of their lungs to magnify the impression of their

Above: The men, Blue and Gray, who battled through the tangled undergrowth of the Wilderness, must have longed for a chance to fight on open ground such as this rare bare patch near the Orange Turnpike.

numbers. They hit with terrible effect, for the Federals had not erected breastworks to protect their flank, and by 11:45 Longstreet was rolling up Birney's flank. Before long Hancock ordered him to withdraw. Providence almost gave the Yankees another chance when, in another repeat of Chancellorsville, an errant Confederate volley struck Longstreet and a party with him. General Micah

Below: Brigadier General Horatio G. Wright commanded the First Division of Sedgwick's VI Corps, and would succeed to its command when Sedgwick was killed by a sniper a few days later. He stands bottom center.

Jenkins went down mortally wounded. Others were killed, and Longstreet himself took a desperate wound in the neck and shoulder that put him out of the war until the end of the year. Field took command of the I Corps temporarily.

But now Lee intervened. Longstreet had been preparing another, more massive flank attack that he believed could achieve Hancock's rear and turn the whole tide of the battle. Lee now cancelled that movement, no doubt lacking confidence that it would succeed without Longstreet to manage it. Instead, in a desperate move reminiscent of the order for Pickett's charge at Gettysburg, Lee decided upon a major assault on Hancock's center. What he hoped to achieve is debatable, for he never explained his intentions. Even if successful in breaking through the works in Hancock's front, Lee had little to gain, and certainly less than from a continuation of the flank attack on Birney. But at 4:15 the gray chieftain sent his men forward only to see the attack broken up without great difficulty by the II Corps veterans, and with appalling losses to the Rebels. It would be Lee's last open field infantry assault of the war.

By the time Lee's attack failed, Grant had given up pursuing the battle further. The Wilderness and Robert E. Lee were too formidable as opponents. But unlike his predecessors, he was not going to turn around and give up the campaign. This indecisive fight was just a check, not a defeat. Even an early-evening attack on Sedgwick's right flank by the brigade of the enterprising John B. Gordon – an attack that actually put Grant's right wing in near panic – came to nought. Grant simply sent more troops from elsewhere, and the reserves and darkness put an end to it.

They put an end to the Battle of the Wilderness as well. The armies entrenched and stared at each other the next day, but then Grant started the series of side-stepping flank marches that were to take him through Spotsylvania, to the North Anna, Cold Harbor, and eventually to Petersburg. Though he could not decisively defeat Lee in the field he forced the wily Confederate to dance to his tune until almost a year later the two great generals finally met in peace at Appomattox. Neither they nor their battered veterans ever forgot the horrors of fighting in the Wilderness. Nor did they forget the more than 17,000 casualties left behind by the Army of the Potomac, and the 8,700 or more of Lee's men who would not fight again. Their sacrifice had decided little. Lee could not stop Grant, and Grant could not yet destroy Lee. The final decision would have to wait for other fields. The Wilderness, in the end, like the ground that hosted the battle, was just a tangled, confused, and painful, stopping place along the way.

References

1 Horace Porter, *Campaigning with Grant* (New York, 1897), p. 39.
2 A.L. Long, *Memoirs of Robert E. Lee* (New York, 1886), p. 327.
3 O.R., I, 34, pt. 2, p. 403.
4 William J. Seymour Memoir, May 5, 1864, William L. Clements Library, University of Michigan, Ann Arbor; William H. Powell, *The Fifth Army Corps* (London, 1896), pp. 609-10.
5 William L. Royall, *Some Reminiscences* (New York, 1908), pp. 279-80.
6 Edward Steere, *The Wilderness Campaign* (Harrisburg, Pa. 1960), p. 330; Theodore Lyman, *Meade's Headquarters, 1863-1865* (Boston, 1922), p. 94.

CHAPTER TWELVE

MOBILE BAY

AUGUST 5, 1864

The Wilderness would be only the beginning for Grant and Lee. Ahead of them lay a solid year of conflict. From the Wilderness to Spotsylvania, to the North and South Anna Rivers, on to Cold Harbor, and finally to Petersburg, and the siege that neither wanted, the epic struggle of these two titans was destined to continue.

Thanks to Grant, there was so much else going on now as well. The attempt to take the Shenandoah failed in May at New Market, when Breckinridge and a scratch force of Confederates turned back a Yankee column and saved the Valley for a time. Then Lee launched his own campaign, sending Jubal Early and the II Corps on a raid out of the Shenandoah toward Washington that actually came within sight of the Union capital. For the next three months, Blue and Gray would conduct an epic battle for that beautiful valley, though in the end the Rebels had to lose.

And down in Georgia, Sherman, Grant's strong, swift right hand, moved toward Atlanta. In a succession of engagements, he pushed the hesitating Johnston back and back again, until in July he stood at the very gates of Atlanta.

There were other places as well where the Yankee juggernaut pressed the beleaguered Rebels. For three years the blockade had squeezed the life out of the Confederacy, until there were only a handful of ports left open to receive the dwindling blockade-running trade. By late summer, it was time to start closing those too.

T HE CIVIL WAR was very much a family affair. More than a dozen sets of brothers became generals on one side or the other, and some families had a son on each side who reached high command. The Lees of Virginia produced four generals, as did the McCooks of Ohio. But no single family on either side so dominated the land war the way the sons of David Porter virtually defined the battle for control of the waters. Porter had been one of the young nation's first naval heroes, and he spawned a generation of seamen whose careers came to fruition just as North and South went to war. William Porter, though not well-liked and nicknamed "Dirty Bill" by associates, was an early influence in the gunboat engagements that won the upper Mississippi and the opening of the Tennessee and Cumberland Rivers for Grant. His more brilliant brother David Dixon Porter became Grant's most dependable captain in the move down the Mississippi to take Vicksburg, and went on to distinction through the balance of the war, though like "Dirty Bill" he had a scheming, dark side that could turn against anyone who threatened his ambitions, including even yet another brother, the best "Porter" of them all – David Glasgow Farragut.

He was an orphan boy from Tennessee only a few years old when the first David Porter adopted him and took him to sea. There he quickly became a favorite, alternately admired and resented by the Porter stepbrothers who later followed their father into the Navy. When the war came, some doubted his loyalty, especially since when on land he had made Virginia his home. But there was never a doubt about allegiance in his own mind, and less than a year after the start of the war he was the Union's premier naval hero after capturing New Orleans, exciting the admiration of all the North, and the jealousy of his stepbrother David, who spent much of the rest of the war writing back-stabbing letters about Farragut to the Navy Department in Washington.

By 1863, Farragut was a rear admiral commanding the West Gulf Blockading Squadron, and charged with controlling the entire Gulf coastline of Texas and Louisiana. His task, on paper, was a simple one – to stop all blockade-running traffic into and out of Rebel ports. New Orleans was already safely in Yankee hands, thanks to Farragut, but there were a few other places where blockade-running traffic continued at a brisk clip as the sleek, fast steamers slipped through the cordon of Union warships. By January 22, 1864, when Farragut returned after a five-month absence, the most significant remaining rebel-held ports were Brownsville and Galveston in Texas, and Mobile Bay, Alabama.[1]

The lighthouse at Fort Morgan, guarding the entrance to Mobile Bay, served not only as a beacon to warn approaching blockade runners, but also as a target for Yankee sailors who knew they must silence the fort and take Mobile.

Of them all, Mobile was the prize, and for a host of reasons. It provided a vital link in the Confederacy's fragile transportation network. At least twice, once before Stones River, and again prior to the Battle of Chickamauga, significant numbers of Rebel troops had been shifted from Mississippi to Tennessee, and the only way for them to go was by rail from Jackson to Mobile, and then up the Alabama River to Selma, to pick up the rails again to go on to Atlanta. Take Mobile, and that link was broken. Mobile was also a key to holding nearby Pensacola, Florida, and its navy yard just fifty miles east, which had only recently been retaken from the Rebels. Then, too, with Sherman expected to launch a drive for Atlanta in the spring, Mobile would be a vital point of origin for supply for Joseph E. Johnston's defending Confederates. Deny him that, and his only other seacoast link for receiving succor by rail would be Charleston, South Carolina, itself under threat. More than that, of course, by those same links, vital war materials coming through the blockade into Mobile could reach other parts of the South. Closing that port would help to starve the vital center of the Confederacy.

Mobile had been one of Farragut's objectives even before he was ordered north back in August 1863. Indeed, before leaving the Gulf he had left instructions on how an attack on the harbor should be carried out should the time and circumstances prove auspicious, but in his absence, nothing more could be done than to maintain fewer than a dozen warships patrolling off the harbor mouth. But by late December, disturbing intelligence came to Washington of a massive new ironclad, the *Tennessee*, being built by the Confederates to defend the bay. Worse, there were mounting reports that this ship and other

Above: His visage always concealing a half-smile, Admiral David Glasgow Farragut was the Union's premier captain. New Orleans, Vicksburg, Baton Rouge, were all on his record, and Mobile was to be next.

Below: The gun deck of Farragut's mighty flagship, the U.S.S. *Hartford*, showing some of the rows of powerful guns that turned their fire on the fort, and then on the Rebel ironclad.

lesser gunboats being built by the enemy intended to sally out of the bay to attack the blockading fleet, composed almost entirely of wooden vessels. After the experience of the C.S.S. *Virginia* virtually destroying two wooden warships at Hampton Roads on March 8, 1862, and only being prevented from wrecking more the next day by the arrival of the U.S.S. *Monitor*, Union naval authorities were understandably worried.[2]

Farragut was the man to send. He reached New Orleans and his flagship *Hartford* on January 22, and found a report there that largely confirmed what he had seen himself two days before when he steamed past Mobile and made a personal reconnaissance. In order to control Mobile Bay, he would first have to run a deadly gauntlet of fire. The entrance to the bay was a little more than half a mile wide, and guarded on the left, or west, by Fort Gaines on massive Dauphin Island. The main ship channel lay on the opposite side of the opening, and wooden pilings had been driven into the bottom extending out from Fort Gaines to force any traffic to run very close to the opposite side, right under the guns of massive Fort Morgan on Mobile Point. The fort mounted at least 48 cannon, 28 of them great monsters ranging from 10-inch seacoast mortars and columbiads to terrible imported British guns that threw shells weighing 160 pounds. Farragut had run past forts before, of course, and knew it could be done as he had done it on the Mississippi. However, he had never run a fleet past this kind of armament, nor right under the guns, at 1,000 feet or less – virtually point blank range.

Should he get past the forts, he then would have to contend with the enemy vessels. Of the *Tennessee*, reports came in constantly, and though they varied, still they presented a picture

of a dreadful adversary, with an iron ram, protected by at least six inches of ironcladding, and mounting four 10-inch smoothbores and two massive 7.5-inch rifles. Then there were its consorts, the steamers *Morgan* and *Gaines* and *Selma*, the ram *Baltic*, the ironclad *Tuscaloosa*, and a sister ship, the *Huntsville*. All told they mounted 47 cannon of varying description, and though they were all of the ersatz kind of armoring that was all the Confederacy could produce, still they were better protected than any of Farragut's vessels. In a battle, his fleet could throw more weight of iron, but against some of these ships it might not be effective.[3]

Farragut went to work preparing for an attack as soon as he arrived. Unlike others, he had no fears of the enemy attacking his blockading fleet, but he believed from the start that he needed ironclads of his own to neutralize the *Tennessee* and the other ships that his own vessels could not meet on equal terms. Unfortunately, he found it difficult to get Washington's attention as the opening of the spring overland campaigns approached. "I fear that I shall lead a life of idleness for a month or two", he lamented in February 1864, "as the Government appears to plan the campaigns, and Mobile does not appear to be included just yet." Meanwhile, "I shall have to content myself going along the coast and pestering all the people I can get at." By March he had some fourteen vessels off Mobile, but none of them other than the *Richmond* were truly powerful ships. "I am expecting ironclads from the North," he said sadly, "but God knows when they will arrive." "I only ask for two, and will go in with one." Alas, he would have a long wait.

Part of the problem was the absence of land forces, for any attack on Mobile to be effective it would have to be undertaken as a combined operation. Farragut could steam into the harbor past the forts any time he chose, but there would be little point. Even if he took out the enemy fleet, he could not batter down those forts. They must be attacked from their landward sides, with Farragut bombarding them from the water. Then the garrisons would be forced to capitulate. Unfortunately, he could not get Washington to assign even a few thousand to the task as every available man was sent to Grant or Sherman, or Nathaniel Banks' soon-to-be-launched expedition up the Red River of Louisiana. And so the long, weary months wore on. By May, with Banks' campaign a shambles, Sherman locked with Johnston in north Georgia and Grant and Lee battling through Spotsylvania and on to the North Anna River, the frustration really told on Farragut. "I have written to the Department time and again about the ironclads," he complained, "but it is of

Rear Admiral David Farragut, U.S.N.

On the face of it, there were considerable similarities between the uniforms of Union Admiral David G. Farragut and his Mobile Bay opponent Admiral Buchanan. It should hardly be surprising, since both were largely patterned after the old pre-war United States Navy costume. Where Buchanan's outfit was Confederate gray, Farragut's was dark blue. The latter also had sleeve insignia to show his rank, though it was more intricate than Buchanan's. He also wore the same Model 1852 sword and beltplate a carry-over from pre-war service.

Farragut's shoulder straps differed, however, in that his rank was indicated by two stars, with a fouled anchor in between, which denoting that his rank was equivalent to that of a major general in the Army. No other Union naval officer would rise as high as he, and he would become America's first full rank admiral. He was promoted to this high rank on July 25, 1866, and died four years later.

no use; they will listen to nothing until the fight is over at Richmond."[4]

It should have come as some little compensation to him to know that his opposite number in Mobile got little more out of his government, and for the same reasons. While all eyes focused on the epic battles in Virginia and Georgia, Major General Dabney Maury commanding land forces in and around Mobile, and Admiral Franklin Buchanan commanding its fleet, were almost entirely on their own. In February, Maury had about 10,000 of all arms scattered among his fort garrisons, but feared that they were barely more than half of what he needed. His counterpart Buchanan, however, faced a much more difficult situation. The enemy may have been dreadfully afraid of him, but he knew the truth of the situation, one that ran throughout the Confederate Navy – not enough of everything, and poor quality to what there was.

A veteran of years of service in the US Navy and a friend of Farragut before the war, Buchanan was one of the South's few naval heroes, exclusively because of his command of the C.S.S. *Virginia* during her near-destruction of a Federal fleet in March '62. Badly wounded during the fight, and not a young man to start with, the Marylander remained a rugged, determined fighter, whose enemies through most of his Mobile Bay command were shortages. He worked constantly with the overburdened Naval Gun Foundry and Ordnance Works at Selma, Alabama, to produce the armor and guns for the *Tennessee* and the other vessels at his command. They could turn out no more than one cannon per week, the first one for the ironclad coming in mid-January 1864. The armament for each of his vessels had to be made specifically for its ship, for matters of weight, length and caliber could make a cannon suitable for one vessel quite unsuitable for another, all of which further slowed the manufacturing. Still, the *Tennessee*'s armament was almost complete by the end of January, and one of her lieutenants could boast that "she is a splendid vessel," though he feared an old hand

Union and Confederate Naval Artifacts

1 Union naval flag from the U.S.S. *San Jacinto* commanded by Captain Richard W. Meade, which ran aground in the Bahamas, January, 1865. Earlier in the war, on November 8, 1861, this vessel, then commanded by Captain Charles Wilkes, stopped the British ship *Trent* which was carrying Confederate Commissioners.

2 Confederate States Navy Ensign of the Second Pattern. This flag was captured on the C.S.S. *Florida*, after that ship was seized under cover of night in the neutral harbor of Bahia, Brazil. The capture by U.S.S. *Wachusett* took place in October, 1864.

3 Confederate floating torpedo (a device now known as a sea mine).

These weapons usually consisted of a metal cylinder with a powder charge and buoyancy chamber. The charge could be detonated by means of a spring mechanism connected to a trigger wire floating on the surface, although more sophisticated designs were set off with electric or chemical detonators

Artifacts courtesy of: The Civil War Library and Museum, Philadelphia, Pa: 1,2; U.S. Army Ordnance Museum, Aberdeen Proving Ground, Md: 3

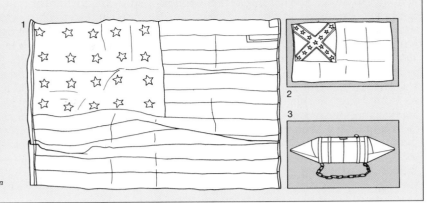

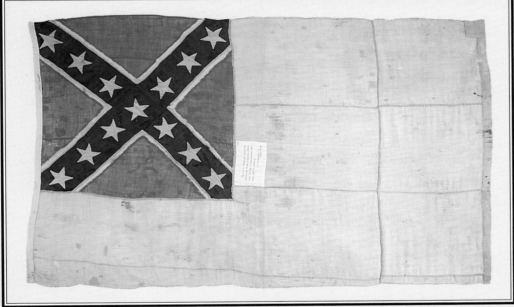

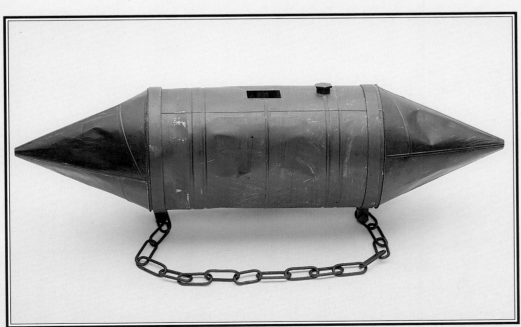

"will almost swear about some of her arrangements." Buchanan would almost swear at his inability to get good officers. He had only two young lieutenants, no midshipmen, and only two "green" mates, a doctor, and a young inexperienced paymaster. "So we go", he lamented.[5]

Buchanan even had to procure his own propellers from Selma, and shells for the guns, though the demand for cannon was so great that Selma had no time to manufacture projectiles. Besides shortage of officers, Buchanan also had not nearly enough seamen for his fleet, and begged Maury to release to him any infantrymen who might have had riverboat experience. Army-Navy rivalries arose, and only an order from the secretary of war at Richmond finally got Maury to turn loose the men Buchanan needed. Then came the cladding of the vessels with plates of iron, two inches thick and about 10 inches square, another time-consuming and laborious process. Also, it tended to make all of them very heavy and low in the water, especially the massive *Tennessee*. When that ship came down the Alabama River and tried to enter the harbor for the first time, it was found that she rode too low in the water to get over the sand bar at the river's mouth. Not until late May did Buchanan get her across. Meanwhile he kept his men busy, some of them being occupied with taking small copper barrels and filling them with gunpowder to make "torpedoes" – actually mines detonated either on contact, or else by an electrical spark sent from batteries on shore. During the spring and early summer, Confederate engineers placed these in several locations in the main ship channel to await any Yankee vessels that dared to enter.

Once equipped, not everyone shared Buchanan's optimism for his fleet. An officer on one of his vessels, the *Baltic*, complained to a friend that "between you and me, the *Baltic* is as rotten as punk, and is about as fit to go into action as a mud scow." The old admiral almost worked himself to death performing all the functions of midshipmen, lieutenants, on up to flag captain, in the absence of good officers. Every day he worked from dawn until 3 p.m. in his office, then made the rounds of the navy yard to see to the armoring

of the *Nashville*, the building of "camels" to help float the *Tennessee* over the bar, the mounting of guns, stockpiling of ships' stores, and more. He even had hopes that a mysterious new "torpedo boat," designed by John P. Halligan and built at the Selma works, would be able to join him. Just thirty feet long, it held a crew of only five, and was powered by a very small and compact steam engine. But when it chose, the boat could submerge. Powered underwater by the crew turning a crank propeller, it could steal up to a vessel,

attach a torpedo to the hull, set a timed fuze, then move away before the explosion. Perhaps thinking of the old myth of the monk who drove the reptiles from Ireland, its builders called it the *Saint Patrick*. It would never see action.[6]

By late June, one of the local army commanders believed that Buchanan looked "humbled and thoughtful" by the frustrations he had faced and the state of some of his ships. The *Baltic* was considered a waste, and others thought the *Nashville* a failure. By late July, the latter still did not have its armor cladding, and on the last day of the month Buchanan was still struggling to have a defective steering apparatus repaired, while relations with Maury remained icy. But time for preparations had run out.[7]

Through a spring and early summer as frustrating as Buchanan's, Farragut finally began to see his plans for an attack approach fruition. "I have long since given up all hopes in the Department of the Gulf", he wrote on June 13, in some excitement at the army's continued lack of interest in cooperating against Mobile and its forts. But at least his own strength grew gradually. And then, following the failure of Banks' campaign in Louisiana, he looked to others for army cooperation, and got it. Major General Edward R.S. Canby, commanding Federal forces at New Orleans, began gathering troops to send for the operation. He hoped to send almost 5,000, commanded by the gallant Gordon Granger of Chickamauga fame, and rapidly the general and the admiral evolved their plans. Finally, too, the long-awaited ironclads began to arrive. By June 1 the *Galena* appeared, an ungainly experimental model that still carried heavy armor. By the end of the month the monitors *Chickasaw* and the *Winnebago* came off Mobile, while the monitors *Manhattan* and *Tecumseh* were on their way.[8]

By the end of July Farragut had his attacking fleet almost intact. He was also getting up-to-date

Above: Admiral Franklin Buchanan was easily Farragut's counterpart, the Confederacy's most distinguished seaman. Only the *Virginia* and the *Tennessee* really challenged Union warships, and Buchanan commanded both.

Below: Farragut delayed his attack for the arrival of several ironclads, among them the double-turreted *Chickasaw*, one of a new class of light draught monitors ideal for river and harbor combat.

intelligence on enemy obstructions and torpedoes planted in the harbor. "Things appear to be looking better", Farragut confessed on July 15. As soon as the monitors all arrived, he would "take a look at Buchanan." Indeed, he believed that with the seas calm, the days long, and the weather excellent, Buchanan was even then passing up his golden opportunity to bring the *Tennessee* and the others out to attack the wooden fleet before the arrival of the ironclads. "If he won't visit me, " said Farragut with typical mock humor, "I will have to visit him."

No later than July 18, Farragut formulated his attack plan, a variant of the tactic he used in running his fleet past Port Hudson the year before. He would take fourteen of his gunboats moving in a column two abreast, into the main channel and past the forts. The four monitors would steam alongside in a single file, between the rest of the fleet and Fort Morgan, acting partially as a screen, but chiefly with a view to turning due east past the fort, to attack the *Tennessee*, moored behind Mobile Point. Once past the forts, the rest of the gunboats would divide their attention between taking on the rest of Buchanan's fleet, and assisting the infantry landed on Dauphin Island to attack Ford Gaines.

During the last week of July, a flurry of orders issued from Farragut's headquarters aboard the flagship *Hartford*, as he awaited only the last of his two monitors and the troops steaming from New Orleans for the land attack. The soldiers, only 2,400 in the end, with Granger in command, left New Orleans on July 29 and arrived three days later. This left only the last monitors to arrive, and Farragut awaited anxiously for the *Tecumseh*, the last one due to come. He had expected her on August 1, and had planned to make his attack on August 3. Bitterly disappointed when he lost that opportunity due to her failure to show, he resolved to give her two more days to

come, or else he would go without her. "When you do not take fortune at her offer," he said, "you must take her as you can find her." With or without that last ironclad, August 5 would be the day.[9]

Happily the *Tecumseh* arrived the next day, and Farragut could make the attack at full expected force. He landed Granger's troops on August 3, to give them two days to move up Dauphin Island and occupy the defenders of Fort Gaines. That would remove most of the pressure from one fort, leaving only Fort Morgan to worry

about. Orders went out through the fleet. "Strip your vessels and prepare for the conflict", he said. All unnecessary spars and rigging must be brought down and stowed. Splinter nets must go up on the starboard side, showing that he expected the worst fire to be that from Fort Morgan, which would pass on their right. The nets were to catch bits of flying debris that shell explosions could turn into deadly missiles. The positions of the men at the wheels were to be barricaded to protect them from being hit while steering the vessels, and the ships' engines were to be given some cover by laying sandbags and chains on the decks above them, to inhibit plunging fire from the fort. Other chains were to be draped over the sides amidships to stop shots from enemy vessels from penetrating to the boilers and machinery. They would get as close as they could to the fort before opening fire, but when it opened on them, they would return fire immediately. And they must watch for the torpedoes, which were believed to be marked by a line of black bouys on the west side of the channel. There were no definite reports of torpedoes on the east side.

The *Brooklyn* would go in first, with the *Octarora* lashed to her port side. Behind them, left to right in pairs, would come: *Metacomet* and *Hartford*; *Port Royal* and *Richmond*; *Seminole* and *Lakawanna*; *Kennebec* and *Monongahela*; *Itasca* and *Ossipee*; and *Galena* and *Oneida*. On their right the monitors would go in with *Tecumseh* leading, followed by *Manhattan*, *Winnebago*, and *Chickasaw*.[10]

Farragut wanted to go in just as dawn was breaking and the flood tide's force would add to their speed. The night before everyone in the fleet knew what was coming. "God grant that we may have good luck," wrote a marine corps private aboard the *Hartford*.[11]

The boatswains called everyone out of their hammocks at 3 a.m. on August 5. By 5:30 the vessels were lashed together and with steam up, they began their run toward the mouth of the harbor under a dark and cloudy sky, a light breeze behind them. They moved slowly, the progress of the faster gunboats retarded by the heavy ironclads beside them. The Rebels in Fort Morgan

Below: A part of the battered face of Fort Morgan, showing not only the impressive depth of the masonry fortifications, but also the effects of the Federal shelling prior to its surrender.

Above: Major General Gordon Granger was one of the Union's most dependable commanders. After saving Thomas at Chickamauga, he now commanded the land forces for the joint attack on Mobile.

first saw them about 6 a.m., and fifteen minutes later opened fire. The *Tecumseh*, last of the fleet to arrive, was the first to return fire. Well before the first of the Yankee vessels came close to the fort, Morgan's guns were delivering a rapid shelling, concentrating chiefly on the gunboats and ignoring the ironclads. Inside the harbor, orders immediately called all hands to quarters, and the *Tennessee* began to get up steam as its gun crews prepared their pieces.

Brooklyn drew the first fire, as Captain Alden, her commander, and his men expected they would. Indeed, this is why they virtually demanded of Farragut that their ship, and not his, should lead the line. They would not risk their commander, and he could not refuse them, especially viewing the fact that *Brooklyn* was the only ship that had a torpedo boom or "catcher" affixed in advance of her bow, expected to explode the devices harmlessly before they came in contact with the ship's sides. If the other vessels followed behind in line as ordered, *Brooklyn* would sweep a clean path for all of them. As soon as his guns came to bear on target, Alden returned fire with his starboard battery, firing grapeshot that appeared to nearly silence Fort Morgan's batteries for a time, by driving their gun crews below to safety.[12]

Then it happened. *Tecumseh*, leading the line of ironclads, got about 300 yards ahead of Brook-lyn. In such a position, and moving to starboard of her, the monitor gained no advantage at all from the other ship's torpedo boom. She was only about 150 yards from the beach, having just passed the tip of Mobile Point, when she struck a torpedo directly amidships. The water rushed in in a torrent, and Captain T.A. Craven immediately ordered the crew to abandon ship. But there was no time. Only twenty-two men and officers managed to get out of the hatchway at the top of the turret. Craven himself stood at the foot of the ladder to safety about to climb, when he stepped aside to let one of the crewmen go first. That was the last man out. In helpless horror, the men on the rest of the leading vessels saw the *Tecumseh*

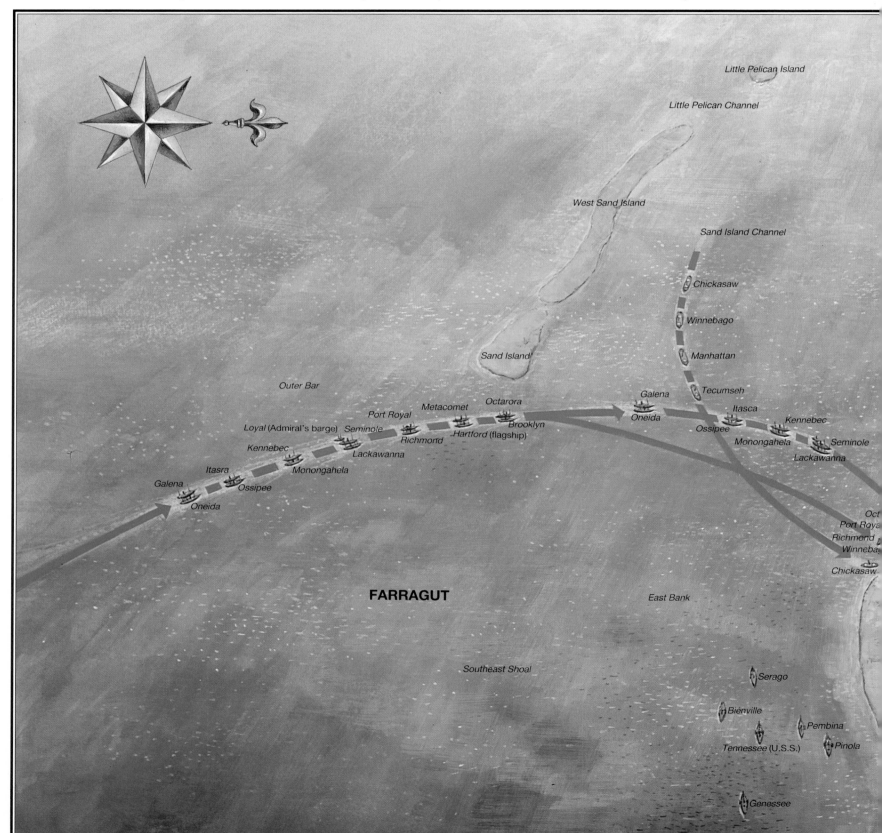

The Battle of Mobile Bay, August 5, 1864

A battle on the water gave an entirely different aspect from one taking place on land, its shape and course defined only by its borders and its depth. Mobile Bay offered an excellent theater for one of the grandest fleet actions of the war.

The entrance appeared formidable at first glance. Mobile Bay, jutting into the water from the east, was protected by massive Fort Morgan, lower center, one of the most intimidating forts to be found anywhere. Almost directly opposite it, jutting out from the west, sat Dauphine Island, with rather less powerful Fort Gaines at its tip, top right center. Obstructions and a line of torpedoes further endangered any Yankee passage, as they jutted out from Dauphine Island and restricted all access to the bay to a narrow channel directly under the guns of Morgan.

Undeterred, however, Farragut counted on his ironclads to pass between the fort and his double-line of conventional ships, thus taking the brunt of the fire. On the morning of August 5 he steamed northward toward the harbor mouth, left, being joined by the line of four ironclads, upper left center. With the *Tecumseh* in the lead on the east flank, and the *Brooklyn* and *Octarora* leading the main line, the fleet approached Fort Morgan. Soon after the firing began, the *Tecumseh* hit a torpedo and quickly went to the bottom, center. Farragut continued steaming into the bay despite considerable damage done by Morgan's guns. Once past the fort, Farragut sent some of his ships after the fleeing Rebel fleet, though the *Tennessee* stood her ground under the protection of the fort. Farragut then took the balance of the fleet to anchor.

This initial phase of the battle had gone remarkably according to plan, except for the tragic loss of the *Tecumseh*. Indeed, Farragut was somewhat surprised that resistance from the Confederate warships had been so feeble. In fact, of course, Buchanan was so terribly outnumbered and outgunned that resistance was destined from the first to be all but futile. The Rebel admiral counted on the fire of the forts, combined with the torpedoes, to slow or stop at least some of the Yankee fleet, while he hung back hoping to pounce on injured or straying vessels.

Before long, however, combative Buchanan brought *Tennessee* back into the fight, steaming straight for the enemy fleet. He hoped to ram one or more of the Union vessels, but heavy fire from them, and their own ramming of his ship, reduced her to an unmanageable hulk. After another half hour of continual battering, the gallant *Tennessee* struck her colors.

It had been one of the largest naval engagements of the war, yet its outcome was never seriously in doubt from the moment that Farragut gave the signal to "go ahead".

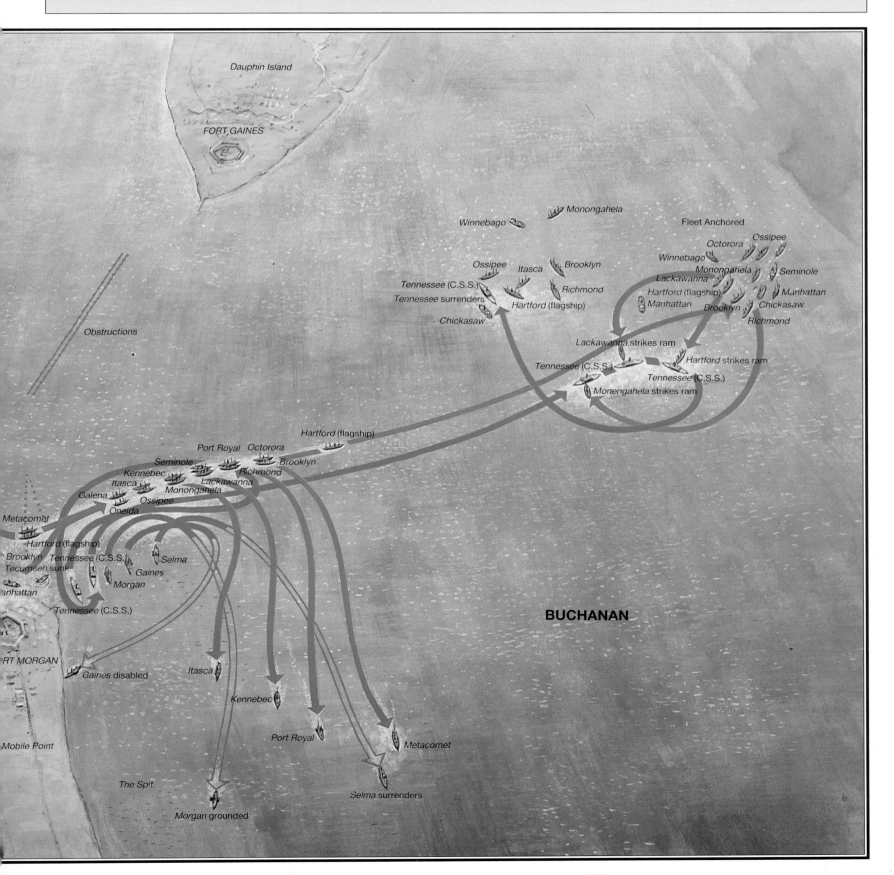

simply roll over and go down, as one described it, "like a shot." With her went Captain Craven and the rest of the crew.[13]

Even as he watched in horror, Captain Alden saw that his own ship had somehow gotten alongside what appeared to be a line of other torpedo buoys, and at once he stopped advancing and started to back out of the presumed danger. This stalled the whole line of ships, and an impatient Farragut did not want to wait. Besides, while they stalled, the *Brooklyn* and *Hartford* were coming under a renewed and heavy fire from Fort Morgan. The admiral himself, to have a better view of the action, actually stood high in the rigging near the top of the flagship's main

mast, reportedly lashed in place so that he would not fall to his death if wounded. It was a foolish way to expose himself, but it gave him a sight that few others could enjoy. "I witnessed the terrible effects of the enemy's shot and the good conduct of the men at their guns," he said with pride a week later, "and although no doubt their hearts sickened, as mine did, when their shipmates were struck down beside them, yet there was not a moment's hesitation to lay their comrades aside and spring again to their deadly work."

In an instant Farragut ordered *Hartford* to take the lead. He sent his ship steaming right between the buoys that had worried Alden. There is more

of legend surrounding this event than any other episode in the entire story of the Civil War at sea. Farragut believed that the buoys did, indeed, mark torpedoes. Certainly the destruction of the *Tecumseh* showed them to be in the vicinity. But he also had intelligence suggesting that the devices may have been in place for longer than the assumed six months it took before water seepage or other causes inactivated them. It was worth the risk to get his attack going again. Exactly what he said in ordering the movement forward will never be known precisely, but from all accounts the meaning was clear. "Damn the torpedoes" was what he meant, and exactly what he proceeded to do.

Confederate Imported Armstrong Rifle

This large garrison piece was made in 1864 in England and imported from there. It is also known as an Armstrong 8-inch Iron Rifle. The weapon weighs 15,737lb excluding the barbette carriage, which allows for easy traverse and a controlled recoil when the Rifle is fired.

The piece was one of those which armed Fort Fisher, the fortress which protected Wilmington, North Carolina – by late 1864 the last major port still open to the Confederacy. An initial assault by Federal forces during December 24-25, 1864 ended in failure, but a second attempt was begun on January 13, 1865, led by Admiral David D. Porter and General Alfred Terry. A massive naval bombardment from over 600 guns from 59 warships heralded a land assault by 8,000 Federal troops, which cut the Fort off from Wilmington and any possible chance of relief from 6,000 Confederates under the command of Braxton Bragg, who had been sent to the area in October 1864. Faced with continued bombardment and attacks from two sides, the Fort's commander, Colonel William Fisher surrendered his garrison of 1,900 men on January 15, 1865. Wilmington fell five weeks later

Artifact courtesy of: West Point Museum, West Point, N.Y.

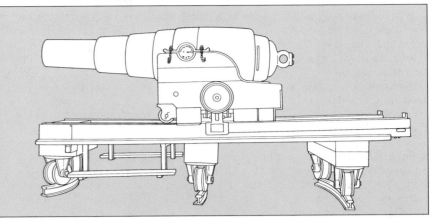

It was about 7:50 when Farragut and the first ships got past the fort. Now they could see the Rebel fleet bearing down on them, the *Tennessee* coming from its moorings on the east, and the *Morgan*, *Selma* and *Gaines* directly ahead. Farragut ordered *Metacomet* to cast loose her lashing and go after *Selma*. *Morgan* and *Gaines* immediately made for the protection of Fort Morgan's guns, however, and the ships coming in behind Farragut were not quick enough at separating to go after them. As a result, *Morgan* escaped later that day, but the fire from Yankee ships succeeded in disabling the other vessel, forcing her crew to run her aground, where they would later destroy her to prevent capture.[14]

The *Metacomet* chased the *Selma* back up the bay, taking fire from the Confederate's stern gun, and giving it back for more than an hour, until at about 9:10 the Rebel vessel's wounded captain struck his colors and surrendered his vessel. Meanwhile, Farragut ordered the rest of his fleet, once safely past the guns of the fort, to come to anchor north of Dauphin Island and about a mile distant from Fort Gaines. And then, about 8:45, they saw the *Tennessee* steaming toward them, Buchanan in command.[15]

There had been no real close action thus far in the morning for the Rebel ironclad. Buchanan traded shots with the Yankee boats as they passed the fort, and hovered nearby apparently with the intention of ramming one or more if the opportunity presented itself, but the ponderous weight and inadequate engine of the ship made her too slow to maneuver competitively with the faster Federal ships. Buchanan tried first for the *Hartford*, seeing Farragut's flag flying over her. She evaded him, and so did the *Brooklyn*, and before long the Confederate admiral simply saw the enemy line of ships steam away from him, content to leave him behind while going after the easier marks of his other gunboats.

But old Franklin Buchanan was every bit as much a fighter as David Farragut. Shortly before 9 a.m., he turned his prow toward the anchoring fleet and asked for full steam from his engines. As

Confederate and Union Heavy Artillery Projectiles

1. 5.3 inch Mullane shell for a rifled 18 pounder Confederate gun
2. 5.8 inch Selma shell for a rifled 24 pounder Confederate gun
3. 6.4 inch Brooke solid bolt for a 100 pounder Confederate gun
4. 6.4 inch Read shell for a 100 pounder Confederate gun
5. 6.2 inch Read shell for 100 pounder Confederate gun
6. Federal 7 inch Schenkl shell for 7 inch rifle. Note that the papier mache sabot is still in place over the projections cast onto the base of the shell to fit the rifling. The fuse on top of the shell also appears to be in position
7. Federal 5.1 inch Stafford shell for a 50 pounder Dahlgren gun
8. Federal 5 inch Whitworth shell for an 80 pounder Whitworth rifle
9. 6.4 inch stand of grape shot for a 100 pounder rifle. Grape was used for close 'soft' targets such as formations of infantry or cavalry
10. 5.8 inch Sawyer shell for a 5.8 inch rifle. Note the cast-on projections designed to fit the rifling in the gun's barrel. The fuse also appears to be in position in this example

Artifacts courtesy of: West Point Museum, West Point, N.Y.

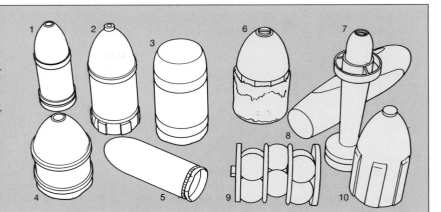

they lumbered on their way toward battle, some of his seamen stood atop the iron-plated casemate of the vessel. "It looked to me that we were going into the jaws of Death," one young engineer, John C. O'Connell wrote that night.

Farragut realized at once that Buchanan's intent was to ram the *Hartford* if he got the chance. Immediately he signalled to the remaining monitors and *Ossipee*, *Kennebec*, *Monongahela*, and *Lackawanna* to join the flagship in attacking not only with their guns, but also with their bows, hoping that by ramming they could bring Buchanan to a halt. *Monongahela* struck first, to no effect, and when *Lackawanna* rammed the ironclad she almost crushed in her own prow, without doing significant damage to the enemy. *Hartford* took the next turn, glancing alongside the *Tennessee* without inflicting injury, but then pouring a withering broadside into her iron sides as they passed. "It was the warmest place that I ever got into," a Rebel seaman recalled, and as the minutes wore on, it got worse.

Steaming around to the ironclad's port side, *Monongahela* rammed her once more, and then a 15-inch shot from *Manhattan* broke through the enemy's iron plating and almost penetrated to the interior of the ship. It was enough to send young O'Connell below with an iron splinter in his shoulder and more in his leg. Then another shot followed hard upon the first. Buchanan had ordered his executive officer, Commander J.D. Johnston, to steer for Fort Morgan once more since the ship was taking on water, but Johnston found that the exposed steering chains that operated the rudder had been damaged and the *Tennessee* could not answer her helm. He went to report this to Buchanan, only to find that another shot, probably from *Manhattan*, had struck the aft gun port while a seaman tried to get the port stopper unjammed. The seaman was literally blown to pieces, another man took mortal wounds, and a fragment of shell or iron hit Buchanan in the leg. The admiral immediately ordered Johnston to take charge.

There was not much left to command. *Tennessee* could no longer steer. Then her smokestack was shot away, so reducing the necessary draft for the furnaces that Johnston could barely keep up steam to make headway. Though his men loaded the guns valiantly amid the din of Yankee iron exploding on the outside of their casemate, defective primers prevented most of the guns from firing, while the aft pivot gun could not fire at all thanks to all three of its portstoppers being jammed shut by Yankee fire. The temperature inside the ship rose to 140° or more, and then Johnston saw the enemy vessels crowd around and literally surround his ship. For the

Admiral Franklin Buchanan, C.S.N.

Admiral Franklin Buchanan was a Marylander with a long and distinguished career in the old Union Navy to his credit before his sentiments took him across the line to the Confederacy. Once in the gray, he saw an extent of service unequalled by any other naval commander. He commanded the C.S.S. *Virginia* in her epic contest with the Federal fleet at Hampton Roads on March 8, 1862, and later the C.S.S. *Tennessee* in Mobile Bay. Thus his was a familiar figure at Confederate navy yards and in Richmond. The regulation naval uniform was a splendid, understated, gray woollen affair with only gold braid to indicate rank on the sleeves, and bullion-rimmed blue shoulder straps. His cap also revealed his rank with its four stars and fouled anchor. He carries the Model 1852 naval officers sword and with it wears the pre-war two-piece naval officer's belt plate. No better commander trod a deck, North of South.

next half hour the Yankees poured a constant fire into the casemate shield, while Johnston rarely if ever returned fire at all. Then the *Ossipee* backed off to get up a full head of steam for a devastating ramming. Looking out at this, Johnston ran to Buchanan to ask if he should yield.

"Do the best you can, sir," answered the admiral, "and when all is done, surrender." Johnston returned to the gun deck, looked out to port, and saw the *Ossipee* bearing down on him. That was all he needed. He sent a white flag up the staff of the brave and battered *Tennessee*, though not quite in time for the charging *Ossipee* to slow down, and she struck a final, glancing blow that fortunately caused no further injury. By 10 a.m., the Battle of Mobile Bay was over.[16]

Almost immediately Farragut put the badly wounded Buchanan aboard a ship and sent him off to Pensacola with many of the other wounded from Union and Confederate ships alike. Buchanan's leg would be saved. All told, thanks to her heavy iron sheathing, *Tennessee* lost only two killed, and nine wounded. Aboard *Selma* another eight were killed or mortally wounded, and seven others injured. Thus total casualties in the Confederate fleet amounted to ten dead and sixteen wounded. By contrast, Farragut took much heavier losses, fifty-two killed and 170 wounded. Hardest hit was the *Hartford* with nearly a quarter of the losses, and the *Brooklyn* the same, being largely the losses suffered by fire from the fort. *Oneida* never got fairly into the fight, taking a shot from Fort Morgan that penetrated her boilers and left her helpless, dependent upon her consort *Galena* to tow her to safety. But none of these figures included the dreadful loss of life aboard the *Tecumseh*. For weeks afterward Farragut would be uncertain just how many men died in her, since no accurate monthly strength report had been filed recently. Yet almost certainly she took down with her at least ninety, including sixteen officers and seventy-four men. None who saw it ever forgot the heart-breaking sight of the proud monitor that struggled so hard to get to the fight at the last minute. Bravely she led the way, firing perhaps the first Federal shot that was also practically her last, going to a watery grave in barely a minute after hitting the torpedo.[17]

The defeat of the Confederate fleet did not, of itself, take Mobile for the Union, but it made the

Above: The gallant *Tennessee*, shown not long after her surrender to Farragut. Against massive odds, she pitted her iron against Yankee wood, but could not overcome poor machinery and a daring foe.

Below: The helmsman who steered the *Hartford* on the day of the battle, stands at his post for a photographer after the fight was done. Like Farragut, he was now a national hero.

rest possible. Granger took Fort Gaines three days later, and Fort Morgan fell on August 23 after a combined land and naval bombardment by Farragut and Granger. Though the city of Mobile itself remained in Confederate hands for several months to come, the loss of the bay put an end to the blockade trade into that port and the taking of the *Tennessee* virtually closed the book on Southern attempts to resist Yankee control of its coastal waters. Never again in this war would mighty ships do battle with mighty fortresses. Never again would ironclad and wooden warships challenge each other for supremacy in this war. Indeed, it would be one of the last fleet fights in all naval history conducted chiefly by wooden vessels. As such, it spelled more than just the end of one of the Confederacy's last sources of supply. While cinching the strangling cord of starvation even tighter around the South, Farragut's victory at Mobile Bay brought down the curtain on the age of wood and sail.

References

1 US Navy Department, *Official Records of the Union and Confederate Navies in the War of the Rebellion* (Washington, 1906), Series I, Volume 21, p.53. Cited hereafter as *O.R.N.*
2 *Ibid.*, pp. 4, 12, 30-31.
3 *Ibid.*, pp. 35-36.
4 *Ibid.*, pp. 95, 121, 122, 141, 300.
5 *Ibid.*, pp. 859, 863, 871, 872, 878.
6 *Ibid.*, pp. 35, 748, 877-78, 886, 896-97, 902-903, 931.
7 *Ibid.*, pp. 903-904, 906, 909.
8 *Ibid.*, pp. 318, 332, 344.
9 *Ibid.*, pp. 378, 388, 400, 403.
10 *Ibid.*, pp. 397-98.
11 C. Carter Smith, ed., *Two Naval Journals: 1864* (Birmingham, Ala., 1964), p. 43.
12 *Ibid.*, pp. 417, 445; Smith, *Journals*, p. 5.
13 Smith, *Journals*, p. 43; *O.R.N.*, pp. 490-91.
14 *O.R.N.*, pp. 417, 419, 444.
15 *Ibid.*, pp. 442, 443.
16 *Ibid.*, pp. 418, 576-77, 580; Smith, *Journals*, p. 5.
17 *O.R.N.*, pp. 407, 445, 492, 578-82.

CHAPTER THIRTEEN

NASHVILLE

DECEMBER 15–16, 1864

Incredibly, given the dire condition that the Confederacy found itself in by the last months of 1864, it still had the spirit and the will, if not the means, to attempt to launch offensives. In the Trans-Mississippi, General Sterling Price, bombastic, preposterous, yet beloved by his men, managed to mount a daring invasion with the aim of reclaiming Missouri and opening another front. He led his army, mostly cavalry, north into Yankee territory, and sent a shock through the Union until he was finally stopped in the biggest battle of the war west of the Mississippi, and the largest cavalry battle of the war, at Westport, site of present-day Kansas City, Missouri.

In the Shenandoah, too, the Rebels tried to retake the initiative. Under the erratic, yet combative, Jubal Early, the Army of the Valley fought three hard battles that fall, and in one, at Cedar Creek, they surprised and came close to defeating Sheridan's forces. But they only came close, and by the end of the year, the Shenandoah belonged incontestably to the Union.

But there was one more offensive in the Confederate blood. It depended upon a tragic, battered young hero, an army that had spent much of the war fighting itself, and the will of an indomitable Yankee general whose own people thought him too slow.

PITY THE poor Confederate Army of Tennessee. No army of the Civil War, not even the hard-luck Army of the Potomac, suffered under such a sucession of sad and inadequate commanders. The statistics of its series of commanders is eloquent. Jefferson Davis created only seven full rank generals for field service during the war. The Army of Northern Virginia used only two of them during the entire course of the war, Lee alone commanding for three of the four war years. But this great army of the west went through no fewer than five. Its brightest hope was probably Albert Sidney Johnston, whose tenure was so brief and ill-starred that none then or later could say with certainty whether or not he might ever have realized his promise. Beauregard gave up any hope of victory at Shiloh, then abandoned his army weeks later, leaving it to its turbulent year and a half in the hands of the hated Braxton Bragg. Many thought that promise returned when Joseph E. Johnston at last replaced Bragg in December 1863, but the following campaign for Atlanta only revealed him to be the Johnston of old – timid, lacking in moral courage, and contentious. Where Bragg spent most of his time fighting his own generals, Johnston spent his falling back and fighting with Richmond. On July 17, 1864, after months of this behavior Davis quite rightly relieved him of his command. Quite wrongly, alas, he replaced him with the newest and youngest of the Confederacy's full generals, John Bell Hood.

That Davis could turn to Hood reveals just how desperate he was for a commander. He could not take Lee from Virginia. Bragg was out of the question. Davis himself so loathed Beauregard that he could not be considered, and the only other available full general, E. Kirby Smith, was already in command of a virtual empire west of the Mississippi and could not be spared. Davis might have looked toward someone like Hardee, but he had declined the army command once before, and now Davis considered that refusal permanent. He needed someone who had the will for a fight, unlike Johnston who seemed only to avoid combat and give up ground. Of all the corps commanders with that army in the summer of 1864, only Hood seemed to have the spirit. The sad-eyed young Kentuckian, just turned thirty-three, also had the stomach for flattery and sycophancy. Recovering in Richmond from the loss of his leg at Chickamauga, he lost no opportunity to ingratiate himself with the president, and once back with his corps, he wrote frequently to Davis to complain of Johnston's hesitation. Finally fed up with the latter, Davis gave his army to Hood, along with promotion to the temporary grade of full general.

At least Hood did fight. He had the courage to use his army, but alas, not the brains. He was not a stupid man. Rather, he lacked the temperament or maturity for headquarters leadership. Removed from his proper position at the head of a division, Hood as a commander could not gain

Just one of the formidable earthwork fortifications ringing Nashville, revealing the strength of the defenses that Hood tried to throw himself against in his attack on the Tennessee capital.

the respect of his corps commanders, and alienated many of the rank and file by intimating that they lacked the courage or fortitude to attack the foe. Certainly he gave them enough opportunity to practise, for he launched a series of offensives as soon as he took command of the army. Yet he also gave them ample cause to fear to attack, for his tactical plans were so fuzzy and undeveloped that time after time his men suffered terribly for it. Indeed, after assuming command, Hood lost every battle he fought, and in September had to give up Atlanta to Sherman to prevent losing his army with it.

Even then, however, the bold commander continued his aggressive tactics. Having lost Atlanta, Hood moved his army quickly north of the city to threaten Sherman's railroad communications back to Chattanooga. As a result, Sherman went after him with most of his army, and gradually they maneuvered their way back almost to Dalton, where the campaign for Atlanta began. Very nearly from the first, Sherman suspected that Hood might try to force him out of Georgia entirely by striking even beyond Chattanooga, to Nashville, and early on he assigned General George H. Thomas to take a division of the XIV Corps back to organize a defense for the Tennessee capital and its vital supply warehouses.

Hood, in fact, did not have very definite plans, one of the consistent drawbacks of his generalship. At first he hoped only to force Sherman to withdraw back to the Tennessee border, and there engage him in battle, perhaps with a view to driving the enemy out of Chattanooga. It seemed to work well at first, encouraging the Confederate commander to think in much bolder terms. Richmond had approved his initial plans. But now, by mid-October 1864, he realized that Sherman simply outnumbered him too overwhelmingly to allow for any hope of success in an attack. Besides, Sherman now began to indicate a disinclination to keep following him once they were in north Georgia. In fact, Sherman had decided not

to pursue Hood any farther. The Federals still held Atlanta, and were still between Hood and Chattanooga. In fact, the confident Sherman chose now simply to pretend that Hood and his army did not exist. Sherman would turn his attention elsewhere and leave Hood at his back. Thomas could take care of him if he posed any kind of threat. Meanwhile, Sherman would march toward Savannah and the sea.

Hood, too, realized that he had achieved nothing by his maneuver. Now he needed to do something far more significant if he was to

Below: Taken from the very portico of the capitol building itself, this image shows downtown Nashville in 1864 at the time of the battle, with cannon virtually on the steps of the capitol.

Above: Photographed in more tranquil times early in the war, these Louisiana boys of the New Orleans Washington Artillery were among those who saw their army all but destroyed by Hood's ill-fated campaign.

achieve anything with his army before the winter came to close down operations. How long the idea gestated in his mind no one knows, but by the end of the second week of October his mind was set. He would take his army quickly across the northeastern corner of Alabama, up to the Tennessee River, cross it, and strike deep into middle Tennessee, driving toward Nashville. If he could meet and defeat Thomas and any others in his way, he could take Nashville with all its supplies, then continue his triumphant march into Kentucky, driving all the way to the Ohio River.

Victorious there, he could then turn east, cross the Appalachians, and either join with Lee, or else strike the Yankee army besieging Petersburg from the rear. It was the sort of grandiose plan that only an immature commander could devise. It might win the whole war, but it depended at a hundred turns on every bit of luck being with the Confederates, on the enemy showing almost no initiative or resolve in countering Hood's movements, on receiving the kind of reception in recruitment and succor in Kentucky that Bragg did not receive in 1862, and on a host of other conditions ranging from the unlikely to the impossible. The planned drive to take Nashville was bold, and just within the achievable if Hood moved quickly. Everything after that was a fool's errand.[1]

Unfortunately, Hood did not move quickly, even though he had himself made speed with one of the essentials of his plan. He moved out of Gadsden, Alabama, on October 22, reaching Tuscumbia nine days later. There he would cross the Tennessee River, but he lost a precious twenty days waiting for supplies that a better prepared commander would have had arranged in advance. Only on November 21 did the Army of Tennessee once more march northward and cross over the line into its namesake state. What lay ahead of him was a hastily assembling variety of commands and commanders. Sherman had sent the IV Corps, under General David Stanley, and the XXIII Corps, led by General John Schofield, to Pulaski, Tennessee, sixty miles south of Nashville. Thomas, meanwhile, had three divisions of the XVI Corps on their way to him, and a host of small garrisons and posts throughout middle Tennessee to draw from. All told, including the force at Pulaski under the overall command of Schofield, he could look for about 60,000 men to meet Hood's 40,000 or more.[2]

Hood's line of march northward would take him several miles to the west of Pulaski, and Schofield at once saw the possibility that the enemy could get between him and Nashville. At once he pulled back toward Columbia, on the Duck River, where his own line of withdrawal and Hood's line of advance converged. It was a near thing, with Schofield just getting to the river crossing first on November 24. Hood's advance elements were there the same day, but it was three days before the entire Army of Tennessee arrived, and by then Schofield was too well entrenched for Hood to risk an attack.

Undaunted, however, and showing a return of the aggressiveness that he had lost for so long at Tuscumbia, Hood concluded to march east a few miles, cross the Duck, and then strike for Spring Hill, fifteen miles north of Columbia. If he got

General John Bell Hood, C.S.A.

General John Bell Hood gave up about as much of himself as any officer who sacrificed for the Confederacy, short of losing his life. He lost the use of an arm after a wound at Gettysburg, then suffered amputation of a leg following a desperate wound at Chickamauga. Any uniform for one who suffered what he had, would most certainly need to be tailor-made.

Hood wears the regulation general officer's frock coat, with the buttons in parallel groups of three's, denoting rank above that of brigadier. His sleeve braid and the trim on his kepi show the four rows of gold required for a general, and though his collar cannot be seen, it would reveal three stars surrounded by a wreath. If his belt could be seen over its yellow sash, his belt plate would be a distinctive rectangular Confederate battle flag design, and extremely rare.

This portrait shows him prior to Chickamauga, when he still had his right leg.

there first, he would once more have a chance to cut Schofield off from his line of retreat, and out-numbering the Federals by 40,000 to about 26,000, he could hope for a conclusive victory.

Once more it turned into a race as Schofield realized Hood's intentions and himself made for Spring Hill. Once again it came down to a whisker of a difference between survival and disaster for Schofield. On November 29 Hood himself led most of his army across the river and on the road for Spring Hill, coming in sight of it by mid-afternoon, with some 25,000 men at hand or nearby. In Spring Hill there were but 5,000 Yankees. But then came an afternoon and even-ing of mistakes, confusions, and poor coordinat-ing command at many levels, most notably Hood himself. The result was that the attack was stopped when only fairly begun, and Schofield escaped. Hood would ever-after blame General Cheatham for the failure to catch the enemy. Un-charitably, he would also blame the army itself for being "unwilling to accept battle unless under the protection of breastworks." If the army felt that way, it was due more to Hood than anyone else, for his fruitless waste of men in his attacks around Atlanta.[3]

In the end, Schofield marched almost all of his army up the road to Spring Hill and beyond, within sight and sound of the waiting Rebels, and once of the most spectacular opportunities of the war went wasted. The next morning there was nothing for the Confederate to do but follow. About 3 p.m. that day, November 30, Hood's advance ran into Schofield, now positioned in and around the town of Franklin, and this time the two armies neither missed nor evaded each other. A ferocious battle ensued, lasting for five hours or more, and well after nightfall. If he needed any further proof of his error of thinking about the valor of his army, Hood here saw it dis-proved. With a fury that can only come from courage, his divisions hurled themselves at Scho-field's lines time after time. To inspire the men, their generals risked themselves with absolute recklessness, and it took a heavy toll. Five fell

Western Variant Confederate Battleflags

1 Flag of the 1st and 3rd (combined) Regiments, Florida Volunteer Infantry, issued to the unit in 1864
2 Flag of the 13th Regiment Louisiana Volunteer Infantry, issued to the unit, April 1864
3 Flag of the 57th Regiment, Georgia Volunteer Infantry. At one time a unit of Bragg's corps, Army of Mississippi/Army of Tennessee. The shape, the large pink border around

the flag, and the twelve, six-pointed stars, indicates that this is the second pattern flag of Bragg's corp. The first had been smaller and squarer of the type introduced into the western theater by General Beauregard, after he was transferred from Virginia in February, 1862. These flags follow the same basic pattern as those first issued to the Army of Northern Virginia in

November, 1861. These flags were usually made of wool bunting rather than silk, which was found to wear too quickly under campaign conditions
4 Unit flag of the 57th Regiment Mississippi Volunteer Infantry. At one time a unit of Hindman's Division, A. S. Johnston's Army of Tennessee. Issued to the regiment sometime after March, 1864

Artifacts courtesy of: The Museum of the Confederacy, Richmond, Va

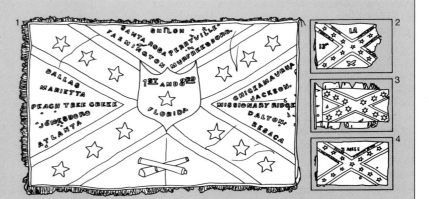
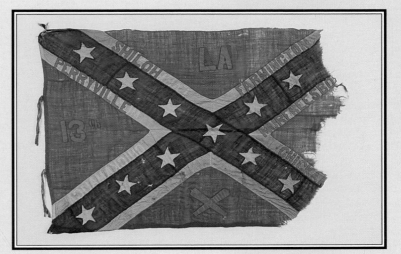

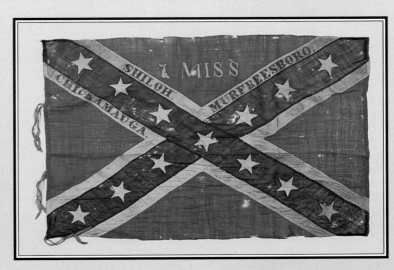

killed or mortally wounded, among them the irreplaceable "Stonewall of the West", Patrick Cleburne. In no other battle of the war did either side suffer such a fearful loss of general officers. It gained nothing. Schofield held out, withdrawing toward Nashville during the night while Hood tried to reorganize his battered legions, intent on pressing on.

By now, Thomas was frantically assembling forces from all quarters to defend Nashville. The XVI Corps reinforcements arrived on December 1, joining the 10,000 infantry and 6,000 cavalry already there. Thomas armed clerks and carpenters and every customarily non-combatant soldier at his command in the quest to defend the city. Men returning from furlough, many of whom had gone North to vote in the recent election, arrived just in time to be commandeered, along with ambulatory hospital patients, and anyone else capable of holding a rifle. By the time all of Schofield's men reached Nashville, Thomas' whole force totaled close to 60,000.

Thomas gave them an improvised organization. Thomas J. Wood, the same general whose strict obedience to an order led to the rout at Chickamauga, now commanded the IV Corps and its three divisions. Schofield reverted to command of the two divisions of the XXIII Corps now with the army, while General A.J. Smith led the three divisions of the XVI Corps. Several brigades, including some black units, made up a Provisional Detachment under General James Steedman, and the old Nashville garrison, including a brigade from the old XX Corps and a scattering of other units, followed General John F. Miller. One of Grant's favorites, the young Major General James H. Wilson, commanded four small divisions of cavalry. Quite clearly, they would all have to fight well in defense, but in case of an offensive, Thomas would naturally turn to the old veterans in the three infantry corps.

Nashville sat on the south side of the Cumberland River, much of it within a bend in the stream. In the years since Federal occupation in early 1862, much work had been done in constructing fortifications and earthworks extending across the open side of the bend, while the river protected the city's back. Several roads led on to the city from the southward, most notably the Hillsborough Pike, the Granny White Pike a little more than a mile east of it, and the Franklin Pike about two miles further east. Thomas' actual communications with Sherman in Georgia were via yet more roads southeast of the city. It was a lot to cover defensively, even with 60,000, especially when many of them were untrained and inexperienced, convalescent, or even civilian.

Thomas went about it with characteristic deliberation. Never rash or impulsive, nor inspired, "Old Pap" as the men called him was, rather, rock solid, methodical, and unhurried. When he saw that Hood, now reduced to probably no more than 25,000 effectives, followed Schofield and then commenced setting up defensive lines around Nashville, Thomas had to know that all the advantages lay with the Federals. Nevertheless, he would not be rushed. Besides, he had more in mind than simply pushing Hood back, which he could easily have done as soon as the Confederates appeared in his front. Instead, he wanted to take Hood and his army out of the war for good. He knew that the Confederates were strapped for supply, strung out with heavy losses due to straggling. If he could punch a hole in the gray line facing him and get swift-moving

Above: The Cumberland River was vital to the supply and reinforcement of Nashville and middle Tennessee. As a result, to preserve such lines against Rebel cavalry attack, the Federals fortified many of the railroad bridges like this one spanning the river.

cavalry through it and into the enemy rear, a successful infantry assault could then virtually crush Hood. The trouble was, Thomas had only a few thousand troopers, and very few good mounts. Consequently, he decided to risk waiting until more cavalrymen and horses could reach him before mounting his attack.

This did not sit well with Washington, or with Grant, who never enjoyed a great fondness for Thomas in the first place. He especially feared that Hood would withdraw and get away if Thomas did not attack quickly. The day after Hood reached Nashville, Grant urged Thomas to attack, and on December 6 he sent a direct order to do so. Still Thomas delayed. "There is no better man to repel an attack", Grant said of "Old Pap", "but I fear he is too cautious ever to take the initiative." Others in Washington began to suggest to Grant that Schofield should replace Thomas. Still Grant continued his urging until December 9, when he finally ordered Thomas to turn over his command. But by then a terrible freezing rain storm had blanketed the Nashville area under a virtual sheet of slippery ice, making any movement impossible. Grant relented, recalling his

order, but two days later told Thomas to attack, ice or no ice. Soon thereafter he dispatched Major General John A. Logan to Nashville with orders to take command from Thomas if the army had not attacked by the time he arrived.[4]

As he prepared for his offensive, Thomas arrayed his forces all across his front. On his left, at the eastern line of defenses, he placed Steedman. Schofield and his corps came next, stretching toward the center, where Schofield's right met Wood's left. At Wood's position the line turned back to the right, taken up by Smith's corps and then, on the extreme right, by Wilson's cavalry positioned alongside the Cumberland. Having taken his time to prepare for an attack, Thomas devoted considerable thought to his precise plan for the offensive, and when he met with his commanders on December 14, everything had been as well thought out as any battle-plan of the war. It was quite simple. Along Hood's entire front, the Federals were to make demonstrations in force to pin down the enemy infantry. Then Smith and Wood would deliver a major hammer blow against Hood's left in their front. At the same time, Wilson would attack with his cavalry. Meanwhile, Steedman was to make a secondary attack on Hood's right, which Thomas felt might have been weakened in the last few days. Schofield would remain in reserve, ready to exploit any advantages, and also perhaps as punishment for being Grant's presumed pet. They would advance on the next day, December 15.

Thomas had little idea of the true plight of Hood and his army. Running out of supplies, short of men and material, Hood had led the Army of Tennessee to Nashville because he simply did not know of anything else to do. Thoughts of Kentucky were out of the question now, yet to retreat back into Georgia would mean moving toward Sherman, with Thomas in his rear. Hood never had many practical strategic ideas, and now he was fresh out of any of them. He could only think to wait here at Nashville and hope for reinforcements from faraway Texas, and hope that he could repel a Yankee attack. With this in mind, he put his men to work building a line of defenses and redoubts. Despite the bitter cold, in spite of poor clothing, lack of shoes and tools, the poor Confederates struggled with the cold earth to erect their works.

Hood's army reflected a lot of changes since its last winter battle in Tennessee a year before, even since Atlanta. All of the corps commanders were new since Chattanooga. The erratic Cheatham now led one corps, composed of three divisions, two of which had lost their commanders at Franklin – Cleburne and John C. Brown. Only Breckinridge's old division had the same general it had a year before – William B. Bate. Hood placed Cheatham on the right of his line, between the Franklin Pike and the Murfreesboro Pike. Thinly manned, Cheatham's works could not quite reach to the banks of the Cumberland.

On Cheatham's left sat the three divisions of Lieutenant General Stephen D. Lee's corps, under Generals Edward Johnson, Carter Stevenson, and Henry Clayton. Lee's line extended to the left to the Granny White Pike, and both he and Cheatham took position just below Brown's Creek, which flowed left to right across their front and emptied into the Cumberland. The creek in front of them, and the line of low hills on which they erected their works, gave the outnumbered Confederates at least some advantage of ground. Meanwhile, to Lee's left sat the corps of Lieutenant General Alexander P. Stewart, three divisions led by William W. Loring, Samuel G. French, and Edward C. Walthall. Stewart extended Lee's line from the Granny White Pike to the Hillsboro Pike, then angled southward along the latter, his position protected by a series of earthwork redoubts constructed along the pike.

Above: This 1864 or 1865 photograph shows a portion of the battleground south of the city of Nashville. Little remains to suggest that here a once-mighty Rebel army was nearly crushed by Federal troops.

None of them were finished as yet but at least they afforded some protection. But a three-mile gap existed between Stewart's left and the Cumberland River to the west of Nashville, and all Hood had to fill it was one division and one brigade of cavalry – a woefully inadequate force for such a task.

Indeed, both Hood's army organization as well as his dispositions revealed the pitiful weakness of the Army of Tennessee. Through intrigue and politics, transfer, or death, gone now were the old

experienced corps commanders like Hardee, D.H. Hill, Breckinridge, and others. Cheatham had been at the head of his corps for only two months. Lee took over Hood's old corps in July when Hood took command of the army. Stewart was the senior in corps command, having replaced Polk after his death in June. As for the divisions under them, it was all a mish-mash. On Cheatham's three divisions, only one, Cleburne's old command, had the same organization as a year before. Brown's and Bate's divisions were a

Below: More vestiges of the ruin of war survive in this contemporary image of one of the fortified hills surrounding the city. The remnants of rail cars and buildings testify to Nashville's brush with capture.

patchwork of brigades formed out of Breckin-ridge's and Hardee's old corps. Lee's corps retained far more integrity of organization, Stevenson's and Clayton's divisions being intact since Chattanooga, and Johnson's almost so. Polk's old corps, now under Stewart, had only rejoined the army during the previous campaign, and was itself reconstituted from several other commands. Worse, none of the corps commanders and only two of the nine division leaders had exercised that level of command for a full year. The war and its own internal strife had battered the high command of the army to pieces. Regiments were placed in brigades with others they did not know, men and officers no longer had the

old bonds of long service together, and they were all tired, cold, and hungry. Worse, they were in front of Nashville with nothing to gain, spread too thin to hold their line, and led by a general in whom they had only shaky confidence. That Hood himself realized his weakness is evident in the works he had them build. That line of redoubts running south along the Hillsboro Pike was his tacit admission that he could not hold the left of his line with the great gap covered only by his cavalry. He expected the enemy to seek his weakness and strike him there.

Thomas had seen the same things, and would oblige Hood if he could. "Old Pap" arose early on December 15. Happily a modest thaw in the

weather had come. The ice was gone and the ground, though mired and boggy, was at least passable. Moreover, with the change in temperature, a fog covered most of the field covering the Federal's early movements. Still, the Rebels certainly heard the shrill bugles that called their foes into line well before dawn, and there was no concealing the clank and bustle of tens of thousands of men, animals, and vehicles and guns, moving into position.

Thomas intended to move Steedman's command forward against Hood's right flank first. In the predawn fog, Steedman formed his men and moved them into position. Off to his right, others moved as well, most of Thomas' army leaving

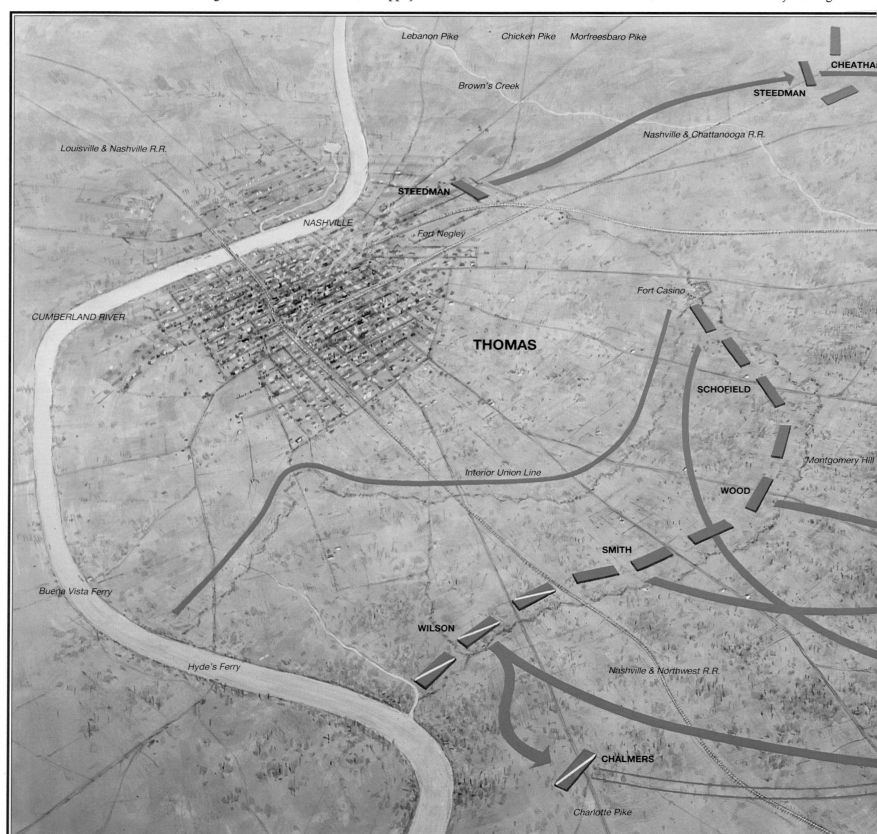

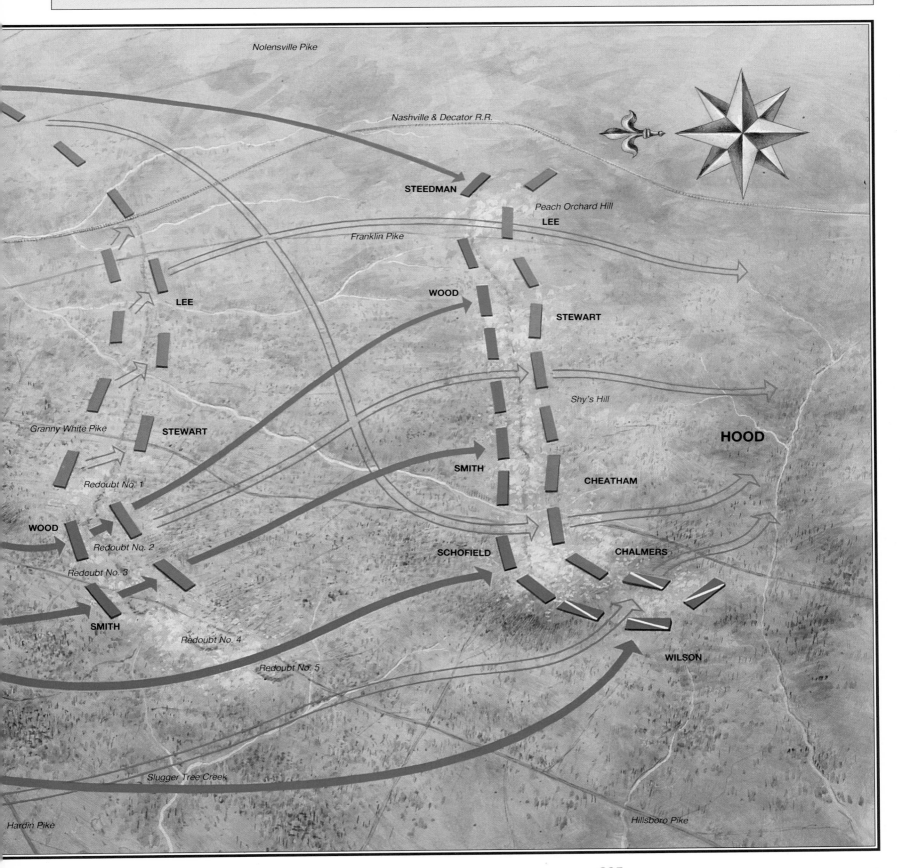

The Battle of Nashville, December 15-16, 1864

Nashville sat inside a bend of the Cumberland River, approached by a series of roads leading west, south, and east. Most important in the Confederate approach to the city, and the subsequent retreat, were the Nashville Pike, top center, the Franklin Pike just below it, the Granny White Pike below that, and the Hillsboro Pike, below that, along which Hood constructed his left flank defenses, when he determined that his intent to

attack was impracticable.

Thomas, meanwhile, had erected his own perimeter of defenses just outside Nashville's limits, left center, and later made a second, outer line of works on his center and right, covering the Granny White and other routes to its west.

When Hood's offensive stalled, and when winter and his own hesitation allowed, Thomas planned an offensive of his own. On December 15 he sent Wilson and his cavalry

out to clear the Charlotte Pike of enemy cavalry, bottom center, and then to start a sweep around Hood's left flank. At the same time, Smith and Wood moved out to strike the Confederate defenses along the Hillsboro Pike, where sharp fighting broke out, lower center. Meanwhile, Steedman advanced against Hood's right behind Brown's Creek near the Nashville Pike. Heavily outnumbering his foe, Thomas had the battle almost all his own way all day, and

Hood was forced to pull back to a hastily erected new line that night between the Franklin and Hillsboro Pikes.

Here Thomas struck him again on December 16, and this time decisively. Steedman swept down on S.D. Lee on Peach Orchard Hill, top right, Wood and Smith struck the center under A.P. Stewart, and Schofield turned Hood's left flank under Cheatham on Shy's Hill, lower right. Wilson, meanwhile, almost completed the envelopment, riding into

Cheatham's rear. Faced with such an overwhelming force on three sides, Hood's army simply started to melt away as the threat of being surrounded became evident.

More troops began to turn and run for the Franklin Pike. A few heroic rearguard actions, particularly one made Hood's small cavalry force, held off the Federals long enough for the battered Army of Tennessee to make good an escape, but the army by that time was in such chaos

that its officers could not even make reports as to how many of their men were left dead on the field. Meanwhile Thomas continued to attack. The Confederate's retreat was one long rearguard action fought out under the most appalling conditions. Only after they had crossed the Tennessee, did Thomas stop the pursuit.

At the end, Hood lost 5,500 or more, forty percent of his army of 24,000. Thomas, by contrast, lost only 3,000 of his 50,000 engaged.

Above: The doleful countenance of General John Bell Hood was noted by many during the war. He lost a leg, use of an arm, and almost an army when he was put in a position for which he simply was not capable.

Above: During the fighting for Nashville, a cameraman captured several images from within the Federal works. Here several Yankees stand gazing at the smoke of battle in the faraway distance.

Right: Men, animals, and artillery support vehicles stand by ready during the Battle of Nashville, though Thomas' men moved so fast during the engagement that their cannon could hardly keep up with them.

their works and marching out to the positions from which they would launch their demonstrations and attacks. It took time in the mist, but by about 8 a.m., they were all ready. Across the way, the Confederates were up, too, and back at work on their redoubts. Lee was even then sending a note to Stewart warning that "I think you may look out for a demonstration on your left to-day." Whether it was precognition or simple military instinct, Lee could not have been more correct. Within minutes he heard the sound of Steedman's guns as the battle commenced.[5]

It went well for Steedman at first. His skirmishers pushed forward and rather handily drove in the Confederate outposts. But then the Federals came up against the first of the earthworks, and their initial assault, though 4,300 strong, was driven back in considerable disorder, and successive charges against Cheatham's line fared no better. Shortly before noon Steedman relaxed his pressure, having he hoped, achieved his goal of diverting Hood's attention away from the main Federal attack shortly scheduled to strike Stewart.

It did not work. Indeed, Hood anticipated what was coming at least two days before, and had already reinforced his cavalry covering the ground from the Hillsboro Pike to the Cumberland. The actual Yankee attempt to get around his flank launched shortly after Steedman's demonstration began. Wilson had about 12,000 cavalrymen, a quarter of them dismounted, and his orders were to move out of the works and strike almost straight south, keeping just in the right of General Smith's XVI Corps infantry, which would move directly against Stewart's lines along the Hillsborough Pike. Unfortunately, the fog and some confusion among Smith's troops delayed the movement until about 10 a.m., but once it started, it operated perfectly. Smith pivoted his left on Wood's right, and swung his corps like a door on a hinge, sweeping across the muddy

ground to strike Stewart. Meanwhile, Wilson moved straight against the pitiful Rebel commands trying to hold that three-mile gap between Stewart and the river. With only a tiny division of cavalry — in fact a mere seven regiments — and a single infantry brigade with only the strength of an undersized regiment, to contend with, Wilson met virtually no opposition worthy of the name.

The Yankee troopers drove the foe straight back from them, and incidentally, further away from the rest of Hood's army. Then they moved for Stewart. No one hurried. Thomas' plans had been deliberate and meticulous, and he instilled this same sense in his subordinates. It was as if they had a wounded bull tethered. There was no need to rush. Rather, they would move slowly and decisively, in order that all their blows might tell, and the destruction of the trapped foe would be complete.

The first of Stewart's works came under fire shortly after the advance began. Smith brought up artillery and started to pound the redoubt prior to a successful infantry charge. By about 12:30, Wood started to bring his right-most brigades into the action as well, and they easily swept over the advanced Rebel defenses. Then the same technique was applied to each of the other redoubts in turn, artillery barrage first, then infantry assault to follow. Redoubt Number 5 fell, at the very left end of Stewart's line on the Hillsboro Pike. Then they turned northward, up the pike to Redoubt Number 4, and though it gave them more resistance, still it, too, fell. Now, as previously ordered, Schofield and his corps in reserve marched across Smith's rear to take position on the extreme right of the Federal army. This put them in position to press Hood's collapsing left even harder, while also freeing Wilson's cavalry to strike out farther south then sweep up and take Hood in the rear. It was all going according to plan.

Now with the two redoubts taken, Smith came up against the infantry of Walthall's division placed behind a stone wall that ran nearly 1,000 yards along the Hillsboro Pike. The Federals shelled them mercilessly, then launched a massive infantry assault that drove its way across the pike, moving northward toward the rear of the balance of Stewart's corps and Lee's. Though Stewart tried valiantly to establish a line to meet the oncoming enemy, the attempt proved fruitless, and Stewart and Lee only barely got their remaining divisions out in time before the Federals swept everything before them. By late in the afternoon, the withdrawing Confederate left had pulled back to the Granny White Pike, almost a mile from where it began the day. Only the coming of nightfall prevented Schofield and Smith from pressing it even farther.

Hood faced a terrible predicament. His left had all but collapsed, and his right had been battered somewhat early in the day. Wood did advance against his center in the afternoon, but it was a half-hearted attempt that Hood ought to have read as a demonstration and nothing more. Now with darkness descending over the field, he faced his army's greatest crisis since Missionary Ridge. His line was too long, even after the day's contractions. He pulled part of Cheatham's corps out of line on the right and moved it to the far left, and later in the evening issued orders for the balance to follow. This corps, still fresh and relatively unblooded, would hold his left now. Hood set it up along the Granny White Pike, with Bate's division place on an eminence called Shy's Hill, somewhat in advance of the pike. The remnant of Stewart's corps was pulled back almost a mile from its position in the center, to form a new center extending from Granny White almost to the Franklin Pike, and there Lee was ordered to put his corps in position to hold the right flank, anchored on Peach Orchard Hill. When the cavalry finally succeeded in rejoining the main

army that night, Hood put it on Cheatham's left to help hold the Granny White Pike, Cheatham's only available line of retreat. Lee and Stewart, in case of disaster, were to withdraw on the Franklin Pike.

Of course, the disaster was already upon them, and it was Hood. Having come this far with nothing to accomplish, he now compounded his folly by refusing to see that his army was in mortal danger of being not just beaten, but destroyed. The only reasonable course open to him was to get out during the night. By swift marching and hard rearguard fighting, he might have gotten the army back to safety south of the Tennessee, though his utter lack of cavalry, and Wilson's 9,000 mounted troopers operating at will on his flank and rear, might have made even that impossible. Instead, Hood doomed his army by deciding to remain for another day. Then, as he imagined it in the fantasy he concocted, he would pull back, only to strike Thomas' exposed right flank the following day, thus retrieving the situation. The very idea would have been laughable but for the Confederates destined to die on December 16 – a cruel joke indeed.

Thomas, too, had much on his mind that night. He wired Washington of the day's events, receiving in return thanks from the War Department and from Grant, who realized that now Logan would not have to supersede "Old Pap." "Push the enemy now," said Grant, "and give him no rest until he is entirely destroyed."[6]

The Federals rearranged their lines that evening, though not, like Hood, with an eye toward lines to retreat. Steedman remained on the far left, and moved forward to face Lee. Wood put his corps in position on Steedman's right, and then Smith took up the line, so that between them they covered the whole center from just east of the Franklin Pike to just west of the Granny Smith. Schofield then wrapped his corps around Hood's left, from just north of Shy's Hill, on

around to the south. Wilson, meanwhile, went into position on Schofield's right, and actually extending beyond and behind Cheatham's depleted flanks. The Federals could literally come at Hood's battered army from three sides at once when the battle commenced again in the morning, and this is about what Thomas had in mind. As before, the Federal center and left were merely to hold the enemy in their front, while Schofield, Smith, and Wilson swept once more around the Rebel left. If they were as successful on December 16 as the day before, they could bag the whole army.

It almost could not have gone better for the Yankees. The advance began at about 6 a.m. as Wood and Steedman moved into the works abandoned by the enemy the night before. Over on the Federal right, Schofield skirmished briskly with Cheatham throughout the morning while the other Federal forces moved forward, but by noon no actual engagement had yet erupted. Once again, Thomas took his time. As the afternoon began, Thomas personally called at all points along his line, making certain that everything was in place as he wished. Meanwhile, throughout the morning he kept up an almost constant artillery bombardment of the whole Rebel line, but concentrating especially on Shy's Hill, where Bate's position formed not only a corner of the Rebel line, but a salient at that.

Once more Thomas ordered demonstrations against Hood's right. Wood sent several strong feints against Lee's position on Peach Orchard Hill, and finally launched a four-brigade attack that the defenders barely beat back. But not before Lee sent Hood a plea for reinforcements. The army commander faced a terrible choice. There was no serious activity in front of Cheatham other than skirmishing. He could pull troops from there to support Lee, but what if Thomas then attacked Cheatham? Of course, it was exactly what "Old Pap" had in mind. At

2 p.m. Hood sent one of Cheatham's divisions to Lee, only to find that by the time it arrived, Wood's attack had been beaten back. Consequently, about 3:30 Hood sent it on its way back across the two miles to its original position. Before they could return, however, Schofield and Wilson finally struck.

The Yankee cavalryman, with his dismounted troopers, ran right around the left flank of Cheatham's line and started to deliver an enfilading fire into the rear of the Confederate line. At the same time, Schofield attacked across a broad front, while Smith struck due south with part of his corps, hitting Shy's Hill. The Rebels under the valiant General Bate were caught in a hopeless predicament. Smith in their front right, Schofield on the front and left, and Wilson in their rear. Wilson's cutting of the Granny White Pike eliminated their only direct line of retreat. If the enemy pushed on to the Franklin Pike, they would have no way out at all.

Seeing this, Bate's old veterans did not have the will to put up the kind of resistance normally expected of these seasoned fighters. Instead, they began to break by pairs and groups, racing for the safety of the Franklin Pike. Others tried valiantly to hold out, but to no avail. Smith broke through to the right of Shy's Hill, and Wilson swept up from the south. Bate's division simply all but vanished. Then the smell of victory swept over to the rest of Smith's corps facing Stewart. Seeing the triumph to their right, they rushed forward without orders. Seeing the debacle on their left, Stewart's corps did not try to stand, but almost as one man got up and ran for safety "in the wildest disorder and confusion." Then the panic spread to Lee's corps on the Confederate right. Some tried to hold out, but the Yankee attackers penetrated the lines so quickly that Johnson's division was cut off and virtually swallowed. Those who could turned to join the fleeing horde on the Franklin Pike.[7]

Major General George H. Thomas, U.S.A.

There was never anything flamboyant about George H. Thomas, though he was unfailingly competent. Somewhat more punctilious about uniform than many Yankee commanders, he usually appeared in full regulation attire, as here. He is obviously a major general, his stars just barely visible on his shoulder straps, and the two stars of his rank quite evident on his regulation saddle blanket.

His uniform is regulation blue, with yellow belt sash, and a flat-brimmed variation of the old Hardee or "Jeff Davis" hat, with yellow cord. He sits on a black leather McClellan saddle, complete with pommel holsters and leather-covered blanket roll at rear.

Thomas was a sturdy man of large, fleshy build, with a grim set of jaw that betrayed his determined, if plodding, nature. No one ever accused him of flourish, but he was unfailingly dependable, and in a Civil War commander, that was a commodity much to be desired.

Only a heroic rearguard action by one staunch brigade managed to hold a position just east of the Granny White Pike long enough for Cheatham to get most of the remainder of his corps past and over to the Franklin Pike. Others managed to cover Lee and Stewart's withdrawal, and before long some modest order re-emerged out of the chaos of the day's fight. But it was clear that this army was terribly beaten, and still in mortal danger. All that saved it was a determined stand by Hood's pitiful little cavalry command. It managed to stall Wilson's advance long enough to prevent him from cutting off all routes of escape. Had this dramatic effort failed, Hood would have been surrounded.

Now on the frantic flight back toward Franklin, Lee covered the retreat, and did so skilfully, while Hood sat that night in his tent alternately crying and pulling his hair with his one good hand. Already the irrepressible soldier wit of Johnny Reb was re-emerging from the bitterness of defeat, and during the terrible days that followed, as the army continued its retreat, boys sang a version of the "Yellow Rose of Texas" that ended with the lines:

You may talk about your Beauregard and
    sing of General Lee,
But the gallant Hood of Texas played hell
    in Tennessee.[8]

"Played hell" he had, indeed. Even though he managed to get his army back to the Tennessee, the march was one long rearguard action, made the worse by the intense cold, the lack of food, and for many Rebels an absence even of shoes. For years afterwards men would tell of bloody

Below: Yankee camps reach off into the distance, with stacked rifles in the foreground. The smoke may be from campfires, or it may be coming from the battle in the distance out of sight.

footprints on the frozen ground. They reached the river on Christmas and crossed the next day, then marched for their ultimate destination at Tupelo, Mississippi. The next day Thomas abandoned the pursuit. Of the nearly 70,000 men at his command, Thomas had actually used about 50,000, and suffered losses that were almost negligible – a mere 387 killed, 2,562 wounded, and 112 missing. Against this total of just 3,061 Yankee casualties, Hood's losses were staggering in proportion. He had no more than 24,000 of all arms at hand. In the aftermath of the battle, he and his commanders were so disorganized that they never compiled or filed reports of their losses. But Thomas recorded at least 4,462 Rebel prisoners taken, including three generals, and fifty pieces of artillery. Killed and wounded can only be guessed at, but Hood's overall loss must have been high, for three weeks later the army mustered only 15,000 men. Thus Hood's killed and wounded must have come to nearly 5,500, and his loss overall amounted to nearly forty percent of his army.[9]

In later years extravagant claims would be made about the Battle of Nashville: that it was perfectly planned and executed – which it was not; that it destroyed the Army of Tennessee forever – which it did not; and even that it was *the* decisive battle of the Civil War, which is patently absurd.[10] Thomas did not execute the battle precisely as planned, and took so long about launching his attacks that he lost some of his great advantage of numbers. Had he done otherwise, he might have eliminated the Army of Tennessee entirely. Of course, it had been not much more than the shell of any army to begin with. But by late January 1865 it could still muster about 15,000, and it would be heard from again before the war was out. As for the decisiveness of Nashville, all it really decided was the continuation of Thomas' tenure in command. The

notion that Hood could have posed a real threat with his shadow of any army, even if somehow he got past Thomas and into Kentucky, is ludicrous. And by defeating Hood, Thomas in no way altered the already overwhelming trend of military events. The real war between the Appalachians and the Mississippi had been over since the fall of Atlanta, awaiting only Confederate realization of the fact.

Nashville, however, did decide the fate of John B. Hood. Humiliated, he asked to be relieved in January, and thereafter saw no active service. The army command then went for a time to Richard Taylor, but very soon Lee's and Cheatham's corps were sent to Georgia to be joined with a command under Hardee. Together they reconstituted the Army of Tennessee once more, and left with no other alternative, President Davis once again put Joseph E. Johnston at its head. Less than four months later, with Johnston commanding, and with Beauregard and Bragg in attendance, it would at last surrender, surrounded by the ghosts, living and dead, of the commanders who had been so much a part of its infrequent triumph and its almost continual turmoil. Of its four commanders still living, only Hood was not there.

References

1 *O.R.*, I, 45, part 1, pp. 654-61.
2 Thomas R. Hay, *Hood's Tennessee Campaign* (New York, 1929), pp. 77-78.
3 John B. Hood, *Advance and Retreat* (New Orleans, 1880), p. 290.
4 *O.R.*, I, 45, part 2, p. 96.
5 *Ibid.*, p. 691.
6 *Ibid.*, p. 195.
7 *Ibid.*, pp. 439, 698.
8 Sam R. Watkins, *Co. Aytch* (Nashville, 1882), p. 229.
9 Hay, *Nashville*, pp. 179-80.
10 Stanley F. Horn, *The Decisive Battle of Nashville* (Knoxville, Tenn., 1956), pp. vi-vii.

Epilogue

Every war must have a beginning, and sooner or later an ending must come no matter what the suffering has been. No-one, excepting a few far-seeing men like Sherman, had predicted that this war would last so long, or cost so much. It was all supposed to have been over in a few weeks or months as they had said in that far-away spring of 1861. Four years later, after 10,000 fights and more than half a million dead, it seemed to some that it would go on forever. If Jefferson Davis and a very few of his leaders had had their way, it would have gone on indefinitely, too, but the will of the South and the ability of her people to sustain the armies in the field were simply no longer equal to the task.

All of which meant that someone, somewhere, had to be that most unfortunate of men, the last to die in a war. Just where the last shots were to be fired, and who were to be the last to die, depended upon the spring 1865 campaigns, now securely in the hands of men who had been seasoned by years of war, North and South, and who had for the most part proved themselves worthy commanders of the greatest modern armies the hemisphere had ever seen. The outcome of that war was no longer in doubt as Blue and Gray faced each other that year; only when, and how.

One of the great errors made by even the greatest of commanders in this war was the notion that there could be such a thing as one, great, decisive battle. It was an obsolete notion, applicable perhaps in the time of Napoleon, but not to a massive conflict on the scale of this one. Individual battles and campaigns may have had decisive influences on the course of the war, but no one engagement could end it, short of actual extermination of one or the other of the participants involved.

And thus it was that the conflict wore on even after such seemingly climactic contests as Gettysburg and Nashville. Indeed, for Grant and Lee, their first meeting in Virginia in May 1864 was only to be the beginning of a year-long confrontation, one that could hardly have disappointed any of those who always wondered what would happen when the Gray chieftain met the Yankee juggernaut. Both men were accustomed to taking chances, both were conditioned to expect victory.

Following the inconclusive meeting in the Wilderness, Grant forced Lee to fall back to Spotsylvania, presaging the approach he would take for fully a month, seeking always to turn Lee's flank and get between him and Richmond. If he succeeded, all was good. If he did not, it could only be because Lee fell back before him to protect his lines of communication. Thus, either way, Grant emerged the victor strategically, if not tactically. Thus their dance of death undulated steadily southward: Spotsylvania, The North Anna River, The South Anna. May passed away, with the armies around Cold Harbor, near where Lee and McClellan had met two years before during the Peninsula Campaign. Here Lee finally stopped Grant cold for the first time. After several days of indecisive skirmishing, Grant launched a massive assault on June 3 that he later acknowledged as the greatest blunder of his career. He sent whole divisions forward in a frontal assault against Lee's entrenched positions, despite Grant's own career-long penchant for avoiding such unsophisticated sledge-hammer tactics. The result was completely disastrous, with thousands killed and wounded in a single hour of combat for no gain at all.

Yet nothing ever deterred Grant from his greater purpose. Stymied at Cold Harbor, with Lee securely between him and Richmond, Grant conceived one of the most daring and skillfully executed movements of the war. Completely fooling Lee, he pulled the Army of the Potomac out of its lines, marched it east and south of Lee, and crossed it over the James River. Lee only discovered what had happened more than a day later, when Grant was moving against Petersburg, south of Richmond. If he took that fortified city, he would have virtually all of the enemy capital's rail links with the rest of the Confederacy at his mercy, a back door to Richmond,

The devastation visited upon the South was unprecedented in American history. Richmond lay in ruins after its evacuation and fall, and the dreadful fire that swept most of the city away.

barely twenty miles away, and Lee in a position where he had nowhere to retreat to. But subordinates bungled the attack. Stout resistance by a scratch force led by Beauregard, and Lee's hurried reinforcements, saved Petersburg through tense days of assault. By then, surprise was gone, and Lee moved most of his army into the fortifications surrounding the city. A distressed Grant realized that he might have another siege on his hands, and when a massive assault launched on July 30 failed to achieve its purpose, he admitted that he had no alternative but to besiege the Confederates. The battered Army of the Potomac, after three months of almost daily fighting, was too weary to do more.

But if Grant was tied down, Sherman certainly was not. Following the capture of Atlanta in September, he decided in the end not to bother about Hood, but to strike out across country through Georgia. His goal was to cut the heart out of this breadbasket of the Confederacy, sever rail lines, destroy industry, and drive to Savannah and the Atlantic. Taking Savannah would deny the South yet another of its ports, and virtually cut off Florida and Georgia from the upper South. There was little to resist him, though the gallant Hardee used what local defense forces he could muster to slow the Yankee advance. Nevertheless, by Christmas 1864, Sherman could wire Lincoln that Savannah was in Union hands.

That done, "Uncle Billy" Sherman turned north, his goal now to drive through South Carolina, ravage its resources for continuing the war, and disperse or destroy the remnants of the enemy left before him. In January Hardee was joined by the corps of Stewart and Cheatham, all once more under the command of Joseph E. Johnston. In the ensuing Campaign of the Carolinas, Johnston futilely attempted to stop Sherman. By the time they met in the one and only real battle of the campaign, at Bentonville, North Carolina, in March 1865, it was obvious that Sherman could go where he pleased and do what he wished. The only hope for the Confederacy in the East now was for Johnston to evade Sherman and

march north, while Lee broke out of Petersburg and moved south. If they could link their forces, they would still have a considerable army of 60,000 or more, and might hope to defeat first Sherman, then turn and drive back Grant and Meade's Army of the Potomac.

The impending fall of Petersburg on April 1, 1865, forced Lee to prepare for such a move. While the government evacuated Richmond, he put his proud old Army of Northern Virginia on the road to join Johnston. Grant though was too numerous, and his cavalry, now led by Sheridan, kept getting ahead of the Rebels, forcing them always to go farther west before they could turn south. Finally on April 8, when Lee camped in

and around Appomattox Courthouse, he saw the glow from Yankee campfires on all sides of him. He could go no further. The next day, in the parlor of the home of Wilmer McLean, who had left his house at Manassas in 1861 in order to get away from the war, the two generals met. The peace that came of their meeting started the slow process of bringing the war to an end.

Sherman and Johnston met next, on April 26. After several days of negotiations, this time involving both Davis' fleeing government, and the Union War Department in Washington, Johnston, too, surrendered. His capitulation would have been easier had it not been for one of the last casualties of the war. Twelve days earlier, Presi-

dent Abraham Lincoln went to the theater, never to return.

There were still other armies, of course. In May Richard Taylor surrendered the last of the Rebel armies east of the Mississippi. The next month the Trans-Mississippi army of Kirby Smith was surrendered in New Orleans. Perhaps the last surrender or organized troops came on June 23, when Confederate Indian cavalry gave up out west of the great river. Yet it was not until many months later, in November, that the last of the surrenders came, when the commerce raider C.S.S. *Shenandoah*, cut off from news and unaware of the collapse of the cause, finally surrendered itself to British authorities in Liverpool.

Left: Following the Wilderness, Grant kept the pressure on Lee, constantly improvising as he sought to turn the Rebel flank. Here, leaning over the pew at left, he conducts a council of war on the road, conferring over a map with Meade.

Right: Logistics won the war for the Union as much as blood, and Grant could command the most and the best that was available. The United States Military Railroad spanned half the continent, taking men and material wherever Grant needed them.

Below: By war's end, when Lee surrendered, Richmond was a Federal garrison, and along the wharves on the James River there were stored the tons of weapons and supplies taken from the Confederates. No more would this artillery bark for the mighty Lee.

It took a long time for Jefferson Davis and some of his advisors to admit to themselves that it was all over. Throughout their long flight south from Richmond, they continued to envision marshalling their remaining forces to continue the fight elsewhere, at first, in the Carolinas, then after Johnston surrendered, off in southern Alabama. Taylor's surrender made that obsolete, and even as he was captured in Georgia on May 10, Davis was hoping to reach the Trans-Mississippi and Kirby Smith, to continue the fight, as long as there was one Southerner with a weapon in his hands, reasoned Davis, the fight could go on.

In the end, he even considered sending the disbanded men from the armies into the hills to continue the conflict on guerrilla terms. Happily, such a form of war, which inevitably must have degenerated into something ignoble compared to the heroic stand made by the Confederate armies, never came to pass. Generals and soldiers alike admitted that it was all over, and Davis finally had no choice.

It had all been more than trauma enough for the Union, which hereafter ceased to be called that in preference for United States. Secession had been tested and failed, and where the former implied unity by consent, the latter clearly proclaimed ineradicable nationhood. It was fitting, considering what bound North and South together as a result of the war. There was all the suffering they had endured, the hardship and privation. There was all the heroism and sacrifice they displayed, not just as partisans, but as Americans.

Even more than this, however, there was the bloodstained land to bind them together. Johnny Reb and Billy Yank had made streams run red as they poured virtual rivers of their blood on the ground for their causes. "Bloody Run", "Bloody Pond", the "Bloody Angle", "Bloody Lane", and a host of other sanguinary new names on the land testified to where they had been and what they had done. The geography of North America would never be entirely the same. Rivers had been diverted, hills leveled, whole forests

THIRD PENNSYLVANIA CAVALRY. C. CO.

B. CO. VETERAN BATTALION,

YORKTOWN, WILLIAMSBURG, SAVAGE JORDAN'S FORD, CHARLES CITY CRO

MALVERN HILL, ANTIETAM, UNIONVI PIEDMONT, ASHBY'S GAP, AMISSVILL

KELLY'S FORD, STONEMAN'S RAID BRANDY STATION, ALDIE,

GETTYSBURG

OLD ANTIETAM FORGE, SHEPHERDSTOWN, CULPEPER COURT HOUSE, OCCOQUAN, NEW HOPE CHURCH, PARKER'S STORE, WILDERNESS,

SPOTTSYLVANIA COURT HOUSE, NORTH ANNA, TOTOPOTOMOY, COLD HARBOR SIEGE OF PETERSBURG, REAM'S STATION, BOYDTON PLANK ROAD, HATCHER'S

FORT STEADMAN, FALL OF PETERSBURG, LEE'S RETREAT, LEE'S SURRENDER AT APPOMATTOX COUR

Union Cavalry Guidons

1 National Guidon of Company C, 3rd Pennsylvania Cavalry, issued at the close of the war and carried by that unit in the Grand Review of the Army of the Potomac, along Pennsylvania Avenue, Washington D.C., May 23, 1865. Note the battle honors in gilt paint on the flag, which begin with engagements in the Seven Days' Campaign and conclude with Appomattox Court House

2 State Guidon of Company G, 1st Pennsylvania Cavalry. This flag is one of 112 manufactured by Holtsmann Brother and Company, Philadelphia, and the only surviving specimen of this pattern. Note the Pennsylvania state coat of arms on the top

3 National Guidon of Company I, 6th Pennsylvania Cavalry, complete with battle honors inscribed. The last battle honor, "Gettysburg", would date this particular cavalry color to some time after July, 1863. This particular company served at General Meade's headquarters during that battle

Artifacts courtesy of: The Civil War Library and Museum, Philadelphia, Pa: 1, 3; Chester County Historical Society, West Chester, Pa: 2

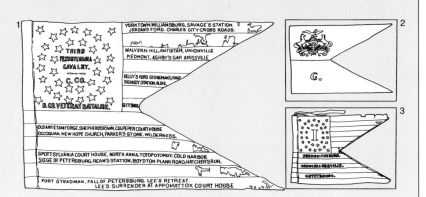

denuded, and the fields of a million farmers sown with lead and iron. Mountain tops once reserved for the wolves, were now to be the haunt of old veterans and men too young to fight, who came as tourists to see where it all had happened. Fields made battlefields, were soon to become parks, there to host the future millions who, like their ancestors, could never entirely turn loose of this war. For too much passion, too many horrors and so many sacrificed lives would ensure that these places would live on in the thoughts of those who came after.

In the wake of the surrenders North and South began the final tally of the cost for their quadrennium of madness. More than 600,000 dead. Over one million wounds inflicted. A whole generation of maimed and crippled men left to live out their days however they could. A section of the country ravaged of all its resources, material, agricultural, and human. An angry scar across the landscape and the soul of its people, a scar that could never be erased. Four million Americans, once enslaved, were now free, with little but their freedom and the clothes on their backs. An old question about the nature of the Union and the relation of the states to the Federal government had been settled at last. The old Union was gone forever, washed away in a flood of blood and tears, but faith still remained in many to create a new America.

And most of all, there remained the nagging question. Had it all been worth the price? Within weeks of Lee's surrender, Union soldiers met on the old Manassas battlefield, where the first full-scale, but still rather haphazard fight had taken place those years before, to dedicate one of the very first monuments to commemorate the dead and what they died for. It was only the first of thousands of such ceremonies, to be repeated on every great battlefield of the Civil War, a continuum that still endures to this day. And always when they come to the bloodied land, now peaceful beneath the sod that covers the dead and their sacrifice, always they wonder. Had it all been worth the cost.

Above: Major General Robert Anderson, the man forced to surrender the Fort Sumter garrison in 1861, raised the flag above that fort exactly four years later amid a euphoric victory celebration.

Above: Charleston, once the seedbed of secession, was largely a ruin at the end, having suffered almost as much as Richmond. Its citizens knew as few others the cost of the war they helped to start.

Below: The great celebration came in Washington, when the Grand Review of the Army of the Potomac and Sherman's armies marched down Pennsylvania Avenue. Grant, Sherman, and others sat in the stand.

Personal Effects and Decorations of Major General Galusha Pennypacker

The youngest general in American history, Galusha Pennypacker was wounded five times during the Civil War, and was awarded the Medal of Honor for conspicious bravery during the Union action at Fort Fisher. He was appointed general at the age of 20, before he could even vote!

1 U.S. Army Model 1904 Medal of Honor complete with case, belonging to General Pennypacker. The Medal of Honor at this time did not feature the neck suspension ribbon

2 Double-breasted frock coat with velvet collar and cuffs, formally the possession of General Pennypacker

3 U.S. Army Model 1896 Medal of Honor, complete with case, formerly the possession of General Pennypacker

4 Model 1850 presentation sword and scabbard, formerly the possession of General Pennypacker. The sword is an import, having been manufactured in Germany by CLAUBERG of Soligen (a town near the Rhine). The sword was shipped to America and retailed by William H. Horstmann & Sons, Philadelphia

Artifacts courtesy of: Chester County Historical Society, West Chester, Pa

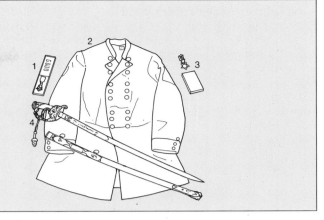

APPENDIX

The specially commissioned color photographs of Civil War weapons, uniforms and personal belongings which have appeared throughout this book, represent perhaps one of the finest collections of contemporary artifacts brought together under one cover. As an additional source of information on each photograph's content and origin, Russ A. Pritchard, Director of The Civil War Library and Museum and technical advisor to the book has compiled this Appendix. Further information can also be obtained from the Bibliography and list of museums and societies on page 256.

Confederate National Flags
pp. 18-19

The Confederate States of America adopted three national flags during its short existence. The first, known as the "Stars and Bars" is by far the most common. The second, the "Stainless Banner", utilized the well-known Confederate battle flag on the canton (upper left hand corner). The last pattern, authorized during the final weeks of the war, saw very limited wartime use, but was more popular in postwar years. Examples of these colors exist in various southern state collections: neglected until recent years, most are in desperate need of conservation.

Federal Uniforms and Headgear
pp. 21-22

Uniforms of both combatants were anything but uniform during the early stages of the war. The state militia system practically encouraged individuality of uniforms and equipment. Uniform regulations allowed the greatest latitude and, in many cases, only the financial resources of the individual militia unit restricted the elaborateness of their clothing. The results were multi-colored, brocaded, plume-bedecked costumes of sometimes comic appearance. The hard realities of war soon relegated such dress to the footlocker, the band or home guard.

Distinctive State Flags of the Confederacy
pp. 28-29

Southern state flags are visual evidence of the strong feeling of state's rights amongst individual Confederate states. Patriotic slogans are quite commonplace on Confederate state flags. A large number of such flags, captured during the war, were returned to their states of origin by the Federal government in the early decades of the twentieth century. The flags usually reside in respective state capitols. Many of these flags, oil-painted on silk or embroidered, are really works of art, and need professional care now, if future generations are to see them.

Union National and Regimental Flags
pp. 36-37

Federal flags face the same acute preservation problem. Many flags are made of silk, and their natural fragility is accentuated by age and pollu-

tants. Many states, north and south, have begun conservation programs, but funding is always a problem. The states of New York, New Jersey, Delaware, Pennsylvania, Ohio, Illinois and Missourri, are all involved in such conservation, and have particularly fine state flag collections – some more accessible than others. Other examples may be found in county historical societies.

Confederate General Service, Naval and State Buttons pp. 40-41

Uniform buttons are another great source of individuality. The Confederate states utilized respective state seals in various configurations, or used other distinctive devices. Various branch indicative letters in several styles: block, script and Old English script were used. Fine collections of Southern buttons may be viewed at the Museum of the Confederacy, the Virginia Historical Society in Richmond, Va, and the Atlanta Historical Society, in that city. In the last two decades, button collecting has become extremely sophisticated and expensive.

Union Hardee Hat and Branch Indicative Insignia pp. 46-47

Branch indicative and unit identifying insignia has always been a popular collecting field. Numerous restrikes during the war's centennial, modern replicas, and outright fakes, however, add to the variety and confusion. Original die variations and unit distinctions make insignia collecting an interesting and challenging field. Caution and expertise are necessary when making acquisitions of such insignia.

Battleflags of the Army of Northern Virginia pp. 54-55

The battleflag of the Confederate States is easily one of the most recognizable symbols worldwide. Many people believe it to be the only Confederate flag, and unfortunately this symbol of a no longer extant military organization has been associated with various radical political groups. This flag, in its myriad configurations, was the rallying point of one of the finest armies of the nineteenth century, and has about it today a mystique like no other. Specimens exist in many southern state collections, and isolated examples may be seen in some northern museums. Without doubt the re-

Jaunty Yankees posing beside their 3-inch ordnance rifle give ample evidence of the indomitable spirit and pride that saw the Federals through so many lost battlefields, to the final triumph.

pository of the finest collection of Confederate battleflags is The Museum of the Confederacy in Richmond, Va.

Federal Uniforms of Enlisted Men
pp. 56-57

Zouave dress, adopted from uniforms of French colonial troops of the period, was extremely popular, although impractical in the field. While camouflage was not yet an official military concern, the common soldier recognized that bright, colorful attire attracted attention and hostile fire. Use of zouave clothing diminished in popularity when the fighting began in earnest. Branch colors, red for artillery, blue for infantry, yellow for cavalry, remained in use throughout the war and into present day service. Excellent Federal uniform collections may be seen at U.S. Army museums at Fort Benning, Ga, Fort Lee, Va, the West Point Museum, N.Y. and the Smithsonian Institution.

Confederate Vandenburgh Volley Gun
pp. 60-61

Various types of artillery, field and siege guns, were primary casualty producers during the war. Various volly and rapid fire guns, forerunners of the modern machine gun, appeared in the arsenals of both sides. While ordnance innovations of the time are crude by today's standards, they were none-the-less quite lethal. Excellent artillery collections may be viewed at the West Point Museum, N.Y., Gettysburg and Petersburg National Military Parks, and the Washington Navy Yard in the District of Columbia.

Union Rockets and Signal Pistols
pp. 64-65

The Signal Corps was a relatively new and small part of the army. Pyrotechnics with multicoloured lights was a rapidly developing innovation in military circles. Flares from the hand-held percussion pistols merely burned in place rather than shooting through the sky. Rocket launchers were not a new idea, and were used without much success during the war. No one could imagine how deadly these contraptions would become 125 years later. An excellent exhibit of this material is on view at the Gettysburg Museum of the Civil War.

Confederate and Southern State Bonds
pp. 74-75

The Confederacy, lacking any meaningful supply of gold or silver, was immediately forced to finance its war effort by means of the sale of bonds, debentures and interest-bearing notes. These had a promise to pay, after the signing of a peace treaty and the end of the war. The central government and the several states concurrently utilized this funding method. Bond issues were many and frequent. Soaring inflation and the eventual collapse of the Confederacy left southerners and many Europeans with reams of worthless paper. These bonds are one of the most common relics of the Confederacy today. These financial documents are available and quite collectible. Most major Civil War museums have representative collections, and there are excellent reference books on the subject.

Excavated Union Artifacts
pp. 82-83

Relic hunting, the use of electronic metal detectors to locate battlefield relics, has progressed greatly since the use of surplus mine detectors in the late 1940s and early 1950s. Such hunting has become so popular that Federal lands have been closed off to it for the last three decades. Nevertheless, hunting continues on private land, and discoveries, while no longer as numerous or as large, continue to draw avid hunters to the fields and woods. The discovery and excavation of a longarm or handgun today is unusual.

U.S. Model 1841 6-Pounder Smoothbore, James Rifle pp. 90-91

Bronze field guns were the most popular artillery pieces of the mid-nineteenth century. The bulk of these historical objects were lost in scrap drives during World War I and II. Surviving specimens may be found unexpectedly in small traffic circles and isolated veteran's cemeteries. Large collections may be seen at Shiloh, Antietam, Vicksburg, and Gettysburg National Military Parks, as well as West Point and the Washington Navy Yard. Live firing of these weapons is becoming an increasingly popular hobby.

Distinctive Unit Flags of the Confederacy
pp. 94-95

All Confederate flags have been popular collectibles, even during the war. In recent years this mystique has assumed considerable stature. While the bulk of surviving Confederate flags now reside in public collections, in the states of their origin, some are still in private hands, trophies of war passed down in families or acquired in various ways. The flags are so sought after by collectors, that substantial numbers of outright fakes have appeared, and other altered or or modified flags have been passed off as 'Confederate'. Extreme caution should be exercised when considering the purchase of such a flag. The extensive collection at The Museum of the Confederacy can provide numerous specimens for comparative study.

Excavated Struck Plates, Identification Badges and Stencils pp. 100-101

Personal items such as belt plates, identification badges and stencils have always had a particular allure for many amateur archeologists. Such items actually identify the presence of a particular individual or unit on a piece of ground. Such evidence is essential to properly interpret what took place on any given battlefield. These bits of personal history are particularly poignant today, as we revisit these battlefields. Accoutrement plates, mutilated by projectile strikes, are proof positive of the ferocity of combat.

Confederate First and Second National Flag Variants pp. 108-1109

Variations of the First and Second Confederate National Flags are extant in many of the public collections; indicative of how prevalent they were. Modern media recognizes the battleflag as the 'Confederate flag' which has added to the confusion of younger generations. Some of the rarer patterns of flags, particularly those of the western

theater, are hardly recognizable as Confederate except by the students of these colors. Some unusual variants exist in the Mississippi, Missouri, Texas, Kentucky and Arkansas state collections.

Union Button Dies, and State and Service Buttons pp. 110-111

Northern buttons, like their Southern counterparts, provide a vast number of state seal variations and branch indicative buttons. Of the buttons unidentified in the key: Figure 51 has a New Hampshire state seal; Figures 52 and 53 have New Jersey state seals; Figures 54-55 have New York state seals; Figure 56 has an Ohio state seal; Figure 57 has a Pennsylvania state seal; Figures 58-59 have Rhode Island state seals; Figure 60 has a Vermont state seal; Figure 61 has a Wisconsin state seal; and Figure 62 displays the badge of the U.S. Sanitary Commission. Federal buttons, in most cases, are readily available in quantity at relatively reasonable prices. Representative collections exist at most museums that have a Civil War period exhibit.

Federal Cavalry and Confederate Infantry Uniforms pp. 126-127

Surviving uniform variations of either Federal or Confederate forces are not common. Scattered specimens do exist here and there, in public and private collections. The collections at West Point, the Smithsonian, and Museum of the Confederacy are truly outstanding, although many examples are not on exhibit due to space and conservation considerations.

Excavated Confederate Artifacts
pp. 134-135

Excavated Confederate used or manufactured objects are considered rarities. Such items are hard evidence of what material was actually in use on the field. The location of such discoveries can often give real corroboration to documentary evidence of the issue of a specific type of weapon, or caliber of ammunition for a particular weapon. While Federal laws prohibiting excavation on Federal property are understandable, the loss of historic objects to natural deterioration is not an acceptable solution. This group of objects is an excellent example of the diversity of Confederate arms and equipment, carefully documented and preserved by dedicated relic hunters and preservationists.

Union and Confederate Coehorn Mortars with Projectiles pp. 142-143

Light, mobile mortars were an integral part of siege operations. Confedrate and Federal forces utilized them in large numbers. Their high angle of fire and surprising accuracy allowed opposing forces to fire projectiles into carefully fortified and entrenched positions with devastating effect. Excellent examples mnay be found at Gettysburg, Petersburg and West Point.

Modified Officer's Grimsley Saddle, Harness and Tack Box (of then) Major General Ulysses S. Grant p. 147

U.S. Grant was known as an excellent horseman, and noted for his conservative dress in the field.

The saddle shown here is probably indicative of the man; plain and sturdy, with little ornamentation: a purely functional piece of equipment. Grant memorabilia is scattered in a number of repositories, including the Smithsonian Institution, West Point, Civil War Library and Museum, Philadelphia, Pa, and the Quartermaster Museum, Fort Lee, Va.

Federal and Confederate Handgrenades
pp. 150-151

Handgrenades were another weapon, still in the early stages of development, that did service with both combatants. Particularly in this case, in siege operations at Port Hudson, Vicksburg and Petersburg. They were not judged to be very effective because of the unreliability of their fuses, which made them at times more dangerous to the thrower than to the recipient. Nevertheless, their development foretold of their place in present day ordnance. Examples may be seen at the Gettysburg Museum of the Civil War, West Point, and Petersburg, Va. Specimens are rare, few having survived, and all are quite collectible.

The Whittier Group of Uniforms and Personal Effects pp. 158-159

Large groupings of one individual's uniforms and equipment are almost unheard of today. Almost all such collections are housed in institutions. Once in a great while, a persistent, dedicated private collector will discover a long-forgotten field chest, placed in an attic or barn shortly after the end of the war. Such is the case with the Whittier Collection. The type of uniforms in the collection are all rare; the labels place the individual pieces in their historical perspectives. Such a discovery happens once in a lifetime.

Union Camp Colors and Field Markers
p. 160-161

Field or flank markers were necessary to designate the extremities of a unit's position on the field, particularly in the early stages of establishing a battle line. Constructed of both silk or bunting, a surprising number have survived. Their small size, usually less than 2 feet × 3 feet, made them excellent souvenirs then, and this small size has made them equally as popular with present day collectors. The state capital in Harrisburg, and the Civil War Library and Museum in Philadelphia have excellent examples of Pennsylvania unit field markers. Specimens also exist in other state collections. Confederate forces did not appear to utilize such a system to any extent.

Excavated Union Cavalry Relics
pp. 168-169

Cavalry relics are particularly difficult to locate because of the nature of cavalry engagements. Rarely was the action concentrated in a confined area, but rather carried on over the countryside in a large area. Thus, literally days may be spent hunting a cavalry site with maybe a horeshoe or dropped bullet to show for a tremendous amount of effort.

Headquarters Designation Flag, Army of the Potomac, July, 1863 p. 171

Flags that are documented as having actually been present at a crucial moment in history are very rare. General Webb's Headquarters Designating Pennant is one of those colors. Soon after the Battle of Gettysburg at which the flag was present, it was placed in safe keeping at the Union League of Philadelphia.

Variant Pattern Confederate Unit Flags
pp. 180-181

With such a large number of different patterns and variations, the study of Confederate flags has developed into a field of its own. Of particular interest are western theater flags that would not even be identified as Confederate by the layman. To have any grasp of the subject, the reader is refered to the definitive source on the subject: *The Battleflags of the Army of the Tennessee*, by Howard Michael M. Madaus and Robert D. Needham. Fine specimens exist in the Tennessee, Mississippi and Arkansas state collections, as well as the Museum of the Confederacy. There have been very dangerous fakes manufactured in the last decade, in some cases using period bunting. Extreme caution is advised.

Union Mountain Howitzer
p. 186

The rough topography of some battlefields required the use of special light artillery pieces. Defense of bridges and entrances to fortifications also required a field piece that could easily be shifted out of position to allow free passage. One result to fill this need, was the 12-pounder mountain howitzer. Such guns were used in sally ports of Washington fortifications as well as on campaign in mountainous territory. Specimens may be seen at the Gettysburg Museum of the Civil War, U.S. Ordnance Museum. Aberdeen Proving Ground, Md, and West Point.

Field Artillery Projectiles and Fuses
pp. 198-199

The general acceptance of rifled artillery, and advances in projectile design and technology, made artillery of both sides much more effective. The importation of advanced English guns and projectiles by Federal and Confederate ordnance also furthered these advances. The 1980s saw an enormous increase in interest in the collecting of projectiles. At least six excellent books have been written on the subject, and are listed in the bibliography. The projectile collection at West Point is excellent, and representative specimens may also be found at various National Park museums.

Excavated Battle-Damaged Union Relics
pp. 206-207

The ferocity of combat in many cases can be determined by the damage on excavated objects. The perforated canteens; the struck and misshapen weapons, all attest to the horror of fighting. Such relics are highly prized by collectors, but are seldom found and recovered today. A number of the older battlefield museums had such relics, but these collections have long been dispersed into private hands. The Atlanta Historical Society and the Gettysburg Museum of the Civil War, have extensive holdings of such objects, some of which are on exhibit.

Union and Confederate Naval Artifacts
pp. 214-215

Naval flags generally are so large as to restrict their display to large institutions. An excellent reference on Southern naval flags is *Rebel Flags Afloat*, by Howard M. Madaus. Such artifacts may be seen at the United States Navy Museum at the Washington Navy Yard, and the United States Naval Academy Museum, Annapolis, Md.

Confederate Imported Armstrong Rifle
pp. 220-221

Imported English ordnance was some of the best in the Confederate arsenal. Armstrong and Whitworth guns were considered to be the most modern types. Range and accuracy were exceptional. Captured trophies such as this large Armstrong rifle may be seen at West Point. The Washington Navy Yard, and Forts Sumter and Moultrie at Charleston S.C. also have captured English ordnance.

Confederate and Union Heavy Artillery Projectiles pp. 182-183

With the advent of armored vessels, projectiles capable of penetrating or crushing such armor had to be developed. The breaking of masonry forts was also accomplished by the use of similar projectiles fired from large bore rifled guns. Many of these projectiles had specially hardened noses designed to punch through armor. Excellent examples are to be seen at West Point and the Washington Navy Yard. The result of bombardment by such projectiles can be seen at Fort Sumter, Charleston, and Fort Pulaski, Savannah, Ga.

Western Variant Confederate Battleflags
pp. 230-231

Variations of the St. Andrews Cross were used as the battleflag of many units serving in the western theater. In general these flags were rectangular in shape, unlike those of the Army of Northern Virginia which were generally square. These flags are avidly sought, but few are in private hands. The premier collection is to be found at the Museum of the Confederacy.

Union Cavalry Guidons
pp. 244-245

Federal swallowtail cavalry and battery guidons are colorful, relatively small, and therefore displayable in a private home setting. They were sometimes available in collector's circles until about the end of the 1980s. They were never inexpensive, and all require conservation. The state of Pennsylvania and the Civil War Library and Museum have excellent specimens, as do other military museums, and the National Park Service at several battlefields.

Personal Effects and Decorations of Major General Galusha Pennypacker p. 247

The personal effects of any ranking general officer of the Civil War are extremely rare. In Major General Pennypacker, we are fortunate that his career began so early in his life. Many of the decorations on show were awarded to him long after the war between the states had finished. This particular selection of artifacts is held at a county historical society, though most other generals' uniforms are held in institutional collections.

INDEX

References for illustrations are listed in *italics*

BIBLIOGRAPHY

Further reading to the Appendix subjects

Albaugh, William A., Benet, Hugh Jr., Simmons, Edward N., *Confederate Handguns*
Albaugh, William A., *Confederate Arms*
Albaugh, William A., *Confederate Edged Weapons*
Albert, Alphaeus H., *Buttons of the Confederacy*
Allen, Glenn C. and Piper, Wayne C., *The Battle Flags of the Confederacy*
Belden, Bauman L., *War Medals of the Confederacy*
Brown, Rodney Hilton, *American Polearms 1526-1865*
Brown, Stuart E. Jr., *The Guns of Harper's Ferry*
Burns, Z. H., *Confederate Forts*
Caba, G. Craig, *United States Military Drums*
Cannon, Deveraux, D. Jr., *The Flags of the Confederacy*
Coggins, Jack, *Arms and Equipment of the Civil War*
Criswell, Grover C., *Confederate and Southern State Bonds*
Cromwell, Giles, *The Virginia Manufactory of Arms*
Crown, Francis J. Jr., *Confederate Postal History*
Daniel, Larry J. and Hunter, Riley W., *Confederate Cannon Foundries*
Davis, Rollin V. Jr., *U.S. Sword Bayonets, 1847-1865*
Davis, William, C., *The Image of War*, Vols. 1-6
Davis, William C., *Jefferson Davis, The Man And His Hour*
Dickey, Thomas S., and George, Peter C., *Field Artillery Projectiles of the American Civil War*
Dorsey, R. Stephen, *American Military Belts and Related Equipment*
Elting, John R. (ed.), *Military Uniforms in America*
Fuller, Claud E. and Stewart, Richard D., *Firearms of the Confederacy*
Fuller, Claud E., *Confederate Currency and Stamps*
Fuller, Claud E., *Springfield Shoulder Arms 1795-1865*
Fuller, Claud E., *The Whitney Firearms*
Garofalo, Robert and Elrod, Mark, *A Pictorial History of Civil War Era Musical Instruments and Bands*
Gluckman, Arcadi, *United States Muskets, Rifles and Carbines*
Govt. Printing Office, *Uniform Regulations for the Army of the United States, 1861*
Hardin, Albert N. Jrs., *The American Bayonet, 1776-1964*
Hazlett, James C., Olmstead, Edwin and Parks, M. Hume, *Field Artillery Weapons of the Civil War*
Hopkins, Richard E., *Military Sharps Rifles and Carbines*
Huntingdon, R.T., *Hall's Breechloaders*
Jangen, Jerry L., *Bayonets*
Keim, Lon W., *Confederate General Service Accoutrement Plates*
Kennedy, Francis H., *The Civil War Battlefield Guide*

Kerksis, Sydney C., *Field Artillery Projectiles of the Civil War, 1861-1865*
Kerksis, Sydney C., *Heavy Artillery Projectiles of the Civil War, 1861-1865*
Kerksis, Sydney C., *Plates and Buckles of the American Military 1795-1874*
Laframboise, Leon W., *History of the Artillery, Cavalry and Infantry Branch of Service Insignia*
Lord, Francis A., *Civil War Collector's Encyclopedia*, Vols. 1, 2, 3 & 4
Madaus, H. Michael and Needham, Robert D., *Battleflags of the Confederate Army of Tennessee*
Madaus, H. Michael, *Rebel Flags Afloat*
McAfee, Michael J., *Zouaves . . . The First and The Bravest*
McKee, W. Reid and Mason, M. W. Jr., *Civil War Projectiles, Small Arms and Field Artillery*
Miller, Francis Trevelyan, (ed.), *The Photographic History of the Civil War*, 10 Vols.
Murphy, John M., *Confederate Carbines and Musketoons*
Phillips, Stanley S., *Bullets Used in the Civil War, 1861-1865*
Phillips, Stanley S., *Civil War Corps Badges and Other Related Awards, Badges, Medals of the Period*
Phillips, Stanley S., *Excavated Artifacts from Battlefields and Camp Sites of the Civil War*
Pitman, John, *Breech-Loading Carbines of the United States Civil War Period*
Rankin, Robert H., *Small Arms of the Sea Service*
Reilly, Robert M., *United States Military Small Arms 1816-1865*
Riling, Ray (ed.), *Uniforms and Dress of the Army and Navy of the Confederate States*
Ripley, Warren, *Artillery and Ammunition of the Civil War*
Sellers, Frank M. and Smith, Samuel E., *American Percussion Revolvers*
Stamatelos, James, *Notes on the Uniform and Equipments of the United States Cavalry, 1861-1865*
Steffen, Randy, *United States Military Saddles, 1812-1943*
Thomas, Dean S., *Cannons*
Symonds, Graig L., *A Battlefield Atlas of the Civil War*
Thomas, Dean S., *Ready . . . Aim . . . Fire! Small Arms Ammunition in the Battle of Gettysburg*
Todd, Frederick P., *American Military Equipage, 1851-1872*, 4 Vols.
Wise, Arthur and Lord, Francis A., *Bands and Drummer Boys of the Civil War*
Wise, Arthur and Lord, Francis A., *Uniforms of the Civil War*

LOCATIONS OF MAJOR CIVIL WAR COLLECTIONS

Ancient and Honorable Artillery Company Armory
Faneuil Hall
Boston, MA 02109

Atlanta Historical Society
3101 Andrews Drive, N.W.
Atlanta, Ga 30305

Augusta-Richmond County Museum
540 Telfair Street,
Augusta, Ga 30901

Casemate Museum
Fort Monroe, Va 23651

Chicago Historical Society
Clark Street at North Avenue, Chicago, Il 60614

Chickamauga-Chattanooga National Military Park
Fort Oglethorpe, Ga 30742

Civil War Library and Museum
1805 Pine Street
Philadelphia, Pa 19103

Confederate Museum
Alexander Street,
Crawfordville, Ga 30631

Confederate Museum
929 Camp Street,
New Orleans, La 70130

Confederate Naval Museum
201 4th Street
Columbus, Ga 31902

Fredericksburg and Spotsylvania National Military Park
120 Chatham Lane,
Fredericksburg, Va 22405

Fort Ward Museum and Historic Site
4301 W. Braddock Road,
Alexandria, Va 22304

Gettysburg National Military Park
Gettysburg, Pa 17325

Grand Army of the Republic Memorial Hall Museum
State Capitol 419 N.
Madison, WI 53702

Kentucky Military History Museum
Old State Arsenal
East Main Street
Frankfort, Ky 40602

Milwaukee Public Museum
800 W. Wells Street,
Milwaukee, WI 53233

Smithsonian Institution
National Museum of American History, 900 Jefferson Drive, S.W.
Washington, DC 20560

South Carolina Confederate Relic Room and Museum
World War memorial Building, 920 Sumter Street, Columbia, SC 29201

Springfield Armory National Historic Site
1 Armory Square,
Springfield, MA 01105

State Historical Museum of Wisconsin
30 North Carroll Street
Madison, WI 53703

The Confederate Museum
188 Meeting Street,
Charleston, SC 29401

The Museum of the Confederacy
1201 E. Clay Street,
Richmond, Va 23219

U.S. Army Military History Institute
Carlisle Barracks, Pa 17013

Virginia Historical Society
428 North Boulevard,
Richmond, Va 23221

V.M.I. Museum
Virginia Military Institute
Jackson Memorial Hall,
Lexington, Va 24450

Warren Rifles Confederate Museum
95 Chester Street, Front Royal, Va 22630

War Memorial Museum of Virginia
9285 Warwick Blvd.
Huntingdon Park
Newport News, Va 23607

West Point Museum
United States Military Academy, West Point, NY 10996